International Conference on Climate Nexus Perspectives: Toward Innovative, Resilient and Sustainable Solutions for Natural Resources and Biodiversity Management (I2CNP 2021)

E3S Web of Conferences Volume 337 (2022)

Online
6 June 2021

Editors:

Lahcen El Youssfi
Adil Aghzar
Sidi Imad Cherkaoui
Alex Godoy Faundez
Diego Rivera Salazar
Emeline S.P. Assede
Ousmane Seidou
Yassine Ouaggajou

ISBN: 978-1-7138-4279-8

Printed from e-media with permission by:

Curran Associates, Inc.
57 Morehouse Lane
Red Hook, NY 12571

Some format issues inherent in the e-media version may also appear in this print version.

For additional information, please contact EDP Sciences – Web of Conferences
at the address below.

EDP Sciences – Web of Conferences
17, Avenue du Hoggar
Parc d'Activité de Courtabœuf
BP 112
F-91944 Les Ulis Cedex A
France

Phone: +33 (0) 1 69 18 75 75
Fax: +33 (0) 1 69 28 84 91

contact-edps@webofconferences.org

Additional copies of this publication are available from:

Curran Associates, Inc.
57 Morehouse Lane
Red Hook, NY 12571 USA
Phone: 845-758-0400
Fax: 845-758-2633
Email: curran@proceedings.com
Web: www.proceedings.com

TABLE OF CONTENTS

Contribution of geomatics tools to the study of the spatio-temporal evolution of forest stands of the Maamora forest in the face of global changes

Fouad Malki[1-2*], *Jamal* Al Karkouri[1], *Mohammed* Sabir[3], *Mohamed* El Mderssa[2-4], and *Youssef* Dallahi[5]

[1]Laboratory "Environment, Societies and Territories", Department of Geography, Faculty of Human and Social Sciences, Ibn Tofail University, Kenitra, Morocco
[2]Forests and Water Department, 10090, Rabat, Morocco
[3]National Forestry School of Engineers, 511, Sale, Morocco
[4]Laboratory "Valuation of the Environment and Microbial and Plant Resources", Faculty of Sciences, Moulay Ismail University, Meknes, Morocco
[5]Laboratoire "Microbial Biotechnologies, Agrosciences and Environment", Faculty of Sciences Semlalia, University of Cadi Ayyad, Marrakech, Morocco

Abstract. A better knowledge of the land use of an environment is based on the establishment and comparison of various spatio-temporal situations of the different components of this space. This work is part of a research on the dynamics of the Maamora forest. It aims to show the contribution of the use of geomatics tools in the evaluation of dynamics of forest cover and therefore the evaluation of the impact of global changes and previous forest management methods. The approach followed for the assessment of forest cover in 1984, 1994, 2004, 2014 and 2019, was based on satellite images of Landsat 5 and Landsat 8, with a spatial resolution of 30 m. The classification supervised by the “Support Vector Machine” (SVM) algorithm made it possible to develop forest cover maps for the dates selected and subsequently to evaluate the changes. The classification evaluation showed a Kappa coefficient over 87% and an overall accuracy over 90% for all the selected dates. The results showed a great dynamic of the forest cover between 1984 and 2019 with two significant periods. The first one is 1984-1994 when areas of cork oak and acacia fell respectively by 20% and 39% in favor of other introduced forest species, mainly eucalyptus and pines, which increased respectively by 28% and 74%. For the second period 1994-2019, there was a reversal trend of regression of cork oak’s area towards an increase of 11%, while the area of eucalyptus plantations was generally stabilized and that of pines and acacia stands fell by 37% and 33% respectively. These results make it possible to provide decision-makers with tools for the future management of the space.

Keywords: Spatio-temporal evolution, Maamora forest, Global changes, Satellite images, SVM, Decision-makers.

1 Introduction

The Moroccan Forest estate covers more than 9 million hectares. It contributes 2.4% to the National GDP, taking into account the formal contribution and that of the income directly

* Corresponding author: malki.fouad@uit.ac.ma

derived by the local population [1]. It has always constituted a multifunctional space (maintaining ecological balance, conserving biological diversity, satisfying product needs, offering recreation areas, etc.).

Forest ecosystems are very dynamic system [2]. They are continuously experiencing changes in composition and content under the influence of various factors: management methods, human activities, climate change, desertification, parasitic attacks, fires, etc. The combination of these factors directly and indirectly impacts the roles and missions of these forests.

Therefore, the monitoring of the evolution of these environments seems necessary to appreciate their dynamics and responses. It requires the implementation of new evaluation methods including space remote sensing, which is of exceptional interest. Thus, the best knowledge of the composition of these spaces is based initially on the establishment of maps of its various components. The comparison of various temporal situations will then allow the determination of the various changes undergone over time.

The Maamora forest, object of this study, is located on the Atlantic coast. It is considered the most extensive cork oak forest in the world [3]. It is a dynamic system, complex and sensitive to various factors such as climate, substrate, human pressure, grazing, diseases, management methods, etc. which continuously generate changes in its composition that are sometimes difficult to appreciate.

This work is part of a research on the dynamics of the forest cover of the Maamora. It aims to show the contribution of the use of geomatic tools (remote sensing and GIS), for the establishment of maps of the forest composition in different dates. It also aims to characterize the dynamics of the cover of this forest over a period of 35 years and to establish consequently an assessment of the impact of global changes and previous management policy on the forest. These objectives were achieved through a supervised classification of Landsat images via the SVM (Support Vector Machine) algorithm.

2 Material and methods

2.1 Study area

The The Maamora forest occupies a total area of 131,921 ha. It is in the North-West of the Moroccan Meseta bordering the Atlantic Ocean and it is in an elongated form from East to West over a length of 60 km and a width of 30 km from North to South [4]. It is divided, from West to East, into five cantons A, B, C, D and E (Fig.1).

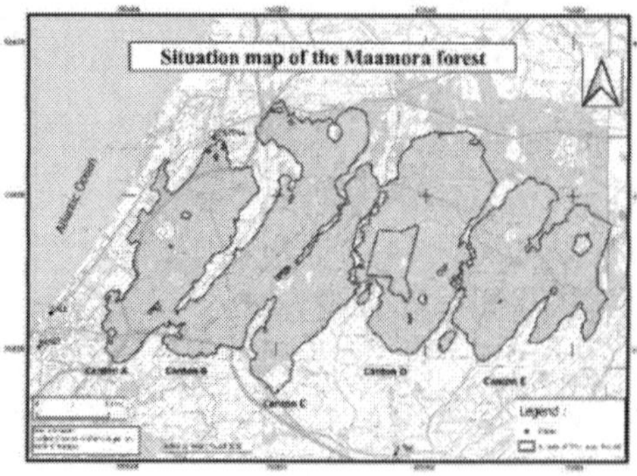

Fig. 1. Situation map of the Maamora forest

Its relief appears flat as a whole [5] and the main soil types that predominate are essentially sand over clay [6]. This forest receives annually between 350 to 650 mm of rain and the average monthly temperatures vary from 12°C to 25°C [7]. The prevailing bioclimate is sub-humid with a warm winter in the West and semi-arid with a temperate winter in the center and in the East [5].

The Maamora forest has been the subject of five management plans since 1918 until now in addition to several interventions initiated in the form of specific projects and specific management plans [7]. Originally a cork oak forest, but for the country's needs for wood and paper; eucalyptus, pine and acacia were gradually introduced [8].

2.2 Methodology

To achieve the objectives set and to have decision-making tools for the future management of the forest, the methodological approach adopted was divided into two stages:

- **Consultation and exploitation of documentation and previous work** in relation to forest management, to obtain an overview of the area and identify the available data necessary for the achievement of the objectives set. The main documents consulted are the various reports and maps published within the framework of the 1972, 1992 and 2014 management plans, the 2001 eucalyptus, coniferous and acacia management plans, as well as other project reports.
- **Acquisition, processing, and classification of satellite images:**

The satellite images employed are Landsat images. The use of this satellite is because it offers recent images and other old with high spatial resolution of 30 m. They concern four images, covering the studied area, of the TM 5 sensor (Thematic Mapper): one for the year 1984, one for 1994 and two for 2004, and four images of the OLI 8 sensor (Operational Land Imager): two for the year 2014 and two others for the year 2019.

The total duration of the study, which is 35 years, is considered sufficient to understand the dynamics of forest stands especially as it covers the period of application of the last three management plans. The 10-years interval chosen between the first four dates (1984-1994-2004 and 2014) will make it possible to assess the impact at the mi-term and at the end of the implementation of previous development plans. As for the year of 2019, it was chosen to provide an updated situation following the application of the first five years of the current management plan. The months in which the images were taken, were chosen between June and October to minimize classification confusions due to the abundance of herbaceous vegetation outside these months and to have images devoid of cloud cover as much as possible.

After the necessary pre-processing of the downloaded images, the isolated bands were combined to obtain multi-band images offering the possibility of carrying out more processing. Subsequently, a supervised classification was performed using the SVM (Support Vector Machine) classification algorithm. SVM methods was officially introduced by Boser, Guyon, and Vapnik (1992) during the 5th Annual Association for Computing Machinery Workshop on Computational Learning Theory [9].

Although initially designed for binary classifications, SVMs have been adapted to the problem of multiclass classification [10]. Their uses in various fields, particularly in the land use classification and in the study of biophysical parameters, as in the case of this study, have been elucidated by Mountrakis et al. [11].

SVMs have often been found to provide higher classification accuracies than other widely used pattern recognition techniques, such as the maximum likelihood and the multilayer perceptron neural network classifiers. Furthermore, SVMs appear to be especially advantageous in the presence of heterogeneous classes for which only few training samples are available [12].

The choice of the SMV method using Landsat images is also justified, in this study, firstly by the monospecific composition of the homogeneous areas and the lack of mixing of the species, and secondly by the differentiated structure of stands (natural stands and plantations) and their botanical classification (leafy stands and resinous stands) thus allowing to give distinct reflectance that facilitate the classification.

The first classification step consisted in identifying the classes to be retained having to indicate the species present in the forest in addition to the other areas. For each of them, a sufficient and representative number of areas of interest was provided, by referring to the previous management documents, particularly stand type maps and action plans, that provide information on the existing forest strata at the homogeneous extents level. In addition, visual interpretation of images and available earth monitoring tools were used to delineate areas of interest. These various documents and procedures were also used to verify and control the results of the classification.

Subsequently, the algorithm learned the areas of interest data and proceeded to separate the pixels then their assignments to the classes selected based on the spectral information contained in these pixels. After completion of the classification operation itself, control and validation tests were carried out based on areas of interest that were not used for the classification, for calculating the overall accuracy and the Kappa coefficient. At the end, the areas of the various forest stands were generated.

Also, to assess and quantify the changes experienced by these stands during key periods, forest cover change maps were created from the classifications made and the statistics of these changes were generated according to change matrices.

The processing and editing of the images were carried out using open-source software (Quantum GIS).

3 Results

3.1 Forest cover maps

The results of the SVM supervised classification of the Maamora forest for the years 1984, 1994, 2004, 2014 and 2019 made it possible to distinguish seven classes of forest cover: Cork oak, Leafy reforestation, Resinous reforestation, Acacia, Empty Land, Water (Dayas) and Others which brings together all the other covers different from those mentioned above, including roads, buildings, wood and cork deposit, etc.

The results as well as the parameters for assessing the quality of this classification are presented in figure 2 and table 1.

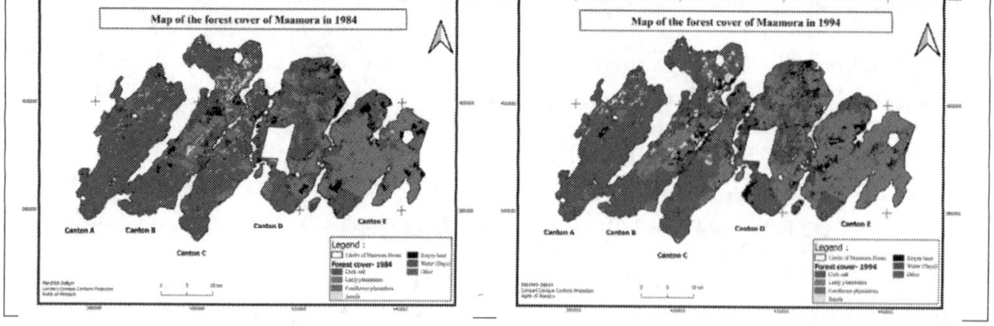

https://doi.org/10.1051/e3sconf/202233701001

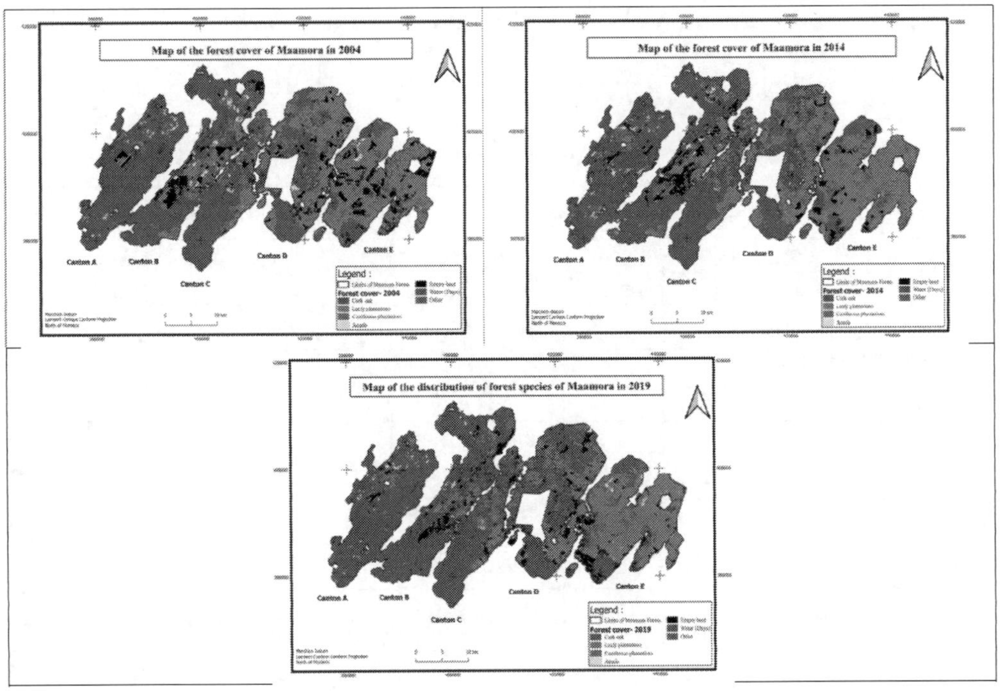

Fig. 2. Maps of Maamora forest covers for the years 1984, 1994, 2004, 2014 and 2019

The qualitative examination of the results presented in figure 2 shows that, for the various dates selected, the forest formations based on cork oak are dominant, occupying the western part of the forest. Leafy plantations, which consist of eucalyptus reforestation, are mainly located in order of importance, in cantons E and D. As for coniferous plantations, made up of pine reforestation, acacia formations and empty spaces, they are in the form of tasks at the level of the five cantons.

Table 1. Overall Accuracy and Kappa Coefficient of the SVM classification

Years	Overall Accuracy (%)	Kappa Coefficient (%)
1984	90.02	87.80
1994	92.01	89.90
2004	91.63	89.47
2014	92.35	91.20
2019	93.77	92.84

The evaluation of the results was done with the objective of assessing the quality of this classification and its ability to differentiate between the previously defined classes. Thus, the values of the overall accuracy and of the Kappa coefficient, presented in Table 1, are very satisfactory and allow to validate the results obtained. They are greater than 90% for the overall accuracy and 87% for the Kappa coefficient for the selected dates.

3.2 Quantitative evaluation of the evolution of the forest cover

The quantitative results of the SVM classification for the various dates selected are presented in Table 2.

Table 2. Evolution of the area of the forest cover of Maamora between 1984 and 2019

Years	1984		1994		2004		2014		2019	
Classes	Area (ha)	%	Area (ha)	%	Area (ha)	%	Area (ha)	%	Area (ha)	%
Cork oak	76098	58	61079	46	64119	49	65487	50	67563	51
Leafy plantations	32855	25	41929	32	41561	32	42231	32	41427	31
Coniferous plantations	5953	5	10340	8	8596	7	9663	7	6544	5
Empty land	8122	6	11985	9	11027	8	7993	6	10626	8
Water (Dayas)	780	1	602	0	1165	1	605	0	761	1
Acacia	5471	4	3362	3	2800	2	3228	2	2269	2
Other covers	2642	2	2624	2	2651	2	2714	2	2731	2
Total	**131921**	**100**	**131921**	**100**	**131921**	**100**	**131921**	**100**	**131921**	**100**

The quantitative evaluation of the evolution of the area of the different Maamora forest cover during the period 1984-2019, showed that the cork oak formations occupied an area of 76,098 ha in 1984 that is 58% of the total area of the forest. This area regressed massively by more than 15,000 ha to reach 61,079 ha in 1994, equivalent to 46% of the forest area. From 1994 to 2019, there has been a reversal of this regression trend towards an increase in the area of around 6,484 ha, i.e. an average annual gain of around 260 ha/year, to reach in 2019 an area of 67,563 ha representing 51% of the total area.

For the eucalyptus plantations, it is an inverse evolution of that of the cork oak which was observed during the period 1984-1994 when an increase in the area of this introduced species was detected, going from 32,855 ha in 1984 to more of 41,900 ha in 1994 i.e. 32% of the area. During the period from 1994 to 2019, the area of eucalyptus has been stable around 32% of the area of the forest.

The area of resinous reforestation composed of pines has known in turn during the period 1984-1994, an increase of 4,387 ha, that is to say an average annual gain of about 440 ha/year. This area subsequently fluctuated between 1994 and 2019 with an overall downward trend going from 10,340 ha at the start of the period to 6,544 ha in 2019.

Acacia formations, for their part, generally lost area during the period from 1984 to 2019 with slight fluctuations, except between 1984 and 1994 when this area declined by more than 38%. In 2019, the area of acacias was estimated at 2,269 ha.

These classification results, as well as the trends in the evolution of the areas for the different species, are confirmed by several studies explaining the extent of these forest formations during the periods of application of the previous development plans [2] [8] [13] [14].

As regards the empty lands and the dayas (Water), they experienced more or less significant changes in their areas during the period 1984-2019. Their areas respectively have increased from 8,122 ha and 780 ha at the beginning of the period to 10,626 ha and 761 ha in 2019, knowing that the area of the dayas is highly dependent on the rainfall of the year and that of the empties is dependent of the removal of plantations especially by regular wood harvesting carried out by the administration of forests.

Finally, for other covers consisting mainly of roads, buildings, and deposits, ... their area has been generally marked by an upward trend since 1984 to 2019, passing from 2,642 ha in 1984 to more than 2,730 ha in 2019.

3.3 Maps and matrices of cover changes

The analysis of the evolution of the areas of the main forest species from 1984 to 2019 made it possible to distinguish two periods marking the significant changes in their areas: the first is that between 1984 and 1994 and the second is that between 1994 and 2019. Figure 3, Tables 3 and 4 respectively show the maps and matrices of forest cover changes during these two periods.

The analysis of changes, which will concern the conversions of the main forest formations, will only consider the areas estimated at the beginning and at the end of the period, without considering the variations that have occurred during this period.

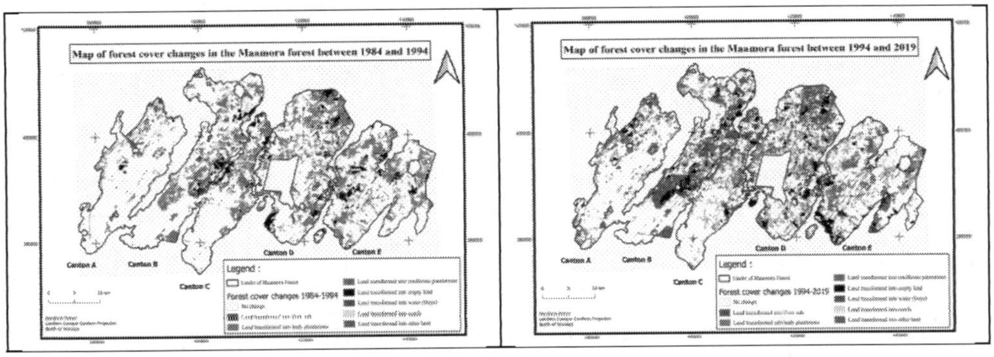

Fig. 3. Maps of forest cover changes during the periods 1984-1994 and 1994-2019

Table 3. Matrix of forest cover changes in % between 1984 and 1994

	Classes	1994: Cork oak	Leafy plantations	Coniferous plantations	Empty land	Water (Daya)	Acacia	Other	Total Area 1984 (ha)
1984	Cork oak	86,4	28,8	32,4	45,0	19,8	54,3	20,9	**76098**
1984	Leafy plantations	5,0	57,1	23,3	25,0	17,3	6,9	4,7	**32855**
1984	Coniferous plantations	1,5	3,3	28,2	2,8	1,2	7,6	5,4	**5953**
1984	Empty land	2,4	7,8	4,2	19,1	13,1	9,2	9,5	**8122**
1984	Water (Daya)	0,3	0,3	0,2	1,4	41,0	0,8	1,2	**780**
1984	Acacia	4,1	2,0	8,9	3,5	4,5	19,3	4,5	**5471**
1984	Other	0,3	0,7	2,8	3,2	3,1	1,9	53,8	**2642**
	Total % 1994	**100**	**100**	**100**	**100**	**100**	**100**	**100**	–
	Total Area 1994 (ha)	**61079**	**41929**	**10340**	**11985**	**602**	**3362**	**2624**	**131921**

The analysis of figure 3 and tables 3 and 4, point out that the forest of Maamora is a very dynamic space that has undergone many changes mainly in the eastern parts of the forest.

During the period 1984-1994, about 14% of the area of the cork oak in 1994 comes from different occupations, especially the eucalyptus; 86% of its area has not changed since 1984, while more than 23,300 ha has been transformed into other occupations.

Leafy plantations have experienced an area gain due mainly to the transformation of cork oak formations and empty lands, which accounted for 29% and 8% respectively of the area of eucalyptus stands when 57% of their area remained unchanged.

As for pine plantations, they have also gained an area of 4,387 ha composed of 72% mainly coming from cork oak and eucalyptus while for the acacia formations, they fell from 5,471 ha to 3,362 ha of which more than 80% of this area comes from the conversion of other species, particularly cork oak (54.3%).

Table 4. Matrix of forest cover changes in % between 1994 and 2019

		2019							
	Classes	**Cork oak**	**Leafy plantations**	**Coniferous plantations**	**Empty land**	**Water (Daya)**	**Acacia**	**Other**	**Total Area 1994 (ha)**
1994	**Cork oak**	69,1	21,4	18,3	27,1	29,2	41,1	11,2	**61079**
	Leafy plantations	14,6	58,4	34,0	42,3	20,0	27,4	13,3	**41929**
	Coniferous plantations	5,0	8,4	32,2	9,1	6,6	9,5	3,8	**10340**
	Empty land	7,7	9,7	11,0	14,1	20,1	7,6	8,5	**11985**
	Water (Daya)	0,4	0,3	0,2	0,3	19,1	0,6	0,4	**602**
	Acacia	2,6	1,1	2,6	5,2	1,6	11,8	5,0	**3362**
	Other	0,6	0,7	1,7	1,9	3,4	2,0	57,8	**2624**
	Total % 2019	**100**	**100**	**100**	**100**	**100**	**100**	**100**	–
	Total Area 2019 (ha)	**67563**	**41427**	**6544**	**10626**	**761**	**2269**	**2731**	**131921**

For the period 1994-2019, the unchanged area of cork oak is around 70% and the rest of the estimated area in 2019 comes from other forest cover, allowing an area gain of around 6,484 ha between the beginning and the end of the period.

Regarding the leafy reforestation based on eucalyptus, they have practically maintained their areas of more than 41,000 ha although the area unchanged between 1994 and 2019 does not exceed 58.4%.

The coniferous pine formations recorded a decrease of about 3,800 ha, of which only one-third remained unchanged.

As for the acacia stands, only 12% of their area in 2019 which remained unchanged while all the rest, i.e. 88%, comes from the transformation of other cover, particularly cork oak (41.1%) and eucalyptus (27.5%).

4 Discussions

The analysis of the general evolution of the area of cork oak shows a regression between 1984 and 1994, while since 1994, a reversal of this trend towards an increase in his area has been observed. On the other hand, in general for the introduced forest species, in particular eucalyptus and pines, it is an inverse evolution of that of the cork oak which was observed.

The increase in their area was noted during the period 1984-1994 after which it decreased. As for the areas of acacias, they continued to decrease from 1984 to 2019.

Trends in the area changes of the different forest species during the two periods confirm the results obtained by different sources [8] [14]. They are mainly due to the forest policies adopted combined with other factors such as climatic conditions and anthropozoogenic pressures. The impact of these factors was manifested mainly, between 1984 and 2019, in the reduction in area and density and the degradation of the state of development of cork oak stands. These impacts were accentuated by the absence of natural regeneration of this species and by the choice of species to plant.

In terms of forest management methods, during the period before 1994, the main objectives of the forest administration were to maintain cork oak where it was economically profitable; and its conversion into introduced species with rapid growth, where the density of the cork oak is low and where edapho-climatic conditions are difficult for its maintenance [8]. This policy has been guided specifically by economic choices aimed at satisfying the local wood market, in particular eucalyptus wood, for the needs of the Moroccan cellulose industrial unit.

Since 1994, following the various evaluations carried out, especially in terms of reduction of area of the cork oak associated with the reduction of the density of this native species, the orientation has thus changed and aimed to densification and reconstitution of the cork oak stands by assisted regeneration. The area of cork oak regenerated only since 1994 until 2010 are estimated at over 14,600 ha [15] and 12,300ha [16]. These actions carried out have essentially made it possible, on the one hand, to extend the surface area of cork oak to the detriment of the surface area of exogenous species, and on the other hand to densify existing low-density stands while ensuring the production of wood by planting introduced species which are eucalyptus and pine.

5 Conclusion

Maamora Forest is a rich and diverse natural ecosystem that occupies an important place among national forest ecosystems. It is a very dynamic space subject to a multitude of factors that impact its state and its future: climatic conditions, anthropozoogenic pressures, management methods, etc.

The present study provides information on the evolution of the areas of the different forest cover of this ecosystem and consequently the analysis of forest dynamics.

The results of this work have made it possible, using spatial remote sensing, to draw up maps of forest composition and maps of changes from 1984 to 2019, to quantify the evolutions and changes that have occurred as well as to analyse this dynamic.

The analysis of the results obtained allowed to highlight two periods oriented essentially by the management methods adopted and the environmental conditions. The first period before 1994, was characterized by a decrease in the area of cork oak in favour of the area of introduced species such as eucalyptus, pines and acacia. The second period is that from 1994 when there was a reversal of the trend of regression of the area of cork oak compared to exogenous species, mainly following the artificial regeneration work carried out.

Finally, the results obtained from this work will make it possible to propose the need to continue the dynamic of rejuvenation and extension of the area of the cork oak, the indigenous species of the Maamora forest due to its better qualities of adaptation to environmental conditions [2].

References

1. FRA, Evaluation des Ressources Forestières Mondiales 2020, Rapport Maroc, 80p (2020)
2. FAO, Plan Bleu pour l'Environnement et le Développement en méditerranée, Optimiser la production des biens et services par les écosystèmes boisés méditerranéens dans un contexte de changements globaux ; Rapport technique : Analyse de vulnérabilité au changement climatique du couvert forestier Forêt de la Maâmora (Maroc) 124p, (2015)
3. J. Natividade, Subériculture. Edition française de l'ouvrage portugais Subericultura, ENGREF, Nancy, France, 303 p (1956)
4. A. Aafi, La Maamora. Encyclopédie du Maroc, N°21 : 7199-7200, (2006)
5. A. Aafi, Etude de la diversité floristique de l'écosystème de Chêne-liège de la forêt de la Maamora. Thèse de doctorat. Institut Agronomique et Vétérinaire Hassan II. Rabat-Maroc. 190 p, (2007)
6. B. Lepoutre, Régénération artificielle du chêne-liège et équilibre climatique de la subéraie en forêt de la Mâamora. Annale de la Recherche Forestières Maroc. Tome 9, 188 p. (1965)
7. H. Oubrahim, Estimation et cartographie du stock de carbone dans l'écosystème de quercus suber de la Mamora occidentale (cantons A et B) et estimation de son stock d'éléments nutritifs, Thèse de doctorat, Université Mohammed V Faculté des sciences Rabat 202p (2015)
8. Haut-Commissariat aux Eaux et Forêts et à la Lutte Contre la Désertification, Etude de base et gestion antérieure de la forêt de la Maamora, révision de l'étude d'aménagement de la forêt de la Maamora, 137p (2014)
9. M. Awad et R. Khanna, Support Vector Machines for Classification, article, 29P (2015).
10. A. Gressin, Mise à jour d'une base de données d'occupation du sol à grande échelle en milieux naturels à partir d'une image satellite THR, Thèse de doctorat, Université Paris-Descartes, 226 p (2014)
11. G. Mountrakis, I. Jungho, C. Ogole, Support vector machines in remote sensing: A review article in Elsevier, 13p (2010)
12. F. Melgani and L. Bruzzone, Classification of Hyperspectral Remote sensing Images With Support Vector Machines, Article in IEEE Transactions on Geoscience and Remote Sensing, 14p (2014)
13. Haut-Commissariat aux Eaux et Forêts et à la Lutte Contre la Désertification, Etat actuel et résultats d'inventaire, révision de l'étude d'aménagement de la forêt de la Maamora, 101p (2014)
14. K. D. Noumonvi, Application de la modélisation spatiale multifactorielle pour l'évaluation de la dynamique et la vulnérabilité des écosystèmes forestiers face aux changements globaux : Cas de la forêt de Maâmora, mémoire de 3ème cycle, Ecole Nationale Forestière d'Ingénieurs, 114p (2015)
15. Haut-Commissariat aux Eaux et Forêts et à la Lutte Contre la Désertification, Etude de régénération, révision de l'étude d'aménagement de la forêt de la Maamora, 38p (2014)
16. H. El Hachimi, cartographie et état des lieux des périmètres de régénération du chêne liège dans la subéraie de la Maamora, mémoire de 3ème cycle, Ecole Nationale Forestière d'Ingénieurs, 77p (2010)

Geographical, Environmental and Anthropogenic Factors Affecting the Occurrence of Wintering Eurasian Spoonbills (*Platalea leucorodia*) in Morocco

Sidi Imad Cherkaoui[1*], *Rhimou El Hammoumi*[2], *Abdeslam Rihane*[3], *Lahcen El Youssfi*[4], *Adil Aghzar*[5], *Mohammed Dakki*[6]

[1] RNDD Laboratory, Faculty of Sciences Kenitra & Kénitra Higher School of Technology, Ibn Tofail University, B.P. 242, 14000, Kénitra, Morocco
[2] Université Hassan II de Casablanca, Av Driss El Harti B.P 7955, Sidi Othmane, Casablanca, Morocco
[3] Department of Life and Earth Sciences, Regional Center for Education and Training (CRMEF) Casablanca-Settat, Casablanca, Morocco
[4] RNDD Laboratory, Faculty of Sciences Kenitra & Kénitra National Higher School of Chemistry, Ibn Tofail University, B.P. 241, 14000, Kénitra, Morocco
[5] Polydisciplinary Faculty of Larache FPL, Abdelmalek Essâadi University, Tetouan, Morocco
[6] Laboratory of Geo-biodiversity and Natural Heritage, Scientific Institute, Mohammed V University, Av. Ibn Battota, Rabat 10, BP 703, Morocco

Abstract. Investigating habitat selection and ecological factors trade-offs is a major avian ecology topic which is closely implicated for conservation purposes. Studies dealing with the impacts of ecological factors on wintering Spoonbills are overall scarce. Here, we used Principal Component Analyses (PCA) and Generalized Linear Mixed Models (GLMM) to test the relevance of geographical, environmental and anthropogenic factors in predicting the occupancy and abundance of the Eurasian spoonbills (*Platalea leucorodia*) during 2008-2011 within 28 Moroccan wetlands. The Eurasian Spoonbill mean annual occurrence was 59.2% (± 1.5% SE; 95%CI: 56.3%-62.1%). Among the occupied wetlands, 15 (83.3%) were regularly occupied. We found that the Eurasian spoonbill occurrence was negatively related to distance to coastline, altitude and human presence, whereas its abundance increased significantly with increasing mudflat areas. These findings highlight the significant effect of human presence in selecting wintering sites, but not in the prediction of abundance. Management strategies should therefore address specific attributes of coastal wetlands and should carefully consider the effects of habitat use especially those related to mudflats. We further suggest guidelines for future studies to understand the dynamic of Eurasian spoonbills wintering in the region.

1 Introduction

When choosing a site at which to settle, individual waterbirds need to simultaneously consider the quality of available sites in terms of their environmental conditions, resource availability and the presence/frequency of predators and/or parasites [1, 2, 3, 4,5]. However, this choice can be also influenced by: (i) geographical and topographical factors, which may either decrease or enhance attractiveness of a site [6, 7, 8, [9], and (ii) human activities, which are known to affect wildlife populations ([10-1]. Patterns of avian habitat use have recently generated numerous interesting questions in community ecology [12-13], and the assessment of avian habitat relationships has become an important part of wildlife, resource management and conservation planning [14, 15, 16], especially within the context of global change [17].

Wetlands are habitats of the highest conservation interest due to their high biological diversity [18, 19, 20]). They also play a vital role in the life cycle of many species, both during wintering, breeding and migration of waterbirds [21]. Among them, the Eurasian Spoonbill *Platalea leucorodia* (hereafter spoonbill) is one of the most relevant species in the Palearctic. It is a widespread wading bird that breeds patchily from Europe to East Asia, India, the Red Sea and Northern Africa [22]. The species is labelled to be of 'Least Concern' globally according to IUCN criteria. The global population is of about 66,000-140,000 individuals [23].

The Kingdom of Morocco, located in the extreme north-west of Africa, is bordered to the north by the Mediterranean Sea, to the west by the Atlantic Ocean, and to the south by Mauritania. It supports habitats ranging from high-altitude moorland through cork-oak forests to wetlands, deltas, arid steppes and deserts [24]. Morocco is located at the crossroads of several avian migratory routes between Europe and Africa. Morocco plays a major role in the migration of waterbirds as it supports numerous wetlands and 70% of its 3500 kilometers of coastline lie within the East-Atlantic Flyway [25].

* Corresponding author's e-mail: imad.cherkaoui@uit.ac.ma

In Morocco, the spoonbill is a common passage migrant, winter visitor and localized breeder in Smir lagoon and Oued EL Maleh [26-27]. In the winter period, the species is mostly found in wetlands along the Atlantic coast, especially at estuaries and lower marshes [28]. The Spoonbill is a protected species according to 29.05 national law.

Most studies on Spoonbils have focused on migration [29, 30, 31, 32, 33, 34], diet [35, 36, 37, 38, 39], breeding [40, 41, 42, 43,44], and molecular biology [45-46]. To date there are few studies on abundance and occupancy of wetlands during the winter period in Morocco and even worldwide [47-48]. Yet, there is a real need to understand the ecological factors and processes affecting these two parameters. Such investigations are fundamental to our understanding of the dynamics of spoonbill communities in North African wetlands and in devising conservation plans. Thus, the major objectives of our study were (i) to identify the most important factors affecting wetland occupancy and abundance of spoonbills during winter period (ii) to disentangle the relevance of environmental versus anthropogenic factors by considering wetlands and habitat features along with human presence. Given the high sensitivity of the Eurasian spoonbill to habitat conditions and its low sensitivity to human disturbance [49], we hypothesized that the distribution and abundance of this species in Morocco are largely dependent on wetland and habitat features rather than human presence.

2 Methods

The 28 wetlands of international conservation importance were studied our of 38 based on their designation as both Ramsar and Important Bird Areas (IBA) sites (www.birdlife.org; n = 20) (Ramsar2021; Ramsar Convention Bureau 2002, Fig. 1) or as exclusively IBA sites ([50]; n = 6; Fig.1). Among these wetlands, four are of national importance since they are designated, by the Moroccan Protected Areas Study [51], as Sites with Biological and Ecological Interest (SBEI). Eighteen wetlands are coastal areas, while twelve are located in inland areas (Fig. 1).

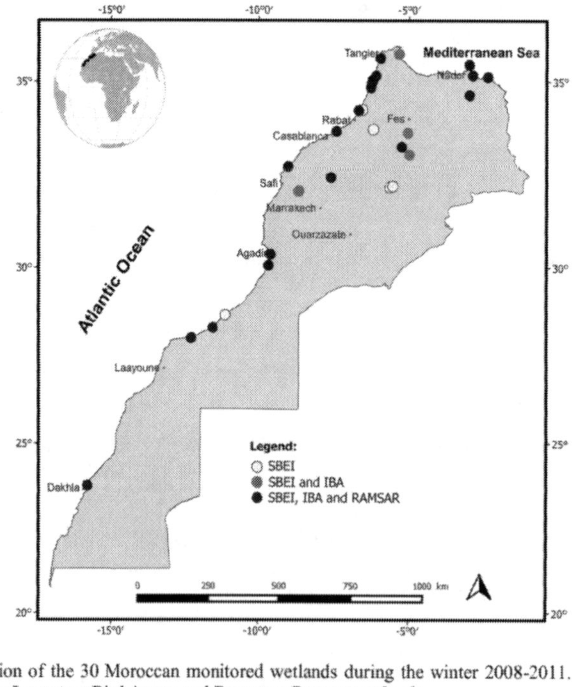

Fig. 1. Geographic location of the 30 Moroccan monitored wetlands during the winter 2008-2011. SBEI = Sites of Biological and Ecological Interest; IBA = Important Bird Areas; and Ramsar = Ramsar wetlands.

2.1 Occurrence data

Eurasian Spoonbill surveys were conducted during four consecutive years, from 2008 to 2011 in December-January because variable used in this study were only collected in this period during the Eurasian Spoonbill censuses, which is in line with the International Waterbird Census (IWC; [52]. All study sites were visited once a year because of limited human resources, shortness of the survey period and the large distances between studied wetlands (up to 1533 km). During each visit, direct counting of spoonbills was carried out using binoculars and a spotting scope from the shore and from accessible vantage points. The number of counting points depended on wetland shape and area size. The number of observation points per site ranged from 3 to 6. Observations were made between sunrise and midday, and in good weather, ensuring that counting conditions were relatively homogeneous.

A wetland was considered to be occupied when at least one individual of Eurasian Spoonbills was recorded. All data were collected by the same author [I. Cherkaoui (IC)].

2.2 Geographical, environmental and human data

For each studied wetland, we collected data related to geographical characteristics, habitat features, and human presence (Table 1). The visual inspection of the resulting proportions revealed no difference between years; data of 2011 were therefore chosen to represent the proportional coverage of habitats.

Table. 1 Ecological variable (grouped into four factors) considered for modeling the occupancy and abundance of wintering Platalea leucorodia in Morocco (2008-2011).

Factors	Variables description	Range	Mean ± SE
Spatial location	Longitude (m)	?	?
	Latitude (m)		
Wetland features	Altitude (m)	0-206	5.16 ± 3.23
	Distance to nearest coastline (km)	0.29-72.02	6.07 ± 1.20
	Wetland surface area (km^2)	0.05-500	77.40 ± 12.31
Habitat features	Shallow water cover (%)	5-51	30.86 ± 1.13
	Mudflat cover (%)	0-42	21.70 ± 1.56
	Vegetation cover (%)	5-45	21.48 ± 1.66
Human variables	Human density (number/km^2)	0-80	2.51 ± 1.24
	Number of recorded humans	0-215	47.11 ± 8.29

The choice of variables was motivated from our overall aim, which was to identify factors driving wetland site occupancy and abundance of wintering spoonbills in Morocco. Some variables have been already correlated to the occurrence of coastal wintering spoonbills [49].

The location of each wetland (latitude and longitude) and its altitude were determined using a Global Positioning System (GPS) device (Garmin eTrex HC) and compared to altitude data from Google Earth to check their accuracy. We took the wetland surface area from [25] for Ramsar sites, from Birdlife datasets (www.birdlife.com) for Important Bird Areas (IBA), and from the Moroccan Protected Areas Study [51] for SBEI.

To estimate the proportion areas covered by the shallow water and mudflat, we used Google Earth Pro (2017 cover), since high resolution images are available for all studied wetlands. The analysis process was realized as follow: (i) for each wetland, and to bypass any difficulties in identification, many available images were explored to draw polygons for all habitat types; (ii)we proceed in the same way to draw lines for low or high tide; (iii) once finished (for all wetlands), we exported all these polygons and lines to Quantum GIS v2.10; (iv) in QGIS, for each wetland, we gathered, for each habitat type, all its polygons in dedicated layer with specific color; and created the intertidal area polygon by taking into account the lowest and the highest exported lines for tides; and (v)the proportion of the area covered by the shallow water and mudflat, at low tide, were then calculated. In addition, the vegetation cover was calculated from the Google Earth Pro and QGIS according the formula: the ratio of vegetation coverage area to the wetland area × 100.

The distance to the nearest coastline was determined in QGIS using the centroid of each studied wetland. To assess the impact of human disturbances, we collected, for each study year and at each wetland, the number of people present. This has allowed us to determine the human density by dividing the number of persons per surface-area unit.

3 Statistical analysis

3.1 Preliminary analyses

Statistical models were implemented in R (version 3.1.0); [53]. Before performing statistical analyses, we checked for normality and homogeneity of variance for all variables. In order to approach residuals from normality, variables that did not conform to the requirements for parametric tests were log or square-root transformed prior to all analyses [54, 55, 56]. We also checked for possible correlations among variables by using Pearson's correlation (r) index. Based on two Principal Component Analyses (PCA) [one for data of occupied and unoccupied wetlands (PCAoc), and the otherone for data of occupied wetlands only (PCAabto model the abundance)], we reduced the total number of studied variables ($n = 9$) to a smaller number by merging different categories, while minimizing the loss of independent information. We used this reduced set of components (four for PCAoc and four for PCAab) as final variables in our further analyses. The Kaiser–Meyer–Olkin measure of sampling adequacy (KMO) indicated that our data were suitable for the two PCAs (PCAoc: $n = 9$, KMO = 0.47; Bartlett-test for sphericity, $\chi2 = 749.18$, $p< 0.001$; PCAab: $n = 9$, KMO = 0.59; Bartlett-test for sphericity, $\chi2 = 315.35$, $p< 0.001$, [57].

3.2 Model development

Two types of models were used to analyze the relationships between the four PCA components and wetland occupancy and abundance of wintering spoonbills. First, we applied generalized linear mixed models (GLMMs) with binomial error distribution and a logit link function using the lme4 package (version 1.1-7; [58] to test the relationships between wetland occupancy (response variable) and the four explicative PCAoccomponents (independent variables).Second, to determine whether our PCAab components are related to abundance of spoonbills, we used another Generalized Linear Mixed Models (GLMMs) with a negative binomial error distribution and log link function using the lme4 and MASS packages [59]. In both sets of models, we included year and site as random effects to statistically control for between-year variation and potential non independence of multiple observations at the same site.

We limited the number of variables included in our models due to the small sample size of wetland sites. Models were then ordered by increasing Akaike Information Criterion (AIC) and AIC weights were calculated [60]. Model with the lowest AIC was selected as the best fitting model. We corrected AIC for small sample size using AICc [60] using theMuMIn package [61]. The top ranked models (within two points of AICc) were then averaged as implemented in the MuMIn R-package [62]. Model weights were used to define the relative importance of each explanatory component across the full set of models evaluated by summing weight values of all models that included the explanatory component of interest.

3.3 Spatial autocorrelation

One of the assumptions of parametric statistics is that observations are independent of each other. This assumption is often violated with spatial data, so it is important, when present, to test for and subsequently address spatial autocorrelation in data prior to data analysis. The presence of spatial autocorrelation was tested using semi-variograms of the residuals of the best model in terms of AICc value using the "sp", "gstat", and "lattice" packages [63]. Nugget to total Sill Ratio (NSR) was expressed as the percentage of total semivariance and was used to define for spatial dependency: $NSR < 0.25$ indicated strong spatial dependence, $0.25 < NSR < 0.75$ indicated moderate spatial dependence, and $NSR > 0.75$ indicated weak spatial dependence [64].

4 Results

4.1 Wetland occupancy

During the years 2008–2011, the mean annual occupancy rate of wetlands by wintering spoonbills was 59.2% ± 1.5% (95% CI: 56.3%-62. 1%). Among the occupied wetlands, 15 (83.3%) were regularly occupied by the wintering spoonbills.

Strong correlations were found between some of the spatial, wetland, habitat and human variables. The PCA summarized these studied field variables into four independent axes accounting for 82.5 % of the variance of the original dataset [28.6 % (eigenvalue = 2.576),22.2 %(eigenvalue = 1.997), 18.2% (eigenvalue = 1.638), and 13.5% (eigenvalue = 1.211) respectively].The varimax rotation revealed a first gradient (PCwetland features) characterized by high loadings of variables related to wetland features as it was positively correlated with distance to the nearest coastline ($r= 0.983, p< 0.001$), and altitude ($r= 0.964, p< 0.001$). The second gradient (PCgeographic position) represents an axis of the spatial positionof wetlands as it was positively correlated with longitude ($r= 0.955, p < 0.001$), and latitude ($r=0.950, p< 0.001$). The third axis represents an axis of increasing habitat features as it was positively correlated with the shallow water cover ($r = 0.801$, $p< 0.001$). The fourth axis represents an axis of increasing human presence as it was positively correlated with human density ($r = 0.864$, $p< 0.001$).

The GLMM testing for effects of the four principal components showed that the two models were equally good (Table 2).The most important predictors that explain the occupancy of Moroccan wetlands by wintering spoonbills (betas, SE and 95% CI given in brackets are model-averaged estimates across all possible models) were PCwetland features($\beta= -36.57 \pm 18.65$, 95% CI:−73.124 to −0.016; importance =0.999), and PChuman presence($\beta= -3.73 \pm 1.52$, 95%CI: −6.71 to −0.75, importance =0.940) (Table 2). The PChabitat features ($\beta= -2.07 \pm 2.26$, 95% CI: −6.50 to 2.36, importance =0.579) contribution, although present in the top model, is not sufficiently important [$\Delta AICc = 1.47$ (< 2)].

https://doi.org/10.1051/e3sconf/202233701002

Table 2. Set of candidate models to assess the effects of ecological factors on the spatial distribution and abundance of wintering Platalea leucorodiain Morocco. Log-lik is the maximized log-likelihood, K is the number of model parameters, AICcis the Akaike information criterion for small simples size, ΔAICc is the difference in AICc value from the top model, and AICcwi is the Akaike model weight.

Dependent variables	Models	Log-lik	K	AICc	ΔAICc	AICcwi
Occupancy	**$PC_{wetland\ features}$ + $PC_{habitat\ features}$ + $PC_{human\ presence}$**	**-10.90**	**4**	**32.38**	**0.00**	**0.56**
	$PC_{wetland\ features}$ + $PC_{human\ presence}$	**-12.73**	**3**	**33.85**	**1.47**	**0.27**
	$PC_{wetland\ features}$ + $PC_{geographic\ position}$ + $PC_{human\ presence}$	-12.49	4	35.57	3.20	0.11
	$PC_{wetland\ features}$	-16.03	2	38.30	5.92	0.03
	$PC_{wetland\ features}$ + $PC_{habitat\ features}$	-15.71	3	39.81	7.43	0.01
	$PC_{wetland\ features}$+ $PC_{geographic\ position}$	-15.94	3	40.27	7.89	0.01
	$PC_{wetland\ features}$ + $PC_{geographic\ position}$ + $PC_{habitat\ features}$	-15.38	4	41.36	8.98	0.01
	Null	-31.95	1	68.01	35.63	0.00
	$PC_{human\ presence}$	-31.90	2	70.04	37.66	0.00
	$PC_{habitat\ features}$	-31.94	2	70.11	37.73	0.00
	$PC_{geographic\ position}$	-31.95	2	70.12	37.74	0.00
Abundance	**$PC_{habitat\ features}$ + $PC_{human\ presence}$**	**-352.56**	**3**	**716.14**	**0.00**	**0.42**
	$PC_{habitat\ features}$	**-353.93**	**2**	**716.54**	**0.40**	**0.35**
	$PC_{habitat\ features}$+ $PC_{geographic\ position}$	-353.87	3	718.78	2.64	0.11
	$PC_{habitat\ features}$+ $PC_{wetland\ features}$	-353.88	3	718.81	2.67	0.11
	$PC_{human\ presence}$	-359.31	2	727.30	11.16	0.00
	Null	-360.48	1	727.37	11.23	0.00
	$PC_{geographic\ position}$ + $PC_{human\ presence}$	-359.30	3	727.64	11.50	0.00
	$PC_{wetland\ features}$+ $PC_{human\ presence}$	-359.13	3	729.29	13.15	0.00
	$PC_{wetland\ features}$	-360.31	2	729.30	13.16	0.00
	$PC_{geographic\ position}$	-360.48	2	729.64	13.50	0.00
	$PC_{wetland\ features}$+ $PC_{geographic\ position}$	-360.31	3	731.66	15.52	0.00

Overall, the wintering spoonbills have especially occupied wetlands located in close proximity to the coastline, at low altitude, and having a low human presence (Fig. 2 and3). With regard to spatial autocorrelation, semi variograms, established from the residuals of the best models, indicate the absence of a spatial pattern in the data (the NSR ratios were equal to 1 for the two top models).

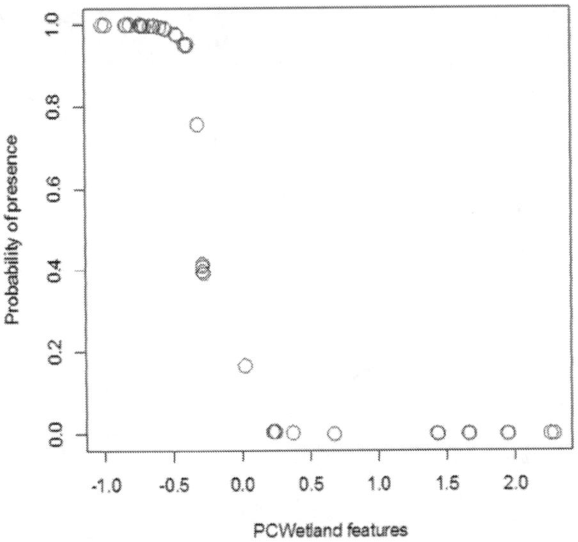

Fig. 2. Probability of presence of wintering Platalea leucorodia according to wetland features in Morocco (2008-2011).

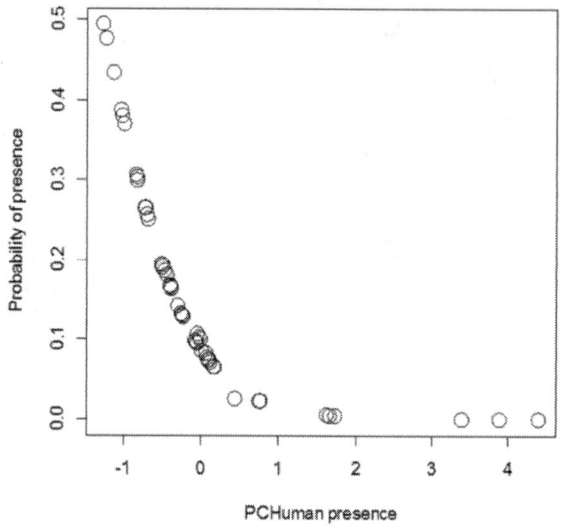

Fig. 3. Probability of presence of wintering Platalea leucorodia according to human presence in Morocco (2008-2011

4.2 Abundance

In evaluating the effect of ecological factors on the abundance of spoonbills, we used the 64 cases of presence as a dependent variable. As for wetland occupancy, strong correlations were found between some of our studied variables. The PCA summarized these field variables into four independent axes accounting for 87.5 % of the variance of the original dataset [31.5 % (eigenvalue = 2.523),28.3 %(eigenvalue = 2.566), 18.2% (eigenvalue = 1.638), and 13.5% (eigenvalue = 1.211) respectively].The varimax rotation revealed a first gradient (PCwetland features) characterized by high loadings of variables related to wetland features as it was positively correlated with distance to coast ($r= 0.923, p< 0.001$), altitude ($r= 0.910, p< 0.001$), and wetland area ($r = 0.703$, $p< 0.001$). The second gradient (PCgeographic position) represents an axis of the spatial position of wetlands since it was positively correlated with longitude ($r= 0.968, p<0.001$), and latitude ($r=0.945$, $p< 0.001$). The third axis represents an axis of increasing habitat features since it was positively correlated with the mudflat cover ($r = 0.956$, $p< 0.001$). The fourth axis represents an axis of increasing human presence since it was positively correlated with human density ($r= 0.972$, $p < 0.001$).

The GLMM testing for effects of the four principal components indicated that the two models were equally acceptable (Table 2). The most important predictor explaining the abundance of wintering spoonbills was PChabitat features ($\beta= 0.78\pm 0.18$, 95% CI: 0.427 to 1.333; importance =0.994). The PChuman presence ($\beta= -0.13 \pm 0.15$, 95 % CI: −0.164 to 0.424, importance =0.427), although present in the top model, its contribution is not significantly important [$\Delta AICc = 0.40$ (< 2)] (Table 2).Overall, wintering spoonbills occurred more abundantly in wetlands with large covers of mudflats(Fig. 4).

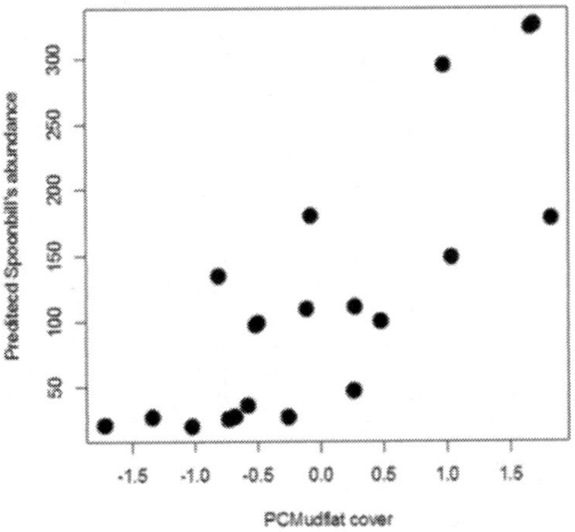

Fig. 4. Predicted abundance of wintering *Platalea leucorodia* according to mudflat coverage in Moroccan wetlands (2008-2011).

Regarding the spatial autocorrelation, the mean NSR ratio was 0.754 ± 0.06 (95% CI: 0.64 to 0.87). This mean value is situated at the lower limit of the third class established by [64] suggesting a weak spatial pattern in the data.

5 Discussion

The aim of this study was to assess the relevance of wetland and habitat features and human presence as predictors of the abundance and occupancy of wetlands by spoonbills wintering in Morocco.

Results of the modelling showed that both wetland features and human presence negatively affect the probability of presence of wintering spoonbills in Morocco. In fact, wetlands in proximity to coasts and located at low altitudes are the most occupied by this wading species. This is not surprising to the extent that, in Morocco, this species is mostly found, during the winter period, in wetlands situated on or near coasts especially at estuaries and at lower marshes [28]. This winter distribution could be dictated by the migratory flyway used by this species: along the Atlantic coast between France and Senegal [65]. Indeed, our results support a scenario where wintering spoonbills are reaching suitable areas in Morocco close to the autumnal migratory flyway. It turns out that these suitable areas are in majority coastal Ramsar sites whose importance for waterbirds is no longer to be confirmed [66-5]. On the other hand, more than 80% of study sites were regularly occupied by wintering spoonbills. Although the present study was carried out over a relatively short period (four years), this rate highlights an important use of these wetlands during the wintering period. Yet, we ignore if the wintering spoonbill are showing fidelity to studied wetlands as we have not tagged birds. But, it seems that this is the case considering this important rate of use. [33] have shown that adult birds are highly site faithful during winter.

The occurrence probability of spoonbills in Moroccan wetlands is also affected, but with a lesser extent, by human presence. This is all the truer that this biotic factor is known to affect the distribution of waterbirds ([67, 68, 49]. Indeed, in recent years, increasing human pressures and activities along coastal areas were at the origin of reducing the quality of intertidal habitats in many regions of the world [69-70]. In Morocco, the occupancy of wetlands by wintering spoonbills seems to be a trade-off between the proximity to coastlines where the migration happens, and the human presence.

Regarding the abundance, statistical analyses showed that the main factor influencing the number of spoonbills wintering in Morocco is the coverage of mudflats. This is consistent with [66], who demonstrated that in general waterbirds abundance was significantly related to habitat characteristics. [22] have also reported the preference of spoonbills to this type of habitat. In a recent study, conducted in the Tunisian Gulf of Gabès [49], have shown that habitat features, especially large mudflats, have a positive effect on the abundance of Eurasian spoonbills. This trend seems to be largely due to higher food abundance [71]. The effect of large mudflats on the abundance is not limited to the spoonbill, but concerns also other groups of waterbirds like shorebirds and wading birds (e.g. [72, 73, 74, 75]. Surprisingly, human presence had no impact on abundance of wintering spoonbills leading to a key question: why the human presence impacted the probability of presence of spoonbills wintering in Moroccan wetlands and not abundance? We suspect that, once a wetland is occupied, the level of human presence remains tolerable by the spoonbills. This is most likely why the human presence has not emerged as a predictor in the abundance analysis. Another possible explanation is that avoidance of human presence is secondary to selection of habitat. This has been also reported by [76], who have stated the importance of habitat selection in comparison with human presence. Many waterbirds are willing to tolerate increased levels of human disturbance in response to the favorable habitat found [76]. This was confirmed by a study in the Gulf of Gabès (Tunisia) where the abundance of wintering spoonbills was also affected by the habitat features and not by human presence [49]. In Morocco, like in Tunisia [49], the spoonbills occurred mainly in large mudflats, which are in general located far away from humans.

Overall, it appears that, for wintering spoonbills, there is a first level of selection, which bears on the choice of wetlands. This latter takes into account both the wetlands position relative to coasts and human presence. Once settled, the number of spoonbills mostly depends on habitat characteristics such as the proportion of mudflats, while tolerating some level of human presence.

6 Conclusions and future prospects

The results of this study increase our understanding of the factors shaping the occupancy and abundance of spoonbills wintering in Morocco. Coastal wetlands most likely represent key areas for wintering spoonbills in this region. At the meantime, a low human presence is the guarantor of the presence of the species. Large mudflats play also a key role in increasing their abundance. Occupied wetlands should receive special priority and protection efforts in preservation programs. It is also crucial to maintain a low level of human presence within mudflats. Indeed, considering the Moroccan protected area context, it is important to integrate, in the management plan and the wetland zoning, mudflats as fully-protected habitats given their importance and their usefulness for feeding wintering waterbirds. In parallel, it is of a primary importance to raise awareness among users of these wetlands on the interest of mudflats, not only for the spoonbill but also for other mudflat-dependent species, such as wading and wader species.

Furthermore, additional detailed investigations are needed to allow deeper insights into the presence of spoonbills during winter in Morocco. Our future challenge is to determine quantitatively, via appropriate statistical analyses, the tolerance threshold of human presence under which spoonbills occupy a wetland, but also the tolerance threshold from which the abundance of spoonbills is affected. These recommendations cannot be implemented without the continuation of counting birds while collecting data on human activities, habitat characteristics, food supply, water quality (limnological data), including the conservation status of sites. The Moroccan wetlands, in majority located at the crossroads of several avian migratory routes between Europe and Africa, offer a great opportunity to document the population dynamics of spoonbills both during the winter period and in the course of spring and automn migration.

References

1. T.E. Martin, Biosci. **43**:523–532 (1993)
2. V. Penteriani, M. Gallardo, P. Roche, J Zool **257**: 365–372 (2002)
3. L. Cardador, M. Carrete, S. Mañosa, Anim Conserv **14**:382–390 (2011)
4. I. Cherkaoui, M. Dakki, S. Lahrouz, S. Hanane, Rev. Écol. (Terre Vie) **68**(2): 167–180 (2013)
5. I. Cherkaoui, S. Hanane, N. Magri, M.A. El Agbani, M. Dakki, Wetlands **35**:913–922 (2015)
6. D. Storch, M. Konvička, J. Beneš, J. Martinková, K.J. Gaston, J. Biogeogr **30**:1195–1205. (2003)
7. D Storch, R.G. Davies, D. Zajíč Storch, R.G. Davies, S. Zajícek,, C.D.L. Orme, V. Olson, G.H. Thomas, T.S. Ding, P.C. Rasmussen, R.S. Ridgely, P.M. Bennett, T.M. Blackburn, I.P.F. Owens, K.J. Gaston, Ecol Lett **9**:1308–1320 (2006)
8. J.Z. Kosicki, P. Chylarecki, Basic and App Ecol **13:** 475–483. (2012)
9. S.I. Cherkaoui,S. Hanane, Jour. Nat. Hist. **52**:37-38, 2457-2472, (2018)
10. R. Woodroffe, Anim Conserv **3:** 165–173 (2000)
11. S.A. Parks, A.H. Harcourt, Conserv Biol 16: 800–808. (2002)
12. M.L. Cody, Biosc. 31: 107-113 (1985)
13. W.M. Block, L.A. Brennan, *The habitat concept in ornithology: Theory and applications. P. 35-91 In: D.M. Power (ed.).* ¬ Current Ornithology, Volume 11. Plenum Press, New York (1993)
14. J. Verner, M. L. Morrison, C.J. Ralph, *Modeling habitat relationships of terrestrial vertebrates. Madison*, WI: The University of Wisconsin Press; 470 p. 1986
15. L. Hoffmann, H. Hafner, T. Salathé, Col. Waterbirds **19**: 12–30 (1996)
16. L. Marone, Lopez, J. De Casenave, V.R. Cueto, Revista Chilena de Historia Natural **70:** 73–81 (1997)
17. T. Newbold, L.N.Hudson, S.L.L. Hill, S.C ontu, I.L ysenko, R.A. Senior, L. Börger, D.J. Bennett, A. Choimes, B. Collen, J. Day, A.De Palma, S. Diaz, S. Echeverria-London, M.J. Edgar, A. Feldman, M. Garon, M.L.K. Harrison, T. Alhusseini,, D.J. Ingram, Y. Itescu, J. Kattge, V. Kemp, L. Kirkpatrick,, M. Kleyer, D.L.P. Correia, C.D. Martin, S. Meiri, M. Novosolov, Y. Pan, H.R.P. Phillips, D.W. Purves, A. Robinson, J. Simpson, S.L. Tuck, E. Weiher, H.J. White, R.M. Ewers, G.M. Mace, J.P.W. Scharlemann, & A.P urvis, Nature **520**: 45-50 (2015)
18. M.W. Weller, *Wetland Birds: Habitat Resources and Conservation Implications.* — Cambridge University Press, Cambridge (1999)
19. P.A. Keddy, Wetland Ecology: Principles and Conservation, 2nd edition. Cambridge University Press, Cambridge, UK (2010)
20. J.C. Garcia, G.E lliott, K. Walker, I. Castro, S.A. Trewick, Jour. Av. Biol. **46:** 001-008 (2015)

21. G.M. Tucker, M.I. Evans, *Habitats for birds in Europe: a conservation strategy for the wider environment.* Cambridge, UK: BirdLife International (1997)
22. P. Triplet, O. Overdijk, M.S. mart, S.Nagy, M. Schneider-Jacoby, E.S. Karauz, C.S. Pigniczki, S. Baha El Din, J. Kralj, A.S andor, J.G. Navedo, (Compilers). *International Single Species Action Plan for the Conservation of the Eurasian Spoonbill Platalea leucorodia.* AEWA Technical Series No. 35. Bonn, Germany (2008)
23. BirdLife International, *Species factsheet: Platalea leucorodia.* Downloaded from http://www.birdlife.org on 10/06/2020 (2020)
24. C. Magin, Morocco. In L. D. C. Fishpool, M. I. Evans, (Eds.) *Important Bird Areas in Africa and Associated Islands: Priority sites for conservation.* Newbury and Cambridge UK: Pisces Publications and Birdlife International (BirdLife Conservation Series n° 11) (2001)
25. M. Dakki, M.A. El Agbani, A. Qninba, Trav. Inst. Sci., Rabat, Sér. Gén. **7:** 238 pp. (2011)
26. M.A. El Agbani, A. Qninba, M. Amezian, F. Cuzin, M. Dakki, Bull. de l'Inst. Sci. **31**(2) : 103-110 (2009)
27. A. Rihane, R.E L Hammoumi,. Go-South Bull. **14**, 29-36 (2017)
28. M. Thèvenot, J.D.R. Vernon, P. Bergier, *The birds of morocco.* British Ornithologists' Union Checklist Series: 20:594 pp (2003)
29. J.G. Navedo, *Identifying stopover wetlands for the conservation of an endangered waterbird species: The role of Santona˜ Marshes for the Spoonbill Platalea leucorodia during autumn migration.* In P. Triplet, & O. Overdijk (Eds.), En: EUROSITE spoonbill network newsletter (vol. 4) (pp. 48–51). 2006
30. J.G. Navedo, J.A.M asero, O.Overdijk, G. Orizaola, J.M. Sanchez-Guzman, Ardea **98:** 3–12 (2010)
31. J. Kralj, A. Zuljevic, T. Mikuska, O. Overdijk, Waterbirds **35**:239–247 (2012)
32. O. Overdijk, J.G. Navedo, Aqu. Cons.: Mar. Fresh. Ecos. **22**: 695–703 (2012)
33. T. Lok, O. Overdijk, J.M. Tinbergen, T. Piersma, Ani. Beh. **82**: 837-844. (2011)
34. T. Lok, O. Overdijk, T. Piersma, Biology Letters **11**: 20140944 (2015)
35. J.H. Kemper, Behaviour **132:** 1285–1299. (1995)
36. E.Aguilera, C.Ramo, C. De Le Court, Col. Waterbirds **19**: 156-166. (1996)
37. E.H.M. El-Hacen, O. Overdijk, T. Lok, HO lff, T. Piersma, Waterbirds **36:**277-286 (2013)
38. E.H.M. El-Hacen, T.Piersma, J. Jouta, O.O verdijk, T. Lok, Jour. of Ornit. **155**: 611-619 (2014)
39. J. Veen, O. Overdijk, T. Teen, Ardea **100:** 123–130 (2012)
40. E. Aguilera, Auk **107:** 416-420 (1990)
41. L. Marion, *Status of the breeding population of Spoonbills in France and relation with Sacred Ibis.* Spoonbill Network News Letter, 4 (2006)
42. F. Halligon, J.C.Beaudoin, A. Fossé, Crex **10**: 67-71 (2008)
43. M.S mart, H. Azafzaf, H. Dlensi, Ostrich **78**(2): 495-500 (2009)
44. T.Lok, O. Overdijk, T. Piersma, Ardea **102:** 181-194 (2014)
45. R.T. Chesser, C.K.L. Yeung, C.T. Yao, X.H. Tian, S.H. Li, Zootaxa **2603:**53-60 (2010)
46. T. Piersma, M. van der Velde, E.M.E l-Hacen, T. Lok, O. Overdijk, Ardea **100:** 131-136 (2012)
47. R. El Hamoumi, M.A. El Agbani, A.Q ninba, M. Dakki, *Les Oiseaux d'eau au Maroc in CMAOT, 2014. Les oiseaux d'eau d'Andalousie et du Maroc.* Consejería de Medio Ambiente y Ordenación del Territorio. Junta de Andalucía. Sevilla (2014)
48. R. El Hamoumi, M.A. El Agbani, , A. Qninba, M. Dakki, *Distribution y evolucion de la Espatula comun durante la invernada en Marruecos.*. In Hortas F. & Ruiz Y. (Eds) 2015. La Migracion intercontinantal de la Espatula (Platalea leucorodia. Grup de Desarrollo Pesquero Càdiz-Estrecho y Sociedad Gaditana de Historia Naturel . Càdiz.Espagna.. 324 pags (2015)
49. F. Hamza, S. Selmi, Mar. Ecol. Prog. Ser. **540:** 251–258 (2015)
50. L.D.C Fishpool, M/I. M.I. *Evans, Important bird areas in Africa and associated islands: priority sites for conservation.* Birdlife International, Cambridge, p. 1144 p. (2001)
51. AEFCS *Plan Directeur des Aires Protégées du Maroc.* Administration des Eaux et Forêts et de la Conservation des Sols/BCEOM/SECA/ISR/EPHE. (Unpublished report). (1996)
52. Wetands International, *Wetlands International State of the World's Waterbirds*, 2010. Wetlands International, Ede, The Netherlands. (2010)
53. R Development Core Team *R: A language and environment for statistical computing. R Foundation for Statistical Computing*, Vienna, Austria. ISBN 3–900051–07–0, URL http: //www.Rproject.org (2014)
54. J.H. Zar, *Biostatistical analysis.* Prentice-Hall, Inc., Englewood Cliffs, NJ, p. 717 p (1984)

55. A.J. Underwood, Ecol Appl **4**: 3–15(1994)
56. G.P. Quinn, M.J. Keough, *Experimental design and data analysis for biologists.* Cambridge University, 2002. Print. (2002)
57. S.V. Budaev, Ethology 116(5): 472–480 (2010)
58. D. Bates, M. Maechler, B. Bolker, S. Walker, lme4: linear mixed-effects models using Eigen and S4. R package version 1.1-7. http://CRAN.R-project.org/package=lme4 (2014)
59. Venables, W.N. & Ripley, B.D . Modern applied statistics with S, 4th edn. Springer-Verlag, New York.
60. K.P. Burnham, D.R. Anderson, *Model selection and multimodel inference: a practical information theoretic approach*. Springer, New York. (2002)
61. K. Barton, *Package 'MuMIn'. Model selection and model averaging based on information criteria*. R package version 3.0.2 (2012)
62. K.Barton, *MuMIn: Multi-model inference.* R package version 1.9.13. http://CRAN.Rproject.org/package = MuMIn. (2013)
63. E.J. Pebesma, *The gstat package* www.gstat.org (2006)
64. C.A. Cambardella, , T.V. Moorman, J.M Novak, T.B. Parkin, D.L.Karlen, R.F. Turco. A.E Konopka, Soil Sci. Soc. Am. Jour **58**: 1501-1511 (1994)
65. T. Lok, O.Overdijk, , J.M. Tinbergen, T. Piersma, Ecology **94:** 2358-2369 (2013)
66. D. Kleijn, I. Cherkaoui, P.W. Goedhart, J. van der Hout, D. Lammersma, J Appl Ecol. **51:**289–298 (2014)
67. J.C ayford, *Wader disturbance: a theoretical overview*, pp. 3–5 (1993)
68. J.D. Goss-Custard, N. Verboven, Wader Stud Group Bull. **68** (Special Issue): 59–66 (1993)
69. H.K. Lotze, H.S. Lenihan, B.J. Bourque, R.H. Bradbury, R.G. Cooke, M.C. Kay, S.M. Kidwell, M.X. Kirby, C.H . Peterson, J.B.C. Jackson, **312**: 1806–1809 (2006)
70. M.L. Kirwan, J.P. Megonigal, Nature **504:** 53-60 (2013)
71. F. Hamza, *Contribution à l'étude de l'hivernage du Flamant rose (Phoenicopterus roseus) dans le Golfe de Gabès*. MSc thesis, Carthage University, Bizerte (2011)
72. L. Zwarts, Ardea **76**: 42-55 (1988)
73. J.D. Goss-Custard, M.G. Yates, J Appl Ecol **29**: 330–340 (1992)
74. B.C. Congdon, C.P. Catterall, Wildlife Res **21**: 507-518 (1994)
75. S. Rosa, J.M. Palmeirim, F. Moreira, Waterbirds **26**: 226–232 (2003)
76. A.H. Traut, M.E. Hostetler, Lands. Urb. Plan. **69**: 69–85 (2004)

E3S Web of Conferences **337**, 01003 (2022) https://doi.org/10.1051/e3sconf/202233701003
I2CNP 2021

Iter-years dynamic of *Juniperus phoencea L.* matorral in the North-Eastern Mediterranen coast of Morocco.

Mehdi Boumediene[1]*,*Nargis*Sahib[1], *Malika*Abid[1], *Mohamed* Addi[1], *Atika*Mihamou[1], *Hana* Serghini-Caid[1],and *Ahmed* Elamrani[1]

[1]Laboratoire d'Amélioration de la Production Agricole, Biotechnologie et Environnement (LAPABE). Departement of biology. Faculty of science. Mohamed I University. BP 717, BV Mohammed VI 60500, Oujda, Maroc

Abstract: In the North-Eastern Mediterranen coast of Morocco, the establishment of seaside resorts built as part of the Plan Azur, substantially degraded the coastalmatorral of *Juniperusphoenicea* L. Nonetheless, the speciesremains an important resource for the ecological balance. The estimation of the changes in land cover is a suitable method to evaluate the degradation rate of the matorral. The aim of the current study is towarn about the degradation of *J. phoenicea*in Saidia coastline by quantifying the inter-years dynamic of *J. phoenicea*matorralin 1965 (before the Plan Azur) in 2004 and 2021(after the Plan Azur). Estimation of changes in matorral cover rate was achieved by photo-interpretation. Analysis indicates that *J. phoenicea*matorral ecosystem of Saidia has been affected by a drop of 63% in land area, over a period of time not exceeding17 years. Currently the picture of the coastal matorral of *J. phoenicea* is rather bleak; the vulnerability of this valuable resource has prompted us to raise the alarmin order to preserve remaining up shrubs to the present time.

*Corresponding author: Boiumedienne.mehdi@ump.ac.ma

1Introduction

After the launch of the Plan Azur2002/Horizon 2010, and its revised and complementary plan Vision 2020 [1]tourism activities in the coastal areas of Morocco are permanently enlarging,butnegatively impactingcoastal naturalecosystems [2], hence leading to loss of habitats and species.Located in the north-eastern Mediterranean part of Morocco in the coastline between the deltaic estuary of the Moulouya river classified as SIBE and the Saidia city, the gigantic tourism project MediterraneaSaidia (SaidiaRessorts: La perle bleu) is one out of the six costal tourist project listed in Plan Azur.Since its launch in 2009, degradation of the coastal matorral ecosystem of *Juniperusphoenicea* L is obvious along the shoreline (Photo. 1), as the 713-hectares tourist project required the destruction of the coastal dune system, thus the vegetation of the dune system was eradicatedat the same time. (Figure1). The degradation of the matorral was deepened further with the excessive extraction of sand and the urban encroachment of Saidia city. Based on its high biodiversity and the number of found habitats, Saidiacostalineis considered as a high value habitat, it houses the estuary of the longest river in Morocco" Moulouya River", it is a site of biological and ecological interest "SIBE" includingendemic and rare plant species, with roughly 86 endangered species including *J. phoenicea* [4]. At the present time *J. phoenicea*matorral in Saidia coastline area is threatened, and its functions are altered. Neither restoration strategies, nor sustainable management have started so far for its conservation. The aim of the current study is towarn about the degradation of *J. phoenicea*in Saidia coastline throughquantifying the inter-years dynamic of *J. phoenicea*matorral,locating and delineating remaining shrubs of *JuniperusphoeniceaL*in 2021 along the shoreline by a ground data approach. The study will conclude by suggesting strategies for conservation and restoration of the matorral.

Photo 1: (**a**) destruction of vegetation (**b**) uprooting long-lived shrub of *Juniperusphoenicea*L

2Materials and methods

https://doi.org/10.1051/e3sconf/202233701003

2.1Study area

The matorral of *Juniperusphoenicea* L. of Saidia coastline located in the north-easternmediterannen coast of Morocco. The region lies between latitudes 35°5' and 35°6'north and longitudes 2°14' and 2°19 west. (Figure1). Climateof the coastline is of a dry Mediterranean type with marine influences..The average temperature is 15.8° C, with temperate winter (11.1°C in January) and high temperature in the summer (22.3 °C in August). Mean precipitation is 319 mm in Saidia city, highest precipitation occurs in winter (November /January). Summer drought is severe, with no rains during summer (July /August), and scarce rainfall in September.

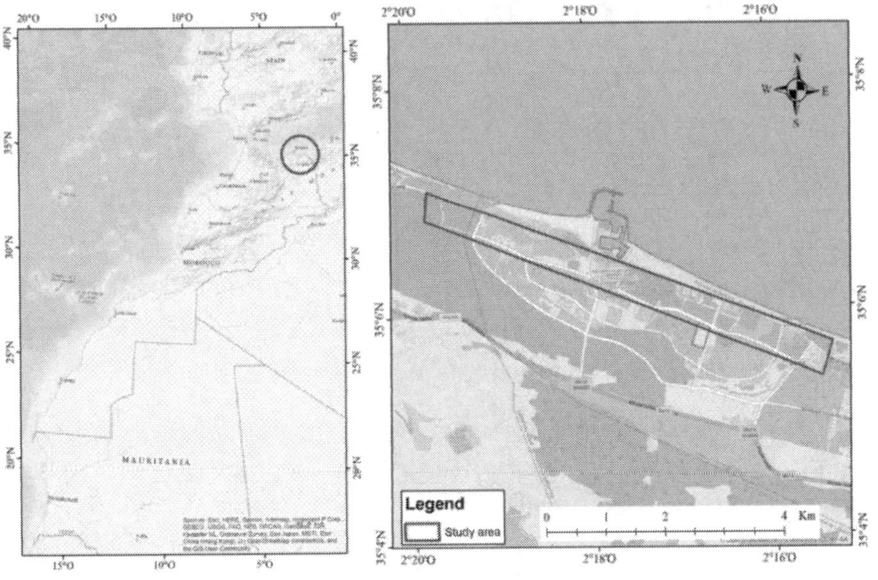

Fig.1.thestudy area.

2.2Data processing

Previousextension of the shoreline vegetation (Tazegraret forest) along coastline of Saidia was based on topographic map dated 1965, (scale: 1/50 000), performed using the software "geographic information system "ARCSIG 10.8" tools. The inter-years dynamic of the matorral, was shown by a comparison of current and past satellite imagery from Google Earth in 2004 (one yearafter Azur Plan) and 2021(11 years after the Azur plan) (Table 1) within the total extend of 31 ha treated with "ARCGIS 10.8" software tools.To map of current distribution of *J. Phoenicea* shrubs along Saidia coastline, was firstly performed with geographic positioning coordinate database within the total extent of the matorral (31 ha area).The shrubs were located and delineated by walking along the shoreline with a GPS receiverusing a "MOBILE TOPOGRAPHY" smart application. Secondly a digital map was performed using the software "geographic information system "ARCSIG 10.8". Was also listed the numbers of regenerations and stumps, spontaneous regenerations young trees, resulting from natural seeding, layering or sprouting from root systems.To assess the vegetation in the ancient remains of the forest a comparison of current and past satellite imagery from Google Earth (in 2004 and 2021) treated with "ARCGIS 10.8" software tools was performed.

Table 1: Details of used documents.

Year and document type	**Source**	**Scale/resolution**	**Uses**
Topographic map dated 1965	Berkan province	1/50000	Old extension of coastline
Image dated 2004	Google Earth	40cm	Inter-years dynamic of Juniper shrubs
Image dated 2021	Google Earth	20cm	Inter-years dynamic of Juniper shrubs

3 Results and discussion

3.1 Most Probablefloristiccomposition & plants associations

Data about ancient floristic composition of the forest of Tazegraretare fragmentary.Investigation about flagship plant species and floristic composition before the establishment of tourist project is deduced from following studies "le Plan Directeur des Aires Protégées du Maroc (AEFCS 1996)[5]; MedWetCoast-Maroc2003 [6], completed by "Ramsar consultative mission 2010[3], Projet FEM/Banquemondiale (GIZC P121271) 2012" [7].None of these studies provides an exhaustive analysis on the ancient vegetation of the coastline of Saidia, the species cited are based on our direct observations, with an extrapolation from the literature.The overall number of Taxa found is 131 representing 43 families classified in (Table 2).They are grouped into 3 large plant formations (sand dunes, Sansouire with *Arthrocnemum*et *Sarcocornia*, ripisylve with *Tamarixcanariensiswilld*and *Tamarixboveana Bunge*), On the Sand dunes was recorded the main following plants association: (a) association of *Otantho-ammophiletumaustralisJ.M. Gehu*, Rivas-Martinez & R.TX ; in [7], occupies small, mobile dunes and white sand dunes exposed to the sea winds.(b) association of *Climatidicirrhosae-juniperetumlyciae*Barbero et al. 1981 [8], which grows behind *Otantho-ammophiletum*often following it directly on the fixed gray dunes.The area is of Thermoméditerranan vegetation série type, it is under semiarid with temparate winter, which corresponds to coastal population of *J. Phoenicea. Juniperusphœnicea*subsp. *lycia*:,Sous-série of *Rhamnusoeloides*: thatcorrespond to*Rhamnorotundifoli-juniperetumturbinatae* [9].The subspecies *lycia* extending from Essaouira to the coastline of Saidia, covers the sandy dunes, semi-arid bioclimate and warm sub-humid to temperate bioclimate on soft sandy substrate [10]

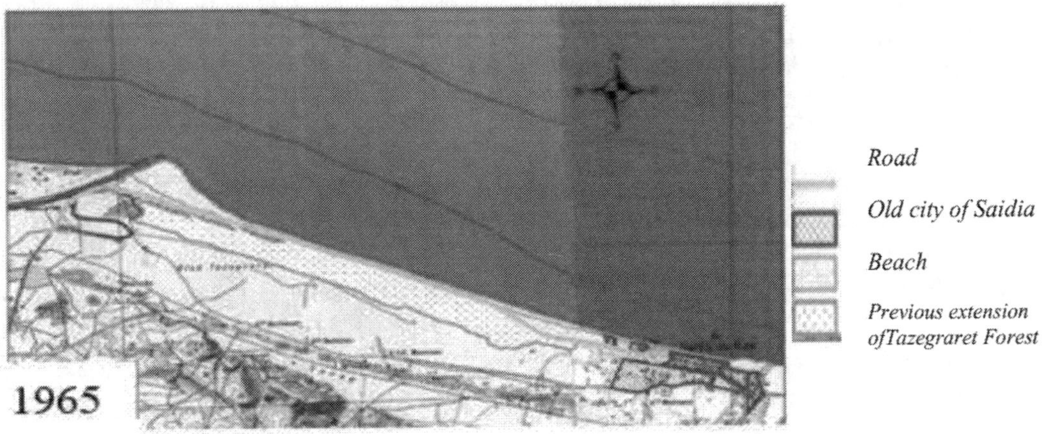

Fig. 2. Topographic map dated 1965, acquired from municipality of Berkane , showing old extension of Tazegraret forest along coastline of saidia, reflects the fatality of the vegetation losses (scale: 1/50 000).

Table 2: Status of conservation of the total taxa recorded

Status of conservation	Number of species	Number of families
Endemic taxa	13	7
Rare taxa	6	5
Threatened	3	3
Remarkable	89	37
Total	111	43

3.2 Analyse of state of *Juniperusphoenicea L.*matorral in Saidia coastline

3.2.1 Inter-years dynamic of Juniperusphoenicea L. matorral

 https://doi.org/10.1051/e3sconf/202233701003

The inter-years dynamic (between 2004 and 2021) of the *Juniperusphoenicea L*matorralin Saidia coastline isof regressive trend (Figure3). Comparison current and past satellite imagery analysis from Google Earth on 2004 and 2021within the total extent of 31 hectares area showees a strong reduction from 155 hectares in 2004 to 57 hectares in 2021, meaning a loss of 98 hectares (red color). Has risen is 63% in a period of time not exceeding17 years.*Juniperusphoenicea* L matorrals represent the late successional stage of the sandy dune of the coastline area in Saidia. In a floristic composition view, it is the most original formation of the shoreline.Currently it corresponds to the last representative matorral of sandy dunes on the Moroccan Mediterranean coast. A major problem it has to cope with is the combined action of anthropogenic pression e.g.urban development, water pumping, and natural pression e.g. high variability of inter-annual precipitation, decrease in precipitation amount, reduction of groundwater recharge rate, aggressiveness of hydrodynamic agents that lead to a huge degradation of shrubs. Threatened after vegetation eradication, a number of 18 species of birds are nowadays threatened of extinction,tourist project has already exterminated thousands of Greek turtles and *Chalcidae*(species closely linked to this ecosystem and listed only in 4 regions of the

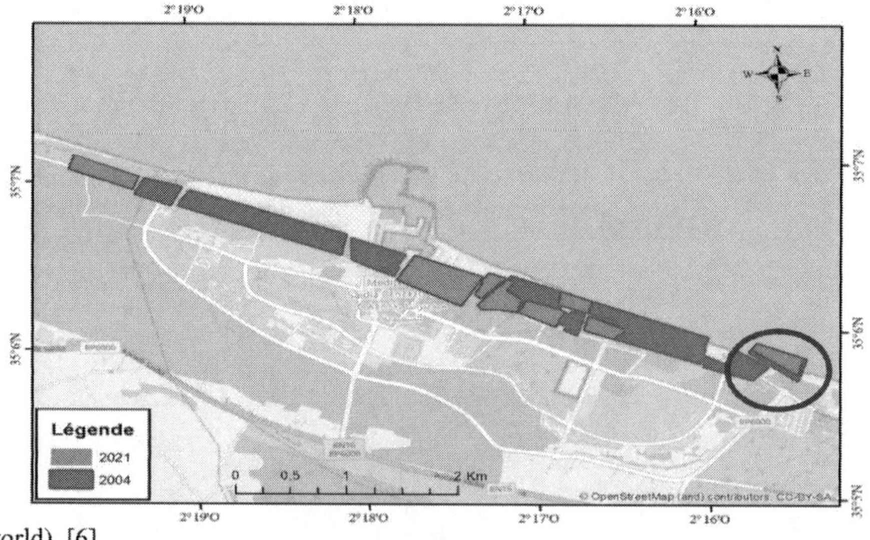

world). [6].

Fig. 3. Inter- years dynamic (year 2004 red color area) and (year 2021 green color) of the *Juniperusphoenicea L*matorral in Saidia coastline. The circled part is the eastern part of the ancient forest subject of Figure. 5 map.

3.2.2 *Current distribution of J. phoenicea along Saidia coastline*

Amongst1233 counted shrubs, 1072 are in good condition when 96 are dead. Shrubs arecurrently covering an area of about 31 ha, spread over a distance of 7km, from the city of Saidia until the site of biological and biological interest "SIBE" at the mouth of MoulouyaRiver(Figure4).Total of 113 new vegetative regenerations was counted, while 48 shrubs found in slops regenerated from seedlings. The mean habitat threat is not only the weak regeneration mechanisms, butdirect destruction of the population and its segmentation in distanced patchs with large disconnections in between (Figure 4).The isolation of the shrubs interrupts ecological processes such as dispersion of seeds. The study enable us to quantify the destruction effects on the Juniper shrubs on Saidia shoreline, given that shrubs fell and uprootingare amongst the most important risks in Saidia (Photo 1). However, the species is rustic and resistant, with little specific ecological demands; it survives at low precipitation 200-400 mm, it occupies a wide range of environments between the coastal sandy dunes to the mountain tops[11].

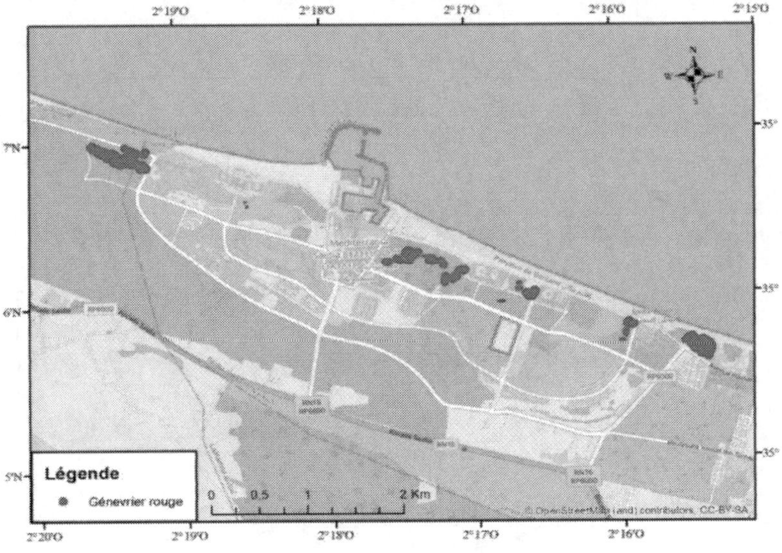

Fig.4.Located and delineated shrubs of *Juniperusphoenicea*Lin 2021 along the shoreline.

In Saidia coastline, the shrub is strongly withstanding the intense effect of coastal environmental stresses, such as salt spray deposition, sand mobility, and low water availability and may be considered stress-tolerant species[12]. The previous cited values give *J. phoenicea* an extreme importance in preserving the ecological balance within Saidia sandy dunes. *Juniperusphoenicea* L, it is indeed the best native species for future restoration programs.The shrub is native to the Northern Mediterranean and North-Africa[13].*Juniperusphoenicea* L is distinguished by the red color of its galbules, by its branched and less powerful port and the bushy appearance of its cylindrical branches *J. phoenicea*isin Morocco is confined in the dunes of the Mediterranean and Atlantic coasts, in the mountains of the Middle and High Atlas and Rif. The shrub is experiencing two types of treats. On one hand human-induced threat such as water scarcity [2];On the other hand, the current climatic changes that have almost affected the shorline zone causing profound changes at the mouth of the Moulouya river manifested by a remarkable retreat of the coastline[4]. The environmental risk predicted with climate change scenarios demonstrated that suitable environmental conditions for *J. phoenicia*are going to change in a way that triggers spatial in species geographic distribution range is expected to occur latitudinally northward and longitudinally eastward[14].

3.2.3Threat includes parts of the matorral out of Azur plan project

The analysis of the circled part (Figure 4),is the eastern part of ancient forest out of Azur plan projectconsidered as less threatened area.The analysis of vegetation dynamics shows an increase in the area of bare land, as it increased from 37,342 m² in 2004 to 61,662 m² in 2021. While the covered area with vegetation reached 54,443 m² in 2004, it decreased in 2021 to 30,103 m², with a rate of deterioration of 45%.This drop is related to an absence of any compensatory measure towards the coastline and its ecological values, the impact of planned facilities e.g. Buildings, golf courses, marina, water withdrawals from the Moulouya; protection channel against floods, bypass road, have accelerated the decline of biodiversity.

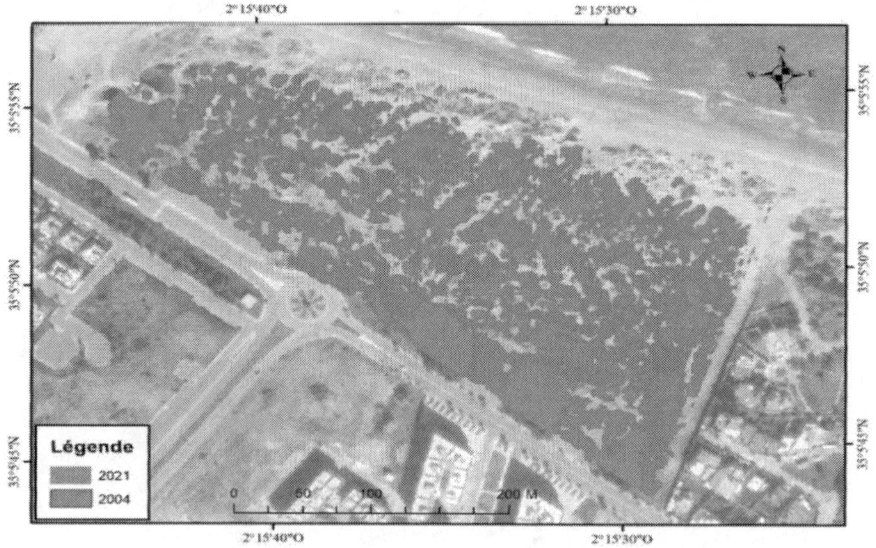

Fig. 5.Inter-years dynamic (2004 red color area) and (2021 green color area) of the vegetation in the eastern end of Saidia coastline.

3.3 Strategies for conservation and restoration of the matorral

The United Nations General Assembly declared 2021 to 2030 as the decade of ecosystem restoration. Hence conservation and restoration measures of coastal dunes in Saidia, are urgent in the current context of growing environmental awareness and concern for the future. When well-managed coastline dunes systems can harbour a great deal of biodiversity, landscapes of heritage value, associated with socio ecological consideration must be included for effective restoration and sustainability (e.g.integrated coastal zone management initiative ICZM). The main targets of coastal dune restoration are: (a) to avoid a loss of biodiversity, especially of rare and endemic species [15] (b) to avoid the encroachment of invasive species [16] that have been considered a major problem to be resolved. Given its ecological values *Juniperusphoenicea*L. is among the native species that can stabilise vegetation recovery of coastal dunes e.g. Conservation Program of Maritime Juniper Woodlands in Andalucia[17]. Planting *J. phoenicea* shrubs is necessary to prevent erosion of sandy dunes, and sustain natural resource management. The high level of degradation of *Juniperusphoenicea*L. is accentuated by a difficult random natural regeneration or natural rejection of sprouts; hence the majority of populations are represented by old shrubs. Yet to cope with is regeneration under controlled conditions as an alternative for conservation and sustainable uses. The real obstacle to regeneration of *J. phoenicea* is the difficulty of seeds germination due to dormancy phenomena. Failures to master regeneration techniques have always been an obstacle.

Acknowledgment

We are grateful to Pr. Sbai A. and Mr. Omar Mouadilifor their remarkable help.

References

1. ELeroux, A. Hmioui... Maghreb - Machrek 2019/1 (N239), (2019)
2. V Tekken, L. Costa,J.P. Kropp, *Increasing pressure, declining water and climate change in north-eastern Morocco.* J.Coa. Cons**17**(2013)
3.P. Triplet, M. Dakki, I.Cherkaoui, M. J de Lope, A. Dufour, *Mission consultative Ramsar* – Rapport n° 71. Moulouya, Maroc. (2010)
4.H. Lasgaa,T. Boumeaza, A Sbai,*Evolution of the coastline of Saidia - Cap Water(Northeastern Morocco).* IOSR J. App.Geol.Geop. **3**,4 (2015)
5.AEFCS.- *Plan Directeur des Aires Protégées du Maroc.* Rapport inédit, Administration des Eaux et Forêts et de la Conservation des Sols-BCEOM/SECA-ISR-EPHE.(1996)
6. MedWetCoast-Maroc. *La flore de l'embouchure de la Moulouya.* Rapp. Inéd.proj.PNUE.Secr. Etat.Envir.(2003)
7. Projet FEM/Banque mondiale (GIZC P121271).*Gestion Intégrée des Zones Côtières/Côtes Méditerranéennesrégion de l'Oriental*"(2012)
8. M. Barbero, A.Benabid, P.Quezel, S.Rivas Martinez, A.Santos,*Contribution à l'étude des Acacio-Arganietalia du Maroc sud occidental.* Documentsphyotosociologiques, nov.ser, Camerino, **6**(1982)
9. A.Benabid**,** *Flore et écosystèmes du Maroc.* Ibis Press, Paris, (2000)
10. A.AAFI, M. Sghir Taleb, M. Fechtal, *Espèces remarquables de la flore du Maroc;* Centre National de la Recherche Forestière, (2002)

11. P. Queze, M. Gast, « *Genévrier* », Encyclopédie berbère [En ligne], document G24, mis en ligne le 01 juin 2011, consulté le 25 septembre 2020. URL: http://journals.openedition.org/. (1998)
12.JP. Grime,*Plant strategies and vegetation processes*. Wiley, New York,(1979)
13. RP .Adams, *the junipers of the world: The genus Juniperus*. (TraPubl, Vict, BC, 2004)
14. A. Arar, Y.Nouidjem, R. Bounar, S.Tabet, Y.Kouba,*Potential Future Changes of the Geographic Range Size of Juniperusphoenicea in Algeria based on Present and Future Climate Change Projections* Con. Pro.Ecol, **13,** 4 (2020)
15. E.Remke, E.Brouwer, AM.Kooijman, I.Blindow, H.Esselink, JGM.Roelefs*Even low to medium nitrogen deposition impacts vegetation of dry, coastal dunes around the Baltic Sea*. Environnement Pollution**157**(2009)
16. H.Marchante, E.Marchante, H.Freitas, *Invasion of the Portuguese dune ecosystems by the invasive species Acacia longifolia* Andrews Willd: effects at the community level. In: Child LE, Brock JH, Brundu G, Prach K, Pysek P, Wade PM, Williamson M (eds) Plant invasions: ecological threats and management solutions. Backhuys Publications, Leiden, Netherlands(2003)
17. M C.Muñoz-Reinoso, S. Azqueta, I. Redondo Morales,*Restoration of Andalusian Coastal Juniper Woodlands,* Restoration of Coastal Dunes, Springer Series on Environmental Management (2013)

Health risks, governance and changes in ecotourism paradigms during Covid19 pandemic, case study of Essaouira Province in Morocco

Mohamed Boukherouk[1,*], *Sidi Imad* Cherkaoui[2], and *Adil* El Filali[1]

[1]Laboratoire de Recherches des Sciences Appliquées à l'Environnement et au développement Durable (SAEDD, Higher School of Technology of Essaouira, Cadi Ayyad University (Marrakech).

[2] Ecole Supérieure de Technologie de Kénitra, Ibn Tofail University, B.P. 242, 14000, Kenitra, Morocco.

Abstract. Tourism merits on natural spaces are not exhausted, namely economic dynamics, development of the local population's life conditions, species conservation and awareness of environmental protection. However, the damage caused by Coronavirus to natural and rural spaces has impacted such merits negatively, for example, suspension of tourism and recreation activities. Further, the rebirth of ecotourism will be slower and necessitates profound changes in travel design and service delivery. Hence, sustainability, sanitary and regulatory measures will probably help manage ecotourism systems. Nevertheless, these will be affected by the increase in exploring relations between nature and wildlife regarding zoonotic and vector-borne infections and the fragility of biodiversity. Because of his mobility, a tourist appears to be a potential pathogen transmission factor of infections to natural areas, local communities, ecotourism destinations, his country and the crossed ones. New procedures would reduce travel and control traveler's behavior, and interactions between travelers and wild animals. The slowdown in demand may thus hinder the development of ecotourism. Sensitivity to epidemiological risk and vector-borne diseases will be greater and the need for health-friendly tourism will be further strengthened.

[*] Corresponding author: m.boukherouk@uca.ma

1 Introduction

Covid19 rise is a moment of rupture [1, 2] in ecotourism as in the various sectors and aspects of human life. In addition to the global pandemic risk, more than a year of stoppage of the activity has profoundly affected the tourism system [3, 4] and transformed the way in which tourism products are consumed, and more particularly those based on interactions with nature. The latter turn out to be the most affected since they call into question the relationship between human and nature [5, 6, 7]. The bitter reality of the birth of the Coronavirus outbreak in bat and/or pangolin populations puts us face to face with the permanent risk of zoonosis and pushes the international community to rethink its relationship to nature, and particularly in ecotourism

In Morocco, ecotourism products were developed steadily from the 1990s. It took the form of hiking activities in the mountains, pre-Saharan areas and the hinterland of traditional tourist poles first (Marrakech, Agadir, and Fes). Then, it spread gradually in other niches developed in more fragile ecosystems such as natural parks and sites of biological and ecological interest. The economic benefits of this activity had limited the rural exodus, diversified the activity of the local population and helped in nature and ecosystems protection and conservations. This is the case of Essaouira. Listed as a Human heritage by UNESCO since 2001, the city has an Archipelago classified as a "RAMSAR" wetland and "Site of Biological and Ecological Interest: SIBE."This last is retained in 1996 within the framework of the Master Plan for the Protected Areas of Morocco, taking into account its rich flora, marine and terrestrial fauna [8]. The area benefits from a great interest in preservation thanks to several assets: coastal and island dune system, vast Argan and cedar forest, protected fauna (Eleonore Falcon) in addition to species, especially in transit since the city is located on the migration routes of birds (Passerines, waders, raptors...) [9,10]. In addition, there are many authentic villages where local life is based on traditional subsistence farming in small plots and breeding of goats and camels in the Argan woodland biosphere reserve [11].

Paralyzed by the Coronavirus pandemic, the ecotourism in Essaouira has been suffering for more than a year now from the absence of tourists and the cessation of activity. Certainly, the coronavirus acts as an element of rupture of the impetus that ecotourism has taken in the territory. The closure of borders, bans on movement within the country and the imposition of travel permits have dampened the desire for travel. Consequently, the actors plunged into a negative spiral and the ecosystems suffered from the cessation of money transfers for the preservation and conservation of species and environments.

If return to normal would seem difficult, given the current economic situation, the beginnings of the recovery will have to be done under the sign of the biosecurity risk for populations and species. Ecotourism should more than ever before take into account health considerations to meet the needs of Eco tourists without endangering the population and the environment. A health friendly approach to tourism is considered a crucial choice for which stakeholders are increasingly committed. The governance of the recovery is important to prepare for the post-covid.

This paper analyzes the new paradigms of ecotourism under the effect of rising health risks and biosecurity considerations. Through the case of the province of Essaouira, it highlights the multiple negative impacts on ecotourism, the ecotourism actors in the province and the governance model proposed for the revival. The results of a field survey make it possible to see how the new measures are applied and how the actors perceive them. It concludes to the difficulty of recovery in an ecotourism environment still under profound transformation and which will need to adapt to the biosecurity and health risk.

2 Methodology

This paper uses a global and systemic vision, combining the deductive and inductive approaches. It tries, through a methodology borrowed from sociology and geography, to suggest a framework of the reality of a tourist destination during the period of Covid19; namely from the beginning of the lockdown until the opening of the Moroccan borders. A review of literature of the concepts of ecotourism, the interplay between human and nature and biological risks were necessary. Then, a documentary study, especially of the grey literature, with an analysis of the data of the official organizations and publications of the Moroccan and international tourist institutions allowed to understand the situation of the territory in relation to the national and international contexts.

The empirical part is made of an analysis of the protocols and guidelines proposed by the Moroccan authorities and which operate a change in ecotourism. In addition a qualitative study on a sample of 30 ecotourism professionals from the province was conducted. This study aimed to assess the impact of the crisis and the perception of the actors of the opening period, after lockdown, as well as the post-Covid period. The questionnaire was drawn up in French and Arabic. The first part of the questionnaire collected data on gender, age, level of education. The second part focused on the Covid19 impact on travel and ecotourism in the studied area, the undertaken actions, and the changes in ecotourism system under the influence of the pandemic and the operator's perception of these changes.

3 Literature review

The economic, social and ecological merits of ecotourism are legion both for territories [12, 13, 14] and natural environments [15, 16]. Based on the principles of sustainable development, it allows both the enhancement and the promotion of environments while aiming at the protection of the environment and its safeguard for future generations [17, 18].

The ecotourism orientation in Morocco was born from the commitment of the country, and the actors in the path of sustainable development, as it's the case in other developing countries [19]. The 2010 Vision, through the strategy of sustainable tourism boosted ecotourism in the country. Perceived, mainly as an opportunity for territories and fragile or marginal areas [20 21, and 22] literature has also developed extolling these different merits [23, and 24]. Moroccan researchers will tackle also the models, actors and sometimes criticize the risks of gentrification [25] and excessive levels of overtourism.

Literature on ecotourism and tourism, in general, has not overlooked the problems that can arise from the interaction of human with natural spaces and fauna. Indeed, Hall [26, 27] insists on the fact that tourism can be a vector of transmission and propagation of diseases, following the same reasoning of other authors which he quoted in his chapter book: ecotourism and biosecurity. He also provides a relevant analysis of the risks and mechanisms and strategies to counteract the biosecurity risks that can be amplified by the tourist activity [28, 29, 30, 32]. Of course, the interaction of humans and natural environments expose humanity to a large number of zoonosis and pandemic risks [31] like SARS [31, 33], H1N1 influenza [34, 35] and Ebola [36]. The question arose sharply putting the bio-security risk at the center of issues related to ecotourism [27].

The impacts of covid19 on tourism are not the only subject to have caught the attention of researchers. Several studies focused on post-Covid19 and changes that will occur in the future. As a moment of rupture [37] states try to respond to covid19 effects in an authoritarian manner [38, 39] while other think about its resilience capacity [40, 41].

Certain affirm that we cannot have a return to normal [42] while others suggest sustainable strategies to ecotourism recovery [43, 44].

For decades, ecotourism has helped to conserve nature and protect endangered species. COVID-19 and the subsequent closure of ecotourism sites has had an indelible impact on wildlife and the communities that protect it [45, 46].

Because of the pandemic, the loss of ecotourism revenue has affected livelihoods and wildlife conservation efforts due to the reduction of internationals travels to Morocco [47].

4 Importance of ecotourism in Moroccan and Essaouira

For two decades, sustainable development has been able to awaken environmental awareness and participate in the reshaping of the tourist offer under the pressure of an increasingly accumulated demand for responsible travel towards the environment and the local communities [48, 49]. While it is still difficult to measure the flows and the impact on the environment, there is no doubt that ecotourism is constantly growing.

In 2000, International ecotourism society [50] published a factsheet in which we can read "Ceballos-Lascuráin (1993) reports a WTO estimate that nature tourism generates 7% of all international travel expenditure (Lindberg, 1997). The World Resources Institute (1990) found that while tourism overall has been growing at an annual rate of 4%, nature travel is increasing at an annual rate between 10% and 30% (Reingold, 1993). Data which support this growth rate is found in Lew's (1997) survey of tour operators in the AsiaPacific region who have experienced annual growth rates of 10% to 25% in recent years (Lindberg, 1997). WTO (1998) stated that ecotourism and all nature-related forms of tourism account for approximately 20 percent of total international travel." Currently, ecotourism seems to be reaching considerable proportions. For Vig & Deshmukh [51] the ecotourism market size was valued at $181.1 billion in 2019, and is expected to reach $333.8 billion by 2027, registering a CAGR of 14.3% from 2021 to 2027.

In 2018 the Center for Responsible Travel [52] estimated between 55% and 66% the average number of travelers willing to pay more for brands committed to sustainability, social and environmental initiatives between 2014 and 2015. According to a study carried out by Booking.com, 73% of international travelers intended to stay at least once in eco-responsible accommodation in 2019. That is to say before the appearance of covid19. According to the same study, this figure has increased for the 4th consecutive year, going from 62% in 2016, to 65% in 2017 and to 68% in 2018 [53].

Moreover, a study by the same website carried out in 2018 reveals the importance of environmental sensitivity without the influence of the travel truce induced by the pandemic. The results show that among the 12 markets surveyed, the vast majority of travelers (87%) say they want to travel while limiting their impact on the environment, and 39% of travelers say they actually succeed in limiting this impact [54].

In Morocco, it is obvious that the development of ecotourism is largely linked to the inauguration of the Central High Atlas (P.H.A.C.) project in 1983 resulting from Franco-Moroccan cooperation. Nevertheless, the Toubkal massif has known a tourist tradition that dates back to the colonial period during which the first climbs to the summit of Toubkal took place between 1880 and 1990.

Since the 1920s, under the French protectorate, the Toubkal massif has experienced tourist and sporting dynamics, in particular with the construction of high-altitude refuges and the Oukeïmeden resort. The Middle Atlas will experience the same process after the opening of Michlifen ski station in 1951. The post-independence period experienced a regression, which was overcome from the 1970s [55]. Subsequently, attendance at Toubkal National Park has increased steadily. According to estimates from the Ministry of Tourism, the number of visitors to the park via Imlil village was around 10,000 people [56], so the latest statistics put forward in the official HCEFLCD website is 40 000 tourists per year in 2004. The statistics on arrivals in rural areas remain interesting but are lacking. On the basis of the estimates of the experts who carried out the study on the Moroccan rural tourism strategy, the ecotourism potential of the country can be estimated to 4,030,566 in 2000. The assumption is that each client arriving in tourist accommodation is potentially a consumer of rural tourism in nearby areas. This allowed them to estimate the number of tourists who passed through rural areas between 2,099,924 or 2,656,142 in 2000 [57]. This study remains pioneer and interesting for several reasons. Although the fact that it's based on estimations, it reveals the potential of natural spaces and therefore the wide possibilities for developing ecotourism in the Kingdom.

In Essaouira province, tourist activity appeared in the shadow of Marrakech. Mainly, day trip city, for tourists from Marrakech, it has gradually developed its own products trying to stand out from the ocher city and from Agadir in the south. Around a nucleus of its intramural city classified as a World Heritage Site by Unesco, the city of the trade winds is surrounded on its west side by an archipelago classified as a site of biological and ecological interest. Inside the country, an Argan grove classified as a biosphere reserve since 1995 populates the plains of the north and east and extends to the Western High Atlas in the south of the province. This endemic tree got its international day recently 2020 (May 10th): the Argania Day. Also found in the region thuja forests and several species of endemic flora and fauna (Eleonora Falcon) and an authentic rural peasant life and ancestral know-how.

In the last two decades, the province of Essaouira has known a tourism dynamic. Thus, according to the local tourism office, the city has 231 accommodation units, mainly made up of small hotels, riads, guest houses and hostels (60% of the whole capacity) as well as 30 restaurants, 32 travel agencies and 34 guides. If the activity has been developed largely only in recent years, it is mainly thanks to the 2010 and 2020 visions which gave more importance to the province. The office estimates at 12,000 the number of direct jobs created in tourism and to 1.4 billion dirhams its incomes. The sector turnover is just over 22% of the province's GDP. The activity is gaining more and more importance both through employment and through its multiplier effect on all other sectors, in particular construction and trade [58].

The hinterland attracts tourists in search of adventure. Hikes with overnights in camps or homestays in the villages of Sidi Mbarek, Ain Lahjar or Idda Ou gourd are the major products. Surfing attracts wave's enthusiasts while ornithology is another niche to develop. Stays in farms and agritourism experiences are emerging, trying to attract tourists by developing a new offer for customers looking for new experiences. The city has only two hotels classified as Clef Verte (Green Key) [59]. Developed in France since 1998, La Clef Verte label is, by the number of labeled establishments and by its age, the first Eco label for sustainable accommodation in France. However, all tourist operators (guest houses, lodges and guides ...) and farmers try to promote and communicate with an ecological and organic orientation. This spirit is also the commitment of the local authority which has made the development of tourism in the hinterland a tool to position strongly and sustainably, their destination. Indeed, the Eco-nature program of the province tourist office in partnership with the provincial tourism council CPT is an example of an integrated development plan of Rural Tourism and Nature that was developed with different stakeholders. It aims the development of the ecotourism potential of this area. Unfortunately, that impetus was stopped by the pandemic.

5 Impact of coronavirus on ecotourism in Essaouira

The covid19 pandemic has paralyzed the tourism sector. The decisions that states took to stop the spread of the pandemic have given a heavy blow to the dynamics in tourist areas, especially those relying on ecotourism. Indeed, without mobility all tourist activity has evaporated and with it jobs and revenues. The world tourism organization estimated the drop in the number of international tourists at 97% compared to 2019 during the period March to May 2020 (i.e. the period of drastic lockdown) [60]. This translates into a fall of 300 million tourist's arrivals and $320 billion of incomes [60]. The effects of this crisis will certainly last until 2024 [61].

In Morocco, economic losses have been estimated at 1 billion dirhams per day of lockdown or one point of GDP every 10 days or the equivalent of 2,000 to 2,500 jobs lost per day. The tourism sector was also a quarantined sector because of the closure of sea and land borders, the suspension of flights, the closures of places of entertainment, etc.). Consequently, professionals were forced to suspend all their activities since mid-March. They quickly found themselves facing cash flow deficit, unable to honor their commitments even those of the short term (salaries, bank and insurance payments, etc.). The National Tourism Confederation (CNT) estimates the impact of the coronavirus in Morocco to $3.4 billion by the end of the year and 70% the reduction for arrivals and more than 60% of overnights at the end of July 2020 [62].

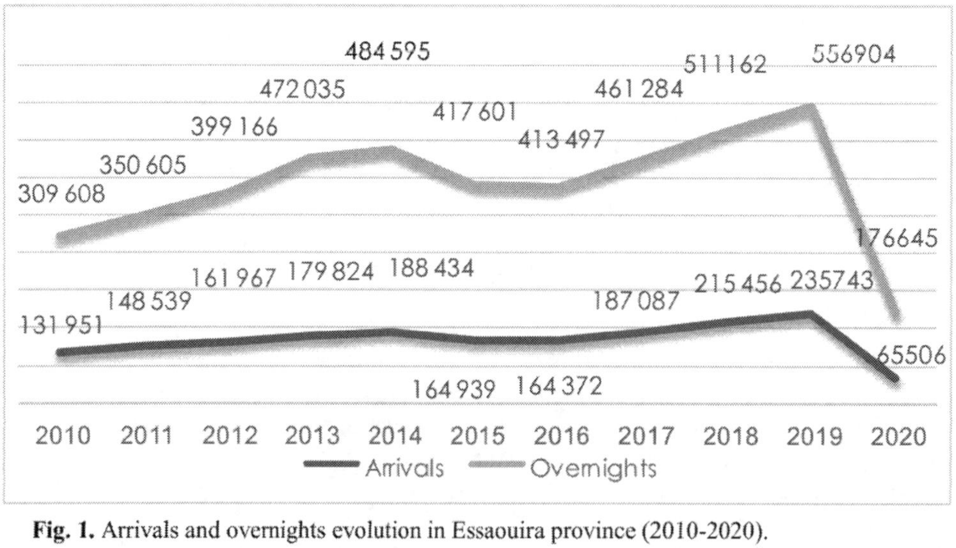

Fig. 1. Arrivals and overnights evolution in Essaouira province (2010-2020).

Locally, and in relation to the province subject of our study, Essaouira seems like a ghost town. Tourism covid19 crisis has impacted all aspects of life and all the tourism system. Hotels, restaurants, transporters, souvenir sellers (bazaars) have been closed since March 2020... Even more serious, our permanent observations confirms the negative multiplier effect on several other activities: agriculture, local commerce and real-estate sector... In other words and according to the province tourism office data: tourist arrivals cut down by 72%, resulting in a 68% drop in overnights compared to the previous year (fig.1). Moreover, 51 accommodations (table.1) located outside the Medina of Essaouira are still closed temporarily, and a capacity of 1653 rooms continues to be closed today. They thus estimated the size of jobs lost at 170 direct jobs, just in the hinterland of Essaouira. The only official guide no longer works since March 2020 and neither do the twenty fake guides when all the Argan oil and Thuja cooperatives are still closed.

Since no museums or Eco Museums exists in the province and no entry fee is imposed for tourists entering the reserve, the only data we have to estimate the volume of ecotourism is the hiking activity involving overnight in camps. It gives us an idea about the losses of ecotourism activity. Thus, after a gradual growth in the number of bivouacs since 2016, this activity has totally stopped because of the pandemic. (Fig.2)

Table. 1. Aaccommodation units closed due to the pandemic.

Accommodation type	Effective
Hostels	1
campings	3
Lodges	10
hotels	6
Guest house	30
Appart hotel	1
Total	51

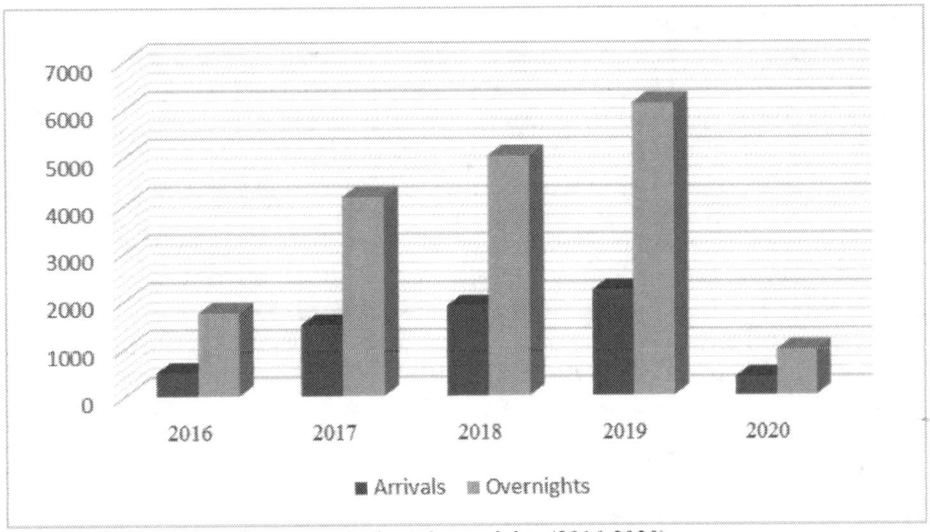

Fig.2. Evolution of camps arrivals and overnights (2016-2020)

6 Environmental Governance of the crisis and ecotourism

Ecotourism in Morocco brings together several actors and communities. However, no specific legal framework manages the development of nature activities. Only National Responsible Tourism Charter of 2006 serves as a basic frame of reference but lacks depth and clarity. The framework law n ° 99-12 [63] on the National Charter for the Environment and Sustainable Development, promulgated in 2014 (Dahir, 2014), The Law of July 16, 2010, on protected areas, the law on environmental protection 2003 and the one on environmental impact studies (Law n° 12-03) [64] provide a more precise framework and general measures but not specifically linked to tourist activities. In these texts, the relationship between human and nature must be respectful. It insists on environment safeguard and its sustainable management and urges the actors to embark on this path. None of the texts mentions the case of health crises and zoonosis.

Nevertheless, the monitoring and control system are rich with legal texts and organizations that participate in the surveillance and prevention of diseases. The modern health control legislation system dates back to 1916. It saw the reorganization of the maritime health police [65] and the organization of health control at land, sea and air borders [66]. They had as main mission the diseases watch (Cholera, plague and yellow fever) and the borders sanitary control in the ports, ports of call and all ships which should dock in the Moroccan ports. They had the power and ability to prohibit the boarding of persons and objects liable to spread transmissible diseases and take precautionary measures against the spread of these diseases (article 15, Dahir 1916) [65]. Since then, the system has been enriched by several texts and institutions.

The governance of the pandemic crisis is essentially dominated by the influence of national health authorities. There are thus central administration institutions such as the Department of Epidemiology and the Fight Against Diseases which provide epidemiological surveillance [67], The National Institute of Hygiene (INH) , The National Virology Laboratory (INV) which deals with the surveillance, isolation and identification of viruses (polio, influenza in particular). These institutions are supported by external services, such as the Tuberculosis and Lung Disease Diagnostic Center (CDTMR) and the Environmental Epidemiology and Hygiene Laboratory (LEHM). At the local level there are the services of the Wilayas and its environmental health services: responsible of the supervision and control of actions related to the monitoring of environmental hygiene. Morocco's recent history of health crisis management has a long and rich experience. Several programs were put in place since mid-2000s. It's the case with zoonosis such as rabies and hydatid/echinococcosis and also those of Environmental hygiene programs (inspection of food industries, food control and restaurants). Fall into this category epidemiological surveillance of animal diseases (avian influenza, foot-and-mouth disease, African horse sickness, bovine spongiform encephalopathy,), and diseases such as blue tongue, sheep pox, tuberculosis, brucellosis, rabies...

The country has a National Animal Disease Epidemiological Surveillance Network (RENESMA) and an epidemiological surveillance program where several institutions collaborate with the Epidemiological Surveillance Service within the Department of Epidemiology and Disease Control. The program aims to regularly collect and analyze surveillance data to feed the alert system to detect early epidemic phenomena and prepare for the response. This program has a history of the epidemiological situation of notifiable diseases, made as early as 1980. The Epidemiological Surveillance Service also manages global alerts and disasters. This was the case of SARS-Cov in 2003 [68]. Indeed, a special attention is paid to influenza surveillance. In fact, avian influenza was at the origin of the creation of the service with a database fed by a network of sentinel centers in different parts of the Kingdom. In the same spirit, the service has a system dedicated to rotavirus gastroenteritis since 2006. Epidemiological surveillance in the Moroccan health system is carried out in partnership with the regional office for the Eastern Mediterranean of the WHO.

The governance of the period of the Covid19 pandemic was managed very early [69, 70]. After the declaration of the state of emergency, a set of institutions were involved in the management of the crisis for which a technical and scientific committee was created and which pilots and coordinates the decisions taken in the matter, in particular with regard to the restrictive and/or easing measures. Thus, coordination between different sectors is ensured by the Interministerial Crisis Management Committee (CIGC). The operation is coordinated by the Central Coordination Post (PCC) [70].

At another level, the Ministry of Health with its organs is strongly represented with its Steering Committee of the response of the health system where we find the Minister or the Secretary General of the Ministry of Health or who represents them. Various departments are also in this committee such as: the Department of Epidemiology and Disease Control (DELM); the Department of Hospitals and Outpatient Care (DHSA); the Directorate of Medicines and Pharmacy (DMP) [67]... In addition, the National Public Health Emergency Operations Center (CNOUSP), at the DELM level, monitors the epidemiological situation at the international level and coordinates the technical aspects of preparation and response. It also monitors the epidemiological situation of the disease at the national level and informs the various stakeholders and partners, the media and public opinion (according to Ministerial Decision No. 013065 of September 16, 2019, on the creation of CNOUSP, CROUSPs). [67].

At the regional level the crisis is managed by the Regional Centers for Public Health Emergency Operations (CROUSPs). It is set up at the level of the Regional Health Directorates (DRS), and is under the responsibility of the Head of the public health service. Its role is to monitor the epidemiological situation at the regional level and coordinate the technical aspects of preparation and response. It is supported in its mission by the regional intervention teams. The Chief and executives of the Public Health Service, the Director of the Regional Hospital Center (CHR) and a clinical doctor, preferably an infectious disease specialist, participate in this crisis ecosystem; the Biologist in charge of the CHR laboratory; the head of environmental health; the head of the Communication Unit; and Any other resource person whose presence is deemed useful [67, 68].. In addition to their mission of monitoring the epidemiological situation and steering the response within their region, the CROUSPs must play a leading role in the organization of training and awareness sessions for health professionals and the various stakeholders in the public and private health care structures. [68].

As far as the tourism sector is concerned, the state of emergency involved the prioritization of health and pandemic issues. The tourist authorities were unfortunately not consulted and remained awaiting announcements from the official authorities in charge of the crisis. The continuous extension of the state of health emergency since March 2020, the ban on travel at the national level and the accompanying health measures (for hotels, restaurants, travel agencies and the other sector companies) have been managed by the Ministry of interior and its regional and provincial bodies, notably the Wali and the governor. But tourism operators have not stopped asking for simplified procedures, at least for domestic tourism. They have repeatedly expressed their fears about the dislocation of the Moroccan tourism system because of the activity shutdown.

At the economic level, the governance of the crisis was based on the creation of an economic watch committee (CVE), at the level of the Ministry of the Economy, Finance and Reform of the Administration. This committee, which includes the economic departments alongside the Wali of the central bank of the Kingdom (Banque Al Maghrib) as well as the ministries of health and labor is responsible for monitoring the development of the economic situation and taking the appropriate measures to support the affected sectors. Since its first meeting on March 16, 2020, the committee has taken a number of measures to cushion the economic shocks due to the pandemic: suspension of the payment of social charges (CNSS contribution), the postponement of incomes declarations and corporation taxes payment, deferral of loan maturities, a loan guarantee and exceptional overdrafts (Damane Oxygène & Auto-entrepreneur Guarantee) for companies affected by the crisis [71, 72]. Other measures aimed either at reducing the burden of debts and receivables or at proposing guaranteed financing mechanisms. The tourism operators benefited from the same measures, in addition to a Covid-19 lump sum, in compensation for lost wages. This allowance was set in the beginning of the crisis and has been extended several times until August 2021.

According to the interviewed operators, the governance of tourism paralysis, has been made under the waiting sign and the domination of measures taken from above. The restrictions lifting during the second part of 2020 was accompanied by new obligations. From now on operators should comply with measures of physical distance, impose barrier gestures, reduce passengers and clients in the same space or transportation and disinfect equipment's and premises permanently (table.2) [72]. The primacy of the management of the pandemic crisis and the risk of propagation become an obligation. To enforce its new measures, regular checks were carried out and led to companies' closure for those which not comply with the established instructions [73].

Table. 2. Summary of changes in tourism services due to covid19 period

General measures: Set up a special COVID 19 HR unit to control staff temperatures, report suspected cases, daily brief the staff: Set up shifts with a fixed staff. Reinforced health and safety instructions. Limit the number of customers present at hotel at the same time. Mandatory and systematic update of all COVID 19 protection procedures. Provide protective equipment: masks, face shields visors, disinfected uniforms. Provide masks for customers and staff; (Compulsory - (to be renewed every 4 hours)) Set up a quarantine procedure in case of a positive case among clients or staff; Notify customers about the possibility of rehousing if required. Compliance with social distancing measures in common areas.	
Hotel Booking, Check-in and Check-out	Priority remote reservations (online or phone). Online check-in and payment, Defer check-ins and check-outs to avoid the risk of overcrowding. Avoid printing (prefer emails to be sent directly to the customer). Set-up a crawl space of 6 hours between reservations. Disinfection of luggage and keys (during the check-in and check-out).
Hotel Front desk	Comply with health and distancing measures in terms of capacity (50%). Ground marking to avoid reception hall overcrowding Equip the reception desks with protective Plexiglas screens and separations. Equip reception with a medical kit suitable: sanitizers, disinfectant wipes for surfaces cleaning, gloves, and disinfectant spray. Display of safety measures (screens or use posters in the reception hall). Regular disinfection (front desk area, desks, lifts and common sanitary blocks).
Hotel common areas, conference centers, pools and beach	Permanent display of the safety and hygiene measures for customers. Redesign the kids clubs, limit reception capacity and hygiene measures. Install sanitizer's dispensers. Reinforced cleaning and disinfection guidelines. Air filters cleaning and regular ventilation. Better use of natural ventilation. Deactivate the electric hand dryers. Disinfect vehicles. Disinfection and cleaning of deckchairs and beach towels. Sports halls, Hammam and Spa must comply with the guidelines; (Closed) Possibility of keeping the seminar rooms open during the events; The installation of parasols must respect social distancing measures. Carry out a site zoning in the beaches.
Hotel rooms & floors	Strict cleaning protocol for the rooms. Train and inform staff on the Covid19 measures; Limit the use of bathroom amenities to single use. Use certified cleaning products for laundry. Cleaning air filters and regular ventilation (Better use natural ventilation).
Restaurant (Hotel and outside restaurants	Train staff on good hygiene and safety practices against COVID 19. Put up awareness posters on prevention and protection measures against COVID 19. Regularly disinfect door handles, stair railings, chair armrests, tables, electronic payment terminals and any other frequently touched work tool. Disinfect equipment and accessories, made available to customers, before and after use. Temperature control for staff, clients and suppliers. Respect of a minimum distance of 1 meter between 2 tables;

	Use separation screens Servers must keep at least one meter from customers when ordering. Suspend operations that require customers to use common utensils or vending machines, such as salad bars, buffets and beverage vending machines. Customer flow management for buffet and dedicated staff for every service; Establish rules with suppliers to verify compliance with health security measures related to COVID-19: delivery and reception staff for deliveries must be trained to properly disinfect, unpack and clean products. Adopted protocols in kitchen: handling of cutlery and utensils. Encourage room service to avoid queues. Room service should consider disposable packaging: boxes, cutlery… Reinforce the cleaning of air filters, air conditioning and hoods. At the entrance, access or high traffic areas (kitchen entrance, delivery area, etc.), place a disinfectant non slipping mat and sanitizers dispensers. Animation activities must respect social distancing measures. Electronic payment. Maximum 4 people per table with minimum distance respected (members of the same family are excluded from this restriction). Dedicate an area to take-away and home delivery service. Practice social distancing during home delivery by opting for "contactless" deliveries: send text messages or call when deliveries are ready.
Travel agencies	At the entrance access or heavy traffic areas, place a non-slipping disinfectant mat; Regular disinfection of the premises. Inform and educate staff and customers on hygiene and prevention rules. Redesign offices to respect social distancing. Limit and control the flow of customers: maximum of two customers Mandatory floor marking outside the agency. Encourage teleworking. Develop online reservation channels. Set up a hotline and social networks presence. Guarantee transparency and inform customers of all the preventive and hygiene measures required when preparing trips. Distribute hygiene guidelines guides and safety practices on the agency's website or on social networks. Establish rules with suppliers and partners to verify compliance with health security measures related to COVID-19. Notify the possibility of rehousing when needed.
Guides	Comply with regulatory recommendation (within the framework of health crisis management) regarding the maximum number of people per group. Wear a mask (Compulsory) Ensure compliance with social distancing rules: before, during after a tour. Respect minimum distance between clients during visits, breaks and stops. For guides in natural areas, used equipment must be systematically disinfected after each use. Have emergency numbers. Take a hygiene kit (masks, gloves, disinfectants, etc.). Use audio equipment for visits. Check if customers wear masks (to be renewed every 4 hours). Stop any tour if the guide or a customer have symptoms of COVID 19. Use electronic payment.

Thus, the collection of recommendations published on June 1, 2021, presents the list of protective measures against COVID 19 that must be put in place: before, during and after every service. The collection was intended mainly for accommodation establishments, restaurants, travel agencies and guides. No mention is made for of transportation. It will be included in other recommendations made by the ministry of interior.

Indeed, after the injunctions for a complete closure of coffee shops, restaurants, Hammams, swimming pools and the ban on all cultural and groups events as well as all travel, during the lockdown. The lifting of these restrictions was done in three different phases and zones. 50% gauges, openings from 8 a.m. until 11 p.m. and curfews have been established. With the curfew still in place, several establishments were not authorized to open, depending on the zone where they are: Hammams, Spa, swimming pools... and events were banned, then banned with a maximum capacity of 50 people. The carriers for their part had to comply with the gauge of 50% during the first stage and 75% before going back to50% of the capacity with the respect of the fundamental obligations: physical distance, mask and regular disinfection. The deterioration of the health situation is behind the permanent change in the given recommendations.

At the environmental level, if a relaxation was felt during the first months of the pandemic. The recommendations and action of the Ministry of the Environment aimed at accentuating controls in fragile and protected areas and raising awareness of the repercussions of the pandemic on ecosystems. The management of new wastes linked to health care or prevention (tests, masks) was the subject of an awareness campaign of the Ministry of the Environment [75].

In short, the governance of the crisis was carried out under the sign of the state of emergency [76]. This situation framed by the Decree-Law No. 2-20-292 of March 23, 2020 [76]. In such situations, the authorities have full prerogatives to take all necessary preventive, urgent and immediate measures, decisions, laws and recommendations to fight the pandemic. The consultation of the ecotourism and tourism actors was thus very limited, but the involvement of all the stakeholders was vaguely requested. It has certainly made it possible to control the spread of the virus, despite the big economic and social losses. It has also initiated the first new transformations that ecotourism will undergo in the coming years.

7 Results

The field study was conducted on a sample of 30 ecotourism professionals. In choosing sample elements we favored a strong presence of the private actors (96.66%), since it is the most affected by the pandemic. SME offering accommodation services are the most important (66.66%) part followed by transportation companies and cooperatives. Each of these last companies represents 13.33% of the sample (table.3).

Table. 3. Field study sample composition

	Sample		Men		Women	
Actors	**N**	**%**	**N**	**%**	**N**	**%**
Guide	1	3,34%	1	3.34%	-	-
Homestay	3	10%	3	10%	-	-
Guest houses	11	36,66%	8	26,66%	3	10%
Lodges	6	20%	6	20%	-	-
Cooperatives	4	13,34%	1	3,33%	3	10%
Provincial tourism Office	1	3,33%	1	3,33%	-	-
Transportation	4	13,33%	4	13,33%	-	-
Total	**30**	**100%**	**24**	**80%**	**6**	**20%**

The typology of the hosts is made of guest houses, lodges and homestay providers. A situation that we can justify by the absence of other types of structures working in the area. Indeed, Essaouira has only one official guide of natural spaces. Likewise, the agencies that offer activities in the hinterland of Essaouira are based in other cities. Providers of activities such as horse riding, surfing, windsurfing... exist, but their activity is more recreational. In addition, the territory does not offer entry rights to the reserve and does not have an Eco museum outside the city. This is what justifies our choice of Argan and goat cheese cooperatives, which remains the only establishments that offer free tours in a formula that is somewhat similar to agro tourism. The propensity of men is higher (80%) compared to women (only 20%). The latter is either managers of guest houses (10%) or cooperatives (10%). The sector is strongly dominated by men, at least for owners and managers. Only cooperatives have a strong presence of women.

Generally speaking, all the questioned people are unanimous in confirming the negative impact of covid19 pandemic on ecotourism locally (table.4). All the hosts confirmed that they have being obliged to cancel their booking, from mid-march 2020 to the study date. Only two guest houses are preparing their re-opening for 2021 summer season, and the rest has remained closed. The only guide working in the area had lost his business and the same for transporters. This situation was confirmed by the tourism office delegate as suggested above in the official statistics about the impact of covid19.

Table.4. Themes and most listed elements during the interviews

Themes	Most listed elements
Negative impacts	Bookings cancellation Stoppage of the whole activity Deposits and fully paid bookings reimbursements Declaration of financial difficulty and inability to repay bank loans inability to meet fixed charges due to prolonged emergency measures; Treasurer exhausted by reimbursements and salaries Employees dismissal and lack of staff No clear strategy due to emergency status and borders closers; Difficulty to plan marketing activities; temporary stoppage of promotion, marketing and attendance at fairs and events. New expenses related to sanitary measures and products and to the training of teams to the new processes
Engaged actions	Closure and cessation of activity lower wages Declaration of the cessation of activity Request for CNSS (Caisse Nationale de Sécurité Sociale) aids Awaiting opening decisions Loan request Maintaining links with customers through the phone and social networks Search and diversify clients by seeking new markets (local) Implement new health protocols and prepare for the opening (by following the new measures for welcoming clients) Reorganize the staff activities in respect of the new measures Develop new products or experiences
Measures requested within the framework of State actions	Declaration of the cessation of activities Suspension of payment of social charges (CNSS contributions) request for monthly lump sum compensation for employees; Request for postponement of loan maturities without payment of fees or late penalties Request for postponement of tax declarations and deadlines (corporate tax, declaration of taxable incomes)
Potential changes in ecotourism	A slow return to normal; A demand mindful of the pandemic risk and services; The vaccination obligation The obligation of a health passport; Adaptation of services including health measures;

Because of the lockdown, accommodations were thus forced to stop their activity. A situation that has been prolonged because of the continued extension of the state of health emergency. Consequently, a series of negative impacts followed. Most significant impacts have a financial component. The majority of managers questioned evoke that they had to reimburse deposits and payments received, that they declared their financial difficulty and have been unable to pay bank loans. The extension of the emergency measures has also led to the incapacity to ensure even the fixed costs for 90% of the companies. Some of them were forced to dismiss employees (70%) because of a cash flow burdened by three months of lockdown and a long inactivity charges. The nature of companies, especially composed of SME's, means that they have not been able to plan to reopen after the lockdown lifting nor to be able to prepare an effective marketing strategy, preferring to continue their closure, despite the encouragement of the State to reopen for domestic tourists. Two guest houses which are preparing for the future opening are doubtful as to the strategy to be adopted in terms of price while new charges must be borne, mainly due to sanitary measures and protocols: purchase of masks and sanitizers, permanent disinfection of premises and staff training to new health and service procedures.

In sum, the situation remains complicated for all the questioned operators. First of all because of the state of health emergency and closure of borders for international customers and because of the obligation of travel authorizations and incessantly prolonged curfews for the local market. It is also these reasons which pushed the owners of accommodation, transport and cooperatives to go through emergency actions such as stopping their activity (100%), lower wages (95%) then the declaration of cessation of activity (100 %). All of them have also appealed to the aids for employees and requested for the postponement of credit maturities. Other actions such as searching for new markets, maintaining contact with customers and suppliers as well as reorganizing the work process and developing new products and experiences are also cited by some of them. This is the case of cooperatives that have tried to sell their products through new channels (online) while they were previously dependent on tourism activity. Even while waiting for the lifting of emergency measures, all of the interviewed declared that they took the pandemic risks seriously and made an effort to implement new health protocols. We were able to observe these measures in cooperatives and the tourist office which remained open after the lockdown. The other companies preferred to stay closed.

In any event, all of the respondents appear satisfied with the management and governance of the current crisis. Especially during the beginning of the pandemic. For the sample questioned, the health situation remains a priority and justifies decisions like lockdown and stopping the activity. Ditto for the aid measures which have made it possible to reduce the shock of the period of complete cessation of activity. However, the opinions of operators appear to differ when it comes to last minute decisions to limit travels and curfews. 65% believe that the situation requires that kind of decisions while the rest think that operators should be given more time to be able to open without incurring stealthy and last-minute decisions.

When it comes to ecotourism return to normal and the changes that will affect the sector, the interviewees are pessimistic. For all of them, the return to normal will take a long time because of the pandemic constraint and the accompanying measures (state of emergency, border closures, curfews, compulsory quarantine, etc.).These decisions will certainly affect the tourists' motivations. They are and will remain increasingly concerned with the travel health risks and constraints. For all the operators, this situation will seriously mark ecotourism in the short and medium terms. Thus, the vaccination obligation and the health passport will be decisive elements in tourist admission in countries borders.

In the same vein, ecotourism will be marked by profound changes in terms of supply. According to all interviewees, generalized and mandatory health protocols will be an integral part of the services they offer. They will be certainly decisive in travel decisions and destinations choices.

8 Conclusion

Ecotourism has been deeply affected by the covid19 pandemic. The catastrophe magnitude is huge, and the states and operators are still under the shock. They all proceed by trial and error to find a solution to the spread of the pandemic and its economic and social impact. There is still a long way to go despite the progress in terms of vaccinations and the efforts made to minimize the economic shock.

This study sheds light on the issues linked to covid19 and the changes in ecotourism in Morocco and more particularly in the province of Essaouira. It is a contextual study conducted during the period of covid19. Through an analysis of the measures which accompanied the attempts to revive the activity and a field study, it reveals the important the impact of the pandemic on Moroccan ecotourism and the difficulties of companies to rethink the post-covid period. It also gave an overview of the profound transformations that the sector will undergo in light of the measures imposed by the Moroccan authorities.

More than ever before, the sector appears to be heavily impacted. The domino effect on the various operators and providers is unequivocal. The closure of establishments, the layoffs, the declaration of financial difficulty have plunged ecotourism operators into a hellish loop of difficulties and deficits. The state's efforts to alleviate the impact of the crisis, although appreciable, do not allow an easy return to normal. All companies are struggling to design their start-up because of the uncertainty associated with the pandemic and the inconsistent recommendations of decision-makers.

In this frame, Covid19 shock is a rupture moment. It foreshadows profound changes in the whole ecotourism system. From now on, health considerations will be an integral part of the service offer of the various providers. Reconstructing the ecotourism supply will involve a strong health component in the service delivery process. Airlines, ground transportation, accommodation providers, museums, guides or restaurateurs all must review their services processes and arguments to be able to reassure their clientele in the future.

References

1. S. Sareen, K. B. Nielsen, P. Oskarsson, D. Remme. Front. Hum. Dyn **3:**636422. (2021)
2. N. G. Uğur., A. Akbıyık. Tour. Manag. Perspect, **36,** 100744. (2020)
3. M. Sigala. J. Bus. Res, **117**. 312–321. , (2020)
4. N. A. Bakar, & S. Rosbi. Int. j. adv. eng. res. sci., **7**(4), 189–193. (2020)
5. JW. Morse, TM. Gladkikh, DM. Hackenburg, RK. Gould. *PLoS One*. **15**(12). (2020)
6. J.A. McNeely. Ambio **50,** 767–781 (2021).
7. P. Girish. Time for rest? TRI, Volume **24, N°2-3**, (2020). 179-184(6).
8. AEFCS. Plan Directeur des Aires Protégées du Maroc. Administration des Eaux et Forêts et de la Conservation des Sols/BCEOM/SECA/ISR/EPHE. (non published report). (1996)
9. M. Dakki, M. A El Agbani, & Qninba, Trav Instit Scienti, Rabat, **7**: 238 pp. (2011)
10. C. Magin. Morocco. Bird Inter. Ser. **11**. (2001)
11. A. Benabid. Flore et écosystéme du Maroc, (Ibis Press. 2000).
12. A. Jurowski, U. Muzaffer, and W. Daniel. **35** (Fall): 3-11. (1997)
13. J. Liu, & V. Turgut (1986). Ann. Tour. Res, **13**: 193-214. (1986)
14. J. R. Ritchie. JTR, **38** (2): 100-10. (1999)
15. R. Schweitz. J for Qual and Parti, **September:** 36-40. (1996)

16. R. E. Owen, S. Witt., S. Gammon. T.M, **December:** 463-74. (1993)
17. J. R. MacGregor. Sustainable Tourism Development." In VNR's Encyof Hosp and Tou, in M. Khan, M. Olsen, and T. Var. New York (Eds): Van Nostrand Reinhold (1993).
18. V. Vincent, W. Thompson. JTR, Vol. **41**, November. 153-160 Sage Publi (2002).
19. A. Girard. B. Schéou. M en Dévelo, **157**(3), 67- 80. (2012)
20. R. Sadki, M. El Wartiti, R. Azelmad, and S. Berred, IJIAS vol. **18**-3. 784–793, November (2016)
21. M. Berriane. Z.F.T. **12**(2), 158-183, (2020)
22. Y. El Azyzy. CirVath Inter Jour of T, **10**(3), 123-142. (2018)
23. H. Zaoul, Rev Org et march, **3**.155-182. **:** (2007).
24. H. Faouzi. & W. Affaghrou, RIMEC, **04** |(2020),
25. Ouassini, N. Ouassini. JAAS. **July.** (2020).
26. M. Hall, Ecotourism and biosecurity, in Higham. J., (Ed). Critical issues in ecotourism, understanding a complex tourism phenomenon. Butterworth-Heinemann, Elsevier. 121-135. (2007)
27. C.M. Hall, D. Scott, D. Gössling. Tour. Geogr **22:3**, (2020).
28. C.M. Hall, M. James. Tour. Rev **66**(1/2): 118–126. (2011)
29. C.M. Hall, Biological invasion, biosecurity, tourism, and globalization. In Handbook of Globalization and Tourism. 114–125. (2019).
30. J. Hanrahan, D. Melly. EJTR **22.** 45–61.(2019)
31. B. Zeng, R. W. Carter., Curr. Issues Tour. **8**(4). 306–322. (2005).
32. S. Sönmez, J. Wiitala, Y. Apostolopoulos, How complex travel, tourism, and transportation networks influence infectious disease movement in a borderless world. In: J.T. Dallen, (ed.) Handbook of Globalization and Tourism. (Cheltenham: Edward Elgar Publishing, 76–88. (2019)
33. B. Zeng, R. W. Carter., T. De Lacy. Curr. Issues Tour **8**(4). 306–322. (2005)
34. H. Joo, B.A. Maskery, A.D. Berro, et al. Health Secur. **17**(2). 100–108. (2019)
35. K.K. Hung, C.K. Mark, M.P. Yeung, et al. Glob. Health **14**(1) (2018)
36. P. M. Maphanga, U. S. Henama. *Afr. J.* Hosp. Tour. Leis **8**(3).59. (2019)/
37. S. Sareen, K. B. Nielsen, P. Oskarsson, D. Remme. Front. Hum. Dyn **3:636422**. (2021)
38. S. Thomson, and Ip, E. C. J. Law Biosci.**7** (1), Isaa064. (2020)
39. M. K. Haywood. Tour. Geogr, **22**:3, 599-609, (2020)
40. G. D. Sharma, A. Thomas, J. Paul. Tour. Manag. Perspect**, 37.** 100786. (2021)
41. M. Traskevich, A. Fontanari, Tour. Plan. Dev. **03**-03. (2021)
42. S. Benjamin, A. Dillette, D.H. Alderman. Tour. Geogr. (2020)
43. M. F. Orîndaru, A.P. Popescu, S.C. Alexoaei, M.S. Ca˘escu, A.O. Florescu, Sustainability **13,** 6781(2021).
44. H. C. Goh, Glob.Res, **3**, December (2021).
45. P. Wight. Jour. Tour. Stud. **4**(2):54-66. (1993)
46. J. C. Isaacs. Wildl. Soc. Bull. **28**:61-69. (2000)
47. S. Cherkaoui, M Boukherouk, T. Lakhal, A. Aghzar, L. El Youssfi. *Conservation Amid COVID-19 Pandemic: Ecotourism Collapse Threatens Communities and Wildlife in Morocco*. InProceedingof *the* 2020 International Conference on Climate Nexus Perspectives: Water, Food and Biodiversity, I2CNP, Scopus. **183,** 2020.
48. M. Honey, (1999). Ecotourism and Sustainable Development: Who Owns Paradise?: (Island Press. 1999)
49. J. K. Sharma. Tourism and Development-Design for Ecological Sustainability. (Kanishika Publishers & Distributors. 2000)
50. The International Ecotourism Society, General Tourism Statistics (2000),
51. V. Himanshu, R. Deshmukh, AMR (2021)
52. CREST. The Case for Responsible Travel: Trends & Statistics. (2018)

53. Booking https://news.booking.com/fr/ecotourisme--76-des-francais-souhaitent-que-des-solutions-durables-soient-mises-en-place-rapidement-selon-une-etude-bookingcom/(2019)
54. Booking. https://news.booking.com/87--des-voyageurs-internationaux-declarent-souhaiter-voyager-en-limitant-leur-impact-sur-lenvironnement-selon-une-etude-bookingcom/ (2018)
55. B. Moudoud, Thesis (Doctoral), Université Joseph Fourier-Grenoble I. 2000
56. A. Billand. Rev. Géogr. Alp. **84**-4. Pp 95-108. (1996)
57. OMT-PNUD, Stratégie de développement du tourisme rural, (2002).
58. DPT. Monographie touristique de la province d'Essaouira, internal document. (2021)
59. La clef verte website. (2021)
60. WTO, L'impact de la COVID-19 sur le tourisme mondial apparaît dans toute son ampleur alors que l'OMT chiffre le coût du blocage, 28 juin 2020, in https://www.unwto.org/fr/news/limpact-de-la-covid-19-sur-le-tourisme-mondial-apparait-dans-toute-son-ampleur-alors-que-lomt-chiffre-le-cout-du-blocage
61. WTO, 2020 aura été la pire année de l'histoire du tourisme avec 1 milliard d'arrivées internationales en moins, 21 janvier 2021, in https://www.unwto.org/fr/news/2020-aura-ete-la-pire-annee-de-l-histoire-du-tourisme-avec-1-milliard-d-arrivees-internationales-en-moins
62. B. Berrissoule. L'Economiste, **5721.** 18/03. (2020)
63. General Secretariat of the Government. B.O. **6240**. 20 mars. (2014).
64. General Secretariat of the Government. BO. **5118.** 507. 19.06 (2003)
65. General Secretariat of the Government, B.O. **177**. 13 mars. (1916).
66. General Secretariat of the Government, B.O. **1780**. 06-12-(1946)
67. Ministère de la santé, Programme de la surveillance épidémiologique, report, 2010
68. Ministère de la santé, Plan national de veille et de riposte à l'infection par le coronavirus 2019-nCov, Version Janvier 202
69. A. Ait Ali, A. Bassou, M. Dryef, K. EL Ayanoui, R. El Houdaigui, Y. El Jai, F. Hossaini, L. Jaidi, M. Loulichki, E.Rezrazi, A. Saaf, PCNS. A**vril** 20-07. (2020).
70. XZ DCAF & CEDHD, Gestion de l'état d'urgence sanitaire au Maroc, gouvernance sécuritaire et droit humains, Report, juillet (2020)
71. CRI, Résumé des mesures du comité national de veille économique, Région de Casablanca-Settat, in CRI Website (2019).
72. DRDQ, Website. Recueil de recommandations de sécurité sanitaire liées au Covid19 à destination des opérateurs touristiques marocains, Ministère du Tourisme, de l'Artisanat, du Transport Aérien et de l'Economie Sociale Département du Tourisme. (2020)
73. H. Dari. Challenge Online. (2020).
74. CESE (Conseil Economique Social et environnemental), (2020), Saisine **28.** (2020).
75. Département de l'environnement, (2020), Press release (2020)
76. General Secretariat of the Government, Morocco. B-O. **6870.** 506-85. (2020)

Simulation of natural ventilation of different configurations of a room equipped with a solar chimney under the city of Ouarzazate (Morocco) climatic conditions

Aboubakr Herouane [1*], *Thami* Ait Taleb [1], and *Mourad* Taha Janan[2]

[1] ERME, Department of Physics- Chemistry, Polydisciplinary Faculty of Ouarzazate, Ibn Zohr University, Morocco
[2]Advanced Technical Teachers Training School, Mohammed V University, Rabat, Morocco.
*Corresponding author:herouaneaboubakr@gmail.com

Abstract. Natural ventilation is considered as an effective way to improve ventilation and thermal comfort and reduce energy consumption in residential buildings. In this sense, natural ventilation is done by a solar chimney. It is used to increase the air flow through the room by converting thermal energy into kinetic energy. In this work, we intend to present the results of a three dimensional numerical simulation by ANSYS software of natural ventilation through a room equipped with a solar chimney in semi-arid climatic conditions of Ouarzazate (Morocco). The chimney is modelled by a rectangular channel with two South-East oriented glass faces. The main objective is to investigate effects of room dimensions for a solar chimney with of fixed dimensions. We also investigate effects of air velocity, turbulence of kinetic energy and air-flow rate in each area of room and solar chimney and determine optimal dimensions of solar chimney for each room size.

Nomenclature

U_i velocity magnitude (m/s)
X, Y, Z Cartesian coordinates (m)
ν kinematic viscosity (m^2/s)
λ thermal conductivity ($W \cdot m^{-1} \cdot K^{-1}$)
C_p heat capacity ($J.kg^{-1}.K^{-1}$)
ρ air density (kg/m^3)
k pressure loss coefficient
g gravitational acceleration (m/s^2)
T Temperature (°C)
P pressure (Pa)
P_m pressure in reference state of the fluid (ρ_0, T_0) (Pa)
β coefficient of thermal expansion (K^{-1})
K thermal diffusivity of the fluid ($m^2.s^{-1}$)

δ Kronecker symbol
Z height (m)
ACH Air change per hour

1 Introduction

Natural ventilation is considered as an effective way to improve ventilation and thermal comfort and reduce energy consumption in residential buildings. In order to well improve natural ventilation, we can use the solar chimney. It is used to increase the air flow through the room by converting thermal energy into kinetic energy. It generates air movement under buoyancy forces that hot air rises and exits from the top of chimney cavity, drawing cooler air into building with continuous cycle [1]. In this context Mekkawi et al. [2] simulated natural ventilation through a solar chimney in Egypt. Their study showed that the installation of the solar chimney of size of (0.4 ×1.5m) can decrease the temperature from 29.45°C to 28.64°C in the room and increase the air velocity with 50%. Also, Kasim et al. [3] investigated the effect of different opening positions on natural ventilation. The study is based on nine different positions of two openings located on two opposite walls. The result of this study shows that the air flow rate is maximal when both openings are located at the top of the air flow level. In the same context Zavala-Guillén et al. [4] conducted a numerical study on the optimal width of a solar chimney used for the diurnal cooling of a room, their study showed that the optimal width of the solar chimney for to have an air flow greater than 0.1072 Kg/s is 0.125 m with a height of 2m, and a solar flow of 700W / m2. Some studies [5, 6] suggest a cavity gap of 0.2 to 0.3m for the solar chimney to achieve maximum performance. Jiménez et al. [7] carried out a study on the performance of a solar chimney installed on a room. The results of the study show that the ventilation rate increases with 45% in summer and 24.89% in the winter and the air speeds in summer and winter vary from 0.16m/s to 0.22m/s, thus the temperature at the room level varies between 29.5 °C and 40.5 °C in the summer and between 16 °C and 26 °C in the winter. Cheng et al. [8] conducted a study on optimizing design of solar chimney for natural ventilation and smoke exhaustion, they was developed an empirical model to predict the flow rate through air inlet under both normal and fire conditions, they found that the empirical model can be used for a solar chimney with an up to 0.3 m cavity gap. Kong et al. [9] investigated a CFD based approach for determining the optimum inclination angle of a roof-top solar chimney for building ventilation in different climate zones in Australia, their results show that the solar chimney with 45° to 60° inclination angles results in the maximum ACH on a yearly average basis, And under a given received heat flux, a higher inclination angle results in a better ventilation performance. Villagran et al. [10] also developed the effect of shape and design of Arc greenhouse roof openings on natural ventilation. They investigated three types of greenhouses, the traditional Colombian wood (TG) greenhouse, Gothic and Arc multi-span greenhouses that are characterized by having larger ventilation areas both on the sides and on the roof. Ahmed Abdeen et al. [11] examined numerically and experimentally the efficiency of using an inclined solar chimney to elevate indoor air motion. The solar chimney used is 1.85m height, 2.65m width, 75° inclination angle and 0.28m air gap. The work found that the width is the most significant parameter, and is found as well that the solar chimney in these dimensions is able to induce air motion of up to 0.28, 0.47 and 0.52m/s at mean solar radiation values of 500, 700 and 850W/m^2, respectively. Leila et al. [12] presented a design of solar chimney connected to a wind catcher and spraying water system for use in a 2-story office under hot and dry climate conditions. The results showed that even on a hot and sunny day, the connection comprising solar chimney, wind catcher and water spray was able to provide proper thermal and air flow conditions for both floors

through natural convection. The connected system reduced average temperature by 5.2 °C and provided around 9 m3 air changes per hour during the hottest hours of the day for the mentioned space. The optimal connected system also saved 75% of the overall cooling and 90% of the required ventilation energy during the working time in the peak hours during summer. Nguyen et al. [13] carried out a study of various parameters (absorber surface length, air channel gap, and height and width of the inlet and outlet sections) and heat flux factor effects on the thermal performance of chimney, The results demonstrated that the increase in these parameters resulted in a higher flow rate with negligible impact on the thermal efficiency. It was also understood that a large amount of outlet width resulted in a low flow rate. Samsudin et al. [14] studied the effect of solar radiation on the efficiency of the solar chimney installed in hospitals in Malaysia, the results of their study indicate that the air velocity and the air flow rate increase simultaneously with the increase of solar radiation. Kashyap et al. [15] evaluated the effect of the solar chimney on the natural ventilation of buildings in different areas of India, their results indicate that the optimal dimensions of the solar chimney to have a minimum ACH of 6.20 are 2.5m in length, 2m wide and 0.1m gap, so the solar chimney is efficient in all climatic zones and throughout the year, an energy saving of 89 kWh / year is observed.
Ouarzazate city is among the cities of Morocco that are characterized by their arid climate: In summer, the temperature often exceeds 40°C, which requires continuous ventilation throughout the day. To address this need, natural ventilation may be the best solution to stabilize temperature and regulate thermal comfort in residential buildings in this city. So, this study presents the results of a numerical simulation under climatic conditions of Ouarzazate city by the ANSYS (fluent) software of the natural ventilation of a three-dimensional room equipped with a solar chimney. Thus, we try to study the air velocity, the turbulence of kinetic energy and the air-flow rate in each area of the room and solar chimney, then to determine the optimal dimensions of the solar chimney for each room size.

1.1 Physical model

The physical configuration is shown in Figure 1; it is a square room of variable side (3m, 4m, 5m) with a fixed height of 3m. The window size was kept constant with a dimension of 1 m×1 m. The height of the solar chimney is 1m and the width is 0.3m. The fluid inside the cavity is the air which considered Newtonian and incompressible, since we study the heat transfer that occurs only by natural convection, the Boussinesq approximation is considered.

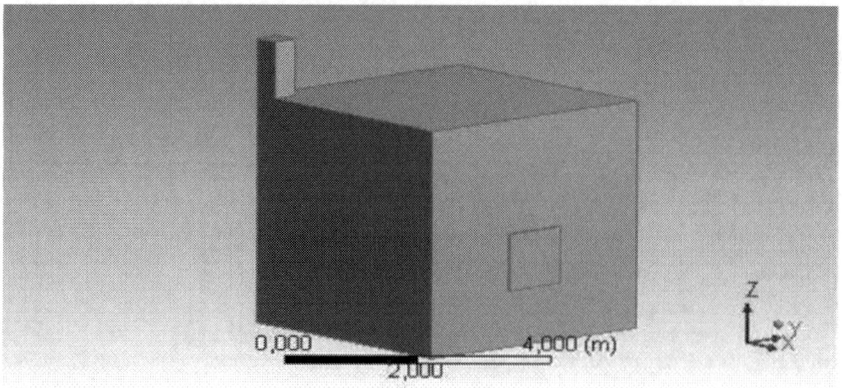

Fig.1. Physical model of the studied configuration

For the simplification of the problem, the following assumptions are taken a count:

- ✓ The fluid is Newtonian and incompressible.
- ✓ The flow is unsteady.
- ✓ The inlet pressure equal to the atmospheric pressure
- ✓ The outside temperature is constant at 313 k.
- ✓ The air flow is considered turbulent.
- ✓ Normal direct solar irradiation is 1423W/m^2

1.2 Mathematical model

- ➢ Equation representing the conservation of the mass (Continuity) is :

$$\frac{\partial u_i}{\partial x_i} = 0 \tag{1}$$

Where u_i
are the velocity components in x, y, z directions

- ➢ Equation representing, the momentum (Navier – Stokes):

Considering the linear variation of the density of the Boussinesq approximation, we obtain the following momentum equation:

$$\frac{\partial u_i}{\partial t} + \frac{\partial (u_i u_j)}{\partial x_j} = -\frac{1}{\rho_0}\frac{\partial p_m}{\partial x_i} + \frac{\partial}{\partial x_j}\left(v\left(\frac{\partial u_i}{\partial x_j} + \frac{\partial u_j}{\partial x_i}\right)\right) + g\beta(T - T_0)\delta_{iz} \tag{2}$$

With:

- ✓ $P_m = p + \rho_o gz$: the pressure corresponding to the reference state of the fluid (ρ_0, T_0), ν the kinematic viscosity and g the gravitational field.

β: the coefficient of thermal expansion at constant pressure. It describes the change of volume at constant pressure as a function of temperature.
Energy equation:
This equation is deduced from the first principle of thermodynamics by neglecting the effect of pressure in front of temperature variation. It is then expressed by:

$$\frac{\partial (T)}{\partial t} + \frac{\partial (u_j T)}{\partial x_j} = \frac{\partial}{\partial x_j}\left(k\left(\frac{\partial T}{\partial x_j}\right)\right) \tag{3}$$

Where:
$k = \frac{\lambda}{\rho C_p}$ is the thermal diffusivity of the fluid (m^2.s^{-1}), C_p is the heat capacity (J.kg^{-1}.K^{-1}) and λ is the thermal conductivity (W·m^{-1}·K^{-1}). [16]
All these equations are taken into account in the program which is part of the software used for the simulation.
In this study, we used a tetrahedron-type mesh with a high number of mesh nodes, access to 99487 nodes, as it is shown in figure 2. After having carried out several attempts and mesh tests in order to choose the optimal mesh which should give a good precision on the speed of the air and on the flow of air, we opted for an optimal choice which is composed of 562130 elements.

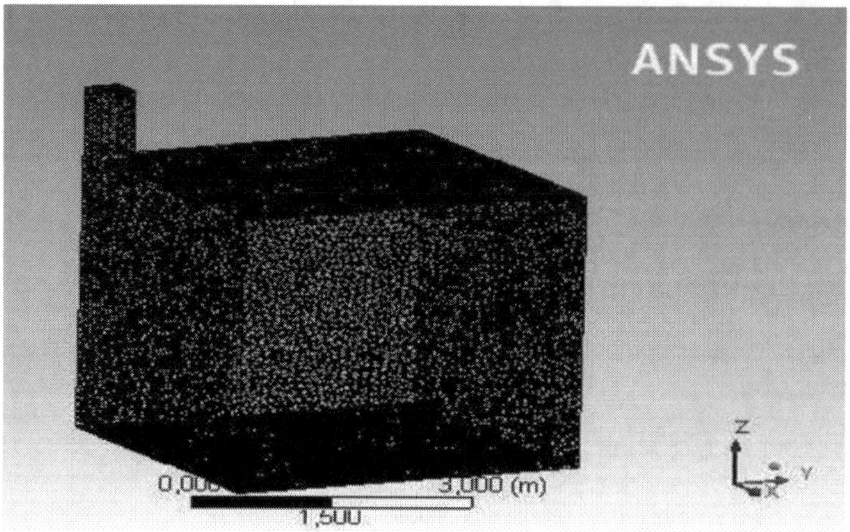

Fig.2. Mesh of the model

2 Results and discussion

- Air velocities, turbulence of kinetic energy and Air flow rate in different plane of the room of 3m in dimensions.

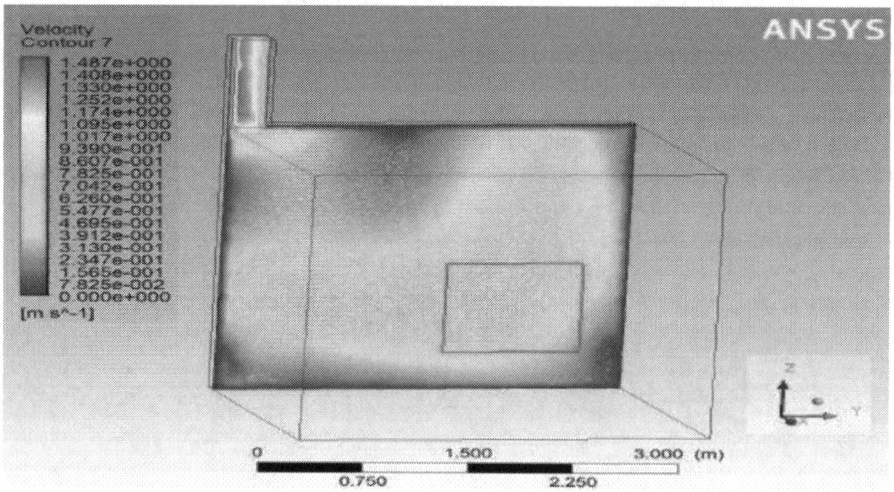

Fig.3. Air velocity in the 0.15m plane

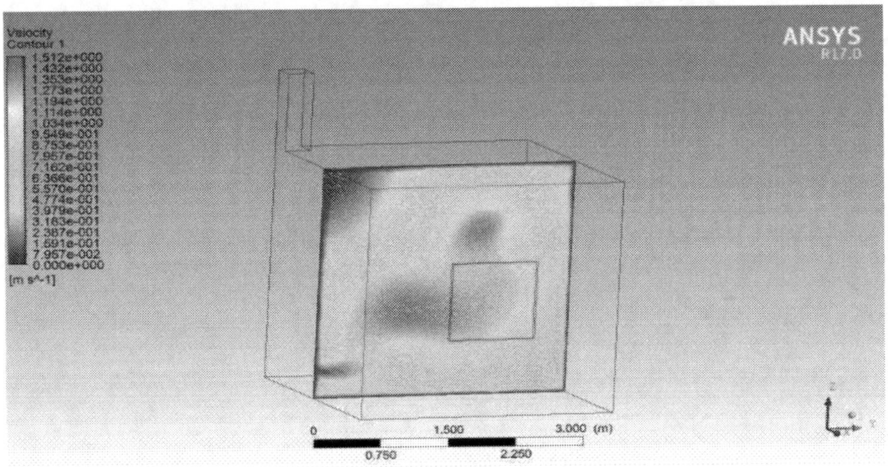

Fig.4. Air velocity in the 1.5m plane

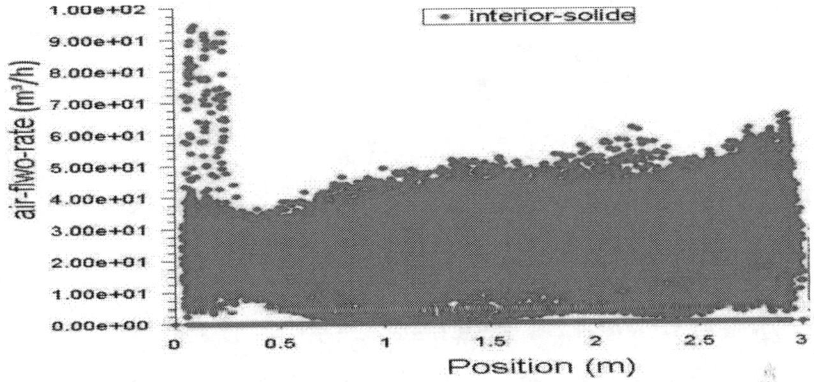

Fig.5. Air Flow rate in the room

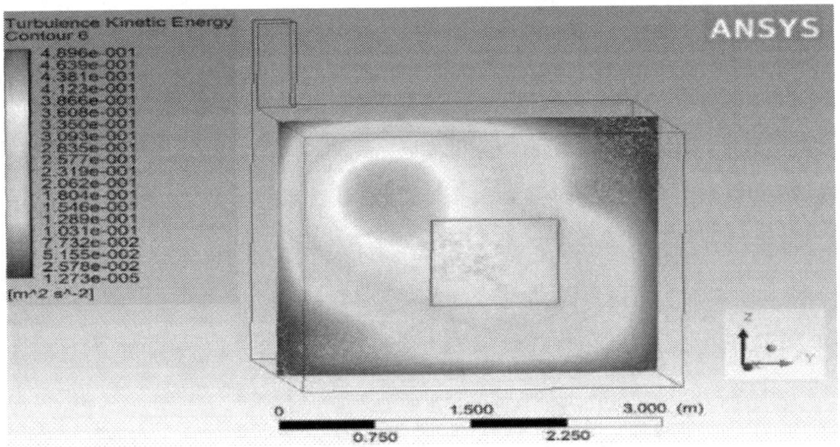

Fig.6.Turbulence of kinetic energy in the 1.5m plane

In these figures we can note that the air velocity increases from the entrance of the window towards the exit of the solar chimney. So it is maximum in the chimney 1.48 m/s. the air velocity is also important in the middle of the room 0.87 m/s, but it's low in the top and bottom of the room. The more important is that the air velocity increases in all direction of the room and create a passive cooling. The air flow rate is greatly increased at the solar

 https://doi.org/10.1051/e3sconf/202233701005

chimney 90m³/h, and also very high at the level of the whole of room 40 m³/h, that makes more effective natural ventilation for the turbulence of kinetic energy its value is low, which allows the room to be naturally ventilated perfectly.

- Air velocities, turbulence of kinetic energy and air flow rate in different plane of the room of 4m in dimensions.

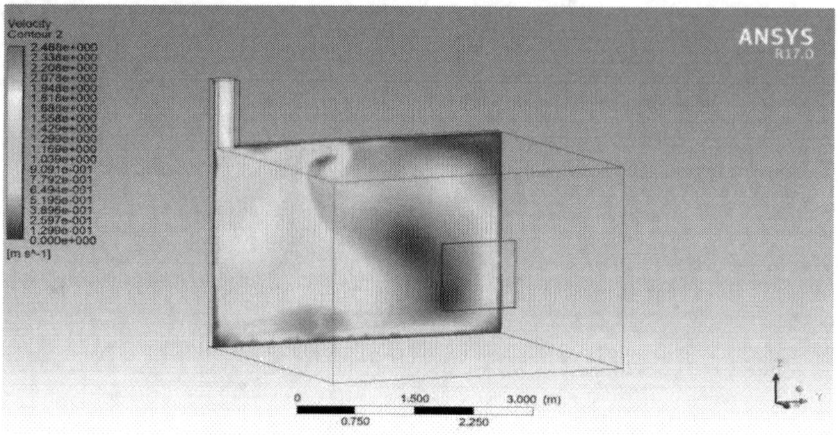

Fig.7. Air velocity in the 0.15m plane

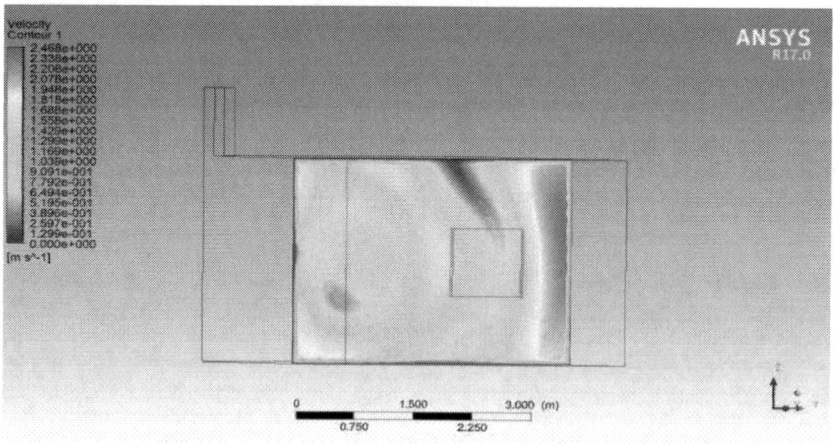

Fig.8. Air velocity in the 2m plane

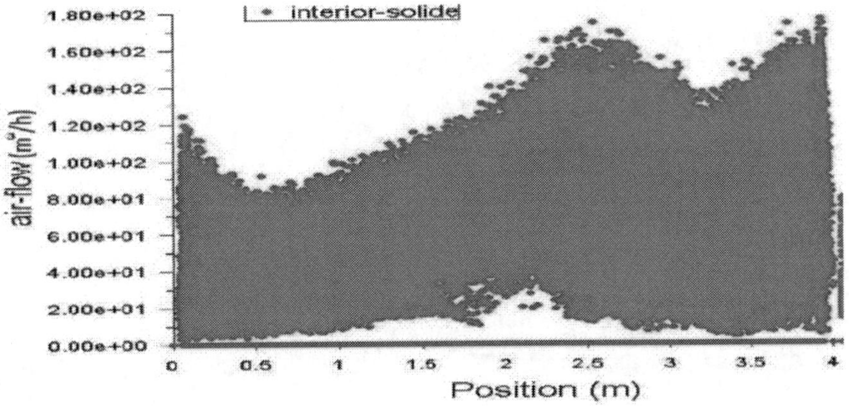

Fig.9. Air flow rate in the room

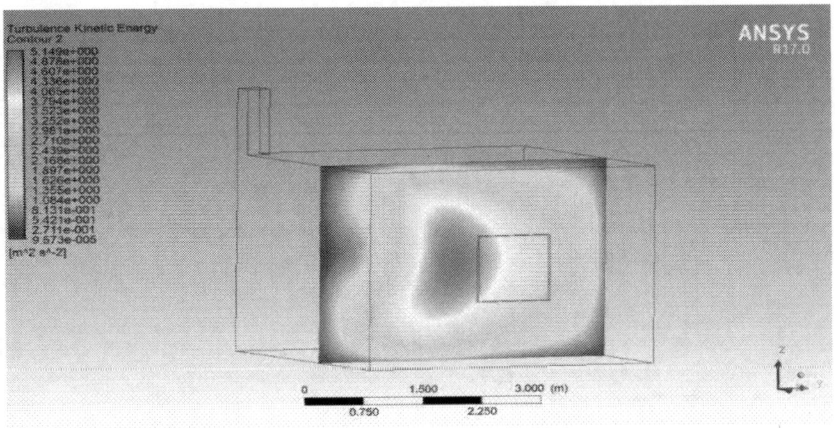

Fig.10. Turbulence of kinetic energy in the 2m plane

This figures show that's the air velocity gradually increases in the solar chimney, which causes the circulation of the air inside the room. We notice that the air velocity is maximum in the middle of the room 2.88 m/s and increases from bottom up to the horizontal plane of the room, and then it decreases to the ceiling. For the air flow rate it is found that it varies in proportion to the air velocity, it is higher in the middle plane of the room 180 m^3/h as well in the solar chimney 120 m^3/h. The turbulence of the kinetic energy is also high throughout the room and it is maximum in the horizontal plane of the window 5.15 m^2/s^2. All that makes the room ventilated naturally.

- Air velocities, turbulence of kinetic energy and Air flow rate in different plane of the room of 5m in dimensions.

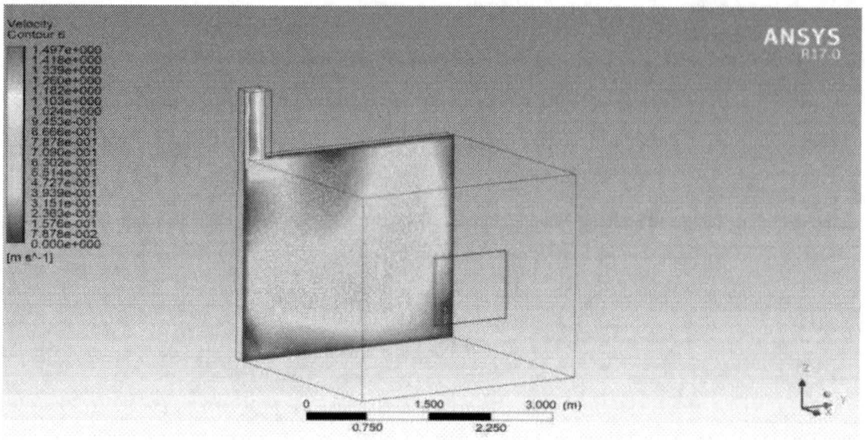

Fig.11. Air velocity in the 0.15 plane

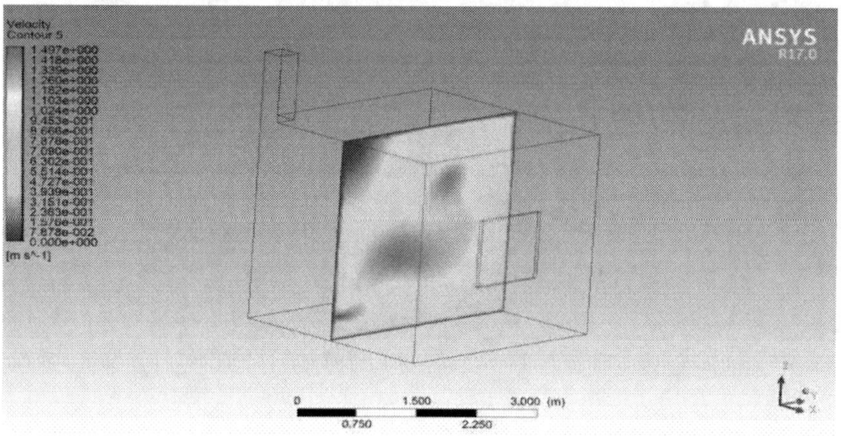

Fig.12. Air velocity in the 2.5m plane

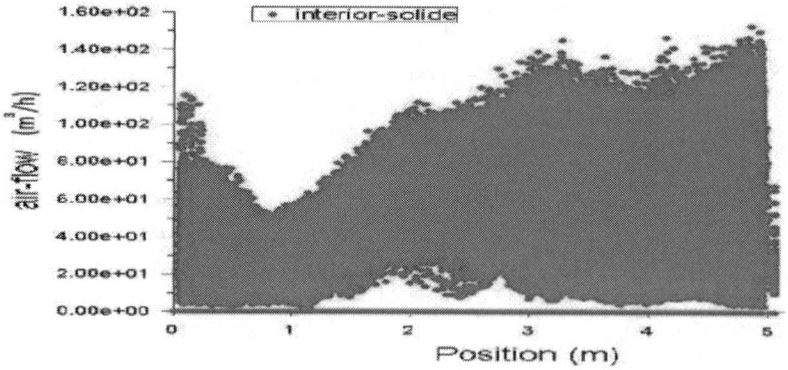

Fig.13. Air Flow rate in the room

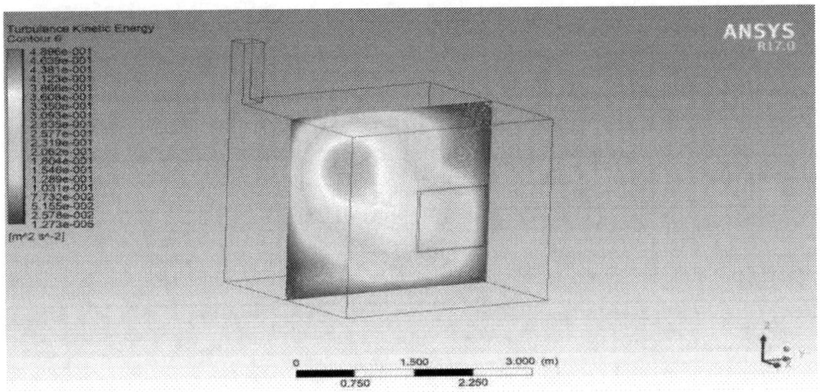

Fig.14. Turbulence of kinetic energy in the 2.5m plane

These results show clearly that the air velocity increase from the inlet of the window to the outlet of the solar chimney for which is maximum 1.49 m/s, it is also medium in all direction of the room and very low in the top and bottom of the room. We notice that the air flow rate is high in the room and the chimney, its maximum value is 140m^3/h. the turbulence of the kinetic energy also is high its maximum value is 48 m^2/s^2. This values show the capacity of solar chimney to improve the thermal comfort in the building by making continuously the movement of air in the building.

By comparing the results of the three simulations, we can see that the air velocity, the air flow rate and the turbulence of the kinetic energy increase simultaneously when we pass from the room of 3m in dimension to the room of 4m, but contrary when we pass from the room of 4m to the room of 5m we can note that they decrease. These results confirm that the solar chimney of 0.3m of width could be designed to the room of 4m in dimensions.

3 Conclusion

This work present a numerical simulation under climatic conditions of Ouarzazate city by the ANSYS (fluent) software of the natural ventilation of a three-dimensional room equipped with a solar chimney.

The results of the study show that solar chimney has an important role to increase natural ventilation in the room. The Air velocity, the air flow rate and the turbulence of kinetic energy inside the room are important in the three simulations, but in the second simulation (room of 4m), the air velocity and the air flow rate are high and stable, that's allows to cool more the room. This different of values in the three simulations insure the need to specify the dimensions of the solar chimney to each room dimensions. So, in that case the solar chimney with the width of 0.3m is optimum to the room of 4m in dimensions.

References

1. E. Bacharoudis, MG. Vrachopoulos, MK. Koukou, D. Margaris, AE. Filios, SA. Mavrommatis, Study of the natural convection phenomena inside a wall solar chimney with one wall adiabatic and one wall under a heat flux. Appl ThermEng. (2007);27:2266-75.
2. G. Mekkawi, R Elgendy, solar chimney for enhanced natural ventilation based on CFDsimulation for a housing prototype in Alexandria, egypt, International Journal of Advances in Mechanical and Civil Engineering, ISSN: 2394-2827 Volume-3, Issue-5, Oct.-(2016)

3. N.F.M. Kasim, S.A. Zaki, M.S.M . Ali, N. Ikegaya, A.A. Razak, Computational Study on the Influence of Different Opening Position on Wind- Induced Natural Ventilation in Urban Building of Cubical Array, Procedia Engineering 169 (2016) 256 263
4. I. Zavala-Guillén, J. Xamán, I. Hernández-Pérez, M. GijónRivera, Y. Chávez, Numerical study of the optimum width of a diurnal double air-channel solar chimney, Energy (2018)
5. T. Miyazaki, A. Akisawa, T. Kashiwagi, The effects of solar chimneys on thermal load mitigation of office buildings under the Japanese climate, Renew. Energy 31(2006) 987 1010.
6. J. Arce, M.J. Jimenez, J.D. Guaman, M.R. Heras, G. Alvarez, J. Xaman, Experimental study for natural ventilation on a solar chimney, Renew. Energy 34(2009) 2928–2934.
7. C. Jiménez-Xamàn, J. Xamàn,M. Gijon-Rivera, I. Zavala-Guillén, F. Noh-Pat, E. Simà, Asessing the thermal performance of a rooftop solar chimney attached to a single room, Journal of Building Engineering (2020)
8. Cheng, X., Shi, L., Dai, P., Zhang, G., Yang, H., Li, J. : Study on Optimizing Design of Solar Chimney for Natural Ventilation and Smoke Exhaustion, Energy & Buildings (2018)
9. Kong, J., Niu, J., Lei, C.: A CFD based approach for determining the optimum inclination angle of a roof-top solar chimney for building ventilation, Solar Energy, Elsevier (2020)
10. Villagran EA, Baeza Romero EJ, Bojaca CR Transient CFD analysis of the natural ventilation of three types of greenhouses used for agricultural production in a tropical mountain climate. Biosyst Eng 188:288–304. (2019)
11. Laila M, Zandi M, Bidi M, Behroozizade E, Kazemi I. New design for solar chimney with integrated windcatcher for space cooling and ventilation. Build Environ. https://doi.org/10.1016/j.build env.2020.10678 5. 2020.
12. Ahmed Abdeen, A.A., Serageldin, M.G.E., Ibrahim, A., El- Zafarany, S. and Ookawara, R.M., "Solar chimney optimization for enhancing thermal comfort in Egypt: an experimental and numerical study", Solar Energy, Vol. 180, pp. 524-536, (2019) doi: 10.1016/j.solener.2019.01.063.
13. Nguyen YQ, Wells JC. A numerical study on induced flowrate and thermal efficiency of a solar chimney with horizontal absorber surface for ventilation of buildings. J Build Eng. 2020;28:1–31.
14. Sheikh Saiful Azam Samsudin, Abdul Muin Shaari, Kamil Abdullah, Mohd Faizal Mohideen Batcha, Potential of Utilizing Solar Chimney as an Energy Efficiency Measure in Malaysian Hospitals , CFD Letters: Vol. 12 No. 4: April (2020)
15. Kashyap.S, Kumar.G.V, Kumar.T.A, Enhancing Passive Cooling and Natural Ventilation for Houses of Various Climatic Zones of India, 2nd International Conference on Power and Energy Applications, 978-1-7281-1596-2/19/$31.00 ©2019 IEEE
16. B. Brangeon 'Contribution à l'étude numérique de la ventilation naturelle dans des cavités ouvertes par la simulation des grandes échelles. Application au rafraichissement passif des bâtiments' PhD thesis in Mechanics and Building Sciences, University of Reunion (2012)

E3S Web of Conferences **337**, 01006 (2022) https://doi.org/10.1051/e3sconf/202233701006
I2CNP 2021

Evolution of air quality in Morocco before and during the COVID-19 pandemic; comparative study

Loubna Bouhachlaf[1*], *Fatimazahra MOUSLI*[1], and Souad El hajjaji [1]

[1]Laboratory of Spectroscopy, Molecular Modeling, Materials, Nanomaterial, Water and Environment, CERNE2D, Mohammed V University in Rabat, Faculty of Science, Avenue Ibn Battouta, BP1014, Agdal, Rabat, Morocco

Abstract. The epidemic of coronavirus is spreading all over the world, which is not only a health problem, but also affects the social sector, politics, the environment in various ways and the world economy. In response to the rapidly changing international epidemiological situation of COVID-19, several governments have imposed restrictions on the movement of people and vehicles and suspended industrial activities as a safety measure to reduce the spread of the virus and protect human health. This study aims to compare the state of air quality in Morocco, before the pandemic and during the lockdown situation of the Moroccan territory between March 2020 and June 2020.

1 Introduction

The Covid-19 is an emerging pandemic of an infectious new disease, caused by the coronavirus SARS-CoV-2, which appeared in Wuhan on November 17, 2019[1], before spreading around the world. The World Health Organization (WHO) first alerts the People's Republic of China and its other member states and then declares a public health emergency of international concern on January 30, 2020 [2]. The Covid-19 epidemic was declared a pandemic by the WHO on March 11, 2020, which calls for essential protective actions to prevent overcrowding in intensive care units and to strengthen preventive hygiene [3]. This global pandemic is causing a series of cancellations of sporting and cultural events around the world, the implementation by many countries of containment measures to curb the formation of new outbreaks of contagion and the closure of borders in many countries [4]. It also has effects in terms of social and economic instability due to the uncertainties and fears it brings to the global economy [6].

2 The influence of COVID-19 pandemic on air pollution in Morocco

The lockdown due to the health situation in the world and in Morocco linked to COVID-19 has reduced air and road traffic, resulting in lower energy consumption and lower demand for oil. These changes in transport activities and oil demand have a significant impact on the quality of the environment [7], indeed since the start of containment Morocco has

*Corresponding Author: loubna.bouhachlaf@um5s.net.ma

experienced a change in the very significant reduction rate of atmospheric pollutants and an improvement. appreciable air quality, it is clearly improved, a preliminary assessment of the air quality has been drawn up at the level of the city of Marrakech and covering the period from November 2019 to April 8, 2020. The first results of this balance sheet show significant reduction rates. Air pollutants, in particular a decrease of 55% for nitrogen dioxide (NO_2), 70% for carbon monoxide (CO) and 67% for suspended particles [8]. One of the consequences is the establishment of lockdown measures to limit the spread of the virus should widely modify anthropogenic emissions of pollutants, both in terms of mass emitted and temporal variations. The change in these emissions should modify the concentrations of surface pollutants observed in the world and also in Morocco. This has been observed since the start of the lockdown, in particular through the analysis of the measurements resulting from the monitoring of air quality [9].

3 Materials and Methods

3.1 Study area

Morocco is an African country located in the extreme northwest of the continent (fig.1). It has a coastline on both the North Atlantic Ocean to the west and the Mediterranean Sea to the north. The capital is Rabat and the largest city is Casablanca. Morocco spans an area of 710 850 km^2 and has a population of over 36 471 769 inhabitants (2019) [10].
This assessment, based on the analysis of data from air quality monitoring stations, before, during and after the state of health emergency [11]. This assessment will allow for a more detailed analysis of the air pollution situation, including the baseline, to draw lessons and make recommendations to limit post-COVID19 air pollution [12].

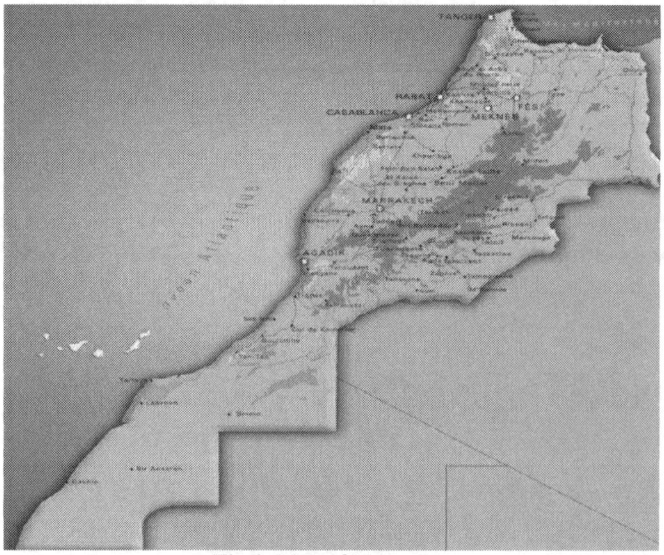

Fig.1. Map of Morocco.

3.2 Methodology

The Ozone Monitoring Instrument is a Dutch and Finnish image spectrometer for ozone monitoring (Table 1). The instrument is designed to distinguish ozone from other atmospheric species. The OMI sensor has a spectral region of 264 - 504 nm, a spectral resolution of 0.42 nm - 0.63 nm, and a resolution of 0.125 × 0.1250. The high spectral and spatial resolution of the instrument is key to the detection of urban-scale air pollution. The acquisition of measurements of the tropospheric and stratospheric levels of the Earth's atmosphere is the main objective of the instrument's mission. MERRA-2 stands for Modern-Era Retrospective analysis for Research and Applications version 2. The MERRA project focuses on historical climate analyses for a wide range of weather and climate time scales and places the NASA Earth Observation System of Systems (EOS) series of observations in a climate context [13].
For the study of total column ozone, the TOMS-like OMTO3e product has been used. It selects the best pixel data from the good quality level-2 total column ozone data (OMTO3). This product data falls in the 0.25 x 0.25-degree global grids. For the study of the tropospheric column, NO2OMNO2d data product has been selected. It is a Level-3 Gridded Product where pixel-level data of good quality are falling into 0.25x0.25 degree global grids [14].

4 Results and Discussion

4.1 Vicissitudes in NO_2 emissions

Nitrogen Dioxide (NO_2) is one of a group of gases nitrogen oxides (NOx). It is a dangerous pollutant in and of itself. Also while all of these gases are harmful to human health and the environment, NO_2 is of greater concern. And also contributes to the formation of $PM_{2.5}$ and ozone, two of the most dangerous forms of air pollution. Therefore, NASA has tracked by satellite and recorded the level of NO_2 in the air (NASA, Global Nitrogen Dioxide Monitoring Home Page, 2020). NASA tracks two cities of Morocco mainly, Rabat, and Laayoune. Therefore, this study shows the satellite detected of the NO_2 in the air across these cities to investigate the decline in the emission that is relatively correlated to the pandemic [15].

4-1-1 NO_2 measurement

NO_2 emission mostly originates after the burning of fuels and the consumption of automobiles. Therefore, to systematically evaluate whether the COVID-19 pandemic has an opposing effect on the concentration of NO_2 in the air in three cities, NO_2 was used as an estimation indicator [16].
NASA is responsible for expressing and realizing research in planetary science by using the ozone monitoring instrument (OMI) to collect the vicissitudes in Morocco's NO_2 level before and after the pandemic.
Figure 2 illustrates the NO_2 level in the air of the two cities over four years. The NO_2 concentrations in the troposphere over cities show that the concentrations as of the end of 2019 are still the same as in 2018

(A)

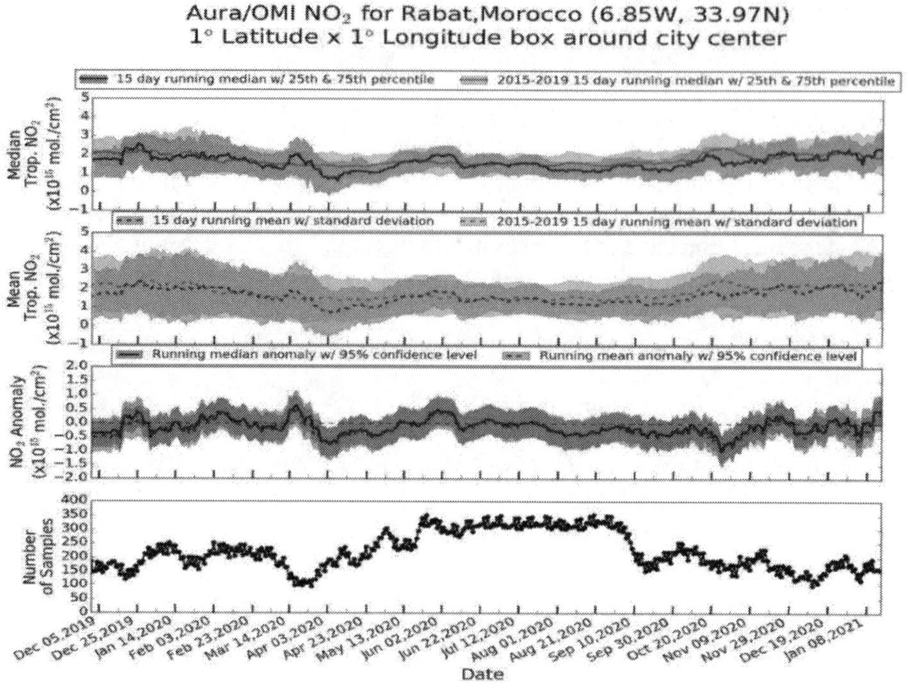

(B)

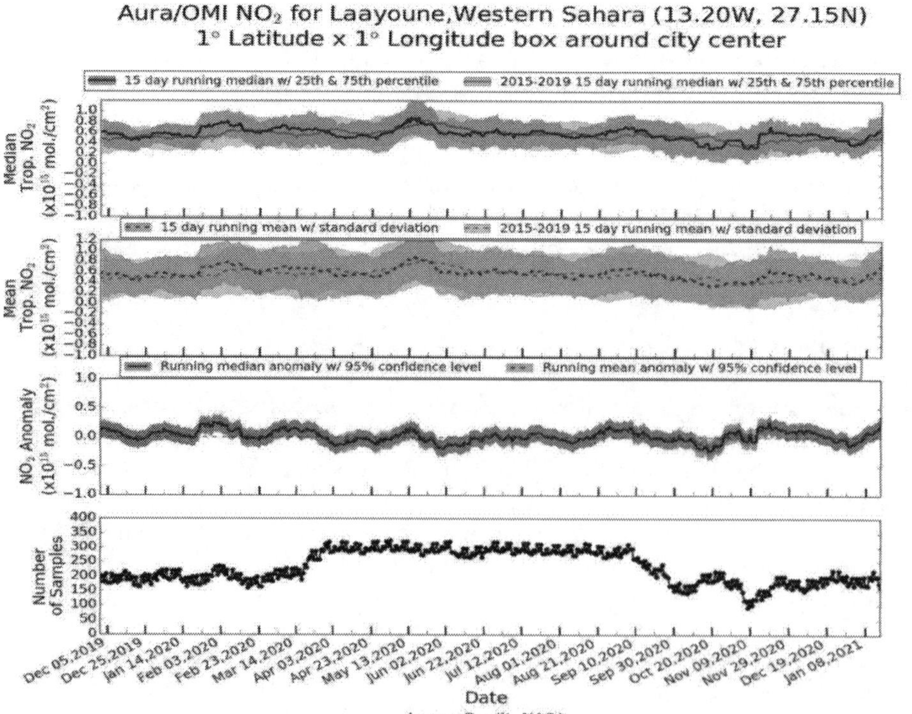

Image Credit: NASA

Fig. 2. Vicissitudes in NO_2 emission levels in **(A)** Rabat, and **(B)** Laayoune, from 2015 to 2019. Source (NASA- Rabat, and Laayoune, Morocco OMI data, 2020).

4-2- NO_2 emissions in Morocco before, during, and the pandemic

NO_2 emissions in Morocco before, during, and after the pandemic gouvernements moroccan applied quarantine to avoid and control the spread of the COVID-19. Figures 3, 4, and 5 show the tropospheric screen on the OMI of NASA's Aura satellite composed of information on the NO_2 level in the air of Morocco between January 1, 2020 and September 15, 2020. Rabat shows the highest NO_2 emission before the pandemic compared to other city laayoune due to the emission from the industriels activities and Traffic. Before the effect of the pandemic in the period between January 15, 2020 and mars 15 , 2020, the figures show that NO_2 emission level in Morocco is high (fig.2.(A)). The effect of the pandemic is apparent on the reduction of NO_2 in Rabat due to stop industriels activities and due to the transportation restriction. One possible reason for this finding was that during the pandemic, the vehicle exhaust and industry emission slowed down. [17]. The results of the examination show that the pandemic is seriously threatened public health; however, in the meanwhile, it contributes to reducing NO_2 emissions significantly. This reduction as Morocco imposed strict restrictions to fight the pandemic, with the entire country closed. These restrictions not only slow local economic progress, nevertheless also efficiently condensed greenhouse gas (GHG) emissions. However, it is early to predict the long term of the contribution of COVID-19 pandemic on air pollution [18].

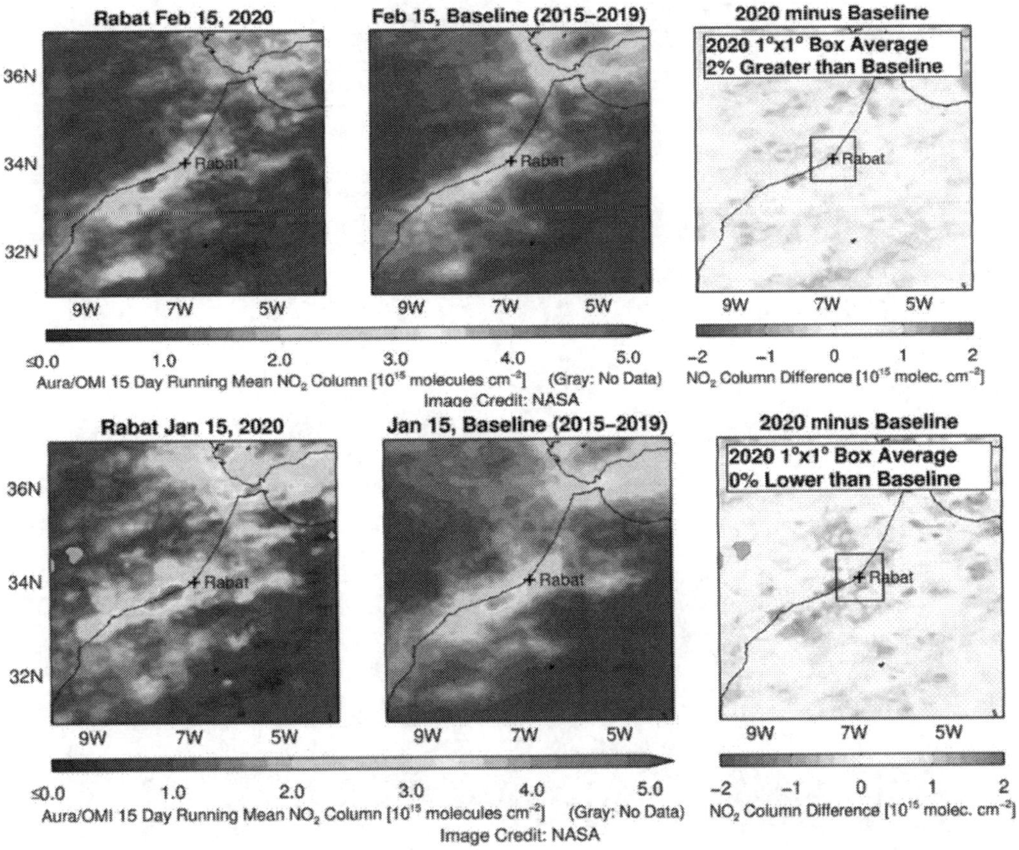

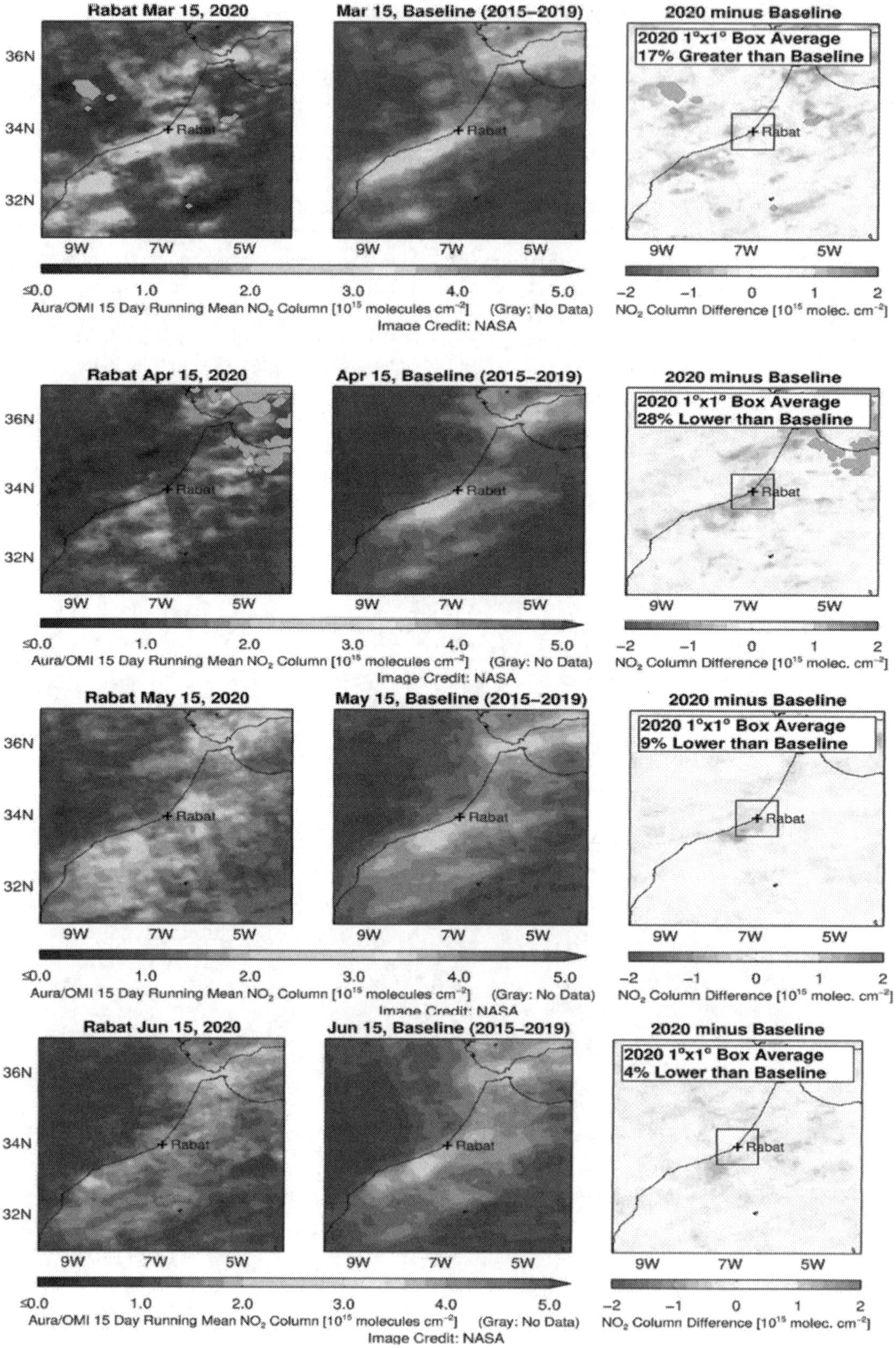
Rabat Mar 15, 2020
Mar 15, Baseline (2015–2019)
2020 minus Baseline
2020 1°x1° Box Average
17% Greater than Baseline
Rabat Apr 15, 2020
Apr 15, Baseline (2015–2019)
2020 minus Baseline
2020 1°x1° Box Average
28% Lower than Baseline
Rabat May 15, 2020
May 15, Baseline (2015–2019)
2020 minus Baseline
2020 1°x1° Box Average
9% Lower than Baseline
Rabat Jun 15, 2020
Jun 15, Baseline (2015–2019)
2020 minus Baseline
2020 1°x1° Box Average
4% Lower than Baseline
Rabat
36N
34N
32N
9W
7W
5W
≤0.0
1.0
2.0
3.0
4.0
5.0
Aura/OMI 15 Day Running Mean NO2 Column [10^15 molecules cm^-2] (Gray: No Data)
Image Credit: NASA
–2
–1
0
1
2
NO2 Column Difference [10^15 molec. cm^-2]

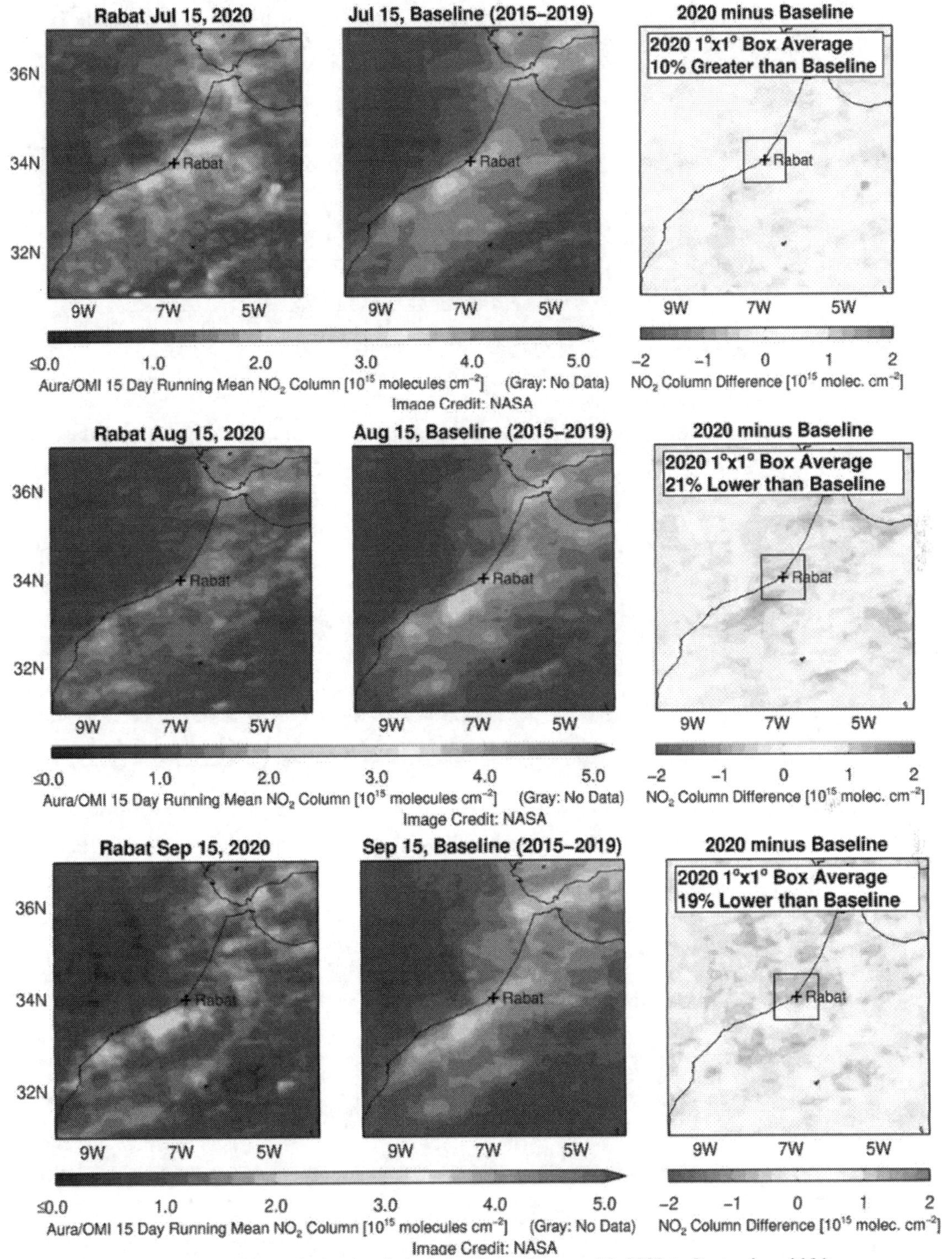

Fig. 3. Vicissitudes in NO_2 emission levels in Rabat from January 15, 2020 to September, 2020. Source (NASA-Rabat OMI data, 2020) [19].

https://doi.org/10.1051/e3sconf/202233701006

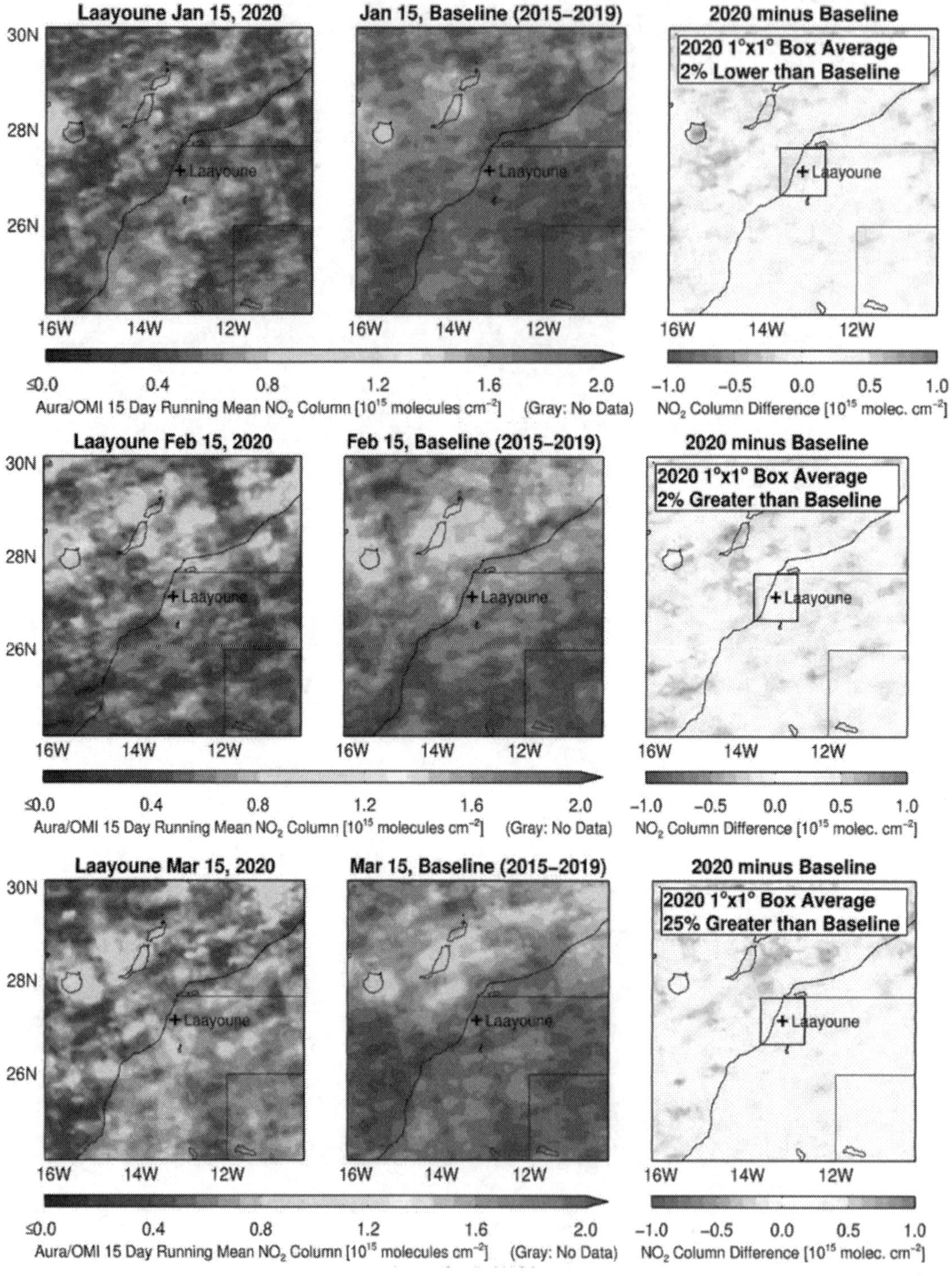

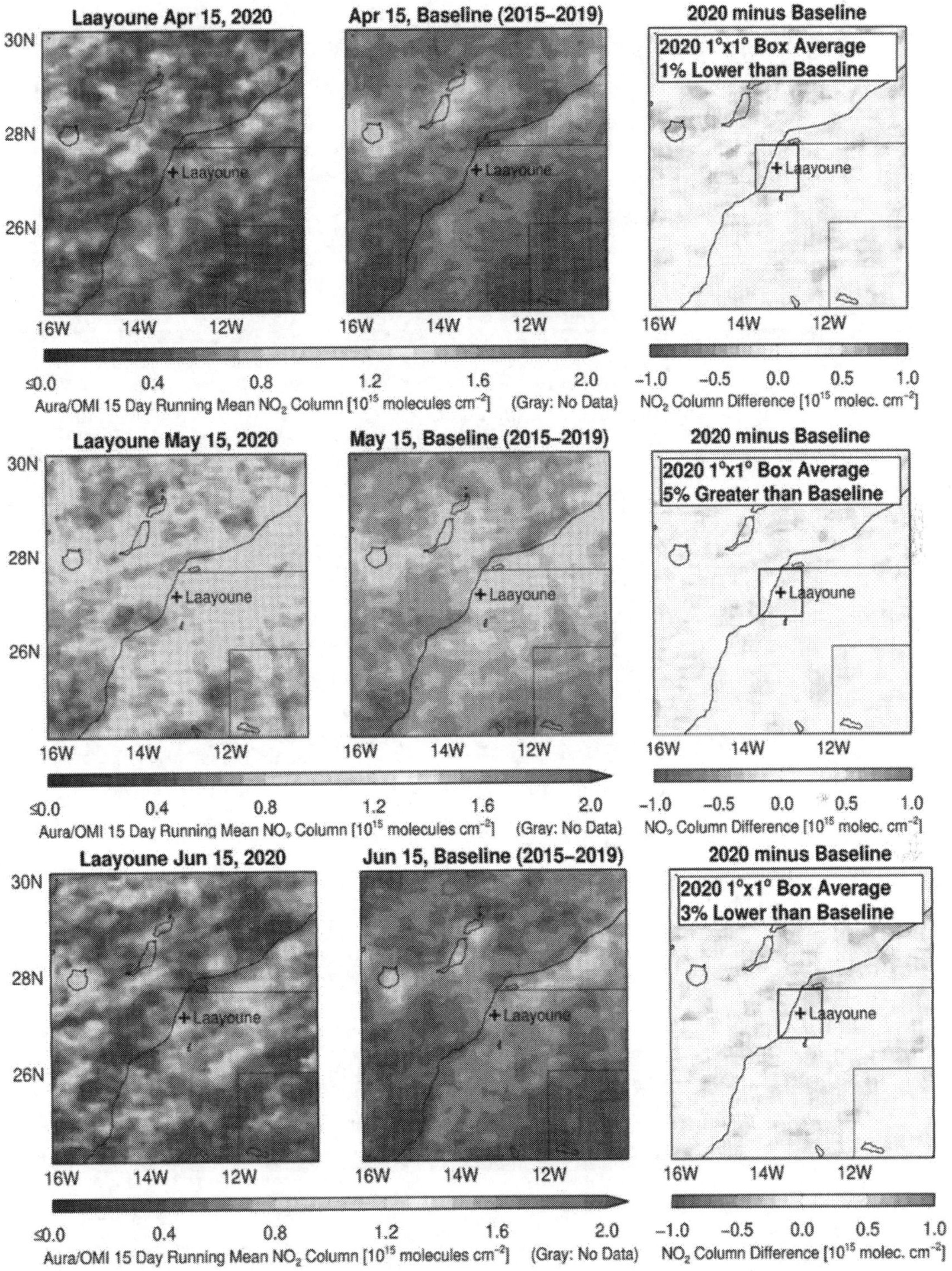
Laayoune Apr 15, 2020
Apr 15, Baseline (2015–2019)
2020 minus Baseline
2020 1°x1° Box Average
1% Lower than Baseline
30N
28N
26N
16W
14W
12W
+Laayoune
≤0.0 0.4 0.8 1.2 1.6 2.0
Aura/OMI 15 Day Running Mean NO2 Column [10^15 molecules cm^-2] (Gray: No Data)
−1.0 −0.5 0.0 0.5 1.0
NO2 Column Difference [10^15 molec. cm^-2]
Laayoune May 15, 2020
May 15, Baseline (2015–2019)
2020 minus Baseline
2020 1°x1° Box Average
5% Greater than Baseline
Laayoune Jun 15, 2020
Jun 15, Baseline (2015–2019)
2020 minus Baseline
2020 1°x1° Box Average
3% Lower than Baseline

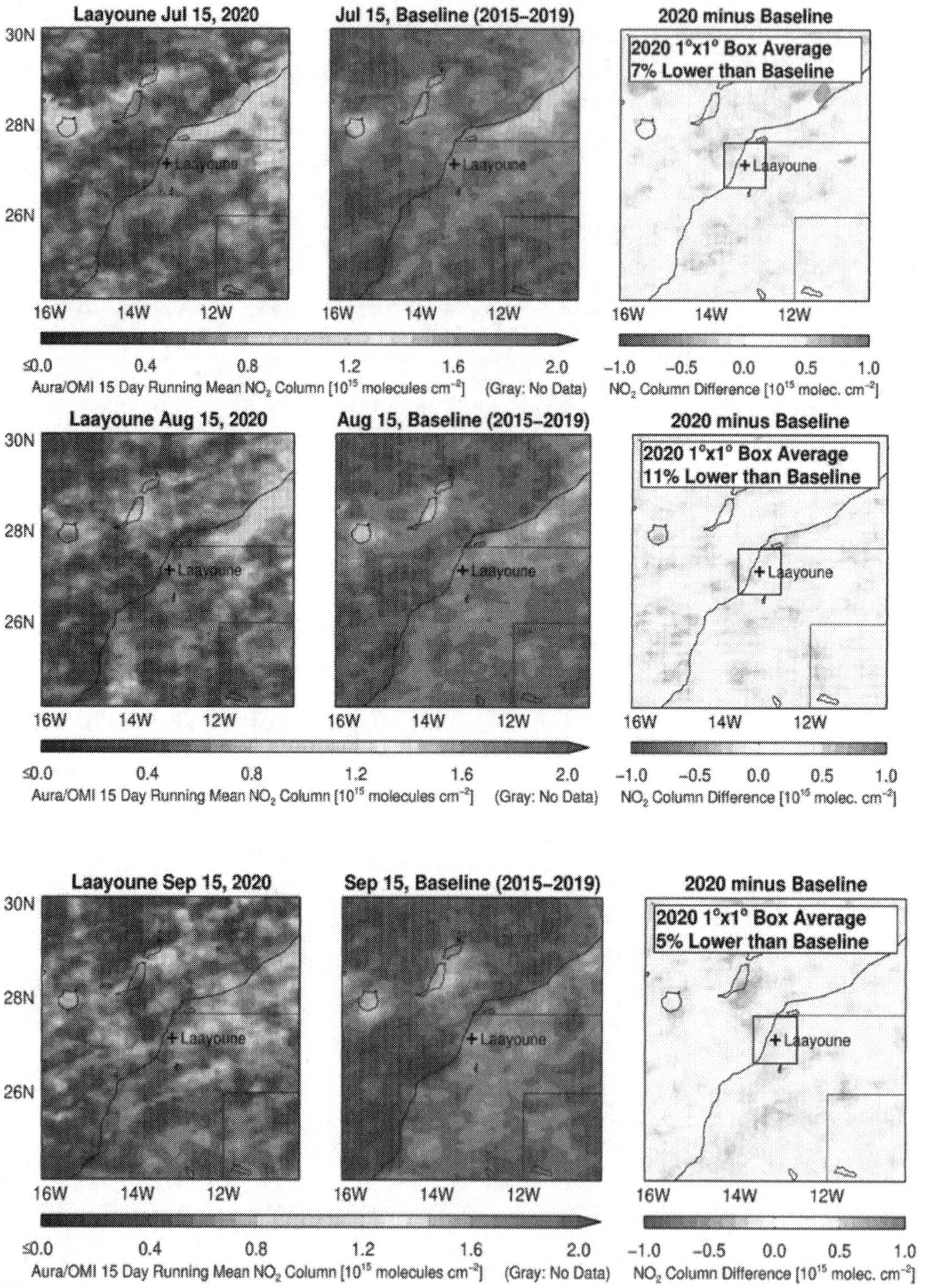

Fig. 4. Vicissitudes in NO2 emission levels in Laayoune from January 15, 2020 to September, 2020. Source (NASA-Laayoune OMI data, 2020).

5 Conclusion

This study provides an analysis of air quality changes in Morocco during the COVID-19 pandemic for the period from March to June 2020. Satellite data comparing the levels of dioxide of azote concentrations after stopping savings indicate steep reductions. Data from the Sentinel 5-P satellite shows that, in confined areas, the average levels of NO_2 between March 15 and May 15, 2020 were lower than the concentrations of June 2020 after lockdown. So we conclude that traffic restrictions applied during the quarantine for COVID-19 pandemic in Morocco were remarkably convincing in reducing emissions of NO_x. This reduction in ozone precursors reduces ozone.

References

1. P. Xu et al., High aspect ratio In2O3 nanowires: Synthesis, mechanism and NO2 gas-sensing properties. Sensors Actuators, B Chem., vol. 130, no. 2, pp. 802–808, (2008).
2. L. Luo, Y. Zhang, B. Pearson, Z. Ling, H. Yu, and X. Fu, On the security and data integrity of low-cost sensor networks for air quality monitoring," Sensors (Switzerland), vol. 18, no. 12, pp. 1–22, (2018).
3. M. A. Zaytar and C. El Amrani, MetOp Satellites Data Processing for Air Pollution Monitoring in Morocco. Int. J. Electr. Comput. Eng., vol. 8, no. 6, p. 4584, (2018).
4. C. Clerbaux and C. Clerbaux, Comité des sciences de l'environnement l'évolution de l'ozone atmosphérique Sophie Godin-Beekmann, no. (Octobre 2015).
5. N. Zhu et al., Morphogenesis and cytopathic effect of SARS-CoV-2 infection in human airway epithelial cells, Nat. Commun., vol. 11, no. 1, (2020).
6. L. Menut, B. Bessagnet, G. Siour, S. Mailler, R. Pennel, and A. Cholakian, Impact of lockdown measures to combat Covid-19 on air quality over western Europe, Sci. Total Environ., vol. 741, p. 140426, (2020).
7. B. El and F. Idrissi, Evaluation et étude des paramètres physico-chimiques de la pollution générée par les unités industrielles dans la région de Fès, no. April, (2020).
8. S. Lokhandwala and P. Gautam, Indirect impact of COVID-19 on environment: A brief study in Indian context, Environ. Res., vol. 188, no. April, p. 109807, (2020).
9. J. McNamara et al., COVID-19, Systemic Crisis, and Possible Implications for the Wild Meat Trade in Sub-Saharan Africa, Environ. Resour. Econ., vol. 76, no. 4, pp. 1045–1066, (2020).
10. J. B. Renard et al., Origins and spatial distribution of non-pure sulfate particles (Nsps) in the stratosphere detected by the balloon-borne light optical aerosols counter (loac), Atmosphere (Basel). vol. 11, no. 10, (2020).
11. M. Akram, M. Amrani, and C. El, "Air2Day: An Air Quality Monitoring Adviser in Morocco," Int. J. Comput. Appl., vol. 181, no. 17, pp. 1–6, (2018).

12. S. Muhammad, X. Long, and M. Salman, COVID-19 pandemic and environmental pollution: A blessing in disguise? Sci. Total Environ. vol. 728, p. 138820, (2020).
13. A. Metya, P. Dagupta, S. Halder, S. Chakraborty, and Y. K. Tiwari, COVID-19 lockdowns improve air quality in the South-East Asian regions, as seen by the remote sensing satellites,Aerosol Air Qual. Res., vol. 20, no. 8, pp. 1772–1782, (2020).
14. S. Haddout and K. L. Priya, Unveiling the causes of reduction in troposphere NO2 in two cities of Morocco during COVID-19 lockdown, Environ. Forensics, vol. 0, no. 0, pp. 1–4, (2020).
15. T. Naciri and H. Hassani. COVID-19 national lockdown in Morocco: impacts on air quality 1 and public health. medRxiv, vol. 8106, p. 2020.07.05.20146589, (2020).
16. NASA earth observatory. Available from https://earthobservatory.nasa.gov/global_maps/MOD_LSTD_M/MOD_NDVI_M.
17. National Aeronautics and Space Administration Goddard Space Flight Center (NASA) 2019. Global Nitrogen Dioxide Monitoring Home Page, Africa OMI data. Available at: https://so2.gsfc.nasa.gov/no2/pix/regionals/Africa/Africa.html
18. KF Boersma, HJ Eskes, RJ Dirksen, RJ Van Der A, JP Veefkind. An improved tropospheric NO_2 column retrieval algorithm for the Ozone monitoring instrument Atmos. Meas. Tech. 4 1905–28, (2011).

Impact of geology and climate change on wetlands: Case of Lake Aguelmam Azegza (Middle Atlas, Morocco)

Abdelaziz Boutsougame[1]*, *Lahcen* El Youssfi[4], *Hlima* Aassine[3], *Soufiane* Khuili[3], *Sidi Imad* Cherkaoui[1], *Hassane* Ouazzani[2], and *Mustapha* Alaoui[3]

[1] EGRN research team, LPSTS Laboraotory, Higher School of Technology of Khenifra, Sultane Moulay Slimane University, Khenifra, Morocco

[2] Faculty of Sciences, Department of Geology, Meknes, Morocco

[3] Sciences and Technologies Faculty of Sultan Moulay Slimane University, Beni Mellal, Morocco

[4] RN2D Laboratory, Kénitra National Higher School of Chemistry ENSC, Ibn Tofail University, B.P. 241, 14000, Kénitra, Morocco

Abstract. The AguelmamAzegza lake is located in the Middle Atlas mountain range, this region consists of carbonate rocks of Lias. It is the best-watered and richest region in Morocco in terms of wetlands, especially natural lakes, rivers and fresh springs. The lake's water supply is ensured by the inflows of water in the form of underwater sources fed by precipitation (rain and snow) following tectonics and the genesis of karstic formations by the dissolution of carbonate rocks. The declining trend in lake levels in recent years results from the decrease in precipitation which decreases during these years.

Introduction

The AguelmamAzegza lake is a natural lake located 30 km east of the town of Khénifra in the heart of the Middle Atlas (Ajdir Causse) at an altitude of 1546m. It belongs to the structural domain of the Middle Atlas Causse, a karstic domain affected by tectonics, formed by carbonate rocks (limestone and dolomites) of the Lias which lies on soft red clay rocks and altered basalts of the Triassic (Lepoutre and Martin, 1967).

The Middle Atlas, to which the AguelmamAzegza lake belongs, constitutes a favourable environment for the genesis of karstic formations following the dissolution of the carbonate rocks that form the major part of the Middle Atlas landscape. This dissolution is due to the intense fracturing that the mountain range has undergone the humid climate, the low temperatures and the accumulation of snow during the winter (Martin, 1981). This karstification allows, on the one hand, the penetration and storage of a very high proportion of meteoric water by infiltration (Bentayeb and Leclerc, 1977), and on the other hand, the filling of the groundwater network.

The specific objectives assigned to this work consist in the synthesis of geological and hydrological data from different sources of punctual observations, analyses and spatialized information in the years 2012/13/15/16/17/18/19/2020 to note the effect of the dissolution of the carbonate geological formations and of the climatic changes on the water levels of lake AguelmamAzegza.

1 Location and geological context

The delimitation of the impluvium of AguelmamAzegza was carried out by means of the maps of Karrouchan 1/50000, Al Hammam 1/50000 and with the help of prospecting field visits. The surface area of the delimited basin is 911 ha.

Lake Azegza is part of the upper basin of the Oum Er Rbia, occupying a deep NW-SE elongated depression with an average depth of 26 m and an area of about 50 ha. It belongs to a small catchment area of about 10.62 km2 (Fig. 1) and is limited to the south by a sheer cliff. It is surrounded by limestone reliefs which are covered by a forest of Quercus *rotundi folia* (holm oak) and *Cedrus atlantica* (Atlas cedar). Its water supply is ensured by precipitation (rain and snow), runoff and underground water inputs (Benkaddour*et al.*, 2008).

This lake is located in a region with a Mediterranean mountain climate, which is characterized by a humid and cold climate (Martin, 1981).

*Corresponding author:a.boutsougame@usms.ma

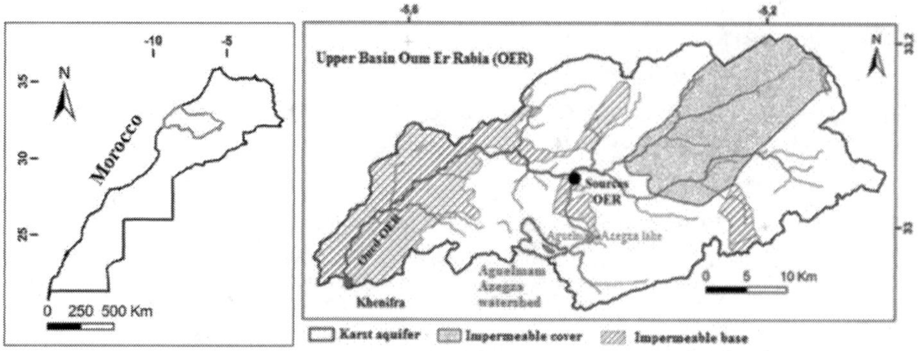

Fig. 1. Geographicallocation of the AguelmamAzegza lake and its catchment area on the hydrogeological map of the Upper Oum Er Rbia basin (Bentayeb and Leclerc, 1977).

Lake Azegza is one of the lakes of tectono-karstic origin (Hinaje and Ait Brahim, 2002). The Ajdir causse, of which the lake is a part, is located between the Aghbal and Srou limestone plateaus and constitutes the western part of the Bekrit syncline. In the western part of the Bekrit syncline, the structuring has generated submeridian faults and folds. The latter are the result of the detachment of the Liassic formations from the Triassic clay-salt. The installation of synclinal basins has favoured the karstification of the Liassic calcareous-dolomitic formations and the development of most of the poljes in this site. The whole area is part of the structural domain of the Middle Atlas causse, which offers the classic landscapes of high limestone plateaus, hence the name of causse. This is the domain of carbonate rocks: limestones and dolomites of the Lias which lie on soft red clayey-marl rocks of the Triassic (Lepoutre and Martin, 1967)(Fig.2).

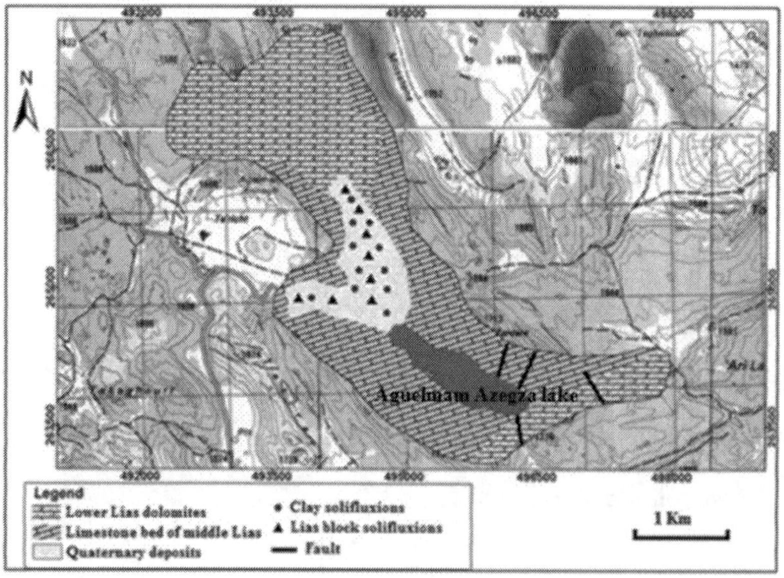

Fig. 2. Lithological map of the lower basin of AguelmamAzegza.

2 Geological conditions

2.1 Tectonics

The AguelmamAzegza lake is fed through the anomalous contact between the crushed and faulted limestone formations that lead the water to the lake (Fig. 3).

https://doi.org/10.1051/e3sconf/202233701007

Fig. 3. Fractures in the carbonate rocks of AguelmamAzegzaLake.

2.2 Karstification

The karstification consists of the dissolution of rocks following an intense fracturing affecting the carbonate formations near the lake. These formations feed the lake by precipitation (snow and rain)(Fig.4).

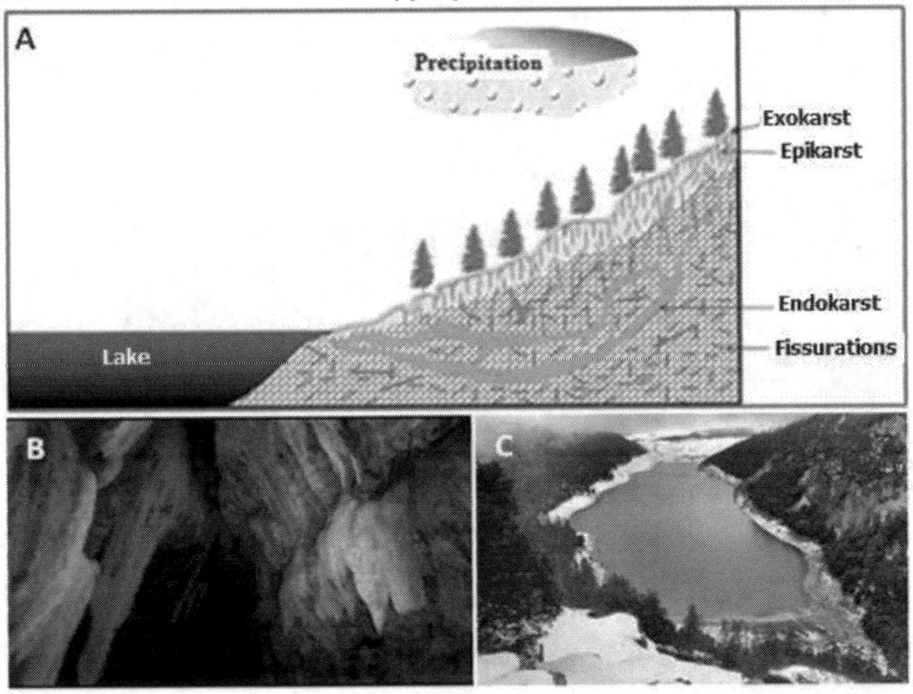

Fig. 4.A: Model of karstification of AguelmamAzegzalake, B: Photograph of karst in the vicinity of the lake, C: Photograph of the lake covered with snow in 2017.

In addition to the caves located to the north-east of the lake, such as Ifri Waman, the IfriWadou mountain on the eastern shore of the lake contains several caves, the largest of which is IfriWadou, named after the mountain. The latter has been studied by a team of Spanish speleologist (Subies and Benedito, 1993)(Fig.5).

 https://doi.org/10.1051/e3sconf/202233701007

Fig. 5. Photos of caves A: Ifri Waman; B:IfriWadou

2.3 Sedimentation of the lake

The study of sedimentation at AguelmamAzegza shows that the rate of accumulation is around 4 to 5 mm/year. The sediments present a granoclassification where the distribution is decreasing from the edges towards the centre of the lake. At the edges, the proximity of the inflow and the strong hydrodynamic energy allow the deposition of coarser materials of millimetre size by excess load. These coarse particles are favoured by the inflow from the gullies, especially the one situated to the SE of the lake (Fig. 6). The winds also exert a significant erosive action on the banks. Towards the centre of the lake, the decrease in hydrodynamic energy favours fine micrometric sedimentation.

Fig. 6. Photograph of the gullies that feed the lake with sediments

2.4 Main chemical elements of the lake water

According to Table1, the lake waters of AguelmamAzegza are characterised by high concentrations of Calcium, Magnesium, Bicarbonates and Carbonates, with moderately low concentrations of chloridesand very low concentrations of Potassium, Sodium, Sulphates and Nitrates. These waters therefore have a calcic and magnesian bicarbonate hydrochemical facies.This facies reflects the lithological nature of the host rock, which consists of limestone and dolomite.

Table1. Concentration of the main chemical elements in lake waters

Chimical elements	Concentration Average	Origin
Carbonates	11,02 mg/l	Alteration of carbonate rocks (limestone and dolomite)
Bicarbonates	281,26 mg/l	Alteration of carbonate rocks (limestone and dolomite)
Sodium	0,16 mg/l	Leaching of forest soil from the lake catchment
Sulphates	1,11 mg/l	Rainfall as a result of the presence of sulphur in the atmosphere
Potassium	1,02 mg/l	Leaching of forest soil from the lake catchment

Chloride	14,24 mg/l	Leaching of forest soil from the lake catchment
Magnesium	43,91 mg/l	Alteration of carbonate rocks (limestone and dolomite)
Calcium	64,12 mg/l	Alteration of carbonate rocks (limestone and dolomite)

3 Climatic conditions

3.1 Precipitations

The annual rainfall measured at AguelmamAzegza over the period 2000 to 2019 has recorded fluctuations that are particularly significant. This annual rainfall was very high in 2009-10 (844mm - 680mm) and after a decrease in rainfall from 2011 onwards becomes lower in 2019-20 (13mm), these results are strongly correlated with those obtained from the study of lake level variation and therefore explain the trend of decreasing lake level in recent years(Fig.7).

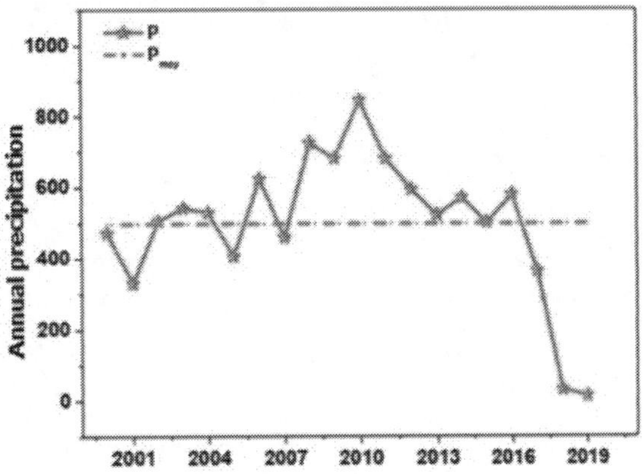

Fig. 7. Annual rainfall in AguelmamAzegza from 2000 to 2019

3.2 Temperatures

The temperature of the lake waters shows strong variations during the year, with a maximum value of 24.8°C recorded during the month of July, and a minimum value of 7.8°C recorded during the month of January, i.e. a difference of 12.64°C. Indeed, lake waters are strongly dependent on external climatic conditions (Adallal et al., 2014). Underground water inputs (~12°C), which are in the form of sub-aquatic springs feeding the lake, also contribute to the decrease in lake water temperatures (Gayral and Panouse, 1954).

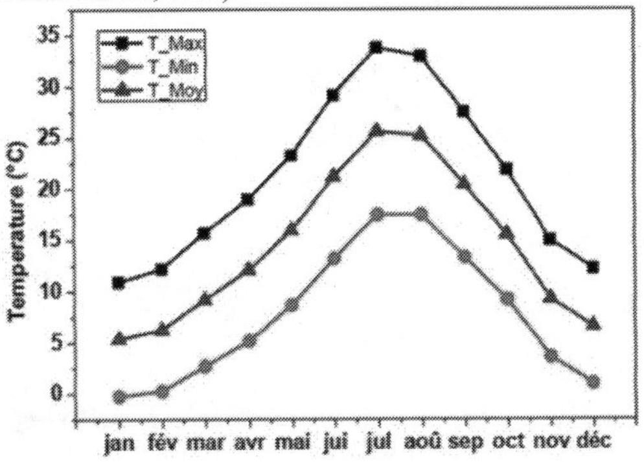

Fig. 8. Average monthly temperatures in AguelmamAzegza from 2000 to 2019

The average temperature values of the AguelmamAzegza station for the period 2000 to 2019 show that January is the coldest month (5°C), while July and August are the warmest (25°C) (Fig. 9). The average temperature is 13.48°C, with annual average values between 12.46°C and 14.25°C.

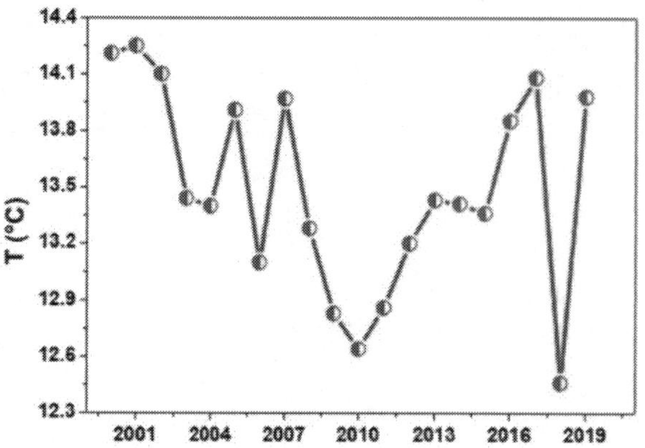

Fig. 9. Average annual temperatures in AguelmamAzegza from 2000-2019.

4 The temporal evolution of the lake edge

The understanding of the recent, past and future fluctuations of the AguelmamAzegza lake level, requires a good knowledge of the physiographic parameters of the lake (height h, surface S and volume V), in order to describe the relationship between the changes of the lake volume and the inflow and outflow of the interconnected lake basin which will be defined as the AguelmamAzegza lake system in an accurate way. These parameters can be determined directly by placing oneself in the site, which requires adequate equipment and instruments, or by using the geographic information system (GIS). The latter alternative will be adopted in this study to achieve our objective. In this context, data from different sources of point observations and spatial information were collected and synthesised in order to clarify the water balance of the lake in a sensitive context of environmental changes.

In order to record the effect of climate changes on the water level of the lake, weproceeded to plot the lake's edges in two ways. The two ways adopted here are based on satellite images corresponding to the last years (from 2012 to 2020). The first one allows delimiting the lake borders using ArcGIS software, while the second one allows having them directly on Google Earth Pro.

According to the results, there is generally a narrowing of these curves representing the level of the lake as a function of time (Fig. 10), which reflects a decrease in the volume of water during these years. In contrast, the expansion of the curve corresponding to the year 2016 implies an increase in water volume during that year.

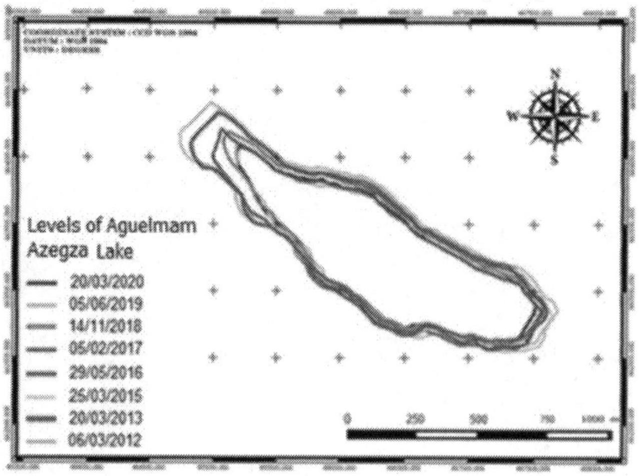

Fig. 10. Mapping the water adges of AguelmamAzegza lake based on satellite images(2012, 2013, 2015, 2016, 2017, 2018, 2019 and 2020)

 https://doi.org/10.1051/e3sconf/202233701007

These results are in good agreement with those found using the second method(Fig. 11). This trend of decreasing lake level has been demonstrated previously by Flower and Foster (1992) in earlier studies and that moreover the lake has varied in depth by 8m during the 20th century.

Fig. 11. Lake AguelmamAzegza levels between the years 2012 and 2020 on Google Earth Pro

Using data on the level of Lake AguelmamAzegza obtained from previous works (Gayral and Panouse, 1954; Flower and Foster, 1992; aerial photos from the mapping service, Rabat, 1964, 1973) and data acquired in the framework of this study, we were able to establish a chronicle of past lake levels based on satellite images and recent photographs(Tab.2). The data are plotted against the reference level established in 2012. All the data indicate that the lake has recorded decametric variations over the last few decades since the 1950s, with a maximum amplitude of fluctuations in its level of about 16 m (between the very high-level of 1456 m a.s.l. in 1965 and the very low-level of 1540 m a.s.l. in 2008 (Flower et Foster, 1992).

Tab.2. Historical fluctuations in the level of Lake AguelmamAzegza

Date	Lake level (m a.s.l)	Perimeter(m)	Surface (Ha)	References
1949	1544	***	***	Gayral et Panouse (1954)
1952	1546.2	***	***	
1953	1545.5	***	***	
1954	1544.7	***	***	
1965	1556	***	***	Aerial photo (1964), Mapping service, Rabat, Rabat
1974	1552	***	***	Aerial photo 1973, Mapping service, Rabat
1979	1549	***	***	Flower et al. (1992)
1982	1547	***	***	
1985	1544	***	***	
2008	1540	***	***	Photo, https://mapsights.com
2012	***	3869,04[b]	52,76	Adallal[a] (2019); current work[b]
2013	1546[a]	3692,63[b]	49,55	
2014	1544[a]	***	***	
2015	1543.5[a]	3355,28[b]	41,43[b]	
2016	1543[a]	3531,02[b]	43,6[b]	
2017	1542[a]	3261,06[b]	39,35[b]	
2018	***	3303[b]	38,2[b]	
2019	***	3151[b]	37,8[b]	
2020	***	2958,99[b]	34,22[b]	

The fluctuation of the level of the lake can only be justified within the framework of the approach of Campy et al (2003) which links this variation of the level to two systems: the first is the catchment basin carrying the material and the second is the sedimentary basin which reacts as a receptor, hence the need for the lake's own water balance which expresses the equilibrium between the inflow and outflow of water.

5 Conclusion

The AguelmamAzegza area is a favourable environment for the genesis of karstic formations through the dissolution of rocks following intense fracturing. The karstification allows, on the one hand, the penetration and storage of a very high proportion of meteoric water by infiltration.

Variations in precipitation and the temperature of the lake water over the last few decades have led to a reduction in the level of the lake. Indeed, the shores of the lake of AguelmamAzegza fluctuate according to the climatic conditions and particularly to the precipitations and the evapotranspiration and also to the dissolution of the carbonate geological formations.

Further studies are needed to assess the impact of socio-economic activities and land use combined with the effects of climate change on Azegza's water resources in order to better predict the evolution of the lake in the coming decades.

References

[1]. S. HinajeandL. AitBrahim. Les basins lacustres du Moyen Atlas (Maroc) : un exempled'activité tectonique polyphasée associée à des structures d'effondrement. Comun. Inst. Geol. E Minerio.**89**. pp. 283-294 (2002).

[2]. B. Lepoutre, J. Martin. Le causse moyen atlasique. Les cahiers de la recherche agronomique, N°24 : Congrès de pédologie méditerranéenne 1966-Madrid. Excursion au Maroc. 2ème partie : description des régions traversées, chap. IV. pp. 207-226 (1967).

[3]. J. Martin. Le Moyen Atlas Central : Etude géomorphologique. Service Géologique du Maroc, Rabat. 482 pp (1981).

[4]. A. Bentayeb and C. Leclerc. Le causse Moyen atlasique. In: Ressources en Eau du Maroc.Service géologique du Maroc, Rabat, pp 37–84 (1977).

[5]. A. Benkaddour, A. Rhoujjati, M. Nour El Bait. Hydrologie et sédimentation actuelles au niveau des lacs Iffer et AguelmamAzegza (Moyen Atlas, Maroc). pp108-118 (2008).

[6]. J.V. Subies, V. Benedito. Exploración de la grangrietaIfriWadou. Grupo espeleológico "la senyera" c/. Monte carmelo, **4**. Apdo. De Correos 216.Valencia. pp. 17-86 (1993).

[7]. R. Adallal, C. Vallet-Coulomb, A. Benkaddour, C. Sonzogni, F. Sylvestre. Water and isotope balances of AguelmamAzegza lake (Middle Atlas, Morocco). 41st IAH International Congress "Groundwater: Challenges and Strategies", Marrakech, Morocco (2014).

[8]. P. Gayral and J.B. Panouse. L'AguelmameAzegza : Recherches Physiques et Biologiques. Bulletin de la Société des Sciences naturelles et physiques du Maroc **36**, 135–159 (1954).

[9]. R. Flower and I.D.L. Foster. Climatic implications of recent changes in lake level at Lac Azegza (Morocco). Bulletin de la Sociéte Géologique de France **163**, 91–96 (1992).

[10]. R. Adallal, Variabilité environnementale des lacs du Moyen Atlas marocain : fonctionnement hydrogéochimique, hydrologique et réponse au changement climatique. Thèse,Université d'Aix-Marseille et Université Cadi Ayyad de Marrakech. 212p, (2019).

[11]. M. Campy, and J.J. Macaire. Géologie de la surface : Erosion, transfert et stockage dans les environnements continentaux. Compte-rendu p.279-280 (2003).

Characterization of climate impacts on a semi-arid agricultural perimeter in Morocco

Abdelhaq BOUNDI [1*] and *Zehor* AIT YACINE[2]

1. Independent consultant, Fkih Ben Salah, Morocco.
2. Environmental Engineering, Faculty of Science and Technology -University Sultan Moulay Slimane, Béni Mellal, Morocco.

Abstract. The manifestations of climate change in precipitation, air temperature and simulated evapo-transpiration for crops in the Tadla perimeter in central Morocco are examined. The study area is characterized by a semi-arid climate, with a potential agricultural vocation. This area is exposed to a scarcity of water resources for crop irrigation. To this effect, it is it is treating of rainfall in this area over 60 years (1958-2018) at 27 measuring stations and for temperature and reference evapo-transpiration (ETo) over 49 years (1969-2018) at 4 stations. Following data analysis by homogeneity tests on the agro-climatic time series (Pettitt test, Lee and Heghinian test, Buishand test and Hubert's segmentation method). It is detecting that there are time breaks on annual rainfall, mean temperature and reference evapo-transpiration at the respective years of 1978, 1995 and 1978. Because of these breaks in time, rainfall is reducing by 36%, atmospheric warming is 1°C and the climate demand for evaporation coupled with plant transpiration increases by 13%. Consequently, the water requirement increases significantly in the agricultural area of crop production.

Introduction

The present research aims to evaluate the long-term manifestations of climate change in a vulnerable semi-arid bioclimatic zone, which is most prevalent in North-West Africa, while characterizing the most relevant agro-climatic indicators, including rainfall, ambient temperature and reference evapo-transpiration of crops. It has been reported precipitation in continental regions has been effectively decreasing during last decade's in West Mediterranean and in West Africa area [1].

In this context, the Tadla perimeter a strategic Moroccan agricultural area was selected as the study area. It is located on semi-arid bioclimatic zone and is part of the important Oum Er Rbia hydraulic. Multiples changes in hydroclimatic variables have been reported in the basin, starting from the late 1970s onwards [2]: gradual rainfall decrease from the late, a relative atmospheric heating and a the gradual increase in climatic demand for evaporation, as well as the higher transpiration of the vegetation affected in this biotope.

1. Material and methods

1.1 Study area

The study area is the 3000 km^2 agricultural perimeter of Tadla, situated in the semi-arid bioclimatic stage of the Beni Mellal-Khénifera region (Morocco). It is located between longitudes West 6° 18' 08" and 7° 04' 39" and latitudes North 32° 10' 16" to 32° 40' 10", and has an altitude of around 400 m.

1.2 Methods used

The major agro-climatic indicators considered are precipitation, temperature and reference evapo-transpiration (ETo). ETo is commonly adopted for its compelling similarity with plant water requirements [3].

The trend of these indicators is conducted using the appropriate statistical methods typically used to test time series homogeneity, which are essentially defined as follows [4]:

-Pettitt test: This test is a modified version of the Mann-Wihtney test. It allows verifying stationarity of rainfall series. It is subdividing into two groups of data with respective sizes "m" and "n", whose values are grouping and are classifying by ascending order. It is following by the calculation of the sum of the ranks of elements in each group; a statistical study allows detecting the existence or absence of a break.

-Lee and Heghinian Bayesian method: It is a Bayesian method. Therefore, that is proposing a parametric approach. It is requiring a normal distribution of serial values. The absence of a break in the series is a null hypothesis.

-Buishand U-Statistics: The procedure has the same concepts and hypotheses considered in Lee and Heghinian's method, with the condition of normality of the series data studied, in particular the position of the possible break point to be identifying by Buishand's U statistics.

-Hubert segmentation: The segmentation is consisting in splitting the time series into segments. For two contiguous segments, a significant difference is then identifying as the break in the series.

The mathematical equations of these four statistical tests of homogeneity of rainfall series are reporting in detail in the publications by the authors concerned [4].

1.3. Agro-climatic data

* Corresponding author: boundhaq@gmail.com

The agro-climatic database coms from a network of 28 climatic stations in the Tadla perimeter. The observations span 60 years (1958-2018) for precipitation and 49 years period (1969-2018) for temperature and reference evapotranspiration.

1.4. Statistical analyses

About analyses at time series trend, cited in paragraph *1.2 Methods used*, it is used "Khronostat" software [4]. Numerical data processing is applied statistical software: "SPSS" and "R".

2. Results and discussion

Following the homogeneity tests conducted on the rainfall, temperature and reference evapo-transpiration (ETo) series, years of time breaks were detected. The implicit analyses by agro-climatic indicator are presenting below.

2.1 Precipitation

In the evolution of the annual rainfall totals, it is noted that the year 1978, a main point of the chronological break. It is illustrated in the graph in Figure 1.

It is noted that in the pre-breakdown sub-period of 1959-1978, the average rainfall is 397 mm/year, versus 254 mm/year during the period after the break (1979-2018). Therefore, the rainfall reduction rate is 36% between these two sub-periods, illustrated in the histogram graph in Figure 2.

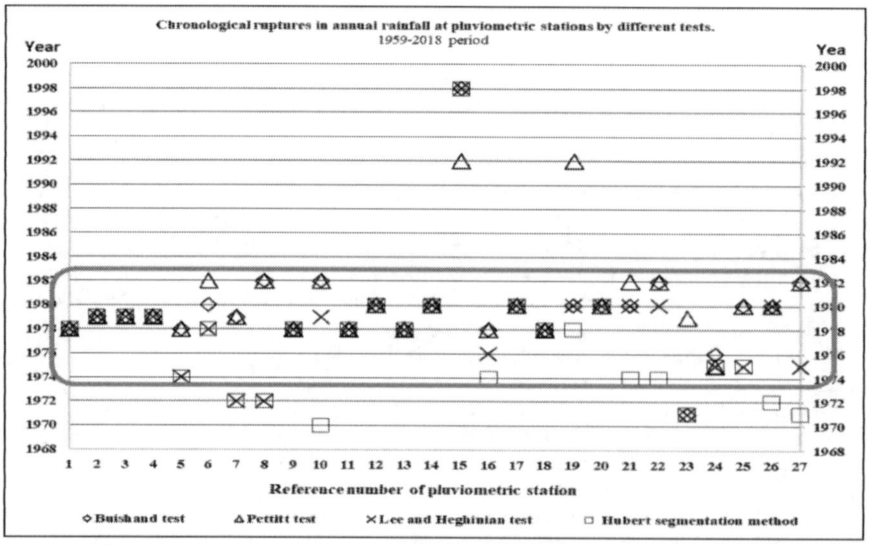

Fig. 1: Chronogram of rainfall ruptures by homogeneity tests. Observation period 1958-2018 in several station. Tadla perimeter (Morocco).

Shift in of the year 1978 is not in line with the findings in recent research works in the eastern part of the Oum Er Rbia hydraulic basin, located near the study area. However, the two studies do not cover the same periods, with

1983-2015 observation period [5]. This is explained, by our study is more recent and with a long observation period (1958-2018).

Furthermore, our results are relatively comparable with those of other authors, who detected shifts in the rainfall regime in the same study area during the decade of the 1980s, others studies detected a shift and a large area of North-West of Algeria between the late 1960s and the early 1970s [1;6].

The amount of reduction in rainfall after the shift is found to be 36% in this study (Figure 2). Shifts between 8% and 15% were found by [2] using a more extended time series of observations.

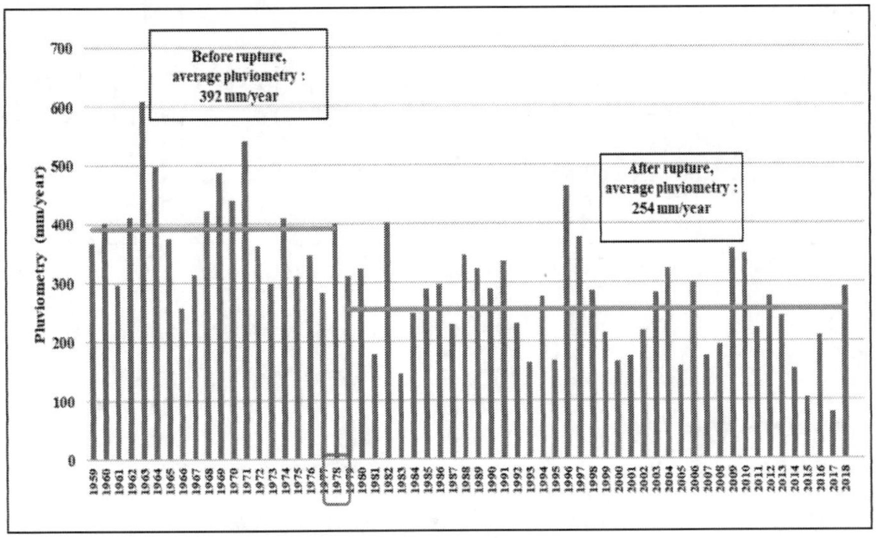

Fig. 2: Annual precipitation evolution before and after rainfall rupture. Observation period 1958-2018. Tadla perimeter (Morocco).

2.2 Temperature

A shift was detected in the year 1995 in the time series of average air temperature (Figure 3).

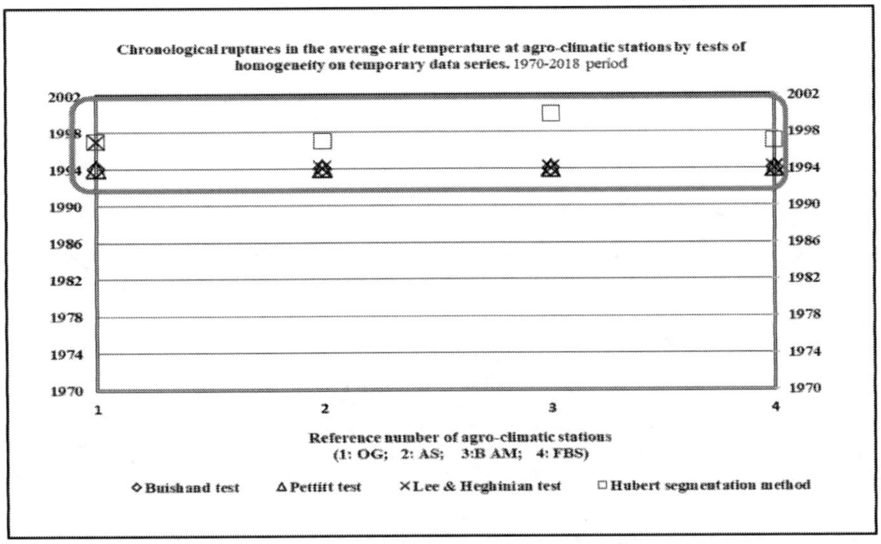

Fig. 3: Shift in mean temperature by distribution homogeneity tests. Period at 1969-2018 in several stations. In Tadla perimeter (Morocco).

In addition, the mean temperature level jumps after the time break from 18.7 to 19.7°C. This shows an atmospheric heating (a positive deviation of 1 °C), which occured two decades ago (Figure 4).

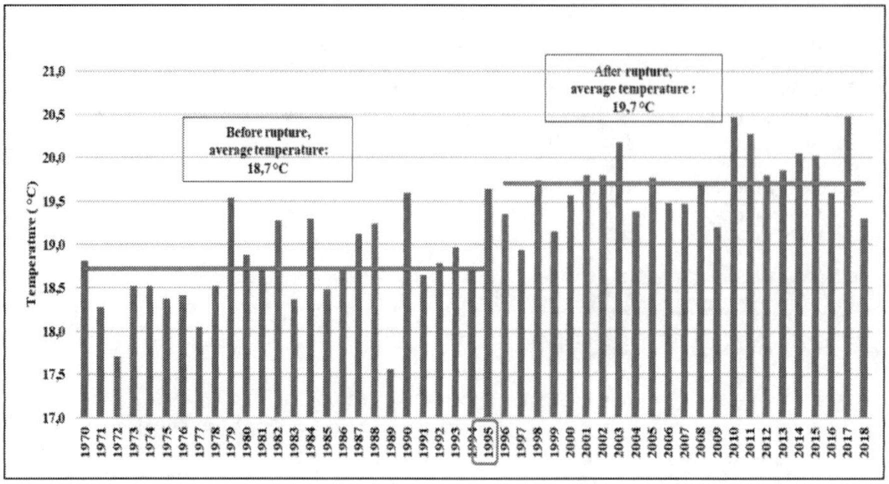

Fig. 4: Average air temperature at the Tadla perimeter (Morocco) before and after the shift.

The observed anomaly in ambient temperature (+1°C) is in accordance with the same shift in at the global air temperature [7].

2.3 Evapotranspiration

A shift was observed in 1978 for ETo (agro-climatic indicator reference evapo-transpiration in the study area (Figure 5).

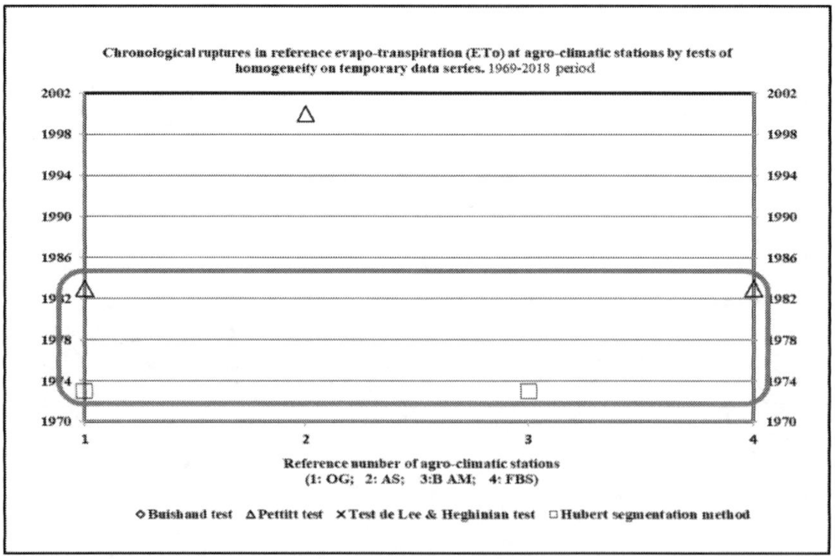

Fig. 5: Dates of time break at reference evapotranspiration (ETo) detected by homogeneity tests. Period at 1969-2018 in several stations. Tadla perimeter (Morocco).

The reference evapo-transpiration (ETo) in the study area increases by 12.5%; after the shift (Figure 6). It is in the same range as 11.4% ETo change reported by [8] in the central Tunisian area with a similar semi-arid climate.

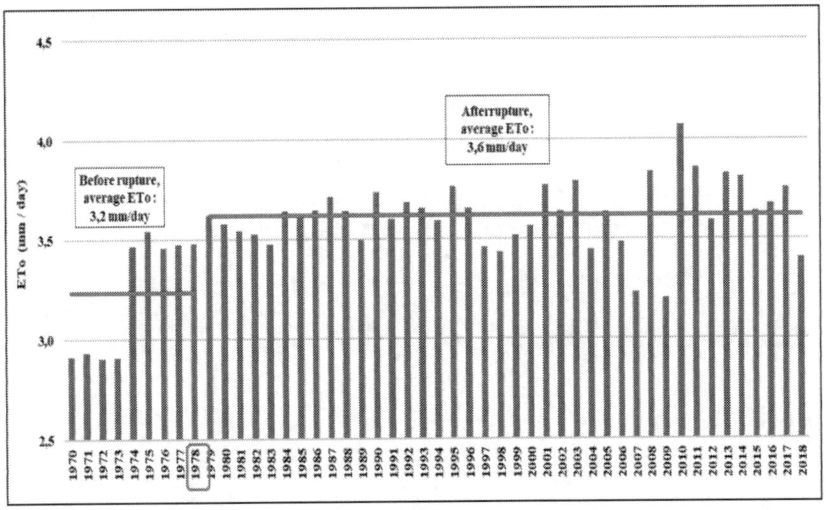

Fig. 6: Average evapo-transpiration at the Tadla perimeter (Morocco) before and after the shift. 1958-2018 period.

3. Conclusion

Climate change in recent decades has been observed in a semi-arid model area of potential agricultural in Morocco. This was achieved through the study of the temporal variability of pertinent bioclimatic indicators, which are of essential used in the agronomic management of crop areas; particularly for cases in the most vulnerable to water scarcity in situation of needed supplemental irrigation. The trend of decreasing rainfall over the last four decades was confirmed by this study along with atmospheric heating over the last twenty years. In addition, the climate's demand for evaporation has been increasing in combination with the increase in transpiration of the plants growing in is this area of bioclimatic stage over the last 40 years. Consequently, the water requirements of the crops in this space are expected to increase.

References

1. TAIBI S., MEDDI M., SOUAG D. and MAHE G. *Larhyss Journal*. Vol. **359**:191–197(2013).
2. SEBBAR A., BADRI W., FOUGRACH H., HSAINE M. and SALOUI A. *Sécherese*. Vol. **22 (3)**: 139-148 (2011).
3. ALLEN R. G., PEREIRA L. S., RAES D. and SMITH M. Crop evapo-transpiration-Guidelines for computing crop water requirements-FAO Ir. & drainage paper 56. FAO, Rome (Italy) 300. **N° 9**: D05-109 (1998).
4. TRAORE V., NDIAYE M. L., MBOW C., GIOVANI M., SARR J., BEYE A. and DIAW A. KhronoStat Model as Statistical Analysis Tools in Low Casamance River Basin, Senegal. **Edit.207**: 10-22 (2017).
5. KRIMISA S., BOUCHAOUI L. and ABOURICHA S. *Larhyss Journal*. Vol. **30** (2017): 7-26 (2017).
6. GHENIM A. N. MEGNOUNIF A. SEDDINI A. and TERFOUS A. *Sècheresse*. Vol. **21 (2)** (2010).
7. LENSSEN N. J. L., GAVIN A. S., HANSEN J. E., MENNE M.J., PERSIN A., RUEDY R. and ZYSS D. Improvements in the GISTEMP uncertainty model. *J. Geoph. Res.: Atmos*. Vol. **124 (12)**: 6307–6326 (2019).
8. HAJRI J. Citrus, pp: 220-225. **29th** International Symposium. 6-9 September 2006-Université Denis Diderot Epernay (France) (2006).

Traditional practices : A window for water erosion management in the Argana basin (Western High Atlas - Morocco)

Mohamed Ait Haddou*, *Belkacem* Kabbachi, *Ali* Aydda, *Youssef* Bouchriti, *Hicham* Gougueni, *Mohamed* En-Naciry, and *Asmae* Aichi.

Geosciences and Environment Team, Department of Geology, Faculty of Sciences, Ibn Zohr University, Agadir 8106, Morocco.

Abstract. Water and soil are two essential natural resources, which their preservation determines the local and national environmental sustainability. The pollution and erosion processes causing deterioration of these resources, which impacting negativaly the environment and economy development. This research aims to suggest the most straight forward and most appropriate control measures for the basins that are socially and economically acceptable for farmers. By consdering Argana basin as study area, the proposed approach consists, first, to analyze the soil loss map and, then, cconducted semi-structured interviews with local farmers at various locations across the basin. The results indicate that the areas of high and very high deterioration encompass 48 % of the basin's surface. In addition, all of the farmers questioned used agroforestry on their managed sites. The study highlight the importtance of the traditional practices in fighting water erosion risk.

1 Introduction

In Mediterranean countries, soil erosion due to rainfall and runoff is a wide spread phenomenon. The soil erosion is the most prevalent type of land degradation worldwide [1, 2]. The negative impacts of soil erosion become critical when it surpasses the optimal rate (10.4 t/ha/year) [3]. In 2014, the cost of land degradation in Morocco, including soil erosion and desertification, was estimated to be over 5.01 billion dirhams or 0.54 % of the Gross Domestic Product (GDP) [4]. The Abdelmoumen dam supplies drinking water to the broader Agadir region (9.5 Mm^3) and irrigation for the Issen periphery (13,000 hectares). This dam is undergoing rapid siltation, with a specific degradation of 646.15 tons/km^2/year [5]. Watershed management and the management of the implications of soil degradation are gaining a lot of interest from policymakers, especially, for the most degraded regions. In this

* Corresponding author: medaithaddou@gmail.com

 https://doi.org/10.1051/e3sconf/202233702002

context, the current study anticipated to provide the simplest and most appropriate method of soil degradation control in the Argana basin (Western High Atlas (WHA), Morocco).

2 Study area

The study area, covering an area of about 1300 km^2, is a basin in the High Atlas downstream section of the Souss watershed (Figure 1). It is part of the Argana corridor and its surrounding valleys. The Issen River is the largest tributary descending from the southern flank of the western High Atlas of Morocco. In general, the Issen basin is a subject to rainfall instability. In addition, orographic variations influence the distribution of climatic settings in this basin [6]. This basin is considered representative of erosion conditions [7,5]. The succession of the crossing lands has a significant supply of detrital materials as a result. Their imperviousness, high relief, and limited vegetation all impact runoff and material transport in this basin. Because of its physical characteristics, it is prone to rapid and torrential floods.

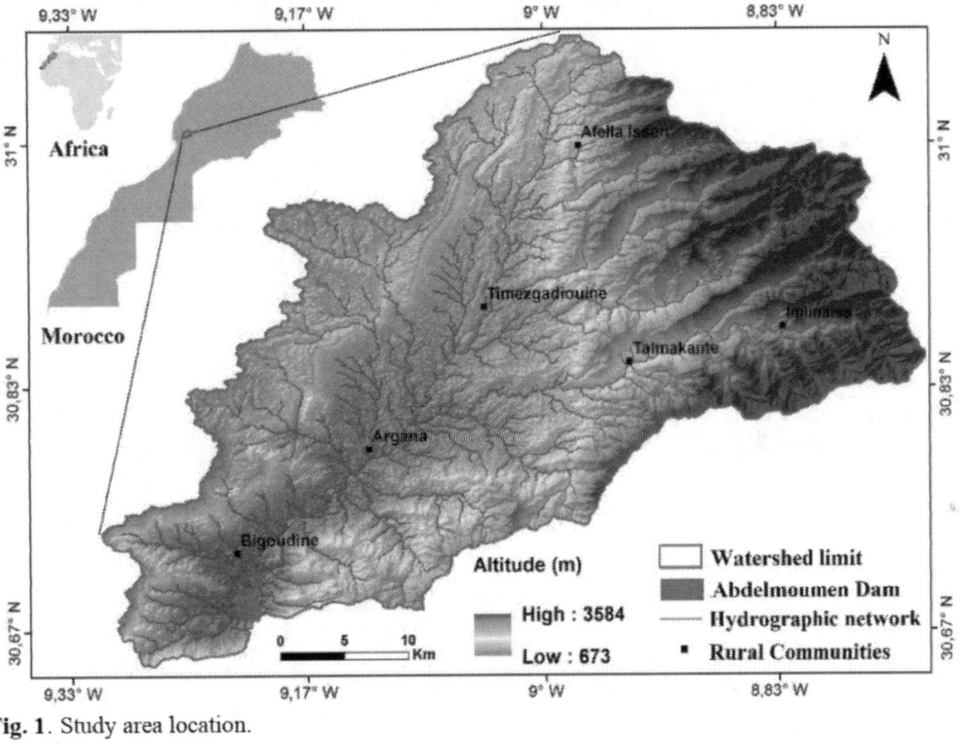

Fig. 1. Study area location.

3 Methods

Integrating erosion mapping and erosion risk management into a GIS creates a comprehensive tool for directing land-use decisions. Several studies have conducted this type of research in the High Atlas watersheds [08]. According to Ancy[9], in order to manage the erosion risk, it is necessary to identify the erosion hazard and conservation practices. The conservation practices (land use) should be properly planned according to the potential of dangers (Fig. 2).

In this study, the hazard map was prepared by integraing 5 factors (rain erosivity factor (R), soil erodibility factor (K), slope factor (LS), land cover factor (C), and erosion control practice factor (P)) in GIS platform using the equation of RUSLE model proposed by[10,11]. In fact, the spatial analysis was utilized to find the regions most impacted by erosion and the most susceptible locations in the Argana corridor. The soil loss risk classes were verified based on in-situ observation. In addition, in order to highlight the conservation practices, semi-structured interviews were done with 30 farmers across the basin, supplemented by spontaneous interactions with a broader group of farmers. Note of that the field surveys were conducted during the priod 2017-2020.

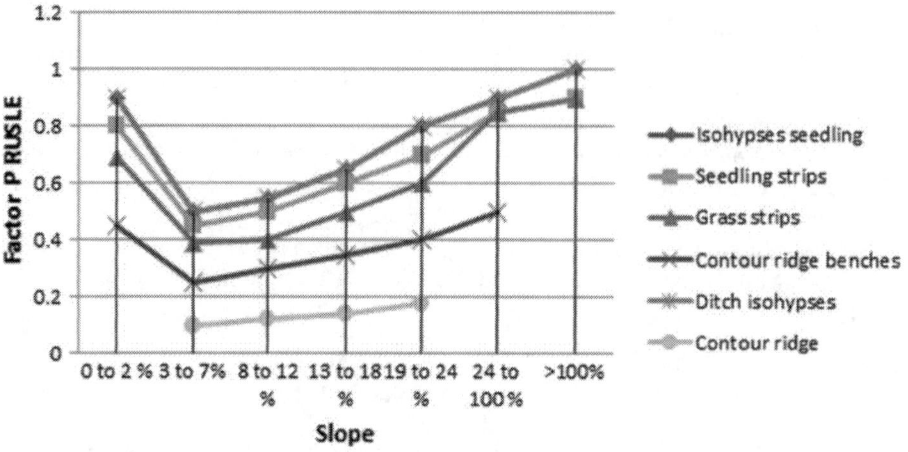

Fig.2.Performance of erosion control methods in relation to the topographic gradient [12].

4 Results and Discussion

4.1 Analysis of the potential erosion map

The obtained erosion risk map reveals that the most part of the study area is threthend by water erosion phenomenon with three degree of risk: very high risk (risk 3 with loss rate estimated by 100 to 430 t/ha/yr), high risk (risk 2 with loss rate estimated by 20 to 100 t/ha/yr), and moderate risk (risk 1 with loss rate estimated by 0-20 t/ha/yr) (Fig. 3).

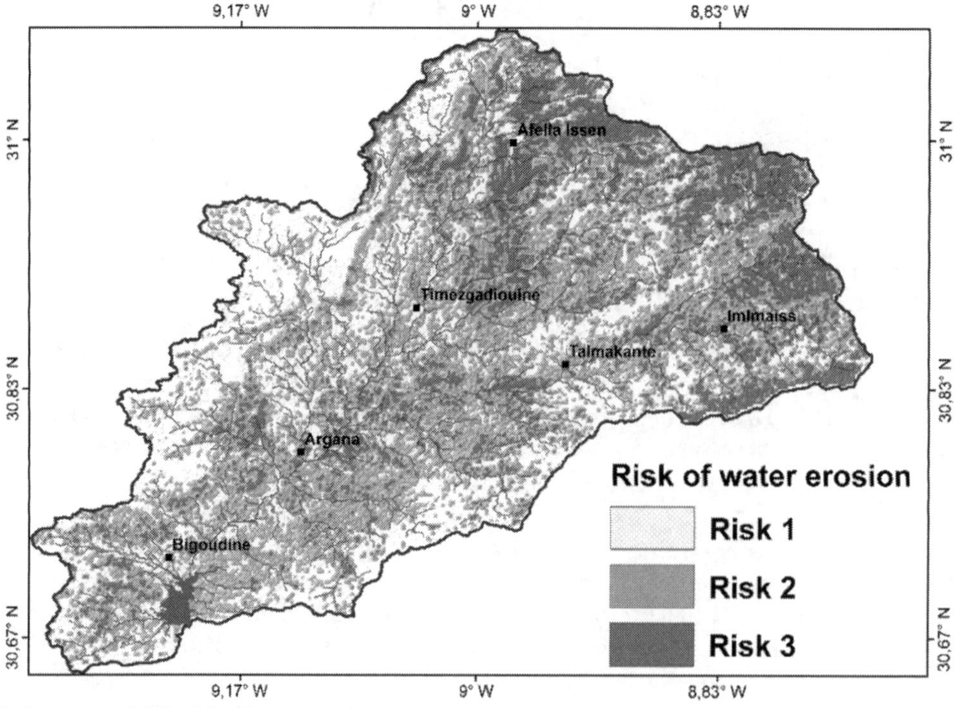

Fig.3. Areas of high risk of water erosion.

Table 1 lists the characteristics of the water erosion risk weighting in the study area. Therefore, huge areas with agricultural potential are sterilized or in the process of sterilization, particularly in locations with fragile rocks, like in the center of the Argana corridor. Because of the large size of the basin and the number of hot spots involved, the sector situated in the center of the Argana corridor was chosen as a priority intervention target.

Table1. Characteristics of the water erosion risk weighting.

Risk of water erosion	Loss rate (t/ha/yr)	Characteristics
1	0 to 20	-Loss rate that is less than or near to the tolerance level: soils on moderate slopes; surface erosion; shallower claws; stripping with bright patches, soil well covered by flora; soil protected by stepping stones; agricultural activities.
2	20 to 100	The loss rate is a concern: evidenterosion; bare soil on steep slopes; bare and badly degraded soil; gullies and ravines; extensive gullying; skeletal soil on rocky outcrops; deserted stockyards; sparse vegetation.
3	100 to 430	- An alarming loss rate necessitates immediate intervention: extremely severe deterioration; weak rocks on rocky slopes; deep gullies; landslides and mass movement; mudslides, bank undercutting, and Badlands; loss of agricultural production.

[13] reported that the success rate of development strategies varies by territory. Under the NABVP, 1.5 Million hectares of managed soils have failed to stop erosion.

4.2 Survey data analysis

All of the 30 farmers and residents interviewed practice agricultural and animal husbandry. 13% depend on beekeeping, mainly in the Ida-Ou-Tanane and Anzerki plateaus. The average age of the interviewes farmers is 43 years. They are all active.7 of the 30 residents are herders under the age of 20. 63.33 % are illiterate, and 36.66 % have only a secondary degree of education. Agroforestry is practiced by all of the farmers in the basin who were interviewed.Growing somewhat demanding fruit trees like olives, almonds, figs, pomegranate, and date palms helped to consolidate and maintain the soil. Grass (barley, wheat, and corn), legumes (fava beans, peas, lentils, and beans), and forage crops, particularly alfalfa *(Medicago sativa),* are all annual crops. Erosion is a severe concern for agricultural management and crops in the area, according to interview responses. Individual plots and farming locations were handled following local customs. Various structures and techniques of development and control of water erosion risk, particularly the building of terraces, banks, dikes, and bunds, are part of cultural heritage competence. However, management is generally collective and applied by all residents. Farmers in such situations use traditional technical skills on a small scale, with relative independence from local authorities. Almost all interviewees stressed the need to replenish what they loosely called small hill dams (SED), created by land dams for surface water supply and groundwater recharge to ensure water use by the rural people in future seasons.

4.3 Proposed support practice

According to the final erosion map, regions of high and very high soil loss encompass around 48 % of the basin's surface. When significant soil loss regions are included, more than half of the entire area, 54 %, comprises degraded soil. This vast region contains practically all of the basin's habitats and bioclimates. As a result, it provides a chance to diversify management strategies and plant species for reforestation. The list of advancements to be proposed is extensive and diverse. However, the most appropriate, practical, and cost-efficient methods for the terrain must be chosen. To preserve the basin's soils against water erosion, two types of interventions are recommended (Fig. 4 & 5).

- Technical measures: contour ridge benches, contour ridge, and concrete or gabion sills.
- Biological measures (revegetation): Argan tree, Berber Thuja, Phoenicean Juniper, Olive tree, Carob, Fig tree and Alfalfa.

Fig.4. Irrigated terraces supported by dry stone walls following the contour lines near the rural commune of Imlmaiss, enclosed in the WHA (Photo a). View of ridges and bunds made only of soil and built exclusively by hand on market gardening plots, with the aim of conducting water from one basin to another (Ait Bkhayr valley) (Photo b).

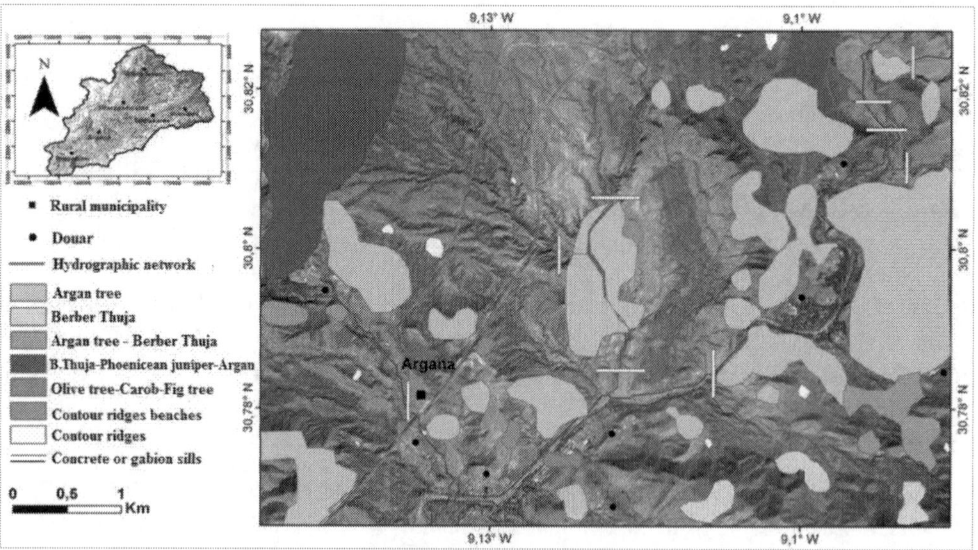

Fig.5. Simulation of the erosion control practices for the Argana site.

5 Conclusions

Direct observations in the Argana basin and indirect views using Google Earth images have indicated that the soils are already degraded. The areas are barren in the bio-geographical sectors most important for development.They are quickly destroyed by fluvio-torrential erosion because they are composed mainly of clay-detritus formations that are sensitive to water erosion, compaction, suffusion, and solifluction.

Erosion control cannot be restricted to the maintenance of existing soils; it must also address soil productivity restauration by enhancing slope water management and soil physical characteristics.

The anticipated management strategy must encompass the whole basin, not only the sections threatened by very high or high soil erosion risk. It should encompass other regions, such as forests and rangelands, to increase pastoral output. These industries are likewise deeply worried about a potential intervention for soil and water conservation. Further, farmers' water and land management accomplishments should not be overlooked, particularly in the mountains. Only a meeting of a watershed's intervening agents, as well as integrated water and soil management strategy, can currently lessen the detrimental effects of water erosion and successfully combat it.

It enables the inclusion of soil erosion rate as a decision-making criterion in the land use planning process, notably in its integrated management.

More persistent monitoring of farmers is required to mobilize and instruct them on the many procedures and circumstances to consider when selecting acceptable species for planting and reforestation initiatives. This observation is common in most of the High Atlas's surrounding watersheds.In any way, the situation remains serious, given the acceleration of deforestation consequences and the limited resources of conservation measures.

As a response, it provided foundation for delivering a set of possible soil and water conservation activities and building a local development plan for erosion risk management.

References

1. W. Maetens, J. Poesen, M. Vanmaercke, Earth-Sci. Rev **115**(1-2), 21-36 (2012b) https://doi.org/10.1016/j.earscirev.2012.08.003
2. J. M. García-Ruiz, S. Beguería, N. Lana-Renault, E. Nadal-Romero, A. Cerdà, Land Degrad Dev, **28**(**1**), 5-21. (2017) https://doi.org/10.1002/ldr.2641
3. Y. Ostovari, A.A. Moosavi, H.R. Pourghasemi, Land Degrad Dev **31**,15 1–12. (2020) https://doi.org/10.1002/ldr.3597
4. L. Croitoru, M. Sarraf, Environ. *Nat. Resour. Global Practice Discussion Paper. 5.* World Bank Group Report Number 105633-M (2017).
5. M. Ait Haddou, B. Kabbachi, A. Labbaci, S. Ezaidi, Y. Bouchriti, Int. J. Sci. Res. Innov. Std., **2**,1 (2020a) https://ijarimss.org/publications/volume2-issue1/
6. M. Ait Haddou, B. Kabbachi, A. Aydda, H. Gougueni, Y. Bouchriti, E3S Web Conf. **183**, 02003. (2020b) https://doi.org/10.1051/e3sconf/202018302003
7. A. Elmouden, N. Alahiane, M. El Faskaoui, Z.E.A. El Morjani,. In :Choukr-Allah R., Ragab R., Bouchaou L., Barceló D. (eds) *The Souss-Massa River Basin, Morocco.* Handb. Environ. Chem., 53 (2016) https://doi.org/10.1007/698_2016_70
8. É. Roose, M. Sabir, et A. Laouina, *Gestion durable des eaux et des sols au Maroc : Valorisation des techniques traditionnelles méditerranéennes*. Marseille : IRD Éditions, 343 p. (2010) https://doi.org/10.4000/books.irdeditions.294

9. C. Ancey, *Notes de cours : Risques hydrologiques et aménagement du territoire*, Laboratoire Hydraulique Environnementale, Ecole Polytechnique fédérale de Lousanne (EPFL). Version 13.1 du 11 septembre, Lausanne, Suisse (2018) http://lhe.epfl.ch/hydrologie.php
10. K. G. Renard, *A guide to conservation planning with the Revised Universal Soil Loss Equation (RUSLE) in Predicting soil erosion by water* , DC : USDA ,Washington, USA (1997)
11. A. Labbaci, S. Moukrim, S. Lahssini, S. Laaribya, H.M. Alaoui, J. Hallam, Adv. Sci. Technol. and Eng. Syst. J., vol. **6**, no. 3, pp. 241-247 (2021) https://doi.org/10.25046/aj060326
12. I. Gaubi, A. Chaabani, A. B. Mammou, & M. H. Hamza, Nat Hazards **86**, 219–239 (2017) https://doi.org/10.1007/s11069-016-2684-3
13. A. Laouina, *Peut-on attenuer les risques hydriques dans les piemonts atlasiques ?* Actes du colloque «Colloque International sur les risques naturels et l'aménagement du territoire». 9 et 10 novembre 2018, Oujda, Maroc (2018), Publications de l'Université Mohamed Ier, mis en ligne le 5 juillet 2019. http://www.ump.ma/

Coagulation-flocculation technique for domestic wastewater treatment in the city of Ain Aouda, Rabat, Morocco.

Naif Al-Jadabi [1*], *Mohamed* Laaouan[2,3], *Maria* Benbouzid [1], *Jamal* Mabrouki[1] and *Souad* El Hajjaji[1]

[1]Laboratory of Spectroscopy, Molecular Modeling, Materials, Nanomaterials, Water and Environment (LS3MN2E-CERNE2D), Department of Chemistry, Faculty of Sciences, Mohammed V University in Rabat, Avenue IbnBattouta, B.P. 1014, Rabat10000, Morocco

[2]International Institute of Water and Sanitation, National Office of Electricity and Drinking Water (IEA-ONEE), Avenue Mohamed BelHassan El Ouazzani, B.P. Rabat-Chellah 10002/ Rabat, Morocco

[3]Central Laboratory for water quality control, National Office of Electricity and Drinking Water (ONEE), Avenue Mohamed BelHassan El Ouazzani, B.P. Rabat-Chellah 10002/ Rabat, Morocco

Abstract. The goal of this study was to investigate the performance of coagulation-flocculation, and sedimentation tests to reduce contaminant concentrations in domestic wastewater to ecologically acceptable levels. The tests were performed at the (ONEE) laboratory using a jar test apparatus (GPB FLC- Didatec, France) for enhancing the primary treatment of domestic wastewater effluent from the wastewater plant of Ain Aouda City, Rabat, Morocco. Aluminum sulphate $Al_2(SO_4)_3.18H_2O$ (termed alum) was employed as a coagulant in the study. The impact of coagulant doses and pH at optimum dosage on the process of coagulation was studied and conditions were optimized to obtain the best removal of studied pollutants. The evolution of physico-chemical parameters pre-and post-treatment was monitored to determine the efficacy of the process. The turbidity and total suspended solids (TSS) were the best indicators of the treatment's efficacy of coagulant concentration (150mg/L) within the pH optimum of 7. Laboratory tests have shown 98.4% of turbidity, and 97% total suspended solids removal at initial concentrations of 121 NTU, 220 mg/L respectively. Also, it has shown good removal efficiency for COD, BOD5 of the order 83%, 77% at initial concentrations of 340 mg/L, 150 mg/L respectively. Following the coagulation tests, the examination of the main parameters studied revealed that the degree of contamination had decreased. This is confirmed that Aluminium sulphate as a coagulant has proven its effectiveness in the treatment of domestic wastewater in this study.

Keywords: Wastewater; Treatment; Coagulation; Flocculation; Morocco

*Corresponding author: naif.naif.2017@gmail.com

1 Introduction

The obtained of clean and safe drinking water is a great concern around the world [1].

In addition, the increase in urbanization and population growth and have led to freshwater pollution in developing countries due increase in wastewater generation and insufficient wastewater discharge. Therefore, wastewater treatment has become an important issue for the development of communities [2], [3]. Furthermore, often the effluent discharge which is improperly treated from wastewater treatment plants (WWTPs) result in major detrimental effects on the health of these surrounding environments and communities situated near sewage or downstream due to worsening physico-chemical parameters and increased microbial pathogens [4]. In the early 20th century, failure to provide large areas and adequate for untreated wastewater disposal, especially in megacities, led to the application of more efficient wastewater treatment methods [5], which took increasingly important at the beginning of the 1970s [6], [7]. Morocco has a semi-arid climate country in the Mediterranean facing scarcity and deteriorating quality of water. The total annual freshwater resources are estimated to be 22 billion (700 m^3/inhabitant/year) with variations of seasonal and spatial, this disparity becomes a source of concern [8]. As a result, the recourse to sources of non-conventional water has become important. Also, untreated wastewater discharges have contaminated the surface water in some urban areas [9]. Nowadays, reusing treated wastewater is an option solution by the limit of this shortage, preventing pollution, preserving natural resources, particularly, in the sector of agricultural, which consumes 80 to 90 % of Morocco's total water [8], [10]. In response to that, in 2005, Morocco launched a national sanitation program that aims to treat 60% of collected wastewater by 2020 [11]. Furthermore, in the plants of water treatment considered the coagulation-flocculation process is one of the best options, common, extensively applied as primary pretreatment, in particular, wastewater treatment relatively cost-effective, robust, and benign to the environment which removes organic and inorganic pollutants from wastewaters with reducing most of the TSS and COD in this process [6], [12-15]. This process aims to form flocs through destabilization of suspended particles followed by adsorption and aggregation [6]. Among others that influence coagulation-flocculation are the factors (effluent quality, temperature, pH, coagulant type, and dosage) [6]. Polyaluminum chloride, aluminum salts, and ferric salts, are inorganic coagulants used frequently at WWTPs for coagulation [16]. This paper aims to evaluate the coagulation-flocculation process' efficiency with aluminum sulphate ($Al_2(SO_4)_3 \cdot 18H_2O$) commonly called alum as a coagulant for (turbidity), TSS (total suspended solids), COD (chemical oxygen demand), and BOD5 (bio-chemical oxygen demand during 5 days)removal as pre-treatment options for domestic wastewater treatment. Alum is commonly used in wastewater treatment because of its proven performance, availability with cost-effectiveness and also, perform comparatively better in the removal of suspended organic solids than to other substitutes such as metallic salts of iron [17], [18].

2 Materials and methods

2.1 Geographic Location of Ain Aouda Wastewater Treatment Plant

The plant of wastewater treatment of Ain Aouda is one of the stations under management of the National Office of Electricity and Drinking Water (ONEE). It depends in treatment on the usage of the activated sludge process with flow reaches 10000 m^3/day. This plant is located about 600 m northeast of the city of Ain Aouda center on the right bank of the thalweg of Wadi Akreuch which is currently serving as a receiver of Ain Aouda discharges While it is bounded from the southwest by the RR401 regional road leading to Rabat [19].

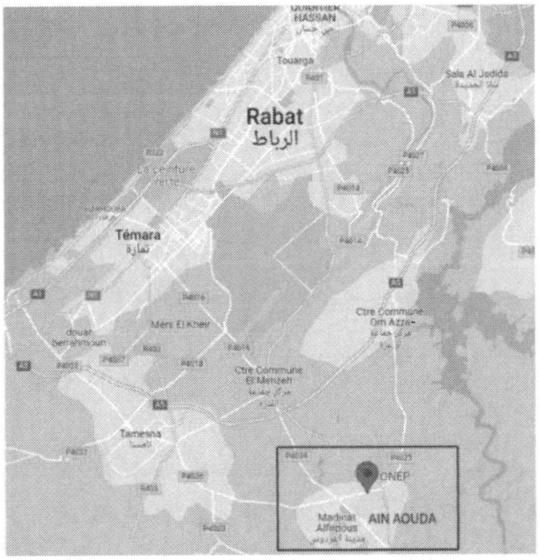

Fig.1. Geographical location of Ain Aouda [Source Google Maps]

2.2 Wastewater Sampling

The wastewater sample used in this research was brought from treatment plant of domestic wastewater of Ain Aouda city, located in the province of Rabat, Morocco. There are a large number of variables that may change the composition of water are numerous, and they can all be potential pollutant sources. The main measurements, carried out on the studied sample, were (temperature, pH, electrical conductivity, turbidity, TSS, COD, BOD5, total

phosphorus, ammonium, nitrates, and nitrites). The samples are taken to the laboratory for analysis and stored at a low temperature (4°C) [20], before use in experimental runs to reduce wastewater decomposition caused by microbial activity [12]. The quality of wastewater

Table 1. Raw domestic wastewater characteristics, Moroccan standards for effluent discharge and analytical devices.

Parameters	Unit	Value	Moroccan standards of rejections	Device analysais
pH	-	8.19	6.5-8.5	WTW Multi-parameter Meter, model 3430
Temperature	°C	20.5	<35	
Conductivity	µS/cm	2200	3000	
TSS	mg/L	220	150	3 pos ramp filtration.
Turbidity	NTU	121	<5	Thermo Scientific, Eutech TN-100 Turbidimeter
BOD5	mg/L	150	120	YSI 52 Dissolved Oxygen Meter, american
COD	mg/L	340	250	Hach Company 45600-00 COD Reactor,

				HachDr 2010 Portable Datalogging Spectrophotometer
TP	mg/L	5.35	10	Thermo Scientific iCAP RQ ICP-MS Spectroscopy Analysis
NH_4^+	mg/L	23.5	0.5	Continuous Flow Analyzer (CFA) (reference 475-424)
NO_3^-	mg/L	<1	-	Ion Chromatography Thermo Scientific™ Dionex™ ICS-5000
NO_2^-	mg/L	<0.5	-	

2.3 Coagulant

Aluminum sulphate $(Al_2(SO_4)_3 \cdot 18H_2O)_{(s)}$ with the purity of between 16-17%. was used as coagulant since this is the salt that is most commonly used in practice [14], [21]. 5-g alum was dissolved in 100 ml of distilled water (ultrapure water) to prepare a five percent alum stock solution.

2.4 Coagulation-flocculation Process

The process of coagulation-flocculation was conducted forwastewater treatment from the plant of Ain Aouda (Morocco), using Jar Test Apparatus set-up with 6 pedals (GPB FLC-Didatec, France). The pH adjustment by adding 0.5 N NaOH and 0.1 N H_2SO_4 solutions). Six beakers were used with one-liter capacity. They were filled with 500 mL of the wastewater sample that would be treated. Next, immersed the propellers into the beakers. In order to determine the effect of the doses of alum, increasing doses were introduced into the wastewater samples under study, with various amounts of Aluminium sulphate $(Al_2(SO_4)_3 \cdot 18H_2O)_{(s)}$(50, 100, 150, 200, 250, 300 mg/L),under rapid mixing, with the speed 180 for 2min. Followed by 40 rpm for 20 min. It's important to note that coagulation-flocculation requires the mixing rapid in order to the colloids meet and neutralize each other while relatively slow speed aim to enhance encounter and aggregation of colloidal without breaking already formed flocs [6]. The samples left to settle for a period extending for 30 min. Once finished of coagulation-flocculation process, the blades of the agitator were carefully withdrawn from the treated water. After settling, the supernatant was withdrawn to determine the optimal dosage required to achieve the highest percentage removal of the most important parameters (turbidity, TSS, COD, and BOD5).

2.5 Analytical Methods

2.5.1 Chemical analysis

Results collection of analysis involved raw and treated samples' initial and final concentrations.The samples were taken from 1 cm below the surface of the supernatant of treated samples using a pipette, and the effectiveness of treatment was based on turbidity, TSS, COD, and BOD5 values. In table 1 shown the characteristics of raw domestic wastewater with the devices used in this analysis according to Moroccan standardized.

2.5.2 Pollutants removal after treatment

The treatment efficiency was calculated based on the percentage removal recorded for the physico-chemical parameters pre-and post-treatment by coagulation process. The removal of each parameter was obtained using the following equation [8]:

$$\textit{Removal rate (\%)} = \frac{C_e - C_s}{C_e} x100$$

where, C_e was the concentration before treatment and C_s was the concentration after treatment

3 Results and Discussion

3.1 Sample characterization

The pollution degree of domestic wastewater is based on the assessment of a number of physicochemical characteristics of this wastewater. Table. 1 shown important information regarding the values of the effluent parameters before the treatment with comparison to the limits of the values fixed by Moroccan standards.

The temperature and Conductivity recorded were 20.5°C and 2200µS/cm below 35 °C, 3000 µS/cm, as limit values of water designated for irrigation according to the Moroccan norms [22]. The pH of this wastewater was 8.19 and usually is pH of wastewater slightly alkaline. Also, this value is within the Moroccan standards range from 6.5 to 8.5 [8], which can be linked to the presence of compounds used in detergents which include a wide range of amphoteric, cationic, and anionic compounds [23]. The turbidity (121 NTU) of this wastewater (Table 1) is very important in this treatment process because it sensitive. This high of this parameter is due to the presence of solid matter, in particular, suspended solids [6]. The COD/BOD5 ratio gives (2.27) that indicates satisfactory biodegradability of this effluent [8]. The suspended solids, COD and BOD5 are the most important parameters because of their high impact generated on the environment when they are released into bodies of water [6].

3.2 Coagulant dose effect

One of the most significant factors to consider when defining the optimal condition of coagulant performance in coagulation-flocculation process is the dose of the coagulant. which is influenced by the TSS or colloid content of the wastewater [7], [18]. Essentially, inadequate dose or overdosing would result in low flocculation results. As a result, determining the optimal dose is significant to reduce cost of dosing, sludge formation, and also, simultaneously achieving the best treatment results [24].

The experiments of coagulation were carried out without prior adjustment of pH with varying the dose range of coagulant between 50–300 mg/L. Fig.2 shows the progression of coagulant dosages from underdosing to optimum dosage and then overdoing.The optimum alum dose was 150 mg/L, which was the lowest needed dosage that resulted in the highest removal. The maximum removal efficiencies of turbidity, TSS, COD, BOD5 were 98.3%, 96.8%, 81.8%, 75% at the optimum dose of 150 mg/L.

3.3 pH variation according to the dose of coagulant $Al_2(SO_4)_3.18\,H_2O$

The pH decreases when the alum dose is increased, as seen in Fig. 3. We note that the progressive addition of this coagulant to wastewater resulted in a pH decrease from 8.19 to 7.6.at optimum dose 150mg/L. This decrease of pH explained due able of alum in released of H^+ ions when dissolution in water [25], and can be interpret the decrease in pH result the reaction of natural alkalinity found in water in water with aluminum [24].

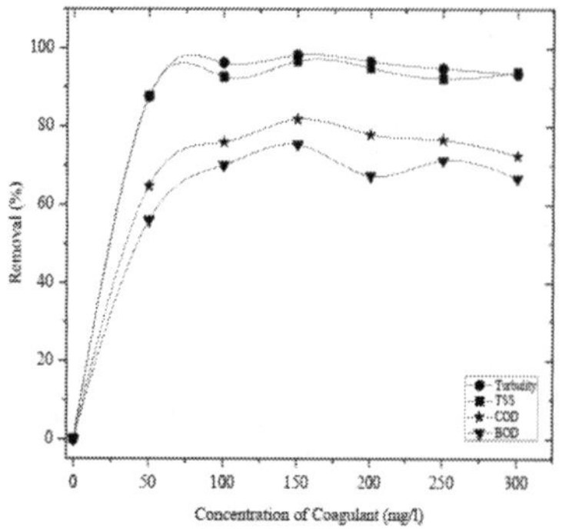

Fig.2. Effect of alum dose on (turbidity, TSS, COD, and BOD5) removal.

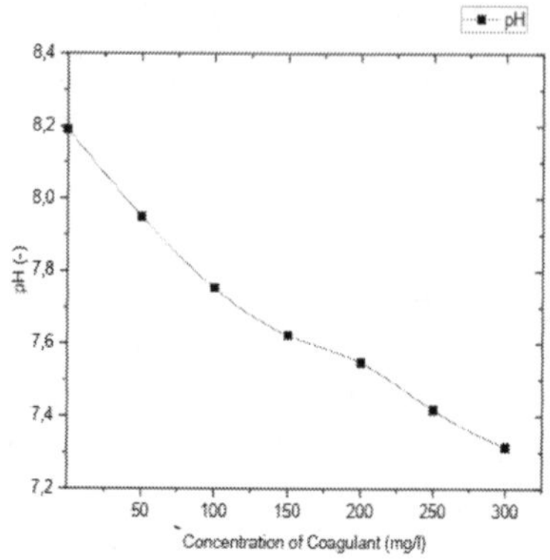

Fig.3. Evolution of pH according to the dose of alum

3.4 Effect of pH

The experiments were performed again with prior adjustment of pH at various values (6-9) for fixed optimal dosage (150 mg Al^{3+}/L to determine the optimum pH. The results are shown in Fig. 4, which plots removals of turbidity, TSS, COD, and BOD5 against pH. The removal efficiency rate of all studied parameters initially increased up to pH 7, where the removal performance was at the maximum 98.4%, 97%, 83%, and 77%, for turbidity, TSS, COD, and BOD5 respectively. Then after this point (pH 7) the rate of removal started to decrease with the pH value increased.At optimal dosage and after 30 min of settling was the amount of produced sludge 26 ml/L with initial turbidity of 121 NTU.

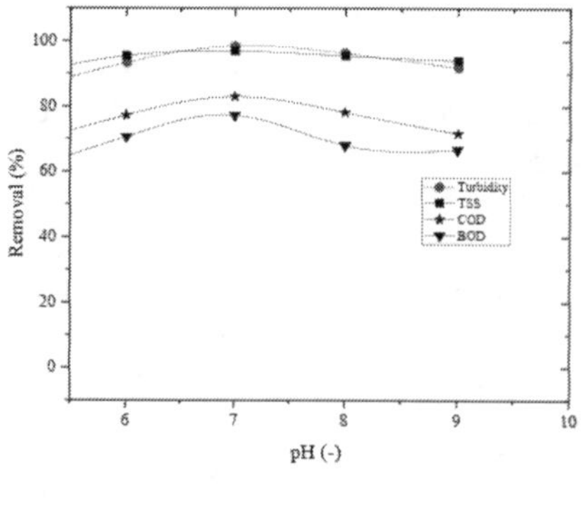

Fig.4. Effect of initial pH on elimination of (turbidity ,TSS, COD, and BOD5) at dose of alum 150 mg/L.

3.5 Effectiveness of the Coagulation Process in Eliminating

Fig.5. Depicts the effects of (alum coagulant dose :150 mg/L and pH at 7) on domestic wastewater effluent (before and after

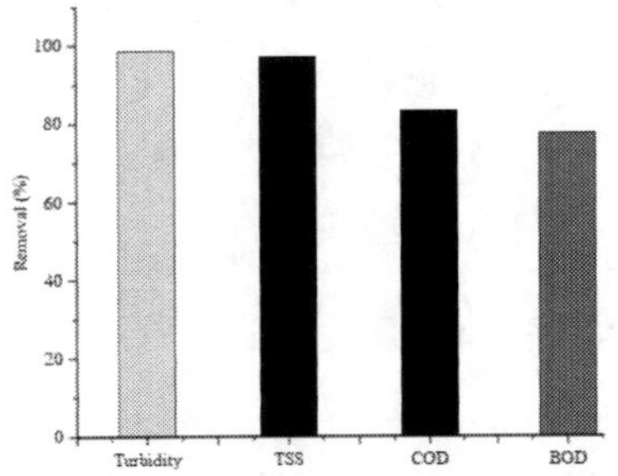

Fig.6. Elimination efficiency of the coagulation process at pH 7 and dose of alum 150 mg/L.

Fig.6 illustrates the results of removal efficiency of the coagulation process for studied pollutants under optimal conditions (optimal dosage 150mg/L, pH=7). The Turbidity and TSS(total suspended solids) were of the best removal in this process that the maximum removalratesreached 98.4%, and 97% respectively. On the other hand, (COD) chemical oxygen demand, and (BOD5) biochemical oxygen demand content removal reach 83% and 77% respectively.

3.6 Comparison of current study with literature

The comparison of results obtained from this study with literature is presented in Table 2. The table isshown various doses of coagulant in determining the effectiveness in pollutants reduction of wastewater.

In our study, it can be described that pollutants reduction using 150 mg/L was best comparable to hight doses with others literature. This study demonstrated good results with 150 mg/L.

Table 2. Comparison of results obtained from current study with literature.

Locality of use	Wastewater Type/ Pollutant	Dose of alum	Removal efficiencies (%)	Reference
Algeria	domestic wastewater 350 NTU (Turbidity)	600 mg/L	Turbidity 99.7%	[15]
Brazil	domestic wastewater 41.1 NTU(Turbidity) 54.2 mg/L (BOD) 223 mg/L (COD)	200 mg/L	Turbidity 93%, BOD 58%, COD 70%	[16]
Algeria	domestic wastewater 90.5mg/L(BOD5)	400 mg/L	BOD5 78.74%	[26]
Iraq	domestic wastewater 834mg/L(COD)	6g	COD 84.41%	[27]

Morocco	domestic wastewater 136 NTU (Turbidity) 220 mg/L (TSS) 340 mg/L (COD) 150 mg/L(BOD)	150 mg/L	Turbidity 98.4%, TSS 97%, COD 83%, BOD 77%	This study

4 Conclusion

This study showed that the performance of the coagulation-flocculation process was efficient in the treatment of domestic wastewater of the city of Ain Aouda plant using aluminum sulfate $Al_2(SO_4)_3 \cdot 18H_2O$ (alum). The experiments of Jar-test were performed to determine the optimum conditions to remove pollution. Turbidity and TSS are considered the best efficiency indicators in wastewater treatment [5]. The efficiency of this process was assessed in terms of its ability to remove in terms of the removal of (turbidity), (TSS), (COD), and (BOD). At an optimum experiment condition (coagulant dose: 150 mg/L, pH: 7 showed 98.4% removal for turbidity, 97% of TSS while having reached 83% removal for COD, and 77% of BOD5. The coagulation-flocculation process has proven it effectiveness using Aluminum sulphate to treat the domestic effluent of Ain Aouda city and adjust to the limits established in Moroccan state regulations and national legislation for the effluents discharge into the aquatic environment.

References

1. O. A. Adesina, F. Abdulkareem, A. S. Yusuff, M. Lala, and A. Okewale, *South Afr. J. Chem. Eng.*, **28**, 46-51(2019).
2. H. Guven, R. K. Dereli, H. Ozgun, M. E. Ersahin, and I. Ozturk, *Prog. Energy Combust. Sci.*, **70**, 145-168(2019).
3. R. S. Al-Wasify, A. K. Mahmoud, S. R. Hamed, I. E. Abdelaziz, and S. R. Hamed, *ARPN J. Eng. Appl. Sci.*, **14**(2), 354-362(2019).
4. S. Naidoo and A. O. Olaniran, *Int. J. Environ. Res. Public. Health*, **11**(1), 249-270(2014).
5. S. H, S. M, N. F, and A. B, *Int. J. Environ. Res.*,**1**(2), 104-113 (2007).
6. F. Z. Choumane, B. Benguella, B. Maachou, and N. Saadi, *Ecol. Eng.*, **107**, 152-159(2017).
7. V. Saritha, N. Srinivas, and N. V. Srikanth Vuppala, *Appl. Water Sci.*, **7**(1), 451-460(2017).
8. I. Mansir, E. Oertlé, and R. Choukr-Allah, *Water*, **13**(7), 14(2021).
9. J. F. Schyns and A. Y. Hoekstra, *PLoS ONE.*,**9**(6), 14(2014).
10. R. Choukr-Allah, A. Nghira, A. Hirich, and L. Bouchaou, Cham: Springer International Publishing, Switzerland, 1-26 (2017).
11. L. Mandi and N. Ouazzani, *Sustain. Sanit. Pract.*, 14,9, (2013).
12. Z. Chaouki *et al.*, *J. Mater. Environ. Sci.*, **8**(8), 2781-2791 (2017)
13. Q. H. Malik, *Appl. Water Sci.*, **8**(1), 40 (2018).
14. N. A. Jadabi, M. Laaouan, J. Mabrouki, and S. E. Hajjaji, *J. Adv. Res. Dyn. Control Syst.*, **12**(7), 147-157(2020).

15. F. Choumane, F. Kandouci, and B. Maachou, *Springer Int. Publ. AG 2018*, 506-515 (2018).
16. P. Vega Andrade, C. F. Palanca, M. A. C. de Oliveira, C. Y. K. Ito, and A. G. dos Reis, *J. Water Process Eng.*, 11(2020).
17. A. H. Jagaba and I. M. Lawal, *Int. J. Eng. Manag.*, **2**(3), 67-75(2018).
18. K. P. Y. Shak and T. Y. Wu, *Chem. Eng. J.*, 256, 293-305 (2014).
19. Assainissement liquide de la ville de Ain Aouda, Lot: station d'epuration des eaux usées, Office National de l'Electricité et de l'Eau Potable (ONEE, Branche eau), Marché travaux, Rabat, Maroc (2017).
20. J. Mabrouki, A. Moufti, I. Bencheikh, K. Azoulay, Y. El Hamdouni, and S. El Hajjaji, in *Advanced Intelligent Systems for Sustainable Development (AI2SD'2019)*, Cham, 200-212(2020).
21. A. B. Şengül, N. Tüfekçi, and S. Aktan, *Desalination Water Treat.*, **57**(53), 25610-25616(2016).
22. Ministry of Land of Morocco. Normes Marocaines, Official Bulletin of Morocco; No. 5062 du 30 Ramadan 1423; Ministry of Land of Morocco: Rabat, Morocco (2002).
23. A. Al-Gheethi, R. Mohamed, A. Wurochekke, N. Nurulainee, J. Mas Rahayu, and M. Amir Hashim, *MATEC Web Conf.*, **103**, 8(2017).
24. J.Mabrouki, M. Benbouzid, M. Dhiba, S. El Hajjaji, International Journal of Environmental Analytical Chemistry, 1-11(2020).
25. N. A. Oladoja and Y. D. Aliu, *Water Qual. Res. J.*, **43**(2-3), 231-238(2008).
26. N. Seghairi, L. Mimeche, A. Bouzid, and Y. Ayachi, *J. Water Environ. Sci.*, **1**, 230-234(2017).
27. L. A. Mokif, O. J. O. Al-Sareji, Z. H. Obaid, N. A. Abdulhusain, and S. S. Mohammed-Ali, *Eurasian J. Biosci.*, **14**(1), 2011-2014(2020).

Quantification and characterization of hydrological drought in the Oued El-Abid watershed between 1978-2019: Statistical and GIS approaches

Elhoucein Layati[1*], *Abdelghani* Qadem[1], and *Mohamed* El Ghachi[1]
[1]LDRH Laboratory, Sultan Moulay Slimane University, B.P: 524, Beni Mellal, Morocco

Abstract. Hydrological drought is considered an environmental phenomenon related to climate changes that affect natural resources in general. The evaluation of droughts is essential to in the management of water resources. This study aims to demonstrate the use of statistical indices (SSFI and NHI) and GIS (remote sensing) to quantify, characterize and identify hydrologic drought. These indices are applied to the watershed of Oued El-Abid using the annual stream flows time series at the three stations studied (Ait Ouchen, Tilouguit, and Tizi N'Isli), spanning from 1978 to 2019. Results showed that dry and very dry years occur more frequently. Because of, the lake of Bin El-Ouidane dam ismarked by a substantial regression of the water storage. Thematic maps of land use using spatial remote sensing (supervised classification) show that the years 1984, 2000, and 2017 are the driest and most vulnerable years.

1 Introduction

Climate variability and changes often manifest themselves with the appearance of drought periods, which influence the water resource, the environment and socio-economic activities [1-2]. Several authors have pointed out that it is very difficult to agree on a precise definition of the notion of drought [3-4]. The international glossary of hydrology defined the word drought as a prolonged absence or marked precipitation deficit. According to the National Institute of Agricultural Research [5], drought is a water deficit compared to average or median conditions. This deficit may affect the atmosphere, the soil, surface runoff, and subsurface water, in a given geographical area. There are several types of droughts: meteorological drought, hydrological drought, agronomic drought, and edaphic drought [6-7]: Meteorological droughts are caused by a deficit of precipitation compared to the normal (median or average) for a given period and region [8-9]; Hydrological droughts is related to the decrease of surface runoff in rivers, which leads to low water levels in rivers, lakes or reservoirs and groundwater. [7-10-11-12], Agricultural droughts are related to the other types of droughts (meteorological and hydrological) has an impact on vegetation development. They can be caused deficits in precipitation, an increase in evapotranspiration or the lack of water in soils and water storages [13]. They are highly dependent on climatic conditions, biological and phrenological characteristics of crops and physical and biological properties of soils. Finally, socio-economic droughts as controlled by the relationship between the supply and demand of certain economic goods and some of the elements of meteorological, hydrological, and agricultural droughts (For example: The

* Corresponding author:houssa8layati@gmail.com

 https://doi.org/10.1051/e3sconf/202233702004

supply of many economic goods, such as water, fodder, food grains, fish and hydroelectric power, depends on weather conditions) [14].

This study aims to study, identify and characterize hydrological drought in the Oued El-Abid watershed. The approach adopted is based on the use of statistical indices (SSFI and NHI) on streamflow time series at three hydrometric stations in this watershed: (Ait Ouchen, Tizi N'Isli and Tilouguit). Supervised classification is used to quantify flooded areas during the study period.

2 Materials and methods

2.1 Study area

Upstream of the Bin El-Ouidane dam, the study area of the present study is part of the Oued El-Abid watershed. It is located in the southeast of the action area of the Oum Er-Rbia Watershed Agency and covers an area of 6405 km^2 (Fig. 1). Geographically, this basin is part of the province of Azilal. It is bounded by the province of Beni Mellal, the province of Fkih Beni Saleh in the North, by the province of Ouarzazat in the South, in the East by the provinces of Errachidia and Ouarzaztin the east and the west by the provinces of El Hauz and El Kalaa des Sraghna in the West. On the hydrological plan: The Oued El-Abid has its sources in the jbel Masker (3 277 m) and flows from E to W. The basin has an elongated shape, extends in the heart of the limestone High Atlas, bounded to the E and S by a line of high ridges above 3 000 m. The highest point of the basin is the jbel Azourki (3 690 m) [15]. The main tributary of Oued El-Abid is Oued Ahansal. The large storage dam of Bin-el-Ouidane (storage of 1.5 billion cubic meters) is located shortly after the confluence of these two Oued(Fig. 1). This dam plays an essential role and meets various uses (hydroelectric production, tourist activities, irrigation of the Tadla plain and Tassaout Aval...).

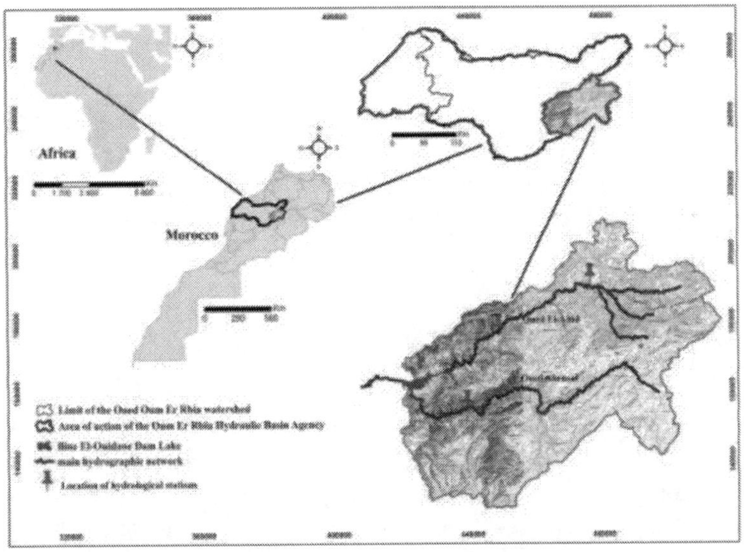

Fig.1. Geographical location of the study area.

2.2 Data and Methods

2.2.1 Available data

a. Hydrological data

The Oum Er-Rbia Water Basin Agency provided the hydrometric data used in this study. The data was collected at three hydrometric stations along the Oued (Tab. 1 and Fig. 1).

Table 1.Hydrometric stations.

Stations	Lambert coordinates			Chronicle
	X (m)	Y (m)	Z (m)	
Tilouguit	422 670	158 500	1100	1978-2019
Ait Ouchen	434 147	185 125	939	1976-2019
Tizi N'Isli	432 300	139 600	1350	1976-2019

b. Satellite data

Landsat TM images from the years (1984, 1985, 1990, 1992, 1993, 1995, 1998, 1999, 2000, 2001, 2002, 2005, and 2007) were used to perform a diachronic study of basin component change (especially flooded aras), from the **T**hematic **M**apper (**TM**) sensor and Landsat 8 **O**perational **L**and **I**mager (**OLI**)from the years (2016 and 2017) images downloaded for free from the website: http://earthexplorer.usgs.gov.

2.2.2 Characterization of hydrological drought

Several indices and methods are used to characterize and identify hydrological drought:

- Standardized Streamflow Index (SSFI):

The normalized streamflow index is similar to SPI and was introduced by McKee*et al.*1993 [16]. SSFI is the difference between the cumulative flow in a year and the mean and divided by the standard deviation of the series equation (1). The value of SSFI can be used to declare different drought management thresholds (Tab. 2).

$$\text{SSFI} = \frac{Fi-\bar{F}}{\sigma} \quad (1)$$

With: F_i: Flow rate in time interval i ;
$\bar{F}$is meanflow rate of the series;
σ is the standard deviation of the series.

- Normalized Hydrologic Index (NHI) :

The Normalized Hydrological Index (NHI) is similar to the Standardized Precipitation Index (SPI); it was developed to quantify the hydrological deficit on several time scales and to show the impact of drought on water resources for a given period [17]. It is expressed mathematically as follows equation (2):

$$\boldsymbol{NHI} = \frac{(\mathbf{D_i - Dm})}{\mathbf{S}} \quad (2)$$

With: Di: Discharge of month or year i; Dm: Mean discharge of the series over the considered time scale and S: Standard deviation of the series over the considered time scale. A classification of the drought is made according to the values of the NHI (Tab. 3).

Table 2. Classification of drought sequences according to SSFI. [18]

SSFI Values	Degree of moisture or dryness
≥2,0	Extreme humidity
1.5 – 2.0	High humidity
1.0 – 1.5	Moderate humidity
0- 1.0	Light humidity
0	Normal
-1.0 - 0	Light drought
-1.5 - -1.0	Moderate drought
-2.0 - -1.5	Strong drought
≤ -2.0	Extreme drought

Table 3. Classification of drought sequences according to NHI. [19]

NHI values	Drought sequence
0.00<NHI<-0.99	Slightly dry
-1.00<NHI<-1.49	Moderately dry
-1.50<NHI<-1.99	Severely dry
NHI<-2.00	Extremely dry
0.00< NHI <0.99	Slightly humid
1.00< NHI<1.49	Moderately humid
1.50< NHI <1.99	Severely humid
2.00< NHI	Extremely humid

2.2.3 Supervised classification:

To extract the years of hydrological drought, we start by downloading the satellite images and proceed with the corrections of distorsions that are due to the sensor and atmospheric effects, as shown in (Fig. 2). Secondly, these images were processed and classified based on auxiliary data (terrain data, topographic maps) to obtain land cover maps. On the other hand, the Maximum Likelihood classification was used to map the different components of the study area. Finally, the results were analyzed to extract statistics of the evolution of water surfaces in the watershed from 1978 to 2019, thus determining the most vulnerable years [20]. Finally, we calculate the areas using the numerical count of each class that is equal to the number of pixels that present it.

The following diagram shows the steps taken to develop maps of the dynamics of the main components in the study area (water surface, vegetation, bare soil, and urbanization).

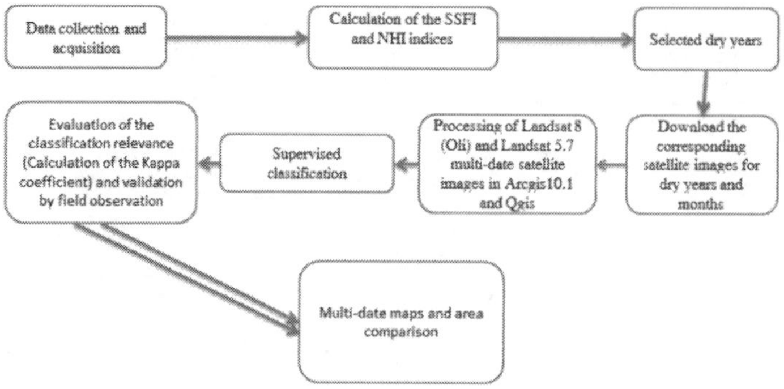

Fig.2. Flow chart of the methodology used to show the driest years.

3 Results and discussion

3.1 Calculation of hydrological drought indices

- Standardized Streamflow Index (SSFI):

The analysis of the hydrological drought was made by the Standardized Stream Flow Index (SSFI) on a series of observations from 1978 to 2019; the succession between dry and wet years was noticed. The results observed for the three stations have an average frequency of occurrence of 57.5% for dry years. It is about 50%, of which 55% and 67.5% are respectively at the Ait Ouchen, Tilouguit and Tizi N'Isli stations (Tab. 4). The dry years that are common to the three stations are: 1980-1981-1982-1983-1984-1985-1990-1992-1993-1995-1998-1999-2000-2001-2002-2005-2007-2016 and 2017. The wettest years correspond to the following years: 1996-2009 and 2010 (Fig. 3).

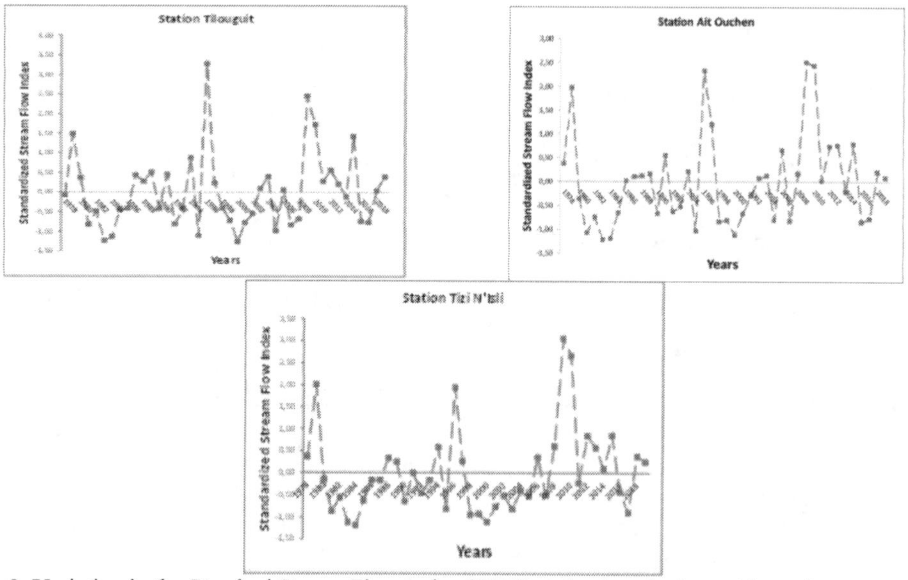

Fig.3. Variation in the Standard Stream Flow Index (SSFI) at the three hydrometric stations.

Table 4. Frequencies (in %) of Standard Stream Slow Index (SSFI) for the period 1978-2019 in the Oued El-Abid watershed.

	Wet period					Dry period				
	Extreme humidity (≥2.0)	High humidity (1.5 to 2)	Moderate humidity (1.0 to 1.5)	Light humidity (0 to 1.0)	Normal (0)	Light drought (0 to -1)	Moderate drought (-1 to – 1.5)	Severe drought (<-1.5 to -2.0)	Extreme drought (≤ -2.0)	Total in (%)
Ait Ouchen	7.5	2.5	2.5	37.5	0	37.5	12.5	0	0	100
Total frequency in (%)	50					50				
Tilouguit	5	2.5	5	32.5	0	45	10	0	0	100
Total frequency in (%)	45					55				
TiziN'Isli	0	2.5	0	27.5	2.5	52.5	7.5	0	7.5	100
Total frequency in (%)	30					67.5				

- Normalized Hydrologic Index (NHI) :

The results obtained by the Normalized Hydrological Index are presented in Tab. 5 and Fig. 4. The three hydrometric stations showed an alternation of surplus and deficit years (figure below).

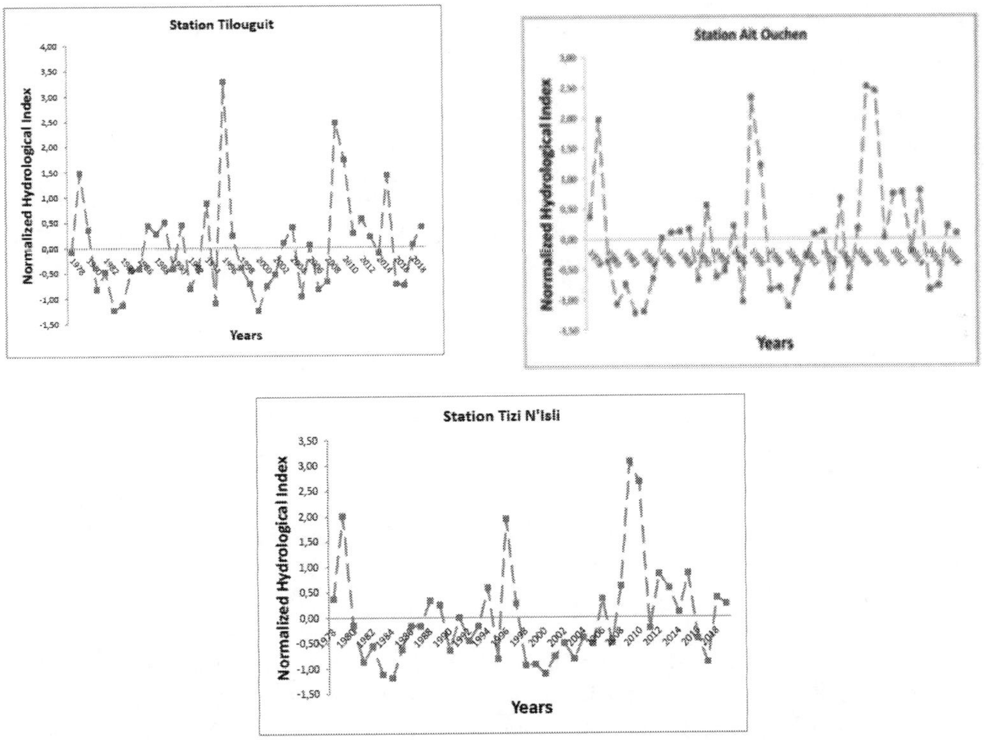

Fig.4. Variation of the Normalized Hydrological Index in the three hydrometric stations.

Table 5. The flow regime's frequency of dry and wet periods in the three stations studied in the Oued El-Abid watershed (1978 to 2019).

	Wet period				Dry period				
	Extreme humidity (2.00<NHI)	Severe humidity (1.50< NHI<1.99)	Moderate humidity (1.00< NHI<1.49)	Light humidity (0.00<NHI<0.99)	Light drought (0.00< NHI<-0.99)	Moderate drought (-1.00< NHI<-1.49)	Severe drought(-1.50<NHI<-1.99)	Extreme drought (NHI<-2.00)	Total in (%)
Ait Ouchen	7.5	2.5	7.5	37.5	37.5	7.5	0	0	100
Total frequency in (%)	55				45				
Tilouguit	5	2.5	5	32.5	45	10	0	0	100
Total frequency in (%)	45				55				
Tizi N'Isli	7.5	2.5	0	30	52.5	7.5	0	0	100
Total frequency in (%)	40				60				

The average frequency of deficit periods is 54% the three stations. The average frequency is of wet periods is 47% (table above). The years of drought common to all the stations are: 1981-1982-1983-1984-1985-1990-1993-1998-1999-2000-2001-2002-2005-2007-2016-2017. The wettest years are 1996-2009 and 2010, which had very high precipitation amounts.

3.2 Supervised classification

The total precision values calculated for our study area are shown in Tab 6. Knowing that the critical value for the acceptance of this classification is 75% [21-22]. Therefore, we can conclude that there is an excellent correspondence between the mapping obtained from the classification of remote sensing data and the field observations.

Table 6. Summary table of classification accuracy.

Selected dry years	1984	1985	1990	1992	1993	1995	1998	1999	2000	2001	2002	2005	2007	2016	2017
Total precision values	100	98.48	100	96.52	99.55	98.48	100	98.99	98.20	99.15	98.70	99.13	99.70	99.50	99.70

The various treatments on the selected images allowed us to generate land use maps from a supervised classification of multiple components of the area studied. The results obtained show that the driest and most vulnerable years are: 1984, 2000, and 2017 when water body surface are as were respectively about 3903.5 ha (Figure 6a), 4156.6 ha (Figure 6i) and 3850.3 ha (Figure 6o). As it is shown [20]. This decline is due mainly to climatic conditions characterized by the regression of rainfall contributions and increase in the needs of the population in different uses: production of electrical energy (industrial and domestic consumption), irrigation (agriculture), drinking water supply, and water recreation (tourism) [23]. Precisely, the hydrological drought infects the filling of the dam lake Bin El-Ouidane. This confirms the results obtained by Qadem et al. [20]. According to which, over the last three decades, the dam has experienced very significant variations in its water height and impoundment, the general trend of which tends to recede with a decrease of almost 40% (Fig. 5) thanks to the occurrence of drought episodes [20-24].

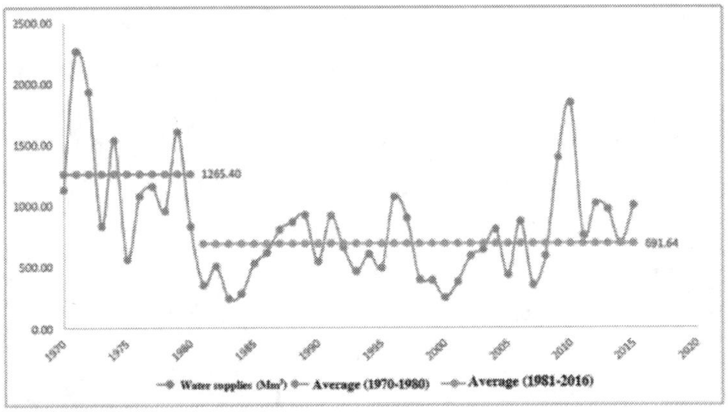

Fig.5. Annual water supply of the Bin El-Ouidane dam.

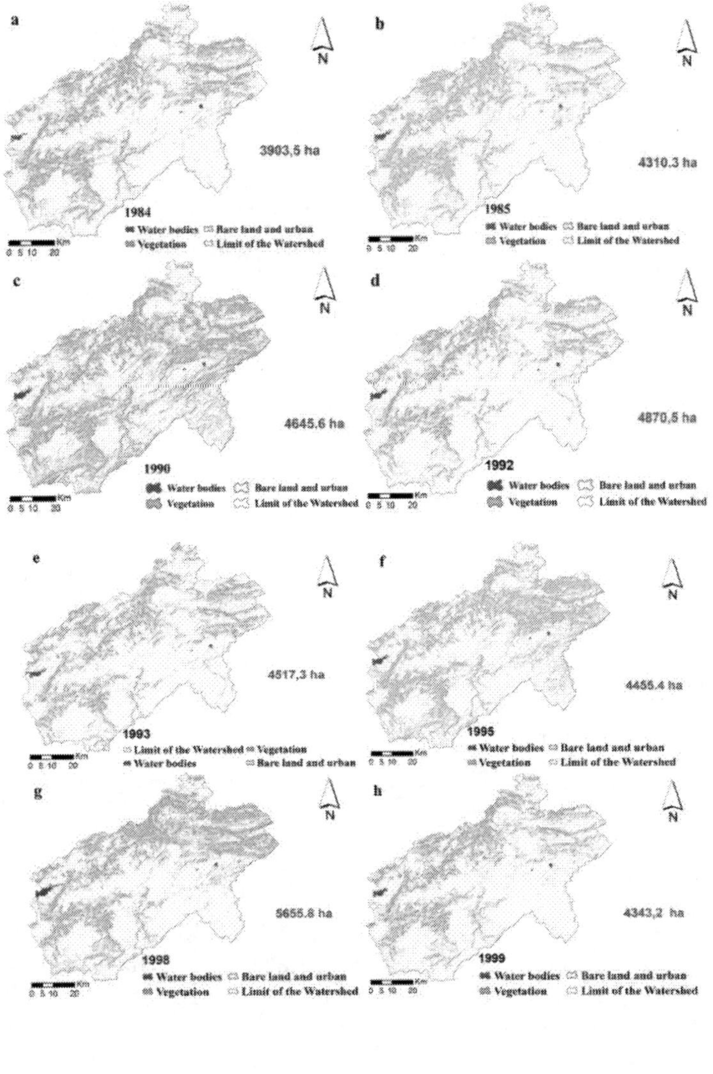

https://doi.org/10.1051/e3sconf/202233702004

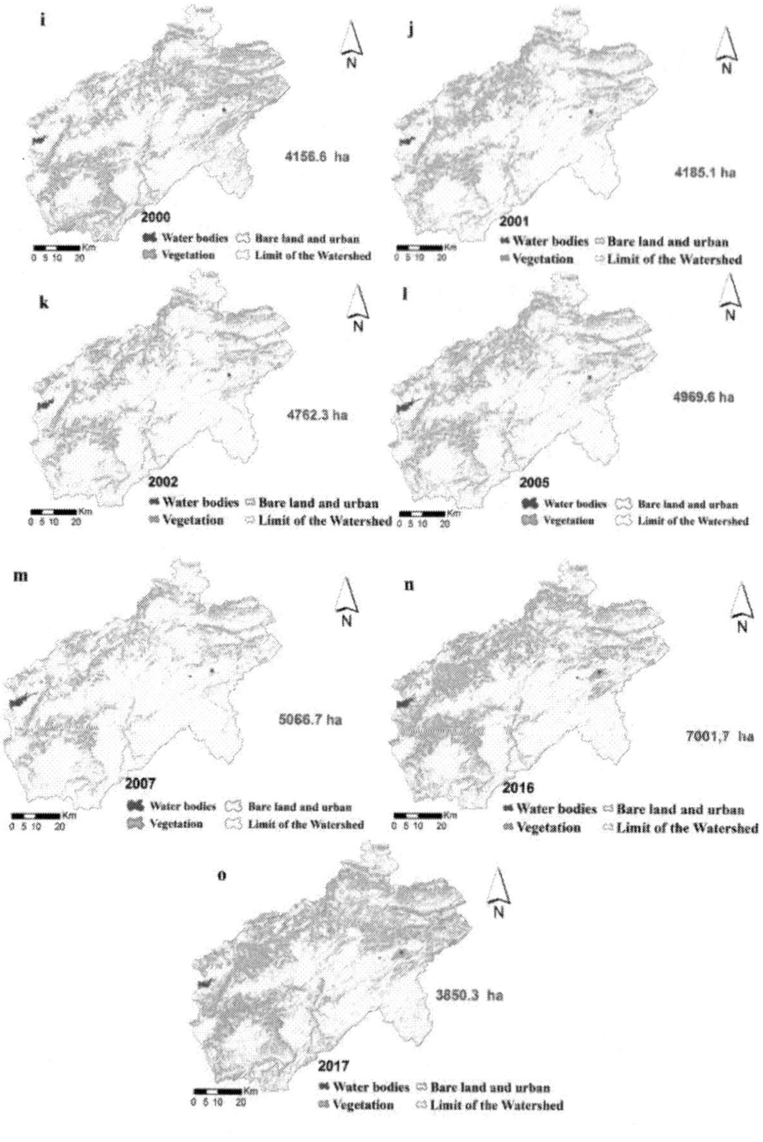

Fig.6. Detection of water body changes in the watershed between 1978 and 2019.

4 Conclusions

In this study, hydrological drought in the watershed of Oued El-Abid during the period 1978-2019was analyzed using two indices SSFI and NHI. Results showed that the study area experienced dry years during 18 years from 1981 to 1985, the hydrological years of 1990-192-1993-1995, 1998 to 2002 and 2005-2007-2016-2017. Spatial remote sensing applied to on Landsat images allowed to make a diachronic study of land use during the periods of drought, and to show the most vulnerable years. It was found that three years (1984, 2000 and 2017) are the driest years in the study period. The changes are due to the annual and interannual variability of flows resulting from rainfall variability.

Acknowledgment

The authors wish to thank the hydraulic basin agency of Oum Er-Rabia for providing data used in this paper.

References

1. S. M. Vicente-Serrano, S. Beguería, L. Gimeno, L. Eklundh, G. Giuliani, D. Weston, A. El Kenawy, J.I. López-Moreno, R. Nieto, T. Ayenew, D. Konte, J. Ardö, G. G. S,Pegram. AppliedGeography, 34, 471–486 (2012)
2. D.D. BA, Sécheresse climatique, dynamique actuelle et perspectives des activités agropastorales dans le Ferlo et la Moyenne Vallée du fleuve Sénégal (Région de Matam), Section de géographie, Sénégal, Thèse de doctorat unique, Université Gaston Berger de Saint –Louis, 196 p (2019)
3. A.Khaldi, Impacts de la sécheresse sur le régime des écoulements souterrains dans les massifs calcaires de l'Ouest Algérien " Monts de Tlemcen - Saida", thèse doctorale, Université D'Oran, pp: 13-18 (2005)
4. D. El Hafid, Z. Zerrouqi, B. Akdim. Larhyss Journal, 31, pp. 83-94 (2017)
5. J.P. Amigues, P. Debaeke, B. Itier, G. Lemaire, B. Seguin, F. Tardieu, A. Thomas, Sécheresse et agriculture. Réduire la vulnérabilité de l'agriculture à un risque accru de manque d'eau. Expertise scientifique collective, Rapport, INRA, France, p. 10 (2006)
6. M. Svoboda, D. Lecomte, M. Hayes, R. Heim, K. Gleason, J. Angel, B. Rippey, R. Tinker, M. Palecki, D. Stooksbury, D. Miskus, S. Stephens, The Drought Monitor,Bulletin of the American Meteorological Society 83(8), pp. 1181–1190 (2002)
7. J. Spinoni, G. Naumann, H. Carrao, P. Barbosa, J. Vogt, Int. J. Climatol. 34, pp.2792–2804 (2014)
8. E.A.B. Eltahir, P.J.F.Yeh, On the asymmetric response of aquifer water level to floods and droughts in Illinois, Water Resour. Res. 35 (4), pp.1199–1217 (1999)
9. R.R. Heim,Bulletin of the American Meteorological Society 83, PP. 1149–1165 (2002)
10. Z. Sen. J. Hydraulics Div., ASCE 106 (1), pp. 99–115 (1980)
11. D.M. Frick, D. Bode, J.D. Salas,J. Hydrological Eng. 116, PP. 733–753 (1990)
12. B. Boudad , H. Sahbi , I. Manssouri,J. Mater. Environ. Sci, 9(1) (2018)
13. D.A. Wilhite, M.H. Glantz, Water Int. 10(3), pp.111– 120(1985)
14. American Meteorological Society (AMS). Soc. 85, pp. 771–773 (2004)
15. M. ElGhachi, H. Ouakhir H. Lameti, Revue Marocaine de Géomorphologie, N° 3, pp 63-45 (2019)
16. T.B. McKee, N.J. Doesken, J. Kleist.American Meteorological Society, Jan 17-23, Anaheim CA, pp. 179-186 (1993)
17. T.C. Sharma, U.S. Panu. Hydrological Sciences Journal, 55 (1), pp. 79-92. (2010).
18. A.Ghenim, LJEE N°18 (2011)
19. Bi. V. H .N'Gnessan, M. B .Saley, M.O. Yeï, A.F. Yapi, B.H.Kouadio, J. Biemi, A. Kouadio,Int J Engineering Science. Vol X Issue X ‖ XXXX: 1-11 (2017)

20. A.Qadem , M. El Ghachi, E. Layati,Geomaghreb 15, pp. 41-47 (2019)
21. M. Girard, Télédétection appliqué, zone tempérées et intertropicales, Rapport, Paris, 260p (1999)
22. E. Layati, A. Ouigmane,M.D.C Alves, M. Bagyaraj, A. Qadem, M. El Ghachi.Papers in AppliedGeography, DOI:10.1080/23754931.2021.1973545 (2021)
23. H. Ouakhir, M. El Ghachi, in *Renforcement de la résilience des hydrosystèmes face aux changements globaux: "de la mesure hydrologiqueaux modèles de gestion"* (Béni Mellal - Maroc), CES'2015, 143p (2018)
24. GIEC, Changements climatiques 2014: Rapport de synthèse. Contribution des Groupes de travail I, II et III au cinquième Rapport d'évaluation du Groupe d'experts intergouvernemental sur l'évolution du climat [Sous la direction de l'équipe de rédaction principale, R.K. Pachauri et L.A. Meyer]. GIEC, Genève, Suisse, 161p (2014)

Impact of Limestone Deposit Operation on the Yield of Surrounding Agricultural Land: Case of the Western Rif, Morocco

Ghizlane Fattah[1,*], *Fouzia* Ghrissi[1], *Jamal* Mabrouki[2]and Hicham Cherifi[3]

[1]Civil hydraulic and environmental engineering laboratory, Water treatment and reuse structure, Mohammadia School of Engineers, Mohammed V University in Rabat, Avenue Ibn Sina B.P 765, Agdal Rabat, 10090 Morocco
[2]Laboratory of Spectroscopy, Molecular Modeling, Materials, Nanomaterial, Water and Environment, CERNE2D, Mohammed V University in Rabat, Faculty of Science, Avenue Ibn Battouta, BP1014, Agdal, Rabat, Morocco
[3]Sidi Mohamed Ben Abdellah University, Sciences & Techniques Faculty, Intelligent Systems, Georesources& Renewable Energies Laboratory (SIGER), B.P. 2202 - Route Imouzzer, Fez Morocco

Abstract. The exploitation of the limestone deposit is carried out in the open air. Its exploitation process has obvious effects on its environment, and in particular on agricultural production in neighboring lands, by suspending fine particles of matter in the air, which can potentially settle in the soil and the water. The latter being essential resources in the agricultural production process. The analysis of the quality of these resources located near the exploited zone as well as in a control zone (reference zone), both located at the level of the mountain of Haouz, of the western Rif, made it possible to assess the 'impact of the exploitation of the limestone deposit on the physico-chemical characteristics of soils and surface water. Soil and water samples were taken in these areas. The results of these analyzes made it possible to assess the level of soil fertility at each identified sampling point, as well as the characterization of surface water intended for irrigation. The results revealed that this exploitation leads to a significant modification of the soil profile of the neighboring agricultural lands, making them less fertile. This exploitation also causes a hardening of the surface layer of the soil, favoring the phenomenon of encrustation of the soil. The only positive point is that surface water is endowed with essential nutrients such as magnesium and phosphorus originating from the dust emitted during the extraction of limestone. An excessive presence of suspended matter could be accompanied by an increase in pH and thus promote the phenomenon of encrustation of soils when they are irrigated. This reveals that the exploitation of the deposits has generally negative physical and chemical impacts on the soil and consequently on agricultural production.

Keywords:agricultural yield, exploitation of the deposit, limestone, dust, soil, water and physico-chemical, Western Rif

1. Introduction

The exploitation of quarries and the production of aggregates constitute a very important sector for the socio-economic development of the country [1]. Indeed, this sector is located upstream of the activities linked to the realization of a large number of urban development projects. The bodies responsible for carrying out these actions have to extract materials and ores of all kinds to carry out their mission [2]. Indeed, quarrying is the basis of all infrastructures. We need only recall its geological definition to understand this [3]. It is an artificial excavation, generally open to the sky, used to extract building materials (limestone, granite, gypsum, sand [4][5]. Furthermore, it is obvious that quarrying requires the occupation of space. However, this occupation is temporary, since at the end of the exploitation, the said space must be restored. If the exploited sites are not rehabilitated, the quarry holes become groundwater recharge areas due to their shallowness and the permeability of the underlying layers [6][7]. Abandoned quarries can also become uncontrolled dumps or unhealthy housing sites.

The expansion of the mining industry has been aided by the development of technology. This in turn contributes to industrial development, which is an important driver of global growth. The performance of the mining sector is thus considered a prerequisite for a country's development and determines its economic position in the world trade [8]. The production of building materials in the Northern Region is ensured by several quarries extracting alluvial materials to meet the local market demand. The number of these quarries is in constant development with a strong positive relationship to the demand for building materials, which is dependent on the downstream activity of the building and public work sector [9]. Generally, an impact is the change, positive or negative, due to a pressure generated by an activity on the Biophysical environment as well as its consequence on man. Indeed, the impact of a career is very

*Corresponding author:ghizlane_fattah@um5.ac.ma

variable depending on the type of material to be extracted, the mode of exploitation and the environment of the site; the eco-system. Indeed, it changes the original physical and geochemical characteristics of the soil [10-11-12-13].

2. Description of the study area

The present study was carried out in the north of Morocco (Figure 1). The deposit in operation is named theHaouzchainof the Western Rif and belongs to the group of limestone geological formations[5]. Pedological studies of the region highlight two types of soil units namely: ferralitic soils on bar land, overlooking valleys and depressions and hydromorphic soils. The climate is humid Mediterranean type under oceanic influence.Temperatures remain mild in winter, mild in summer both on the coasts and in altitude. They rarely reach 0 ° C in January and the most frequent maxima of this month oscillate between 14 ° C and 18 ° C. In summer, the atmosphere warms up noticeably, the most frequent maximum temperatures in July vary between 16 ° C and 26 ° C. The average interannual rainfall is 1,800 mm in the high reliefs.

The study area covers a purely rural area characterized by the presence of towns and villages, generally near the water sources that emerge along the limits of the limestone range. The locomotor activity in the region is subsistence agriculture based mainly on the cultivation of cereals and grazing agriculture.

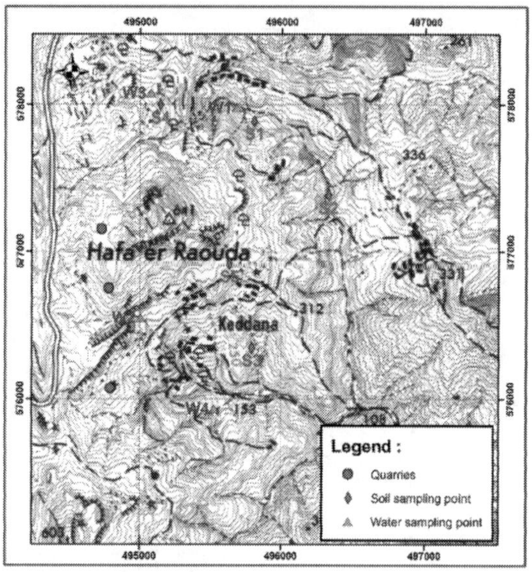

Figure 1 : Site of sample (Tangier topographic map: 1 /250000)

3. Materials and methods

3.1 Sampling

With your hand below the surface of the water, uncork and fill the 500 ml bottle, rinse it and then empty it out of the water. Quickly close the bottle and immerse it in the water to the desired depth. If no precise indication is given, take the sample 30 cm below the surface of the water [14]. At the truck, remove the METALS bag corresponding to the station to be sampled from the bottle set transport cooler and place it with the boxes of long and short polyethylene gloves in the small field cooler designated for trace metals [15].

The soil sampling sent to the laboratory should usually weigh around 400 grams. However, it must be representative of 20,000 tonnes of soil. It is therefore understandable that the sampling must be done carefully [16]. The choice of sampling area can greatly influence the precision of the soil analysis. It is a relatively simple matter to take a sample from each field if the field is small. On the other hand, very large fields should be divided into smaller sampling areas [17]. It is important, as far as possible, to ensure that each sample area is uniform and separate from those that are obviously different.

Soil fertility can vary according to bedrock composition, soil texture, amount of nutrients removed by the crop or topography. Most differences in fertility, However, can be attributed to previous applications of nutrients, in the form of fertilizer or manure [18]. If there is little variability, several soil cores can be pooled into one sample. If the variability is high, it is recommended to take separate samples. If the above field boundaries are known, they can be used to divide larger fields into smaller plots [19]. Samples from plots that clearly differ in composition from the rest of the field should not be pooled into the composite field sample [20]. They should therefore be avoided in ditches, eroded areas, alleys or in the location of old manure or lime piles. If these areas are large enough, it is advisable to take samples from them and have them analysed separately.

3.2 Physico-chemical analyzes

3.2.1 Soil

Soil analysis is an important part of crop production and nutrition management [21-22].Soil sampling is a three-step process: first, taking a representative sample from each field or plot, then analyzing the sample to determine the amount of available soil nutrients, and finally, using the results to determine optimal fertilizer rates [23].

Total porosity is the volume of pores contained in an apparent volume of a given soil. For a soil sample, the apparent volume is the sum of the solid volume (real volume) and the volume of the gaps. A detailed study of porosity variations under the effect of cultivation practices has not been carried out until now. The present study allows a more in-depth discussion of the consequences of no-tillage systems on soil porosity. Indeed, the main objective of this study is to analyses the changes in porosity of a swelling clay soil under different cropping and tillage systems in the semi-arid zone of Morocco.

3.2.2 Surface Water

A number of elements in water can be directly toxic to the crop. Setting toxicity limits for irrigation water is complicated by the reactions that can occur when water reaches the soil. Potentially harmful elements in the water can be inactivated by chemical reactions or accumulate in the soil to levels that are toxic to plants.

Suspended solids, dissolved oxygen, nitrite (NO^{2-}), ammonium (NH^{4+}) and nitrogen (N), Biochemical Oxygen Demand (BOD_5) were determined by the methods approved by AFNOR[6]. The Chemical Oxygen Demand (COD) was determined by the dichromate method. The sediments were dried and then weighed to determine the water loss and the fraction below 63µm. Organic matter was determined by loss on ignition at 550°C.

4. Results and discussion

4.1. Impact onSoilquality

The soil is a powerful medium for water purification and recycling. Clays, hydroxides and organic matter adsorb cations (Ca, Mg, K, Na, heavy metals) and certain organic molecules. Phosphates and nitrogen are well fixed by soil particles; they can however be carried away with these particles (erosion of bare soil) [24]. The microbiological content of the soil can purify the soil of undesirable substances: mineralisation of nitrogen compounds, denitrification, degradation of organic micropollutants (pesticides). Pathogens introduced into the soil are eliminated when the conditions for their survival are unfavourable: UV, competition with native soil microflora [25].

The different elements of soil fertility can undergo degradation processes (OM, nitrogen, ...), leading to a decrease in plant production. In our case the distribution of fertilising elements varies with distance. In cultivated soils, degradation can be caused by excessive tillage and inappropriate soil management practices. Examples include excessive tillage that degrades soil structure and promotes compaction and erosion, too little fertilisation that depletes soil reserves, or salinisation of the soil by irrigation water. The organic matter content of the soil increases when the soil is moved away from the centre, the aggregates are less stable and are more easily dispersed by rain [26]. The decrease in organic matter content and the use of phytosanitary products (pesticides, etc.) and the impact of polluted water lead to a correlative decrease in the biological activity of the soil (figure 2). Earthworms, which are less numerous, no longer carry out the "biological leaching" that ensures the homogenisation and aeration of the soil, while the formation of battance crusts is favoured [27].

In this study, the results indicate an increase in the concentration of conductivity, phosphorus, magnesium, potassium and a decrease in the concentration of organic matter, nitrogen as well as the rate of porosity compared to those of the reference soil. These differences are significant especially for the S5, S4 and S3 stations located near the limestone mining sites and less significant for the S1 and S2 stations(Figure 2).

However, this difference is not significant for the pH values (Figure 2).

This means that the exploitation of the limestone deposit in the study area had an impact on the soil parameters by settling of the suspended matter of the exploited rockin the S5, S4 and S3 stations located near the sites from which it enriched the soil with potassium, phosphorus and in magnesium but it demined the fertility of the soil in organic matter and nitrogen as well as the rate of porosity causing the hardening of the surface layer of the soil.

This modification in physicochemical and granulometric parameters of the soil prevents the emergence of young shoots and influences the hydraulic processes on the soil surface and consequently, the agricultural yield of the land surrounding the quarries decreases.

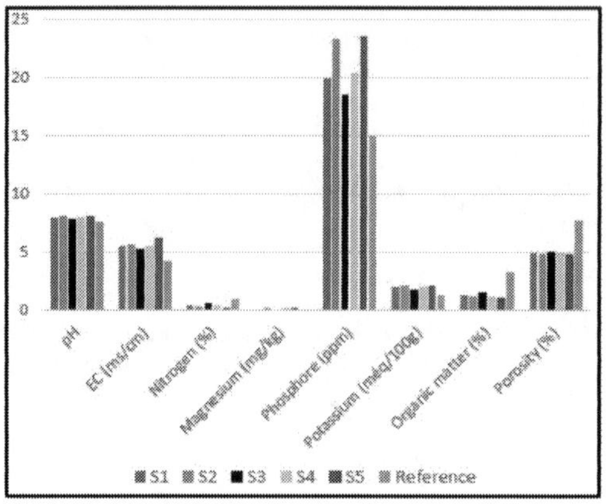

Figure 2 : Characterization of the soil in relation to its distance from the quarries
Sn: Soil sampling station near a limestone extraction area
Reference: Soil far from the mining activity

4.2Impact on surface water quality

Soil particles are also carried into surface waters along with agricultural inputs (fertilisers, pesticides) and pollutants from industry, cities and roads. The input of nitrogen and phosphorus causes eutrophication of the river (or coastal zone). The ecological impact of phytosanitary products is recognised but more difficult to assess because of the multiplicity of these substances and their broad spectrum of action.

The pH of the water conditions the physico-chemical equilibrium, in particular the calco-carbonic equilibrium and therefore the action of the water on the carbonates (attack or deposit). The pH is basic in surface water. It is alkaline in limestone. Suspended matter can be important in water. It causes discomfort in the appearance of the water and its taste (earthy taste). Turbidity peaks follow heavy rainfall; they are frequent in the Norman Chalk aquifer. Runoff water laden with clay particles - and other undesirable material (W2) - rushes into the sinkholes. The speed of water circulation in the underground network during floods does not allow them to be decanted; moreover, the previously deposited particles are torn out of the cavities and increase the suspended load which ends up at the outlet. Any development that increases surface runoff and soil erosion increases turbidity - agricultural consolidation, drainage of surface water to agricultural soils, filling in of ponds storing pollutants in the soil. Manganese generally accompanies iron in rocks. Like iron, its solubility depends on human activities. It is not very abundant in the waters of the region.

The quality of irrigation water is very important for soil fertility management. Irrigation with sodium-rich water leads to progressive salinisation of the soil. Salinisation increases the osmotic pressure needed by plants to absorb water and nutrients. Irrigation of soils with acidic water results in the leaching of nutrients present as ions (Ca, Mg, K, P, S, etc.) and can make toxic elements such as aluminium and manganese bioavailable by solubilising them.

Pollution is widespread in space and time; it is chronic and affects large areas. Indeed, current farming and livestock practices strongly influence the water regime and quality.

On the plus side, the surface water is endowed with essential nutrients such as magnesium and phosphorus from the dust emitted during limestone extraction. Excessive suspended solids could lead to an increase in pH and thus to the scaling of the soil during irrigation. This shows that the exploitation of deposits has generally negative physico-chemical impacts on soils and consequently on agricultural production[28].

Figure 3 indicates a slight increase in pH concentration and conductivity, average oxygen, suspended matter, and interesting phosphorus and magnesium in all sampling stations. The element load becomes more significant in the sites which are close to the quarries.

However, the modification of the physicochemical parameters of the surface water, neighboring active quarries, intended for irrigation comes from the settling of the dust emitted during the extraction of limestone from the rivers.

This change in parameter thus promotes the phenomenon of incrustation of soils when they are irrigated.

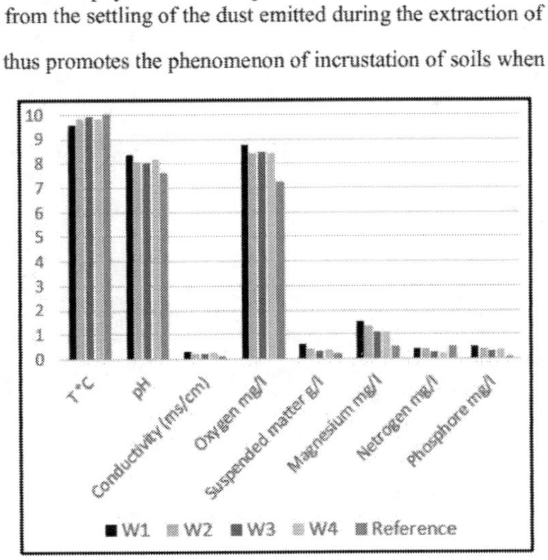

Figure 3 : Characterization of surface water quality in the study area

Wn:Water sampling station near a limestone extraction area
Reference:Water far from the mining activity

5. Conclusions

The assessment of the impacts on water and soil shows that this quarry as a whole has more negative than positive impacts on this ecosystem, bearing in mind that, if the accumulative effect of the other quarries is added, the results will surely be increased. The identification of these negative impacts by phase shows that the local actors and promoters are obliged to take, among others, the environmental protection measures that we have mentioned above in order to develop these quarries which present important socio-economic potentialities and participate in the local and regional development and contribute significantly to the financial maintenance of the territorial municipalitywhere this activity is located. This will allow a coherent integration in the local landscape while respecting the receiving ecosystem.

References

1. Messineo, P. G., & Barros, M. P. (2015). LithicTechnology, 40(1), 3-20.
2. Zhong, Z., Marcel, T. C., Hartmann, F. E., Ma, X., Plissonneau, C., Zala, M., ... & Palma Guerrero, J. (2017). New Phytologist, 214(2), 619-631.
3. Li, W., Wu, L., Xie, Z., Tao, L., Zou, K., Li, F., & Miao, J. (2019). Earth Science Informatics, 12(4), 599-613.
4. Marvila, M. T., Azevedo, A. R., Barroso, L. S., Barbosa, M. Z., & de Brito, J. (2020). Construction and Building Materials, 250, 118786.
5. Fattah, G., Ghrissi, F., Mabrouki, J., &Kabriti, M. (2021). In E3S Web of Conferences (Vol. 234, p. 00018). EDP Sciences.
6. Hamdan, M. H., & Kamel, M. T. (2011). Flow through variable permeability porous layers. Adv. Theor. Appl. Mech, 4(3), 135-145.
7. Mabrouki, J., El Yadini, A., Bencheikh, I., Azoulay, K., Moufti, A., & El Hajjaji, S. (2018, July). In International Conference on Advanced Intelligent Systems for Sustainable Development (pp. 22-33). Springer, Cham.
8. Zhang, J., Hoenig, S., Di Benedetto, A., Lancioni, R. A., &Phatak, A. (2009). Industrial Marketing Management, 38(2), 207-218.
9. Bogus, S. M., Molenaar, K. R., &Diekmann, J. E. (2006). Construction Management and Economics, 24(8), 829-837.
10. Ongley, E. D., Bynoe, M. C., & Percival, J. B. (1982). In Sediment/Freshwater Interaction (pp. 41-57). Springer, Dordrecht
11. A. Aboubakar, A., El Hajjaji, S., Douaik, A., MfopouMewouo, Y. C., Birang a Madong, R. C., Dahchour, A., ... &Labjar, N. (2021). International Journal of Environmental Analytical Chemistry, 1-22.
12. Aranda, V., Ayora-Cañada, M. J., Dominguez-Vidal, A., Martin-Garcia, J. M., Calero, J., Delgado, R., ... & González-Vila, F. J. (2011). Geoderma, 164(1-2), 54-63.

13. Greenman, D. W., Bennett, G. D., &Swarzenski, W. V. (1967). US Government Printing Office.
14. Meador, M. R., Cuffney, T. F., &Gurtz, M. E. (1993). US Geological Survey.
15. Li, Y., &Migliaccio, K. (Eds.). (2010). CRC Press.
16. Carter, M. R., &Gregorich, E. G. (Eds.). (2007). CRC press.
17. Pennock, D., Yates, T., &Braidek, J. (2008). Soil sampling designs. Soil sampling and methods of analysis, [2, 25-37.].
18. Desbiez, A., Matthews, R., Tripathi, B., & Ellis-Jones, J. (2004). Agriculture, ecosystems & environment, 103(1), 191-206.
19. Dennis, P., Thomas, M. B., & Sotherton, N. W. (1994). Journal of applied ecology, 361-370.
20. Palm, C. A., Gachengo, C. N., Delve, R. J., Cadisch, G., &Giller, K. E. (2001). Agriculture, ecosystems & environment, 83(1-2), 27-42.
21. Smith, K. A., & Mullins, C. E. (1991). SMR, 873, 7.
22. Peverill, K. I., Sparrow, L. A., & Reuter, D. J. (Eds.). (1999). CSIRO publishing.].
23. Pueyo, M., Mateu, J., Rigol, A., Vidal, M., López-Sánchez, J. F., & Rauret, G. (2008). Environmental pollution, 152(2), 330-341.
24. Martínez-Zavala, L., &Jordán, A. (2008). Soil Use and Management, 24(1), 108-117.
25. Byappanahalli, M., & Fujioka, R. (2004). Water Science and Technology, 50(1), 27-32
26. Burke, I. C., Yonker, C. M., Parton, W. J., Cole, C. V., Flach, K., & Schimel, D. S. (1989). Soil science society of America journal, 53(3), 800-805.
27. Cheng, W. C., &Kuo, M. T. (2010). Groundwater Monitoring & Remediation, 30(4), 73-80.
28. Dekker, L. W., &Ritsema, C. J. (1996). Catena, 28(1-2), 89-105.
29. Campbell, G. S., Jungbauer Jr, J. D., Bristow, K. L., & Hungerford, R. D. (1995). Soil Science, 159(6), 363-374.

Cost and profitability analysis for a seabass production farm in Morocco

Hassan NHHALA [1, 2, *], *Housni* CHADLI [3], and *Hassan* ER-RAIOUI [2]
[1]National Institute of Fisheries Research (INRH), Aquaculture Centre, M'diq, Morocco (Former researcher).
[2]Faculty of Sciences and Technics, University of Abdelmalek Essaadi, Tangier, Morocco.
[3]Marine Fish Farm « Aquam'diq », M'diq, Morocco.

Abstract. Cost production and profitability analysis was performed for a marine fish farm, located in M'diq Bay, in Morocco. Currently, it is the alone farm in operation on Moroccan Mediterranean cost while three others had been ceased their activity. Created in 1998, it has experienced three main periods during its history, start-up period, followed by a crisis period followed by a recovery period on which the present study has been focused on. Production costs analysis has shown financial results yielding a very low profitability (0.24%). Two main constraints were found to be relevant, one environmental related to the weak local annual seawater temperature profile (18°C in average, 14-21°C in range) and the other technical related to feed and fingerlings management. The use of SGR and TGC tools, based on real growing and production data, have allowed to establish suitable start-up date for annual batches. Feed and fingerlings constraints have analysed in order to contribute in production costs improvement. Results showed that total costs could be reduced and profitability improved following pertinent inputs management optimization, marketing strategies improvement and production plan adjustment. Support measures provided by public powers offer suitable opportunity to improve farm's financial results.

1 Introduction

Despite Moroccan marine fisheries has reached a total annual amount of about 1.5 million tons in recent years, marine aquaculture is still facing development constraints and sustainability problem, although suitable sites availability. The lack of substantial support, mainly in terms of incentive measures, represent a major bottleneck affecting aquaculture development. Investors seemed to be fear about domestic marketing risks because aquaculture products still not welly perceived by domestic consumers. Unlike some European countries, Moroccan Government did not give any subsidies to aquaculture farms.

Shellfish culture has started in Morocco since the fifties in Oualidia lagoon on Atlantic coast and marine fish culture has begun in the late of eighties in Nador lagoon on Mediterranean coast. The first marine fish farm was growing both gilthead seabream,

*Corresponding author: nhhalahassan@yahoo.fr

Sparus aurata, and European seabass, *Dicentrarchus labrax*. Afterward, three others marine fish farms have been established on Mediterranean coast, one in early nineties near Moulouya river estuary and the two others in late nineties in M'diq bay and Martil bay. They all have been growing seabream and seabass with same main targeted market, the European one. However, their contribution to national cultured fish production was low and variable. Maximum total production was 1,224 tons achieved in 2005, followed by a sudden great collapse in 2006.The maximum marine aquaculture production amount, recorded for both fish and shellfish, was 1,500 tons.

Seabass and seabream selling prices collapse in European market, occurred since the beginning of 2000s [1], had negative effects on Moroccan fish farms. Three of the four said fish farms have successively ceased their activity. Selling process of seabass and seabass have collapsed and became lower than these three farms' production cost. Moreover, national market for aquaculture products was very weak and attempts to reduce the production cost have been disadvantaged by the dependence from foreign inputs (fingerlings and feeds) and their high import taxes. If they would sell their products in domestic market, they should first pay these inputs taxes; if not, they could import these inputs in a temporary admission way (legal procedure) but, in this case, they should to export their products to foreign market. The exported quantity is legally based on imported feed amount, using a standard feed conversion rate of 2. This was possible when European market was interesting in terms of selling prices.

Nowadays, only one marine fish farm is still in operation; it is located in M'diq Bay, in the West part of Mediterranean coast of Morocco. Since 2006, it sells its farmed seabass in domestic market while reducing, its production cost and optimising its farming process as much as possible. It has tried using local manufactured fish feed (first pressed then extruded) and locally produced seabass fingerlings through INRH's experimental marine fish hatchery. The later could not satisfy its total need of this farm in fingerlings because of its small scale. So, since 2010, and due to domestic market need in farmed fish, it has been forced to restarted importing small fish from foreign hatcheries, paying import fees and taxes.

In this context, the present study was carry out for and with joint cooperation of this seabass farm in M'diq. The purpose consisted in seeking for opportunities allowing to reduce production cost and optimize selling price of its products to avoid losses and ensure even a minimum profitability allowing first its survival while keeping working for its development. As reported by [2], the profitability of fish producers is mainly a result of two factors: sale price and production costs, and according to [3], the better understanding of the relevant elements and their interrelationships inside the entire production process allow to achieve a sustainable and profitable aquaculture operation. So, considering several studies conducted in different countries, dealing with productivity and efficiency analysis and on structural and economics analysis of fish farming, the present study was focused on cost and profitability analysis of M'diq seabass farm. It is crucial for the development and management of M'diq seabass farm to know its production cost structure and evolution and to search where it could be possible to react to reduce its production cost. For this reason, the present study aimed at providing M'diq seabass farm a clear visibility about options of management improvement in terms of production cost and selling price under which it could be able to ensure a profit margin allowing its sustainable exploitation.

2 Materials and methods

2.1 Site characterization of sea bass farm in M'diq

Seabass farm, subject of the present study, is located in M'diq Bay, in Morocco. The distance between the farm site and the local harbour is about 1.6 km. Currently, this bay still represents the only exploited site for Moroccan marine fish culture. It is located in the west part of Moroccan Mediterranean coast, between Sebta Cape (35°54'N, 5°17'10"W) and Negro Cape (35°40'N, 5°16'40"W). M'diq bay stretches for 23 km in length with a water surface of about 150 km^2 [4]. It is located close to Gibraltar Strait and oriented North-South between two Caps (Sebta and Negro) and limited by Mediterranean Sea in the East and a mountainous hinterland in the West. Current average speed in fish framing site is 0.12 m/s with a maximum of 0.27 m/s. Tidal currents directions are North-West at rising tide and South-East at ebb tide with a tidal range of 20 to 80 cm. The swell could reach a height of 5.5 m with an associated period of 5 s. The seawater temperature regime ranged between 14°C in February-Marsh and 21°C in August-September. The salinity is about 36.0 ‰ to 36.6 ‰ along the year.

2.2 Production system characterization of sea bass farm

M'diq seabass farm has 14 high density polyethylene (HDPE) floating cages of 12 m in diameter with 10 m useful net depth. It has been set up in the South part of M'diq Bay in 1999 with a total investment of 30 million dirhams (Moroccan money (MAD)). Its nominal production capacity is estimated at about 200 tons. The production cycle is variable from 18 to 22 months. Feed and fingerlings are imported from Europe countries, mainly France. Seabass growing culture system is based on the intensive method using a final fish density of about 20 to 25 kg/m^3. Mortality ratio is less than 20%.

Seabass fingerlings have an initial weight of 3 to 8 g. Fish are fed three to twice a day for younger stages and once a day for large stages. Feeding rates are expressed as percentages of fish biomass of each cage, depending on fish monthly average weight and daily seawater temperature. Commercial feed designed for seabass is used, first entirely imported and then partially replaced by locally produced feeds. Fishes are harvested when their mean body weight reached at least 300 g. Commercialsizes are mostly composed of 300/400 g (designed as 3/4), 400/600 (4/6) and 600/800 g (6/8) and also but, sometimes, they include others sizes such as 200/300 g (2/3) and over 800 g. Fishes are sold along the year round, fresh and packaged in 10 kg polystyrene. Their selling prices of seabass are variable, depending on commercial range size and also seasons.

M'diq farm has buildings located inside land area of M'diq fishing port. These buildings comprise a warehouse, cold chamber for harvested fish, room storage for feed, room storage for nets, offices, toilets, changing rooms, handling room, etc. It owns an aquaculture boat and uses loading and unloading dock. The farm staff were educated locally.

2.3 Data source and economic analysis methodology

Data and values used in this study were collected from M'diq seabass farm, covering a period ranged between 2010 and 2016. These data include technical characteristics, zoo-technical parameters, production plan, investment budget and costs of different production components. So, cost structure and the financial and economic analysis were made based real seabass production data. This study was performed based on seabass growing batches with starting farming date recorded from 2010 un until 2015, and fish sales data recorded from 2012 until 2016. It covered production cycle of each growing batch, from stocking fingerlings until harvesting commercial fish sizes.

Economic analysis was based on an average production year, calculated from real farm data to determine production cost and estimate profitability in terms net return and benefit/cost ratio. A comparison of production cost and its breakdown components was to provide a better understanding of cost structure and its relative production efficiency. Total production cost structure includes variable and fixed costs which are related to fish farm operations along the year. Variables costs components include feed, fingerlings, taxes related to feed and fingerlings, marketing and others variable costs. Fixed costs components include labour, repair and maintenance, service provision, premises rental, insurances and taxes, depreciation financial expenses and others fixed costs.

Economic analysis was based on farm-gate prices of harvested seabass and current local markets (MAD). Data on operating expenses including information required for variables and fixed costs calculation were provided. Profitability analysis was performed in static economic study framework, considering average values of income and expenses, based on an average production year calculated for a period of 5 years (2012-2016).

Total income was calculated by multiplying total sales (kg) by their average sale price (MAD/kg). Net return (MAD/kg) was assessed by subtracting production costs from average sale price. The Benefit/Cost ratio (profitability) was calculated as net return divided by production cost; it indicates the net gain generated by the farm for each spent monetary unit.

2.4 Bioeconomic model

Bioeconomic model was carried out with the same data as in the previous paragraph. Many strategic and operational decisions in fish farming can be optimized by mathematical modelling, e.g., stocking strategy, feeding rate, harvesting time, and harvested weight [5]. However, the complexity of the aquaculture production process in open sea makes difficult to find universal optimal solution.

To help making optimization decisions at farm level, analysis should involve general characteristics study of the production function describing relationships between inputs and outputs in the production process [6]. The latter is described over a limited period, by fish weight (W), number of fish (N), time (t), production costs (C) and revenues (R) [5]. For each batch, decision variables are stocking time (ti), initial weight of stocked fingerlings (Wi) and final weight of harvested fish (Wf) [5].

Seawater temperature conditions in fish farming is not able to be controlled. It is a constraint that should be considered deterministic. Feeding rates are managed according to feed quality, rearing site temperature and fish size. Stocking time and initial weight of fingerlings and stocking densities are also managed according to farm production strategy and farming site conditions, mainly seawater temperature profile. Number of fish growing batches per year is defined by fish farmer taking in consideration production cycle length, harvesting time and fish harvested weight.

Among pertinent factors influencing fish growth, seawater temperature, initial fish weight and feeding quality and quantity are essential for bioeconomic analysis. However, considering [1, 7] works, respectively for gilthead seabream and sharp snout seabream farming in sea cages in Spanish Mediterranean coast, a growing simulation is made based on initial and final average weight of cultured seabass (5-400 g) and rearing site temperature conditions (14-21°C), recorded in seabass floating cages farm in M'diq bay.

An estimated production plan was prepared based on fish growth performed according to growth model [8] or Thermal Growth Coefficient (TGC) model [9, 10, 11, 12, 13], taking into account the effective temperature or the minimum temperature for growth. The TCG calculation formula is as follows:

$$TGC = (W_f^{1/3} - W_i^{1/3}) / \sum (Tef) \quad (1)$$

- W_f: Final Weight;
- W_i: Initial Weight;
- ΣTef: sum of effective temperature = 15 x (Tef biweekly average - 12°C)

Effective temperature of seabass is 12°C [14]. On this basis, a seabass growing model was developed for each month of the year, based on an initial mean juvenile weight of 5 g until a mean marketable fish size of 400 g, to determine the best possible fry stocking periods for better yields as a function of temperature. So, three annual batches should be chosen among the ten monthly possibilities simulated, in which the cycle was relatively shortest and allowing both commercial sizes throughout the year. M'diq farm has introduced fingerlings in each month from December to September since 2010 until 2015, based generally on three, or sometimes two, introductions per year. Introduction date choice varies according to empty cages availability. So, in each year, subsequent three new batches of juveniles should be added to the existing rearing fish stock in the fish farm in the same months that have been selected and which could be obviously taken into account in calculating feed consumption [7].

An average TGC values was calculated from real growth results in M'diq farm, for each of the ten monthly batches (i.e. fingerlings stocking month) and the effective temperatures were calculated from daily temperature data recorded during 2010 until 2016.Starting with 20 batches corresponding to the first and second fortnight of all the ten months of the year (designated as Jan-d1 for the first fortnight in January, Jan-d2 for its second fortnight, and so on the others months), the growth of seabass was estimated by elaborating a summary table of weights in order to choose three among them, so that sales of each batch could last for four months in such a way to cover continuously fish demands throughout the year. For each of selected batches, a production and feeding plan could be elaborated by the fish farm, through determining a series of parameters such as number of fish, biomass, total feed consumption and cost, as well as cages number needed to carry each seabass batch until cultured fish reach 400 g, following [14, 15] methodology.

3 Results and discussion

3.1 Seabass sales evolution of M'diq fish farm

Considering total cage size, the said farm has a total rearing volume of 15,825.60 m^3 which allow a yearly production capacity of about 200 tons. Figure 1 shows annual sales evolution since the starting year (1998) until 2016.

https://doi.org/10.1051/e3sconf/202233703001

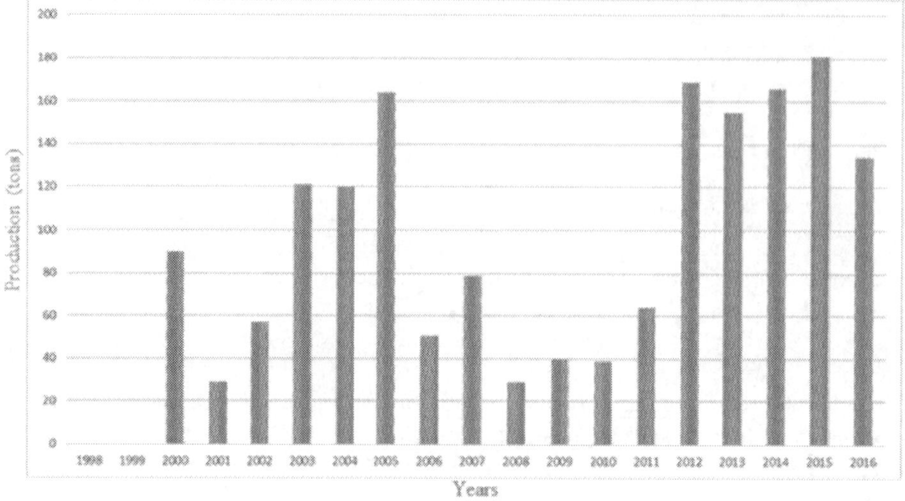

Fig. 1. Annual sales (in tons) evolution of M'diq fish farm (1998-2016).

This figure shows three main periods characterizing the production (considered through sales) evolution of the M'diq farm, including start-up period, followed by a crisis period and then a recovery period. The first one, lasted from starting year until 2005, was generally characterized by a gradual increasing sale totally destined for export, to Spanish market. Then, after the shares held by foreign shareholders were sold and purchased by Moroccan shareholders in 2006, where M'diq farm was forced to change its targeted market by domestic market during the crisis period, lasted from 2006 to 2011. Its production has been reduced less than 80 tons because of limited availability of locally produced seabass fry. Also, sales were mostly intended for a single customer, a supermarket. Furthermore, when recovery period has begun in 2012, after a shareholding of foreign shareholders in 2010, M'diq farm restarted to operate at its full capacity, by importing seabass fingerlings from foreign hatchery and increasing its fish sales, reaching 160 to 180 tons/year. The targeted market remains domestic with diversification of customers, composed both of supermarkets and restaurants through wholesalers.

During its life history, sales of this farm have recorded an average value of 99,3 tons/year, oscillated between a minimum of 29 tons recorded in 2001 and a maximum of 181 tons recorded in 2015. Considering recovery period, mainly from 2012 until 2016, annual average sale was about 161 tons, ranged between a minimum of 134 ton recorded in 2013 and a maximum of 181 tons recorded in 2015. Seabass sales vary also inside each year; monthly sales evolution recorded in this recovery period is shown in Figure 2. Overall monthly average sale was 13,634 kg ranger between a minimum of 211 kg recorded in December 2013 and a maximum of 31,955 kg recorded in November 2016.

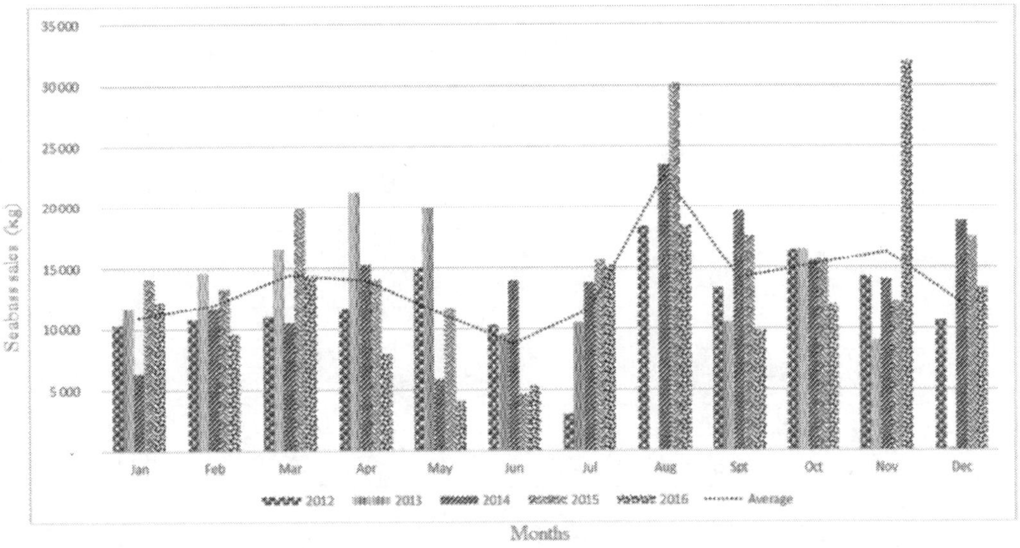

Fig. 2. Monthly sales (in kg) evolution of M'diq fish farm (2012-2016).

Fish sales are performed throughout the year with a monthly variability characterized by a seasonality usually composed of two main periods of high sales, the first in March and April and the second in June, July and September. This monthly sales variability is mostly influenced by fish orders importance, ordered commercial fish sizes availability in fish farming stock and losses level caused by storms and pathologies. Storm losses have been occurred in March 2012 (all fishes (61,559) from one cage after being grown for 21 months), in November 2012 (all farmed fishes (64,288) from one cage and a part of farmed fishes (21,667) from another cage after being both grown for 8 months), in February 2013 (parts of farmed fishes from four cages) and in February 2015 (all farmed fishes (83,983) from one cage after being grown for 10 months). Main pathologies losses were happened in 2012, first in April in two cages containing farmed fishes aged of 17 months and the second in following month (May) on other two cages containing fingerlings introduced in the previous month (April). Also, monthly sales were affected by Ramadan (fasting month) occurrence during which sales of farmed fish are reduced, mainly because of orders lack from restaurants witch are closed during Ramadan month. It has been occurred during July and August in 2012 and 2013 and during June and July in 2014, 2015 and 2016. It lasted 29 to 30 days and is advanced 11 days from year to another. Ramadan influence on sales is very important and is welly took in consideration by farm manager in his marketing strategy.

3.2 Analysis of production costs

Although M'diq farm's production has been significantly increased since 2012, profitability remained a critical. It has been facing technical and financial difficulties. Production costs analysis was based on realistic data related to recovery period. However, production costs were calculated based on an average year data calculated from recovery period's real data to serve as a basis for this study purposes. Table 1 shows production costs of this average year.

Table 1. Production costs of seabass average year production in M'diq's farm.

Items	Average operating year		
	MAD/kg (± standard deviation)	**% (partial)**	**% (total)**
Feed cost	27.41(± 5.85)	54.52	37.06
Fingerlings cost	9.91(± 2.40)	19.71	13.40
Feed & Fingerlings Taxes cost	5.16(± 0.95)	10.27	6.98
Marketing cost	3.32(± 0.56)	6.60	4.49
Others variable costs	4.47(± 0.56)	8.90	6.05
Total variables costs	**50.27**(± 3.83)	**100.00**	**67.98**
Labour cost	11.74(± 2.18)	49.59	15.88
Repair and maintenance cost	1.58(± 1.14)	6.67	2.14
Service provision	1.54(± 1.11)	6.51	2.08
Premises rental	0.17(± 1.12)	0.72	0.23
Insurances & taxes	1.11(± 0.80)	4.68	1.50
Depreciation cost	1.53(± 1.11)	6.46	2.07
Financial expenses	2.10(± 1.52)	8.86	2.84
Others fixed costs	3.91(± 2.82)	16.52	5.29
Total fixed costs	**23.68**(± 6.88)	**100.00**	**32.02**
Total costs	**73.94**(± 4.48)		**100.00**

Economic analysis of M'diq's fish farm production system revealed that seabass cost production is 73.94 MAD/kg for an average year production. Variable costs were a little bit more than the twice of the part shared by fixed costs in the total costs, having respectively 67.98% and 32.02%. These data are similar to those reported by [3] for seabass farms in Turkey, respectively 67.49% and 32.51%.

Variable operating costs are very influenced by two major elements, feed and fingerlings; they both represent 84.50% including their taxes fees. The latter is composed 61.62% of feed taxes and 38.38% of fingerlings taxes. In fact, imported feeds and fingerlings are subjected to customs duties (CD) and value added tax (VAT). They are respectively of 25% (CD) and 20% (VAT) for imported feeds and 20% (CD) and 10% (VAT) for fingerlings. Since 2016, VAT has been reduced to 2.5% for two years in case of imported feeds and cancelled (0%) for imported fingerlings.

So including these taxes, feed showed the highest part in variables costs with 60.85% followed by fingerlings with 23.65%. These values are higher than those reported by [3] in Turkey for sea bass farms with respectively of 47.73% and 10.43%. Therefore, any effort of optimization of these feed and fingerings costs, allowing even a little reduction level of the hole operating costs, would have a positive effect on the fish farm profitability. Their respective part in total costs are respectively 37.06% and 13.40% without taxes and 41.37% and 16.06% including their taxes, totalling both a common part of 57.45%.

Fixed costs seemed to be greatly influenced by one major element, the labour; it represents almost 50%. On the overall operating costs, it is the third most influencing, after feed and fingerling, with a part of 15.88%. The latter value is lower than the one reported by [3] in Turkey for sea bass farms with 22.59%. In M'diq case, feed, fingerlings and labour both account for about three-quarters of production costs, with 73.32%. These factors, and particularly feed and fingerlings, influence a lot the farm profitability. The latter is presented in the Table 2.

Table 2. Profitability of sea bass average year production in M'diq's farm.

Items	Values	Standard deviation
Total sales (kg)	160,463.67	17,828.12
Average sales price (MAD/kg)	74.12	0.93
Total income (MAD)	11,893,567.22	128,088.25
Total production (kg)	171,647.78	22,693.08
Production cost (MAD/kg)	73.94	4.48
Net return (MAD/kg)	0.18	0.19
Benefit/Cost ratio (%)	0.24	0,28

The total income of M'diq farm is of about 11.9 million MAD for an average year, with a net return of 0.18 MAD/kg and a benefit/cost ration of 0.24%. According to data reported by [3], Turkish seabass farms have shown values of net return and benefit/cost ratio respectively of 0.48 US$ and 1.10 (i.e. 10%). Environmental conditions in Turkey sea farming sites are not same than those in Morocco; Turkish conditions, mainly annual temperature profile, allow shorter seabass production run cycle.

Considering its variable contribution margin, which resulted from subtracting variable production costs from revenue. It is important tool because it gives an idea about available margin for the payment of fixed costs and most useful for making incremental pricing decisions in such a way to allow setting prices high enough to cover fixed costs and generate profit. For the M'diq farm, variable contribution margin is of 23.85 MAD/kg (average sales prices (74.12 MAD/kg) - variable costs part (50.27 MAD/kg)). This margin is so small that it just covers fixed costs without generating sufficient profit. This situation is critical and does not give the fish farm enough means to develop its activity and ensure its durability.

The main constraints pointed out by the local fish farm in M'diq which could be considered much affecting negatively the production cost are the lack of capital, the highest price of the local feed, the far-away sea bass fingerlings source, the high charges of taxes and the seasonal disease and the lack of sorting fish stock (high heterogeneous sizes in fish stock in each cage). Among these factors, feed and fingerlings, mainly in terms of price, quality and taxes fess, are considered to be the most important constraints for the local fish farm in M'diq, and even for former three marine fish farms that have ceased their farming activities.

Regarding to M'diq farm current production system, it follows intensive seabass farming technique. According to its recovery period's data, it bought annually an average number of 635,117 seabass fingerlings of 3 to 8 g in individual body weight average. The growing cycle of seabass in M'diq Bay conditions is ranged between 19 and 25 months, depending on the period of fingerlings introduction in rearing fish stock. Total average feed conversion rate is 2.1 (1.9 to 2.6) and harvested densities are about 20 kg/m^3. Some factors such as initial stocking densities, the quality and quantity of feed supply, labour qualifications and managements skills are well taken in consideration. The current densities of the farm are however not managed with size to re-establish fish population homogeneity before reaching commercial sizes because of the lack of a suitable fish sorting device.

Sure, even when starting from a homogeneous population of fry, a significant spread in growth and in biomass distribution is usually observed [16, 17]. According to these authors, seabass growth exhibits a slower dynamic below a critical temperature of about 10°C. It is not the case in M'diq Bay where the minimum seawater temperature is much higher than this value.

Considering marketing characteristics, M'diq farm sold basically its seabass fishes when they reached an individual body average weight of 300 g or more. In spite of the fact that sea bass sold price fluctuates with seasons, average wholesale price is about 74.12 MAD/kg (Table 2). The market price of aquaculture products mainly depends on size and weight, quality, seasonality, supply and demand and competitiveness [18].

To try to improve its economic results, M'diq farm should consider optimization of its production plan, mainly on structural and managerial aspects. For the first one, it is planning to increase its total production volume capacity without significant added investment; and for the second one, it is focused on feed and fingerlings optimization uses.

3.3 Improvement through structural aspect

M'diq farm introduces yearly about 630,000 seabass fingerlings to be stocked in 7 cages of individual volume of 1,130 m^3. The decision of these fingerlings introduction depends on the availability of empty cages according to the planned number of batches. It is necessary to have at least 2 empty cages available in order to ensure that the transport costs using a tank truck could be appropriate.

Right now, M'diq farms uses multi-batches production, by stocking fingerlings through three batches a year, over a time period of four to ten months, depending on harvesting time (availability of harvested or empty cages). When fish are actually harvested, space is made available for new fish; this is important, as the availability space in fish farms is limited. Furthermore, the present farm performed partial harvests during two to four or five months, based on a weekly harvest frequency.

The rotation of cages utilization depends upon the fish growing performances witch influences, in its turn, the batches cycle and the harvested time [5]. It is useful therefore to find out an optimal rotation time for the batches of seabass in M'diq bay conditions, i.e., the time for harvesting each fish culture batch and stocking the next. The best economic results will basically be obtained by producing the maximum number of batches in appropriate period, regardless to larger fish size.

So, to overcome this constraint, it has been set up additional small cages, of 7 m in diameter in each, which have been used to stock fingerlings for a short period of pre-growing before having empty cages available for growing. In fact, 7 small cages have been set up in the same farming site without any additional investment in supporting and mooring systems. They were suitable for stocking seabass fingerlings of an average weight of 3 to 5 g. They have allowed appropriate fingerlings introduction schedule and commercial sized fishes sales. They have also allowed to ovoid seasonal outbreak of some diseases in seabass fingerlings when introduced in farming fish stock which occurred in summer period causing high mortalities and reducing the value of harvested fish.

If this working strategy, M'diq fish farm could then invest in expanding its production capacity. Moreover, once the fixed investment has been made, farm production decision should be based on expected return or income above variable costs. Fixed investment is considered as sunk cost and may not be recovered in short-run period of at least one production cycle.

3.4 Improvement through managerial aspect

This aspect is very pertinent. To improve variable contribution margin and to increase profit of M'diq farm, feed and fingerlings management should be improved according to local conditions. The importation of these two inputs from foreign origins increase variable operating costs, both through taxes fees and also exchange ratio of MAD to Euros. Local feed has been used since 2005, first as pressed pellet then as extruded pellet while the use of locally produced seabass fingerlings by a small experimental hatchery could not be maintained for reasons of limited production not satisfying farm's needs. The lack of local commercial marine fish hatchery Regardless of farming system, the lack of a local commercial marine fish hatchery has then forced M'diq farm to import again seabass fingerlings from foreign hatcheries.

Marketable seabass sizes are harvested usually after a growing period ranged between the 18th and 23th month, but, in fact, harvested weights are usually given by market conditions and consumer preferences. Fish growth can be influenced by several factors, including fish weight, water temperature, feeding rate, water quality, diet quality, stocking rates and environmental conditions [6]. In M'diq Bay, where the concerned fish farm has its rearing site, the annual temperature profile is estimated a little bit weak to allow good growth performances. It is characterized by an annual average of 17.6°C with a minimum and maximum monthly average values of 14.2°C and 20.7°C respectively.

The analysis of growth performances of all growing batches performed between 2010 and 2015 has allowed to drawn average growth curves for every monthly batch among the ten, conducting from December to September (Figure 3).

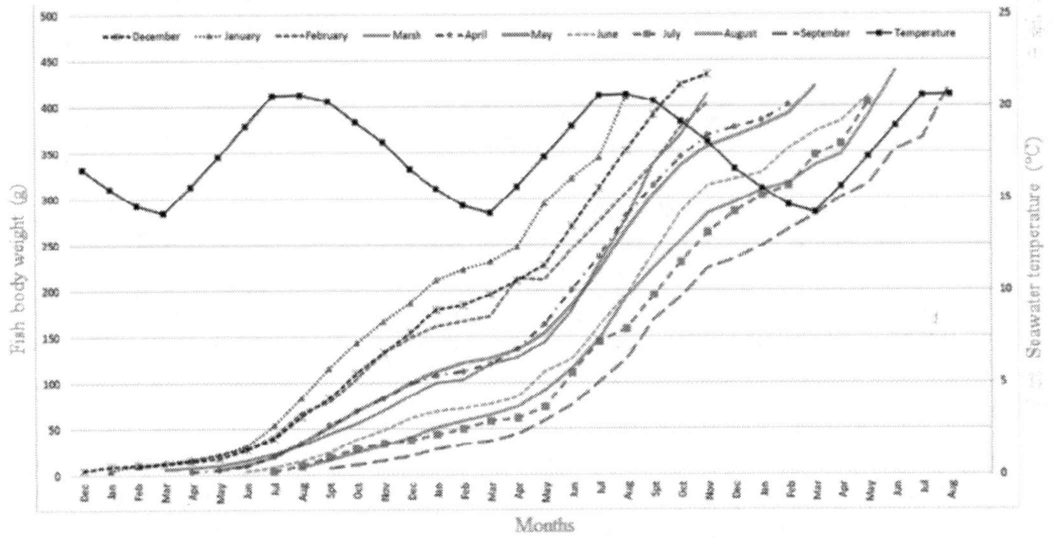

Fig. 3. Average growth curves of the monthly batches conducted by M'diq fish farm during the period of 2010 and 2015.

As showed in Figure 3, temperature influence when fingerlings were first introduced in stock culture is very important. According to these real data, batches started in January showed better results than those introduced in December. The growth during the first months after introduction became a little bit faster as fingerlings date is closer to warm season. The analysis of these growth curves could suggest that three introductions a year, in January, March and May would be important as they would allow harvesting commercial fish sizes (of 400 g in average weight) respectively since July, November and March with sales lasting during four months each.

The analysis based on Specific Growth Rate (SGR) was performed on all monthly batches data, using the following formula [12]:

$$SGR = (\ln W_f - \ln W_i) / days \quad (2)$$

Where (ln) is natural logarithm, (W_f) is final weight and (W_i) is initial weight.

Table 3 shows results relating to SGR analysis, calculations for each monthly batch of seabass lots grown in M'diq fish farm from real data recorded since 2010 until 2015.

Table 3. SGR average data calculated for monthly batches of growing seabass lots in M'diq's farm (2010-2015).

Batches	**SGR**				**Time (months)**
	M6 (%d^{-1})	**M12 (%d^{-1})**	**M18 (%d^{-1})**	**Bc (%d^{-1})**	
December	0.69	0.90	0.70	0.64	22
January	1.00	0.99	0.76	0.73	19
February	0.90	0.82	0.65	0.59	21
March	0.91	0.78	0.70	0.66	20
April	1.49	0.97	0.82	0.68	22
May	1.34	0.86	0.74	0.61	22
June	1.36	0.91	0.80	0.64	23
July	1.22	0.91	0.78	0.63	23
August	1.02	0.80	0.67	0.57	22
September	0.84	0.79	0.66	0.56	23

With: M6, M12 and M18 respectively the 6th month, 12th month and 18th month of the growing cycle and BC means batch growing cycle

In this table, SGR are calculated for each growing period of six months' intervals, mainly at the 6th month (M6), the 12th month (M12) and 18th month (M18) during the growing cycle and also for the whole batch growing cycle (Bc). The analysis of these data showed that SGR in M6 are very low in January's batch and very high in April's batch. M6's SGR data showed a decreasing trend from April's batch toward September's batch then it a little bit increased in January's batch with a relative stabilization in February's and March's batches. Moreover, M12 and M18 showed same trend but their SGR values are very low comparatively to those of M6. Except for the December's batch, all the other month's batches showed a proportional decrease of SGR's values with their fishes ages. This evolution is highly marked for batches started in April, May, June, July and August.

Considering these characteristics showed in Table 3, batches started in January, March and May seem to be very interesting for selection as common practice of regulated seabass fingerlings recruits since they showed relatively highest SGR values and also appropriate batches cycles.

To highlight the importance of the effect of seawater temperature when fishes are introduced and also the subsequent evolution of temperature according to the local pattern of its annual profile, it was judged to use growth curves simulation based on thermal growth coefficient (TGC) (Table 4 and Figure 4).

Table 4. Calculation of TGC data based on monthly batches of growing seabass lots in M'diq's farm (2010-2015).

Batches	Wi (g)	Wf (g)	T-av. (°C)	STef (°C)	TGC ($g^{1/3}$ °C^{-1})
December	5.00	423.70	17.60	3,96	0.00146
January	5.00	410.00	17.50	3,30	0.00173
February	8.00	402.50	17.60	3,76	0.00143

March	6.00	413.00	17.80	3,73	0.00150
April	3.50	401.00	18.00	4,36	0.00134
May	6.00	422.00	18.10	3,68	0.00154
June	4.00	411.00	17.80	3,99	0.00146
July	4.00	404.50	17.70	4,14	0.00140
August	8.00	414.00	17.80	4,05	0.00134
September	7.00	424.00	17.40	3,73	0.00150

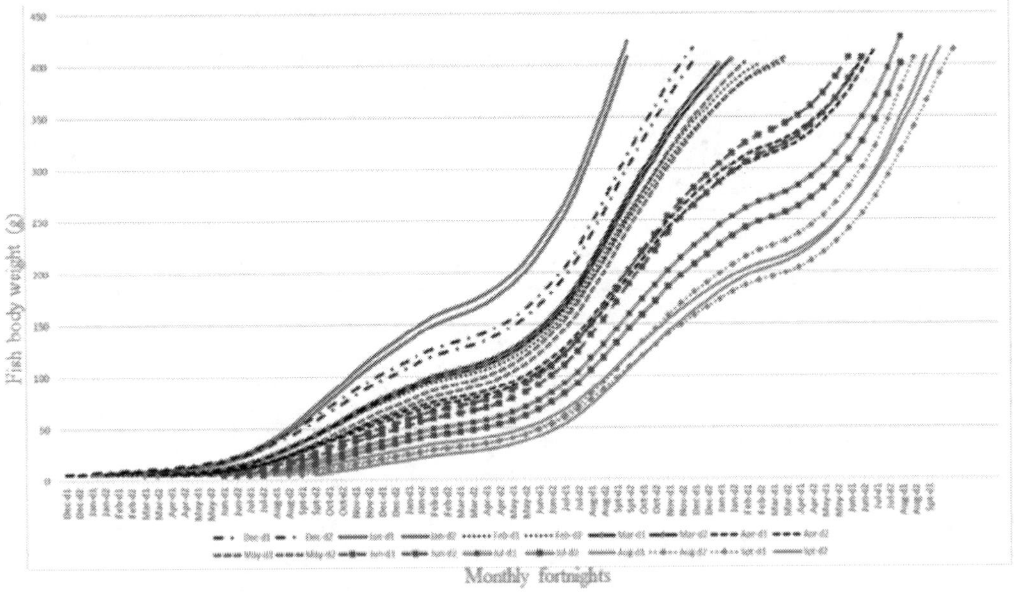

Fig. 4. SGR average data related to growth evolution of monthly batches of seabass growing lots in M'diq farm (with: Jan-d1 for the first fortnight in January, Jan-d2 for its second fortnight, and so on the others months).

In Table 4, the meaning of used terms is initial weight for W_i, final weight for W_f, average temperature of each batch period for T-av, cumulative effective temperature (effective temperature is temperature in degrees' Celsius minus 12) for STef and Thermal Growth Coefficient for TGC.

The comparative analysis of these data showed that preselected monthly batches (January, March and May) have TGC respectively of 0.00173, 0.00150 and 0.00154 (Table 4). Simulated growth curves showed in Figure 4 allowed selecting the three preselected monthly batches as suitable for production planning in seabass farming in M'diq fish farm conditions. This proposal procedure based on small pre-growing cages and introduction plan of seabass fingerlings (may be initial average weight ranged between 5 and 10 g) should allow fish growth improvement and economic results.

Furthermore, Moroccan Government measure related to the exemption of customs duties and value added tax (VAT) came into force since the beginning of 2018 for five years, should allow M'diq farm to improve its economic resultants (Tables 5 et 6). This measure is very important since fishes are sold tax free in Morocco and therefore do not allow aquaculture farms to recover VAT paid for inputs used during the production process. Imported fishes are however subjected to VAT (20%).

Table 5. Production seabass costs in M'diq's farm after exemption of customs duties and VAT since 2018.

Items	MAD/kg	% (partial)	% (total)
Feed cost	27.41	60.76	39.85
Fingerlings cost	9.91	21.96	14.40
Feed & Fingerlings Taxes cost	0.00	0.00	0.00
Marketing cost	0.00	0.00	0.00
Others variable costs	3.32	7.36	4.83
Total variables costs	**45.11**	**100.00**	**65.58**
Labour cost	11.74	49.59	17.07
Repair and maintenance cost	1.58	6.67	2.30
Service provision	1.54	6.51	2.24
Premises rental	0.17	0.72	0.25
Insurances & taxes	1.11	4.68	1.61
Depreciation cost	1.53	6.46	2.23
Financial expenses	2.10	8.86	3.05
Others fixed costs	3.91	16.52	5.69
Total fixed costs	23.68	100.00	34.42
Total costs	68.78		100.00

Table 6. Profitability improvement of seabass in M'diq's farm after exemption ofcustoms duties and VAT since 2018.

Items	Values
Total sales (kg)	160,463.67
Average sales price (MAD/kg)	74.12
Total income (MAD)	11,893,567.22
Total production (kg)	171,647.78
Production cost (MAD/kg)	68.78
Net return (MAD/kg)	5.34
Benefit/Cost ratio (%)	7.76

Exemption of customs duties and VAT, performed since 2018 should then allow reduction of variable costs to reach 45.11 MAD/kg instead of 50.27 MAD/kg, improvement of production cost witch could be reduced to 68.78 MAD/kg instead of 73.94 MAD/kg, improvement of variable contribution margin witch could reach 29.01 MAD/kg instead of 23.85 MAD/kg, sufficiently higher to cover fixed costs (23.68 MAD/kg) and allowing an increase in net return and benefit/cost ratio, respectively to reach 13.24 MAD/kg instead of 0.18 MAD/kg and 7.6% instead of 0.24%.

With regard to mortality, fish losses contribute also to affect fish farm results. One should distinguish between mortality due to particular pathology and a physiological (or natural) mortality. In M'diq farm, mortality is followed and counted once every two or three days and reported on a monthly basis. The recorded data showed an average of mortality (both natural and pathologic) of 14.66% with 5.6% and 47.0% respectively as minimum and maximum recorded values. There are also lost fishes discovered when comparing final harvested fishes number to initial introduced number of fingerlings from witch mortality number is subtracted. This kind of loss accounted on average 17.49% with a minimum and maximum values respectively of 6.4% and 39.1%. The total loss of fishes reached on average 32.2%; the overall survival rate is then 67.8% that should be improved to reduce the impact of this farmed fish loss on the economic farm results.

Pathologies occurred generally when introducing fingerlings during summer months and fish's loss by storms occurred mainly in winter season, following rips of old nets. Furthermore, concerning fish's loss, it seems also that there is may be another reason which could be related to fingerlings quality, particularly in terms of sizes heterogeneity level and deformities level. The analysis of the introduced fingerlings batches showed some different values of coefficient of variation (ratio of the standard deviation to the average weight of fingerlings). It is a statistical measure of the dispersion of data points in a data series around the mean. The recorded values are ranged between 15.65% and 20.27%. So, in case of higher values, cannibalism could happen, mainly between biggest and smallest fingerlings and juveniles or when heterogeneity level increase with fish's age. According delivering hatchery, initial fingerlings average weights are ranged between 2.88 g and 5.21 g (sampling data recorded few days before delivery transport). Also, recorded deformities values ranged between 1% and 6%. The major part of fingerlings batches has deformities level more than 3%. No one could know what should happen to these deformed fishes but they are very vulnerable.

Concerning feed performances, for each fish weight class, the feed amount to supply is computed as a function of water temperature and monthly fish average weight using locally appropriate data, established on the basis to feeding tables provided by aqua-feed suppliers and adjusted following local experience. At the end of fish growing batch, a global value of feed conversion ratio (FCR = Feed supplied / fish weight gained), is computed as a ratio of the provided feed and the harvested fish biomass. There is a common practice of overfeeding when production is undertaken in off-shore cages with the advantages of natural recirculation [6].

The average value recorded for FCR in M'diq farm is equal to 2.1 with a minimum and maximum respectively of 1.9 and 2.6. These values are accounting data, including feed used even for fishes lost or died during growing process. Improvement of quality, in terms of formulation and used ingredients, and quantity by optimizing feeding rationing is a key factor to reduce feed cost and improve growth performances.

Experiences still are conducting to improve local seabass feed, using locally available ingredients and aiming also to reduce as much as possible the used level of fish oil and fish meal.

It is therefore evident that the management benefits depend on the accuracy in the predicted trends for the exogenous variables, especially market and prices. It is also clear that, due to the complex interactions of the many variables affecting the production process, the optimal management strategies cannot be provided by simple rules [16]. Study of the national farmed fishes market and the diagnosis of its regional typology will allow to develop knowledge about on commercial fish sizes' needs of different domestic market branches and to be able to adopt marketing strategic according to the demand evolution. On the average, M'diq fish farm sold different commercial sizes of its farmed seabass (Table 7).

Table 7. Average composition of commercial sizes of farmed seabass sold by M'diq fish farm.

	Average operating year		
Commercial fish sizes	**% (in amount)**	**% (in value)**	**Average prices (MAD/kg)**
200-300	18.90	15.25	60.70

300-400	35.30	34.81	74.21
400-600	39.50	42.07	80.16
600-800	5.80	7.21	93.53
800-1000	0.50	0.66	99.15

As shown in this table 7, the two important commercial fish sizes are 300-400 g and 400-600 g shared both a cumulated part reaching 74.80%, with individual parts of 35.30% and 39.50% respectively. Their average sales prices are respectively 74.21 MAD/kg and 80.16 MAD/kg and are higher than the production cost (73.94 MAD/kg). They both shared 76.88% of the farm income. Large commercial fish sizes (> 800 g) are very interesting in terms of average sales price. The latter is more than 90 MAD/kg. This category of fish size represents 6.30 % in amount and 7.87% in value in the case of M'diq farm. It worth to be developed much but in real proportional trend of domestic market. By the way, several previous studies underwent in others countries have concluded that batches producing larger fish sizes allow to reach higher operating profits due to their higher value in the market [5, 19, 20].

The main fish sales constraint in terms of commercial fish sizes is the part of small harvested fishes (< 300 g). This category is sold on average at 60.70 MAD/kg in the case of M'diq fish farm. This sale price is quite lower than the production cost with lost gap of 13.24 MAD/kg. This category of fish size represents 18.90% in amount and 15.25% in value. To reduce the impact of fish size category lower than 300 g, M'diq farm should to perform fish sorting and include it in its production plan or, if not, increase the average fish weight threshold to begin harvesting and sales from 300 g to more than 350 g. According to [6], the results show that optimal decisions about the ration size approach observed managerial practices. This optimal decision involves a trade-off between biological and economic inputs.

Finally, it is important to explore technical and market feasibility to develop new product alternatives of framed seabass for commercial exploitation. M'diq farm have recently tried it but opportunities of this potential different market is still under investigation taking into account socio-economic aspects and consumer requirements. M'diq farm is very interesting undergoing experimental works aiming at evaluating and improving its farmed seabass market performance through an appropriate supply chain, market-oriented design of diversified or new types of added-value its products for Moroccan consumers. This may contribute to diversify aquaculture products and enhance Moroccan aquaculture farms competitiveness.

4 Conclusion

The present study has examined biotechnical and economic characteristics of a marine fish farm in M'diq Bay, in Morocco. It analysed its seabass production costs and profitability. This farm was created in 1998 and started fish production in 2000. Three main successive periods characterized its history: start-up period, crisis period and recovery period. The present study has been focused on the latter period.

The production system is composed of 14 floating net-cages of 12 m in diameter, placed in open sea in M'diq Bay. Annual farmed seabass sales are ranged between 134 tons and 181 tons. Production costs, calculated over the recovery period (2012-2016), is 73.94 MAD/kg in average among witch variable costs are of 50.27 MAD/kg representing a little bit more than the two thirds (67.98%). Seabass are sold along the year round, at an average price of 74.12 MAD/kg.

The net return and the benefit/cost ratio recorded very low values, respectively of 0.18 MAD/kg and 0.24%. The resulted variable contribution margin is just enough to cover fixed costs. These results show very weak profitability and an economic balance just ensuring financial equilibrium of M'diq fish farm. Considering variable costs, feed and fingerlings fees, including their taxes, shared both a major part of 84.50%. Any steadily increase in these two inputs would affect operating and limit development of this farm. So, how could this farm do to ascertain its sustainability and to be timely able to develop its fish farming activity.

To answer this question, fish production batches data were analysed from 2010 to 2015. Fingerlings stocking procedure has been shown based on two to three introductions per year, performed according to the availability of empty cages. In this period, fingerlings introductions have been performed in ten different months, from December to September. So, growth performances were influenced by stocking date. Results showed that batches started in January, March and May could be more interesting for production plan. Farm's aim is to shorten growing cycle to improve its farming results in terms of production cost decrease and sale income increase.

Productivity and economic results are very influenced by seawater temperature. Growing performance is variable during all the year. The annual temperature profile in M'diq Bay is characterized by a range (14-21°C) with an annual average temperature of about 18°C, recorded in floating cages site. These conditions are a little bit weak, not allowing high growth performance for seabass compared to middle and eastern Mediterranean countries. The latter have higher average seawater temperature in their rearing sites (optimal geographical locations of fish farms) allowing higher productivity. So, M'diq farm should take in consideration this environmental constraint to adjust its production strategies and optimize its operational economic performances through suitable technical solutions that could result in reduction of its operating costs. This was highlighted through analysis of Specific Growth Ratio (SGR) evolution with fish age and growth performance simulation using Thermal Coefficient Growth (TCG) based on real data which showed that growing batches started in January, March and May are more interesting in production planning optimization.

In this regards, improvement of farm's profitability could be generated through optimization of fingerlings and feed management, as major variable costs components having high influence on production costs. Their management could be improved in terms of their appropriate introduction date (January, March and May batches), their initial average weight (to be increased to 5 -10 g) with more relatively suitable homogeneity level, their deformities level (less than 3%), etc. Moreover, it was found that batches started in summer months were more exposed to disease outbreaks. Regarding feed management, it is very relevant to master its conversion ratio (FCR) which should be reduced from the current average value of 2.1 to less than 2.0. Lowering feed's FCR through improving feed formulation and feeding practices could help to reduce feed costs and improve farm's profit margin. In fact, researches on feed quality are undergoing based on locally available low-cost ingredients to improve local produced seabass feed.

Moreover, profit level of this farm could also be significantly improved by the Moroccan Government support in terms of exemption of imported feed and fingerlings from customs duties CD and value added tax VAT. Performed since the beginning of 2018 in Morocco, this exemption should contribute to reduce production cost could to 68.78 MAD/kg, generating a variable contribution margin sufficiently larger to cover fixed costs and allow an increase in net return and benefit/cost ratio reaching respectively 13.24 MAD/kg and 7.76%.The decrease in operating cost is one of major means of increasing profit for M'diq seabass farm. The net income of this farm is then affected by its production cost and also by the level of its production and selling prices. So, increase in farm productivity, reduction in production costs and increase in average farm revenue are major measures to increase net return.

Under the lights of these results, sharing risks by growing others species in terms of cultured species diversification or in multi-trophic model would be economically efficient and may reduce market risks and increasing its profitability. Farmed seabass is still sold in nich market. An economic assessment of seabass production based on the current state of technical and economic knowledge could help overcome its aquaculture development constraints and enhance its social perception.

References

1. S. Merinero, S. Martínez, A. Tomás, and M. Jover, Revista AquaTIC, **23**, pp.: 1-19 (2005)

   Available at: http://www.revistaaquatic.com/aquatic/art.asp?t=p&c=188
2. F. De Benito, F. Maicas, I. Jauralde, S. Martínez, M. Marín, and M. Jover, Revista AquaTIC, **37**, pp.: 123-138. (2012).

   Available at: http://www.revistaaquatic.com/aquatic/art.asp?t=p&c=255
3. M. Bozoglu, and V. Ceyhan, J. Anim. Vet. Adv., **8**(2): 217-222, (2009).
4. D. Nachite, E. Baijot, N. Marchesi, and F. Zyadi, *Plan local GIZC pour la préfecture littorale de M'diq - F'nideq. AT SMAP III: Renforcement des capacités de la cellule locale du littoral de M'diq*. 105 pages, (2009).

   Available at: http://www.academia.edu/6233803/PA_GIZc_Mdiq_morocco
5. I. Llorente, and L. Luna, Aquaculture International, **22**(6), pp.: 1837-1849, (2014).

   C.J. León, J.M. Hernández, and E. Gasca-Leyva, Aquaculture Economics & Management, **5**(3–4), pp.: 147–170, (2001) doi:10.1080/13657300109380284.
6. G.J. García, and G.B. García, Aquaculture International, **18**(5), pp.: 955–971, (2010). doi:10.1007/s10499-009-9311-1.

   C.Y. Cho, and D. Bureau, Aquatic Living Resources, **11**(4): 199-210, (1998). https://doi.org/10.1016/S0990-7440(98)89002-5
7. C.Y. Cho, Aquaculture, **100**, 107–123, (1992).
8. A. Dumas, J. France, and D. Bureau, Aquaculture, **267**, 139–246, (2007).
9. I. Seginer, Aquacultural Engineering, **70**, pp.: 15–32, (2016). doi: 10.1016/j.aquaeng.2015.12.001

   P. Mayer, V.D. Estruch, and M. Jover, Aquaculture, **358–359**, pp.: 6–13, (2012). doi: 10.1016/j.aquaculture.2012.06.016
10. P. Mayer, V. Estruch, J. Blasco, and M. Jover, Aquaculture Research. **39**(10), pp.: 1046–1052, (2008). doi:10.1111/j.1365-2109.2008.01963.x
11. J.C. Martínez, *Acuicultura Avanzada.* Escuela Técnica Superior de Ingeniería Agronómica y del Medio Rural. Universidad Politécnica de Valencia, 63 pages, (2016).

    Available at: https://georgiusm.wordpress.com/universidad-politecnica-de-valencia
12. M. Jover, J. Martínez, A. Tomás, A. Moñino, J.A. Gómez, S. Martínez, J. Villaplana, J.F. Asturiano, and L. Perez, Análisis económico del crecimiento de la dorada en jaulas marinas con diferentes tasas de alimentación. Actas VIII Congreso Nacional de Acuicultura, pp.: 259-260, (2003).
13. G. Rizzo, and M. Spagnolo, Marine Resource Economics, **11**(4), pp.: 267–286, (1996).

M. Besson, M. Vandeputte, J.A.M. Van Arendonk, J. Aubin, I.J.M. de Boer, E. Quillet, and H. Komen, Aquaculture, **462**, pp.: 47–55, (2016). doi: 10.1016/j.aquaculture.2016.04.030.

14. Y.C. Shang, Aquaculture Economics: Basics Concepts and Methods of Analysis. Westview Press, Boulder, Colorado (1981)
15. E. Gasca-Leyva, C.J. León, and J.M. Hernández, Journal of the World Aquaculture Society, **34**, 29–31, (2003).
16. J.M. Hernandez, M. Leon-Santana, and C.J. Leon, European Journal of Operational Research **181**, pp.: 872–886, (2007).

Behavioural pattern in changes to diets and food availability during broodstock conditioning of captive clam *Ruditapes decussates*

Yassine Ouagajjou [1*], *Adil* Aghzar [2], *Meriem* Ahensal [3], *Nadia* Berday [3], and *Mohamed* Idhalla [1]

[1]National Institute of Fisheries Research, Departement of Aquaculture, Casablanca, Morocco
[2]Polydisciplinary Faculty of Larache FPL, Abdelmalek Essâadi University, Tetouan, Morocco
[3]Institute of Agromony and Veterinary Sciences Hassan II, Rabat, Morocco

Abstract. In terms of the nutritional needs, the choice of the appropriate phytoplankton strain is the main criterion during bivalves breeding improvement in captivity. Seven nutritional diets based on tree microalgae species (*Chaetoceros calcitrans*, *Tetraselmis suecica* and *Pavlova lutheri*), and seven feeding rations (0%, 1%, 2%, 3%, 4%, 5% and 6%) were used to assess the nutritional pattern of *Ruditapes decussatus* during conditioning practices. All regimes have shown the highest ingestion rate at low food availability (1% to 3%) with a leaning to diets based on *C. Calcitrans* followed by *T. suecica* and *P. lutheri* respectively.A significant influence of diets quality on absorption efficiency of phytoplankton was observed for both monospecific and bispecific diets ($F = 21.78$, $p = 3.2\ 10^{-7}$ *** and $F = 250.09$, $p = 2.2\ 10^{-16}$ ***respectively). Whereas, low influence of food availability on absorption efficiency among all used rations has been revealed by clams reared under mono, bi and trispecific diets respectively ($F = 11{,}19$, $p = 1.9\ 10^{-7}$ ***; $F = 16{,}63$, $p = 1.03\ 10^{-9}$ *** and $F = 2.78$, $p = 0.0534^{ns}$). Consequently, this study could be handled as a standard protocol for conditioning process in furtherance of yield maximizing and cost minimizing during spat production of bivalves in hatcheries.

1 Introduction

The clam *Ruditapes decussatus* is a common species in the estuarine and lagoon areas of most of the Mediterranean basin [1, 2]. It is globally widespread from the British Isles coastlines to the Mediterranean Sea until the Red Sea and along the Atlantic coasts of Africa from Morocco until Senegal [3-6]. Several studies related to many aspects were carried out on clams in terms of its biology and ecology [3, 7, 8, 9, 10], its ecotoxicology [11-14], its breeding in captivity [15-19] and to physiological energetic and biochemical parameters [20-22]. In Morocco, it is distributed all across the marine ecosystems (bays, lagoons and estuaries) from Alboran Sea to the south Atlantic coasts [23, 24].With its

*Corresponding author: ouagajjou@inrh.ma

excellent organoleptic qualities and economic importance, *R. decussatus* represents a great commercial value as one of the most frequently harvested bivalves in the Mediterranean [25]. Consequently, this clam has been heavily fished in all across its habitats along Moroccan coastlines [23, 25]. Accordingly, a management strategy has been imposed recently to preserve this resource and to regulate its fishery activities. [26] proposed a clam conservation strategy based on morphological variations which were thereafter applied on local populations in order to define their morphometric patterns [27]. In addition, the establishment of management strategy depends on spat production process in captivity inside hatcheries in order to ensure the sustainability production of shellfish aquaculture.

Broodstock conditioning for spawning is a crucial step which allows extending spat production season regardless the short period of natural maturity in wild beds. During breeding process, the major aim of broodstock conditioning is to achieve highest gametes release and maximum fecundity of clams whilst improving gametes quality and viability [28-30]. This achievement depends essentially on feeding practices during conditioning where microalgae are the only trophic source during broodstock conditioning and their production is one of the most important practices in mollusks hatchery [31]. The nutritional quality and cost benefit assessment of the different species of microalgae used in hatcheries for many species of bivalves have been studied during the last decade [32-36]. Therefore, broodstock maturation appears to be more controlled by food quantity and quality rather than other factors such as temperature, salinity and photoperiod [19].

Through the past years, many studies were performed in terms of using bio-deposition method to assess feeding and absorption rates in bivalves [20, 37] in order to optimize feeding regimes and rations during broodstock conditioning. From eco-physiological standpoint, the influence of quality and availability of food on maturation and biomass production is equated to the particular effect of each physiological aspect such as ingestion, digestion and absorption capacities [38]. The aim of the present research was to evaluate the effect of multispecific microalgae diets and suitable food ration on physiological patterns like consumption, ingestion and absorption during locale clam (*Ruditapes decussatus*) broodstock.

2 Materials and Methods

2.1 Clam sampling and broodstock conditioning

All the experiments were conducted during June 2019, at the Station of Shellfish Research, located in the marine area of Amsa (Alboran Sea, Tetouan, Morocco). The biological material consisted of 430 adult specimens of the local clam (weight = 13.78 ± 3.03 g; length = 40.33 ± 3.27 mm) collected from natural populations from Kabila (35°43'04" N/5°20'06" W, M'diq, Morocco),then rinsed and placed in 20 L tanks and held in a closed water system. An average of 10 individuals per tank was used for each experiment under natural conditions of temperature (19 °C), salinity (36 PSU), pH (8-8,3) and dissolved oxygen (6.4 ± 0.5mg/L). Forty adults were randomly chosen in order to calculate their mean dry weight according to [21]. Before carrying out feeding tests, all adults were held to fast during 48h in order to get totally rid of their faeces.

2.2 Feeding process

Three species of live microalgae were used (*Chaetoceros calcitrans*, *Pavlova lutheri* and *Tetraselmis suecica*). Three types of regimes (Diet-A, Diet-B and Diet-C) were held during this experiment with different composition. Three monospecific regimes (**Diet-A1**: 100% *P. lutheri*; **Diet-A2**: 100% *C. calcitrans* and **Diet-A3**: 100% *T. suecica*), three bispecific regimes (**Diet-B1**: 65% *P. lutheri* + 35% *C. calcitrans*; **Diet-B2**: 80% *P. lutheri* + 20% *T. suecica* and **Diet-B3**: 75% *T. suecica* + 25% *C. calcitrans*) and one Trispecific regimes (**Diet-C**: 50% *P. lutheri* + 25% *T. suecica* + 25% *C. calcitrans*). Furthermore, for each diet we established seven feeding rations in triplicate as described by [28] (1%, 2%, 3%, 4%, 5% and 6 % of clam dry weight) in addition to control diet (0%: only natural seawater). For all experiments (except the control), only 0.2 µm filtered seawater was used in a closed rearing system.

2.3 Physiological patterns

The consumption of microalgae in each experiment was handled by counting the number of cells each thirty minutes during three hours from different level of rearing tanks (bottom, middle and surface). By the end of each experiment, early at the evening, the decanted feces in breeding tanks are recovered and quantified according to [21]. In order not to overestimate the feces weight measurements, the pseudo-faeces occurrence was taken into account according to [21] before collecting of the fecal matter.

- Consumption was estimated according to [21]: $C = (N1 - N2) \times Ps/Pi$; (1)
- Ingestion rate was estimated according to [21]: $IR = (N1 - N2/N1) \times 100$; (2)
- Ingestion was calculated according to [21]: $I = C - P0$; (3)
- Absorption efficiency was estimated according to [39]: $AE = (I - PF)/I) \times 100$; (4)
- Absorption rate was estimated according to[21]: $AR = I \times AE$; (5)

Where *N1*: Initial number of microalgae; *N2*: Final number of microalgae; *Ps*: Dry weight of microalgae (g); *Pi*: Dry weight of clam flesh (g); *P0*: weight of the pseudo-faeces (g); *PF*: weight of fecal matter (g).

2.4 Statistical analysis

In order to estimate the effect of each treatment (diet and ration) on each calculated parameter, a two-way analysis of variance (ANOVA) was performed at a 95% level of confidence ($\alpha = 0.05$). When ANOVAs were significant, the post hoc Tukey multiple comparison test was used to determine which treatments differed. All statistical tests were analyzed using Rcmdr interface [40] implemented in R package version 3.1.2 [41].

3 Results

3.1 Ingestion rate and consumption

For monospecific diets, the results reported in Figure 1A show a large difference in terms of ingestion rate (IR) between the food rations used during this study. For instance, clams reared under Diet-A2 (*C. calcitrans*) have revealed the highest IR (approximately 100 %)

for all feeding rations, followed by clams reared under Diet-A3 (*T. suecica*) especially at the first three feeding rations. Diet-A1 (*P. lutheri*) has shown the highest IR only for 1% of feeding rations.

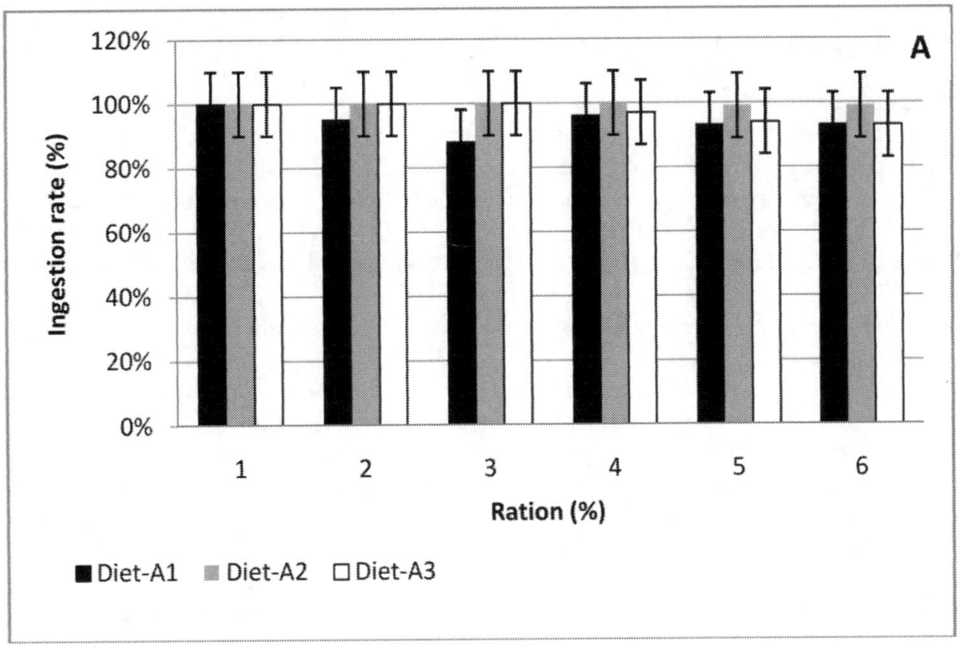

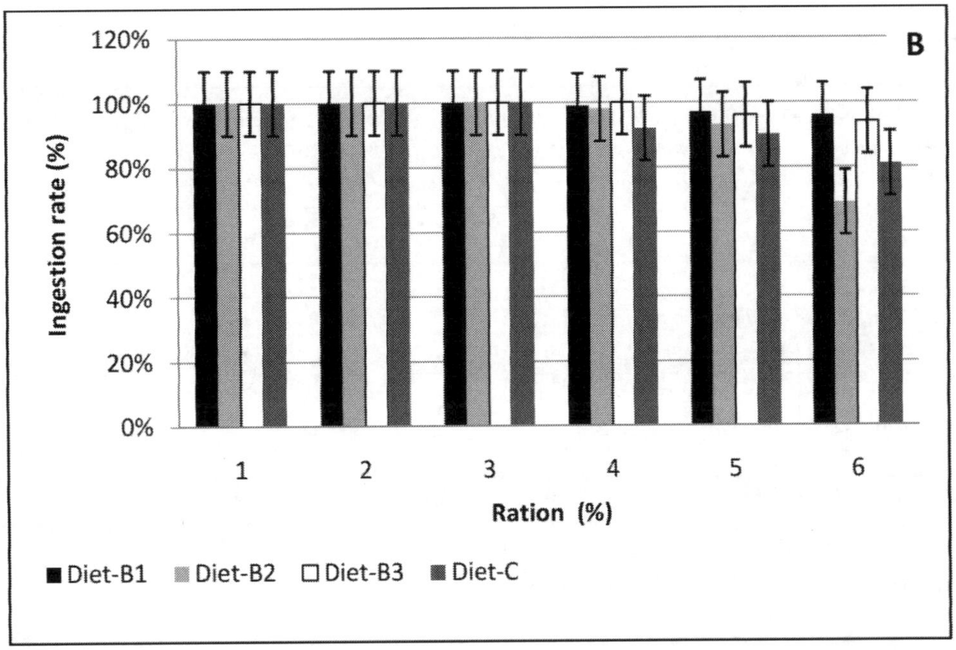

Fig.1. Ingestion rate (%) of clam *R. decussatus* exposed to different microalgae diets (**(A)**: monospecific diets Diet-A1 (100% *P. lutheri*), Diet-A2 (100% *C. calcitrans*) and Diet-A3 (100% *T. suecica*); **(B)**: Multispecific diets (Bispecific diets: Diet-B1 (65% *P. lutheri* + 35% *C. calcitrans*), Diet-B2 (80% *P. lutheri* + 20% *T. suecica*), Diet-B3 (75% *T. suecica* + 25% *C. calcitrans*) and Trispecific diet: Diet-C (50% *P. lutheri* + 25% *T. suecica* + 25% *C. calcitrans*)) and food ration (%) during broodstock conditioning, (Vertical bars indicate SD)

As regards multi-specific diets, the IR varies remarkably from one diet to another and shows an outstanding difference regarding microalgae composition of each diet. All regimes have shown the highest IR at low food availability (1% to 3%).Whereas, at high food availability, clams showed less capacity of IR (Fig. 1B). Taking into account the specific ingestion within each diet, we clearly observed that clams reared under bispecific (Diet-B) have revealed the highest IR toward *C. calcitrans* followed by *T. suecica* and *P.* lutheri, respectively.

For trispecific (Diet-C), the best IR (100%) was also registered at low food availability (1% to 3%, Fig. 1B). Regarding the specific IR, we observed that *C. calcitrans* is thoroughly ingested for all feeding rations, followed by *P. lutherie* which has been fully ingested during almost the feeding rations, while less IR toward *T. suecica* was remarkably observed when multispecific diet is used. However, we note that clams reared under natural seawater (0% ration) have shown 100% of IR during this study. Moreover, the highest effective consumption of microalgae was recorded at 6% of feeding ration across all studied regimes (data not shown).

As regards the amount (post-ingestion) of consumed food (generally after pseudo-faeces lifting),we observed the significant effect of both diet and ration on food ingestion for clams during conditioning process (Table 1). For instance, for monospecific diets, the food ingestion is highly influenced by the quality of microalgae ($F= 126.7$, $P< 0.001$) rather than food availability ($F= 44.6$, $P< 0.001$). Whereas, for bispecific diets, the ingestion is extremely influenced by both ration and diet respectively ($F= 579.7$, $P< 0.001$ and $F= 454.2$, $P< 0.001$). Moreover, the ingestion of clams fed with trispecific diet have been also highly influenced by the ration ($F= 557.8$, $P< 0.001$). In view of the influence of feeding patterns (diet and ration) on clam ingestion, the highest effect was observed across all rations ($F= 298.06$, $P< 0.001$) followed by the effect of diets ($F= 190.53$, $P< 0.001$) and the interaction of diet and ration ($F= 16.44$, $P< 0.001$).

Table 1. Two-way ANOVA of the effect of different regimes and feeding rations on the ingestion. Diet-A: Monospecific diets, Diet-B: Bispecific diets, Diet-C: Trispecific diet, Sum Sq: Sum of square, Df: degrees of freedom, F: value of Fisher test, Pr: probability of error, ***: highly significant at 1 ‰)

		Sum Sq	Df	F value	Pr(>F)
Diet-A	Diet	0.316	2	126.75	$2.2\ 10^{-16}$***
	Ration	0.334	6	44.61	$2.2\ 10^{-16}$***
	Diet-Ration	0.246	12	16.44	3.5410^{-12}***
Diets-B	Diet	0.152	2	454.29	$2.2\ 10^{-16}$***
	Ration	0.585	6	579.78	$2.2\ 10^{-16}$***
	Diet-Ration	0.049	12	24.45	$3.899\ 10^{-15}$***
Diet-C	Ration	0.255	6	557.87	$7.772\ 10^{-16}$ ***
Total	Across diets	0.706	6	190.53	$2.2\ 10^{-16}$ ***
	Across rations	1.105	6	298.06	$2.2\ 10^{-16}$ ***
	Diets-Rations	0.365	36	16.44	$2.2\ 10^{-16}$ ***

3.2 Absorption efficiency and Absorption rate

For monospecific diets, in terms of the evolution of the absorption efficiency (Fig. 2A), clams rise their AE while increasing food availability until ration 3% (54 %, and 83 % for Diet-A2 and Diet-A3, respectively) and until ration 5% (80 % for Diet-A1). In the

https://doi.org/10.1051/e3sconf/202233703002

meanwhile, within monospecific diets, clams fed on Diet-A3 have shown high performance in terms of AE comparing to other diets during the first feeding ration.

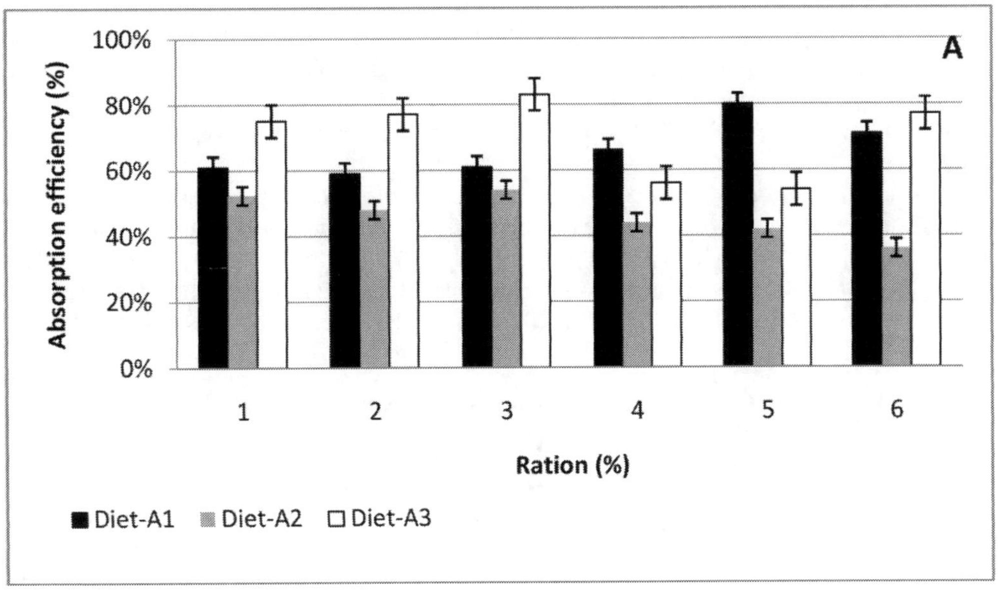

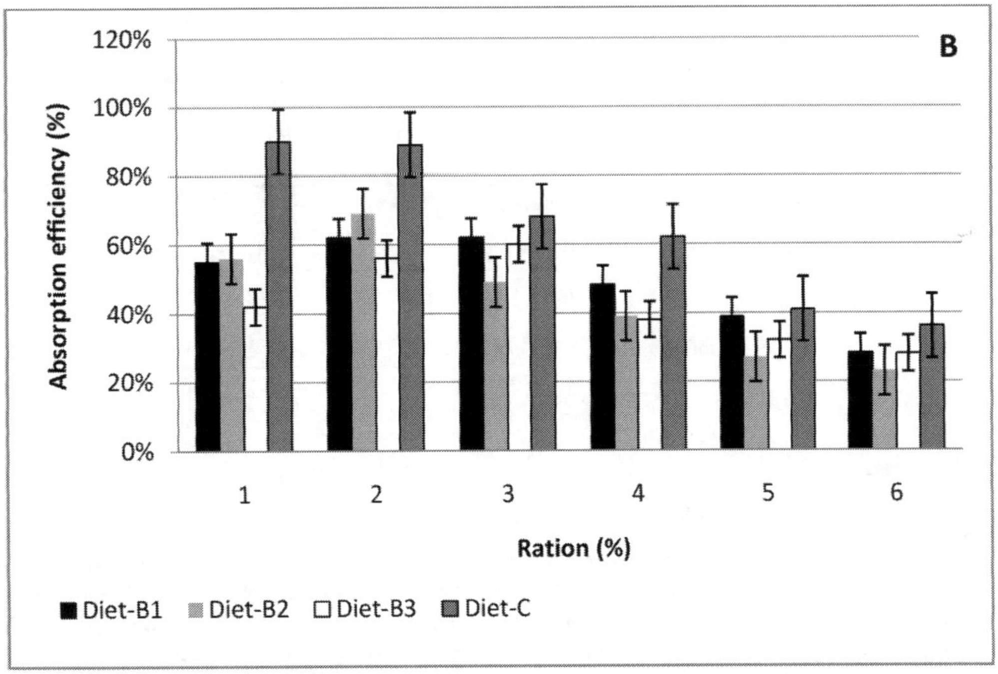

Figure 2. Absorption efficiency (%) of clam *R. decussatus* exposed to different experimental microalgae diets (**(A)**: monospecific diets Diet-A1 (100% *P. lutheri*), Diet-A2 (100% *C. calcitrans*) and Diet-A3 (100% *T. suecica*); **(B)**: Multispecific diets (Bispecific diets: Diet-B1 (65% *P. lutheri* + 35% *C. calcitrans*), Diet-B2 (80% *P. lutheri* + 20% *T. suecica*), Diet-B3 (75% *T. suecica* + 25% *C. calcitrans*) and Trispecific diet: Diet-C (50% *P. lutheri* + 25% *T. suecica* + 25% *C. calcitrans*)) and food ration (%) during broodstock conditioning, (Vertical bars indicate SD)

For multispecific diets (Fig. 2B), we observed that clams rise their AE depending on food availability until a certain limit where this efficiency decrease progressively even though food ration increases. The highest value of AE (90% and 89%) was registered for clams fed on trispecific diets (Diet-C) especially at low food availability (1% and 2% respectively) then decreases remarkably starting from the ration 4%.

According to two-way ANOVA analysis (Table 2), the absorption rate is much more influenced by the regime quality for clams fed with a combination of two microalgae species (bispecific diets) (F= 250.1, df=2, P< 0.001) than clams fed with monospecific diets (F= 21.7, df=2, $P<0.001$). As regards feeding ration, the absorption was also significantly influenced for both clams fed with bispecific and monospecific diets respectively (F= 16.63, $P< 0.001$ and F= 11.19, $P< 0.001$). Whereas, no significant difference has been observed in terms of the influence of the availability of food on the absorption rate for clams fed with trispecific diet (F= 2.7, $P= 0.0534$). In view of the effect of all diets and rations on the absorption rate of clams, the highest influence was observed across all diets (F= 44.10, $P< 0.001$) followed by the effect of rations (F= 17.19, $P< 0.001$) and the interaction of diet and ration (F= 5.45, $P< 0.001$).

Table 2. Two-way ANOVA of the effect of different regimes and feeding rations on the absorption rate. Diet-A: Monospecific diets, Diet-B: Bispecific diets, Diet-C: Trispecific diet, Sum Sq: Sum of square, Df: degrees of freedom, F: value of Fisher test, Pr: probability of error, **: significant at 1%; ***: highly significant at 1 ‰, ^{ns}: not significant)

		Sum Sq	Df	F value	Pr(>F)
Diet-A	Diet	0.006	2	21.78	$3.2\ 10^{-7}$ ***
	Ration	0.010	6	11.19	$1.9\ 10^{-7}$ ***
	Diet-Ration	0.006	12	3.59	0.001**
Diets-B	Diet	0.010	2	250.09	$< 2.2\ 10^{-16}$ ***
	Ration	0.002	6	16.63	$<1.03\ 10^{-9}$ ***
	Diet-Ration	0.003	12	13.65	$7.09\ 10^{-11}$***
Diet-C	Ration	$3.02\ 10^{-5}$	6	2.78	0.0534^{ns}
Total	Across diets	0.019	6	44.10	$< 2.2\ 10^{-16}$ ***
	Across rations	0.007	6	17.19	$1.73\ 10^{-13}$***
	Diets-Rations	0.014	36	5.45	$1.22\ 10^{-11}$ ***

4 Discussion

Different responses in terms of nutritional behavior of *R. Decussates* broodstock have been observed as regards food availability and diets quality during this study. Our results have shown that feeding based on monospecific diet Diet-A2 (*C. calcitrans*) have the best ingestion rate recorded across this experience, flowed by Diet-A3 (*T. suecica*) at only the first three rations (moderated food availability). The highest observed IR of these two strains can be explained by their ease ingestion by bivalves [42, 22]. Beyond the moderated ration, the IR decreases systematically for all diets even though consumption increases (data not shown), which can be explained by the saturation of food-transport mechanism [43]. Inversely, other studies have found that the ingestion capacity is saturated and a functional response occurs decreasing the filtration rates (which is not congruent with our results) in order to keep the IR at maximum level [32, 44, 45, 46, 47, 48, 49, 21]. In parallel, sometimes ingestion can also be regulated by increasing production of pseudo-faeces and remaining the IR constant [45, 50]. In our case, neither regulation nor constant ingestion rate has been recorded. Instead, our results are consistent with [43] where the IR increases when food availability is moderated and decreases at saturated feeding rations [43]. According to [51], maximum feeding rates in *Venerupis pullastra* depends on the natural habitat of the species where particle concentrations are normally low which is congruent with our study at starved situation (100 % of IR at 0% ration: natural sea water).

As regards multispecific diets, the best ingestion rate was generated with bispecific Diet-B3 (*C. calcitrans* and *T. suecica*) during moderated food availability including a specific leaning toward *C. calcitrans* across all rations. However, such result can explain the early maturity of the *R. Decussates* broodstock during conditioning process using *C. calcitrans* alone or mixed with *T. Suecica* [52].Taking into consideration the IR and absorption are the main physiological processes governing spat growth, [53] reports that the combination of *C. calcitrans and P. lutherie* also generates significant filtration and good growth during *Crassostrea gigas* larvae rearing. The maximum ingestion recorded for *C. calcitrans* even under multi-specific diets shows a certain type of specific preference towards this microalgae that might be referred to its high carbohydrate content [52].

Generally, bivalves digest and absorb the ingested microalgae with different efficiencies [54]. At very low food availability, the ingested food is directly digested in the digestive gland with high efficiency and the remaining unassimilated matter is rejected as glandular faeces [46]. Whereas, at high concentrations of particles (cells of microalgae), the absorption efficiency was very low. This could surely due to the formation of pseudo-faeces which generates difficulties in separating normal microalgae cell from pseudo-faeces ones [43]. However, as the seston concentration increases, the excess of phytoplankton entering the stomach bypasses the digestive gland and is rejected undigested as intestinal faeces [55]. This inverse correlation between the ingestion rate and absorption efficiency has been previously described in other species of mollusks [20, 21, 47, 56, 57] and even in fishes [58].

Nevertheless, the extremely low or negative absorption efficiencies might be due to unsuitable, indigestible food, abnormal physiological state of the bivalve or to experimental errors in the measurement of the faeces [20]. Moreover, several difficulties in separating faeces from pseudo-faeces could result in an underestimating of the absorption efficiency calculated using the Conover method. On the other hand, the production of pseudo-faeces means an overestimation of the ingestion rate [43].

Differences in degree of regulation of ingestion and absorption were also achieved herein in terms of food quality (dietary composition). This observation has been previously described by several authors [45, 59] indicating a preferential ingestion of the algal component in the mixture diet. At a certain pseudo-faeces threshold, the absorption efficiency increases with increasing food quality. This interactive effect between food quality and availability has been also reported for *Mytilus edulis* [60] and *M. galloprovincialis* [61]. This mechanism has been explained by [59], by the fact that at high particle concentrations, pseudo-faecal production becomes a significant component of total bio-deposition as quality of food decreases and pre-ingestive selection results in an enrichment of ingested ration; consequently, "true" absorption efficiency tends to increase with increasing filtration in diets of low organic content. So, when feeding on diets of high organic content, bivalves do not achieve a significant enrichment of food upon ingestion and, for rising particle concentration, a reduction in absorption efficiency would take place in spite of partial regulation of ingestion rate. Resulting effects of food availability on absorption are thus minimal. To summarize, feeding behavior of clams (as cockles) exposed to varying concentrations of particles of low food value which suggest that, in turbid environments, bivalves maximize net energy gain by maintaining high selection efficiencies and high rates of pseudo-faeces production. In conclusion, the degree in which pseudo-faeces formation contributes to regulate ingestion varies according to food availability and quality.

5 Conclusion

In conclusion, many uncertainties over the mechanism of seston capture or the role of physiological process, still exist concerning fundamental traits of feeding behavior in suspension feeding bivalves. Therefore, from a shellfish standpoint, a moderated availability of food (generally 3-4 % DW daily), is the target food ration to maximize biomass increase and minimize feeding loss. In addition, this relatively moderated food level was also the optimum ration, since it corresponded to the maximum efficiency of conversion of food into biomass. In the industrial scale, bivalve seed production should attempt to combine both economic and physiological criteria in order to guaranty high growth rate.

References

1. A. Parache, La pêche Maritime, **1254**, 496-507(1982)
2. P. E. Lubet, Haliotis, 14:49-68 (1984)
3. J. M. Poutiers, Bivalves (Acephales, Lamellibranches, pélécypodes). In: Fischer, W., M. L. Bauchot et M. Schneider (ed.), Fiches FAO 'identification des espèces pour les besoins de la pêche. (Révision 1). Volume I (Végétaux et Invertébrés) FAO, pp: 371 – 512 (1987)
4. M. Roméo, M. M. Gnassia-Barelli, Camp. Biochem. Physiol.,**111** C (3),457-463(1995)
5. M. Puigcerver, Aquaculture Research,**27**,399–404 (1996)
6. A. Gharbi,K. Said, A.V. Wormhoudt,J. Shellfish Res**34**, 791–801 (2015)
7. F. Garcia, Oceanologica Acta. **16,** 199-203 (1993)
8. J. Aníbal, E. Esteves, C. Rocha, J. Shellfish Res. **30**(1), 17-23 (2011)
9. R. Moreira, P. Balseiro, A. Romero, S. Dios, D. Posada, B. Novoa, A. Figueras, Dev Comp Immunol **36**,140–149 (2012)
10. A. Derbali, A. Hadj Taieb, W. Kammoun, J. Gouirah, A. Wannes-Ghorbel, N. Zamouri-Langar, M. Ghorbel, O. Jarboui, J. Mar. Biol. Assoc. U. K., **96**(1), 177-184 (2016)
11. S. Chora, M. Starita-Geribaldi, J. M. Guigonis, M. Samson, M. Roméo, M.J. Bebianno., Aquat. Toxicol., **94**, 300-308 (2009)
12. C. Velez, E. Figueira, A. Soares, R. Freitas, Estuar. Coast. Shelf Sci., **155**, 114-125. (2015a)
13. C. Velez, P. Galvão, R. Longo, O. Malm, A.M.V.M Soares, E. Figueira, R. Freitas, Environ. Sci. Pollut. Res., 1-15 (2015b)
14. H. Chalghmi, I. Zrafi, P. I. Gourves, J. P. Bourdineauda, D. Saidane-Mosbahib, Environ. Sci.: Processes Impacts, **18**, 895-907 (2016)
15. M. Delgad, A.Perez-Camacho, Bol. Inst. Esp. Oceanogr. **18** (1-4): 293-300 (2002) (In Spanish)
16. M. Delgado, A.Perez-Camacho, Sci. Mar, **71**, 471-484 (2007)

17. E. M. Gosling, Oxford, OX2OEL, Fishing News Books: 443pp. (2003)
18. J. Ojea, A. J. Pazos, D. Martínez, S. Novoa, P. García-Martínez, J. L. Sánchez, M. Abad, J Shellfish Res **27**(5):1093–1100 (2008)
19. D. Matias, S. Joaquim, A.M. Matias, A. Leitão, Invertebr Reprod Dev **60**(1):49–58 (2016)
20. R. Beiras, A. Perez-Camacho, M. Albentosa, Aquaculture **116**(4):353–365 (1993)
21. R. Beiras, A. Pérez-Camacho, Mar. Biol. **120**, 427–435 (1994)
22. M. Albentosa, A. Pérez-Camacho, U. Labarta, M.J. Fernández-Reiriz, Aquaculture **148**(1):11–23 (1996c)
23. M. S. Shafee, M. Daoudi, Aquac. Res., **22**, 203-216 (1991)
24. A. Kamara, N. Rharbi, A. Berraho, M. Ramdani, Mar. Life. **15** (1-2), 43-50 (2005)
25. A. Kamara, N. Rharbi, M. Ramdani, A. Berraho, Bull. Soc. Zool. France, **133**, 189–198 (2008)(In French).
26. N. Caill-Milly, N. Bru, M. Barranger, L. Gallon, F. D'amico, J. Shellfish Res, **33**, 355–372 (2014)
27. Z. Amane, L. Tazi, M. Idhalla, M. Chlaida, AACL Bioflux**12** (5), 1623-1634 (2019)
28. S. D. Utting, P. F. Millican, Aquaculture. 155, 45–54 (1997)
29. F. da Costa, A. Cerviño-Otero, Ó. Iglesias, et al. Aquacult Int **28**, 1675–1708 (2020)
30. Y. Ouagajjou, T. El Aloua, M. El Moussaoui, E. M. Ait Chattou, A. Aghzar, Y. Saoud, AACL Bioflux **13**(6):3664-3673 (2020)
31. P. Coutteau, P. Sorgeloos, J Shellfish Res **11**, 467–476 (1992)
32. P. R. Walne, J. Mar. Biol Assoc. U.K. **52**, 345–374 (1972)
33. C. E. Epifanio, Aquaculture, **18**, 1 – 12 (1979)
34. CT. Enright, G. F. Newkirk, J.S. Craigie, J. D. Castell, J. Exp. Mar. Biol. Ecol., **96**, l-13 (1986)
35. 1. Laing, P. F. Millican, Aquaculture, **54**, 245-262 (1986)
36. A. Aghzar, M. Minambres, P. Alvarez, P. Presa, Thalassas, **29**,9–16 (2013)
37. J. I. P. Iglesias, M .B. Urrutia, E. Navarro, I. Ibarrola,J. Exp. Mar. Biol. Ecol. **219**, 71–86 (1998)
38. I. Martínez-Pita, C. Sánchez-Lazo, F.J. García, Aquacult. Nutr., **22**, 202-216 (2016)
39. R. J. Conover, Limnol Oceanogr**2**,338-354 (1966)
40. J. Fox, M.Bouchet-Valat, Rcmdr. R Commander. R package version **2**, 3-1 (2016).
41. R Development Core Team. R: A language and environment for statistical computing. R Foundation for Statistical Computing, Vienna, Austria, ISBN 3-900051-07-0, UR (2008)
42. M. RBrown,. Nutritional value of microalgae for aquaculture. In: Cruz-Suárez LE, Ricque-Marie D, Tapia-Salazar M, Gaxiola-Cortés MG, Simoes N (eds) Advances en Nutrición Acuícola VI. Memorias del VI Simposium Internacional de Nutrición Acuícola, 3–6 Septiembre 2002. Cancún, Quintana Roo, México (2002)
43. M. Albentosa, A. Pérez-Camacho, R. Beiras,AquacNutr**2**(4):213– 220 (1996b)
44. E. H. Schulte, Mar Biol**30**, 331-341 (1975)
45. R. L. Foster-Smith. J. exp. Mar Biol. Ecol. **17**, 1-22 (1975)
46. J. Widdows, R. Fieth, C.M. Worrall, **50**,195 –207 (1979)

47. J. M. Navarro, J. E. Winter. Mar Biol, **67**, 255-266 (1982)
48. D. Gerdes, Aquaculture **31**, 195- 219 (1983)
49. V.M. Bricelj, R.E. Malouf,Mar Biol**84**, 155 – 165 (1984)
50. E. Navarro, J. I. P. Iglesias, M.M. Ortega. J exp mar Biol Ecol, **156**, 69-87 (1992)
51. K. R. Tenore, W. M. Dunstan, J. Exp. Mar. Biol.(1973)
52. A. S. Abbas, E. El-Wazzan, A. R. Khafage, A. F. M. El-Sayed, F. A. A. Razek. Aquacult Int **26**, 1297–1309 (2018)
53. E. Ponis, G. Parisi, G. Chini Zittelli, F. Lavista, R. Robert, MR. Tredici. Aquaculture, **282**, 97-103 (2008)
54. B. L. Bayne, R.C. Newell,. Physiological energetics of marine molluscs. In: The Mollusca. Vol 4. Physiology, Part I (Saleuddin, A.S.M. & Wilbur, K.M., eds), pp. 407-515. Academic Press, New York(1983)
55. P. B. Van Wee1,**1**, 245-252(1961)
56. P. Calow, Oecologia, **20**, 33-49 (1975a)
57. J. Widdows, B.L. Bayne, D. R. Livingstone, R. I. E. Newell, P.Donkin, Comp. Biochem. Physiol. (1978a)
58. J.M. Elliott, J. Anim. Ecol.(1976)
59. J. I. P. Iglesias et al., J. Exp. Mar. Biol. Ecol. (1992)
60. B. L. Bayne, A.J.S. Hawkins, E.Navarro,**111**,1-22 (1987)
61. E. Navarro, J. I. P. Iglesias, A. Perez Camacho, U. Labarta, R. Beiras, Aquaculture, **94**, 197-212 (1991)

Ecological risk analysis in marine fish farming: a case study of a seabass (*Dicentrarchus labrax*) farm located in Moroccan Mediterranean coast

Hassan NHHALA [1,4,*], *Abdeljallil* BAHIDA [4], *Imane* NHHALA [4], *Housni* CHADLI [5], *Azeddine* ABREHOUCH [1], *Benyounes* ABDELLAOUI [3], *Mohamed* ID HALLA [2] and *Hassan* ER-RAIOUI [4]

[1]National Institute of Fisheries Research (INRH), Aquaculture Centre, M'diq, Morocco (Former researcher).
[2]National Institute of Fisheries Research (INRH), Department of Aquaculture, Casablanca, Morocco.
[3]National Institute of Fisheries Research (INRH), Regional Centre, Tangier, Morocco.
[4]Faculty of Sciences and Technics, University of Abdelmalek Essaadi, Tangier, Morocco.
[5]Marine Fish Farm « Aquam'diq », M'diq, Morocco.

Abstract. The present study aimed at studying ecological interaction between a fish farm cages located in M'diq bay, in western Moroccan Mediterranean coast, and its local environment. It was carried out in 2013 and 2016 through seabed analysis in terms of organic matter and some nutrients parameters such as Nitrogen and Phosphorus, as well as through monitoring macro-fauna assemblages. The results showed recorded values relatively stable in three years' interval and have a well-integrated fish farming activity within its ecosystem. Sediment was sandy to sandy shelly types and macro-benthic fauna were composed of six groups, Gastropods, Bivalves, Scaphopoda, Amphipods, Nematoda and Polychaeta. Organic matter and Total Organic Carbon were respectively less than 3.2% and 1.87% and very low comparing with others Mediterranean fish farms sites. Estimation of Nitrogen, Phosphorus and Carbon potential releases of M'diq fish farm gave low data related to its production capacity not exceeding 200 tons by year. Influenced with open sea location and high current state, dispersion pattern of these fish farm discharges seemed to play a key role in keeping lower values. ANOVA analysis showed no significant difference between studied parameters in the five sampling stations comparing with the reference station both on geographical and temporal levels (seasons and years).

1 Introduction

The development of aquaculture production is related to consumers' perception, knowing that aquaculture products are registering a higher growth ratio, almost 5.41% per year (3.02 million tons/year) [1]. Consumers' perception is depending on many criteria [2, 3] among

*Corresponding author: nhhalahassan@yahoo.fr

which the durability of aquaculture, relating to potential massive impacts on the environment, is the dominate negative perception of coastal community. As [4] reported, fish farming was, and still is, viewed as a potential polluter of the marine environment.

In marine fish farming, the main environmental impact is widely thought to be coming from organic matter discharges, resulted from cultured species metabolism (faeces, pseudo-faeces, excretion…) and uneaten food (likely generated by excessive distribution of food or driven away by current). The abundance of organic matter discharges may create a change in chemical and physical characteristics of the environment and may have a potential effect on the biological chain composition of the receiving environment. Then, seabed could represent a memory reflecting the effect of anthropic activities on the environment. Seabed sediment is characterised by variable granulometry and physico-chemistry patterns depending on geomorphological and hydrodynamic conditions and by holding a multitude of benthic species. For these reasons, it could be considered as a mirror providing information on the local environment health state. Therefore, impacts are clearer on the seabed sediment composition [5-13].

In Morocco, even if marine aquaculture activity has started in fifties, mainly with marine shellfish culture, namely in Oualidia lagoon (in Atlantic coast), marine fish farming has only begun in 1986, in Nador lagoon situated in Mediterranean coast [14]. The first established fish farm in this lagoon was producing two fish species (Gilthead seabream, *Sparus aurata*, and European seabass, *Dicentrarchus labrax*). Since then, three others marine fish farms have been also established in Moroccan Mediterranean coast, one in 1991 in its Eastern part and the two others in 1998, in its Western part. They have been all growing seabream and seabass and selling them in the European market.

Moroccan fish farms were not so competitive than European ones; seabass and seabream produced by European farms were sold at low prices [15] in comparison with the ones produced by Moroccan farms. Due to difficulties in selling their cultured fishes in local markets, Moroccan farms have faced a critical situation leading to the activity cessation of three of the fourth existing marine fish farms.

Since then, only one marine fish farm is still operating in Morocco. It is located in its Western part of its Mediterranean coast, in M'diq Bay. This farm has been well adapted to the national context market through creating its own cultured fish marketing network. Moreover, it has been concerned by its environmental farm consideration to be able to sustain its activity. Indeed, Moroccan legislation has settled up, in 2003, measures for environmental aquaculture farms monitoring based on results of the environmental impact study. The implementation of this monitoring has become mandatory. The purpose is to ensure sustainability and proper local integration of aquaculture activity with respect to local environment conditions.

The present study was conducted in this context, in join cooperation with the M'diq fish farm, the Aquaculture Centre of the National Institute of Fisheries Research (INRH) in M'diq and the Scientific and Technical Faculty of Abdelmalek Essaâdi University in Tangier. Knowing that aquaculture activities are depending on its local environment conditions to develop and persevere, the present work focuses on some sensitive parameters which are used in ecological approach to assess aquaculture impacts. So, the main objective is to characterise the current situation of the interaction between the said fish farm and its culture site environment and to evaluate ecological risks associated with the fish farming activity through some pertinent geochemical and biological parameters (nitrogen, phosphorus, total organic carbon and macro-benthos), providing information on potential impacts on local and surrounding area, particularly on sediment and benthos.

2 Materials and methods

2.1 Study area

The study area is located in M'diq bay, in the north-western part of Morocco, particularly in western part of its Mediterranean coast (Figure 1). It is situated between Sebta Cap in the north (35° 54' N, 5° 17'10" W) and Negro Cap in the south (35° 40' N, 5° 16'40" W). It has 23 km in length and 130 km² in surface area.

This bay is bounded by Rif mountain chain in the west and opened towards to the East. The hinterland is represented by a large watershed, situated between the sea (the bay) and the mountain chain, generating during the winter a great flow and runoff of the rainy waters, led by two main rivers (Smir river and Negro river) [16, 17].

The hydrodynamic characteristics of this bay can be summarized as follows:

- The tide is semi-diurnal;
- The mean tidal range at spring tide is of 0.8 to 1m [18];
- The swell direction is from East to North-East, generated by local winds;
- The coastal current is dominant and bears Southeast and the offshore current bears East. The tidal current direction is northwest at the swell tide and south-east at the ebb tide [19].

The fish farm culture facility is composed of 14 circular Polyethylene high-density floating cages. They have each 12 m in diameter and 10 m in net depth (with approximately 1,100 m3 in rearing volume). Cages are set on a bathymetry range of 18 and 22 m. Farming site extends over an area of 5.2 ha, located 2.3 km east of the M'diq beach (35°41'16.3"N, 5°19'23.1"W). The annual fish capacity production of this farm is reaching approximatively 200 tons. The cultured fish is seabass and it is fed manually every day on conventional extruded commercial diets.

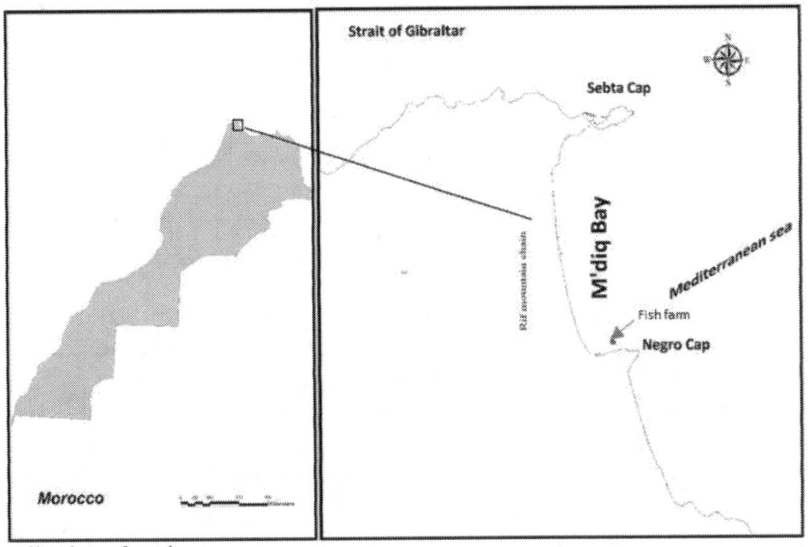

Fig. 1. Localisation of study area.

2.2 Water column investigation

Main oceanographic parameters investigated in fish farm water column are dissolved oxygen, pH, temperature and salinity. They were measured in-situ (field data) using a portable multi-parameter (ELMETRON, CX-401; WTW, Multi 340i). Remote sensing data (satellite data) was also collected by using satellite imagery. A statistical correlation test was performed on field data and satellite data and it has shown an important correlation coefficient (R=0.76) between them. Nitrate (NO_3) and Phosphate (PO_4) were measured through seawater sampling which were first frozen and then analysed later according to the experimental protocols described by [20] and compared with their satellite data. Therefore, and basing on the important correlation coefficient field data and satellite data, only the latter were employed in this study because of their availability for all the year round. Chlorophyll-a data was only collected using Remote sensing for each station position. Field data were measured during six times in 2013 and same in 2016 while satellite data were collected once a week in each month along each year of 2013 and 2016.

2.3 Sediment sampling strategy

Sediment sampling (Table 1) was performed in 2013 and 2016 based on the following strategy. Sampling stations number has been set at six falling into three following categories: one station located in the central point of the cages (Central Station (S-C)), four stations located each in of one of the fourth edges of the farm's site (designated as S-1, S-2, S-3 and S-4) and the last one is located outside of the farm's site at a distance of about 500 m from its limit and considered as a reference Station (S-R) (Figure 2). The latter station was selected in an offshore area, within a depth and bottom type comparable to the two site's farm edges stations oriented towards open sea (S-2 and S-3). In similar study, particularly in Turkey [21], Greece [22] and Italy [23], reference stations were respectively set at 500-700 m and 300 m and 900 m away from fish farming farms locations.

Table 1. Geographical coordinates and depths of sampling stations.

Sampling Station	Depth (m)	Latitude (N)	Longitude (E)
S-1	19.2	35.690314	-5.296541
S-2	21.7	35.691611	-5.297545
S-3	21.2	35.691228	-5.298485
S-4	18.8	35.689961	-5.297382
S-C	20.8	35.690811	-5.297404
S-R	23.5	35.695463	-5.295067

Sampling operations are each based on three replicates taken from each sampling station. They were carried out with a small and a little bit heavier Van Veen grab (fitted with weighting blocks). Its sampling surface area is of 250 cm^2. The mode of its operation to retrieve sediment samples is very easy. It is composed of two jaws which are kept opened by a hook, just before sampling. Then, the grab is steadily and carefully sunk in the sea. Both jaws are fitted with holes allowing air to escape during the sinking. As soon as the jaws touch the bottom, the hook loosens its grip and when hoisting the rope, the jaws shut tight because of the leverage by the rods. The drawn sample amount depends mainly on the nature and the compactness of the sea bottom. However, this heavier grab catches more sediment than another lighter one that has been locally tested. Once surfaced, the grab is emptied and cleaned before being reused.

Moreover, in spite of the heavy closing force, it sometimes happened that a pebble was caught between the buckets. In such case, the sample was not considered because it could not be representative; the smaller parts of sediment might have been lost during hoisting. Then, the sampling operation is systematically repeated.

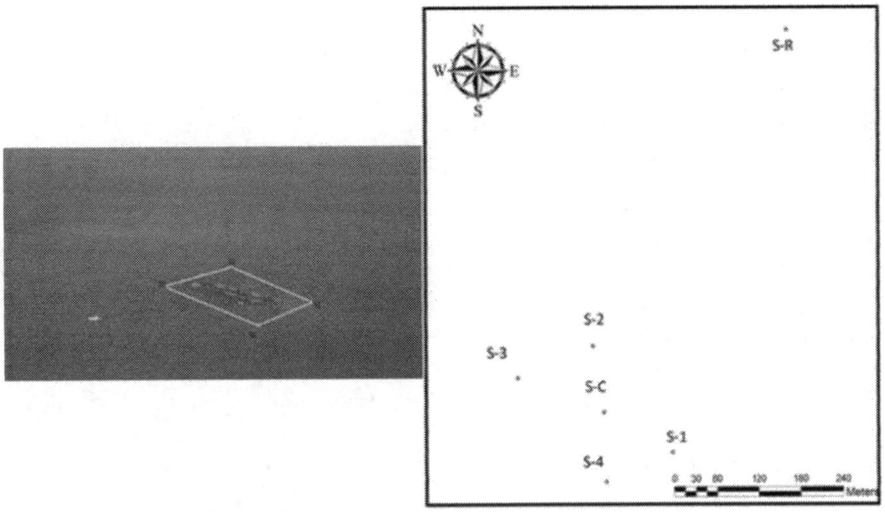

Fig. 2. Localisation of sampling position.

2.4 Sediment grains sizes

Granulometric sediment study was carried out on the six sampling stations. Granulometric characterization was performed applying the international classification used in geology for which Bellair and Pomerol have proposed a simplified terminology [24]. This classification provides three categories of particles according to their size: Rudite "gravels" (< 2mm), Arenite "Sand" (between 2 and 0.063mm) and Lutite "silts and clays" (<0.063mm).

2.5 Organic Matter (OM) and Total Organic Carbon (TOC) in sediment

Two different methodologies of assessing TOC in sediment were used both in 2013 and 2016. The first method dealt with Organic matter measuring using the ignition loss method (at 475-500°C). Samples were first sieved to remove particles sizes over 2 mm and then, the remaining grains sizes, lower than 2 mm, were subjected to calcination in an electric furnace in order to create optimal condition for total oxidation of organic matter, destruction of carbonates and elimination of a maximum of inter-foliar water and water of the clay borders. Organic matter is further expressed as a percentage of the sample mass during the first heating at 220°C (M_1) and the second heating at 450°C (M_2) and the initial weight of the crucibles (M_0):

$$OM\ (\%) = (M_1 - M_2) / (M_1 - M_0) \times 100 \quad (1)$$

The second method was based on Walkley and Black [25] wet oxidation method. Organic matter was determined in sediment as follows: first allows to evaluate the TOC and then calculate OM percentage using the following approximation formula:

$$\%OM = \%C \times 1.72 \quad (2)$$

2.6 Sediment Nitrogen and Phosphorus

Nitrogen and Phosphorus measurements in sediment were also both subjected to two different analysis methods in 2013 and 2016. Total Phosphorus was measured by the Dabinmodified Method [26] and by the scanning electron microscopy analysis.

Total Nitrogen was performed by Colorimetric Methods and also by the scanning electron microscopy analysis.

2.7 Potential chemical elements release

In order to have, on an indicative basis, an idea about the potential Carbon, Nitrogen and Phosphorus fluxes produced by the cultured fish stock in the studied farm, an estimation of total discharges of these chemical elements (resulted from mineralization of uneaten feed and cultured fish faeces) have been made, based on results reported by [27]. These authors reported that 1 ton of produced seabass can generate 45 kg of carbon, 8 kg of Nitrogen and 0.3 kg of Phosphorus, released as faeces and 284 kg of Carbon, 45 kg of Nitrogen and 4 kg of Phosphorus realised as wasted feed.

Therefore, the produced fish biomass and the feeding quantity data were collected from the studied fish farm to estimate its potential Carbon, Nitrogen and Phosphorus fluxes, that could be produced local area in M'diq bay and likely deposed in its site seabed. For this, and taking into consideration local high hydrodynamic conditions (reaching a max of 1.1 m/s) and the seabed slop (about 2.2%), potential chemical elements release have been estimated and expressed according first to the cages area and then to the exploited concession area. These latter have an area surface of 5,600 m^2 and 52,000 m^2 respectively.

2.8 Macro-benthic data

Ecological impacts of M'diq fish farm were assessed through macro-benthic variability assessment in sediment retrieved in beneath and surrounding area of fish cages. Samples of sediment were collected by scuba diving, using a circular piece of PCV pipe of 20 cm in diameter and 10 cm in height. Three replicates were operated in each of the same six sediment sampling stations. Collected sediment samples were first sieved through a 1 mm mesh and retained parts were checked for organisms visible to the naked eye. The smallest ones were kept in a petri dish filled with few seawater. They were examined under a magnifying glass, a profile projector or a stereomicroscope, depending on the organism size. Photos were taken for each kind of collected organisms. They were sorted and identified mainly by major groups. The total individuals number of each group was determined for each station's sample and reported in terms of abundances (Ind.) and percentages (%) per m^2 for each group in each sampling station in 2013 and 2016.

2.9 Statistic data analysis

Results of chemical, physical and biological water parameters obtained for each season and station position were statically analysed using one-way ANOVA analysis of variance. Statistical significant differences ($P<0.05$) between stations position and sampling years were performed following ANOVA Test for organic matter (OM), Total organic Carbone (TOC), Nitrogen (N) and Phosphorus (P). A Correlation Coefficient Test was also applied for parameters water column: Chlorophyll-a (Chl-a), Temperature (T), Nitrate (NO_3), Phosphate (PO_4), potential of Hydrogen (pH), Salinity (S) and dissolved Oxygen (Oxy).

Also, significant changes in macro-benthic community were assessed between sampling stations and sampling years, using one-way ANOVA analysis of variance in terms of each group of organisms and organisms' numbers found in 2013 and 2016.

3 Results

3.1 Seawater column parameters measurements

Biological, chemical and physical seawater parameters were calculated for the two samplings years, 2013 and 2016. Data are presented according to seasons (Table 2). A significant difference ($P<0.001$) was found between seasons while non-significant difference was found between sampling stations (Table 3). This result can be explained by the importance of the local hydrodynamic leading a well renewal and mixing of waters and the great capacity of water flushing of this opened bay allowing dilution and dispersion of fish farm discharges. The local current speed can reach a maximum 1.1 m/s coming from the SW and the NE (source: SIMAR buoy) with a respective frequency of 10% and 8% per year since 1997 until 2016.

Table 4 shows a matrix of correlation coefficients, determining relation between seawater column parameters. Negative correlations were found between Temperature and dissolved Oxygen (R= -0.89), pH (R= -0.90), phosphorus (R= -0.61) and Chl-a (R= -0.65) in 2013 and 2016 samples. In the other hand, an important positive correlation was found between dissolved oxygen and pH (R= 0.92), nitrate and phosphorus (R= 0.89) and phosphorus and Chl-a (R= 0.78).

Table 2. Results of sampling stations parameters (Mean ±standard error, Minimum and Maximum).

	Winter	**Spring**	**Summer**	**Autumn**	**Range (max-min)**
Chl-a (mg/l)	0.569±0.162	0.289±0.138	0.168±0.060	0.406±0.151	0.776-0.115
NO_3 (mg/l)	0.017±0.014	0.004±0.002	0.003±0.001	0.017±0.011	0.049-0.002
PO_4 (mg/l)	0.012±0.003	0.004±0.002	0.002±0.000	0.009±0.004	0.017-0.002
pH	8.097±0.013	8.082±0.025	8.028±0.017	8.072±0.008	8.200-8.000
Oxy (mg/l)	7.836±0.153	7.784±0.232	7.237±0.177	7.540±0.162	8.120-6.930
S (‰)	36.287±0.071	36.152±0.023	36.235±0.042	36.221±0.036	36.400-36.000
T (°C)	16.482±0.730	17.309±1.079	20.373±0.947	18.014±0.727	22.600-15.500

Table 3. Values of one-way analysis of variance for water column measurement.

	Stations			**Seasons**		
	d.f.	F	P Level	d.f.	F	P Level
Chl-a (mg/l)	5	0.0005	Ns	3	87.8712	***
NO_3 (mg/l)	5	0.0001	Ns	3	20.9992	***
PO_4 (mg/l)	5	0.0000	Ns	3	68.9110	***
pH	5	0.0000	Ns	3	165.6104	***
Oxy (mg/l)	5	0.0001	Ns	3	120.2864	***
S (‰)	5	0.0002	Ns	3	76.9153	***
T (°C)	5	0.3532	Ns	3	317.7456	***

* $P<0.05$, ** $P<0.01$, *** $P<0.001$ and ns, non-significant.
d.f.: Degrees of freedom

Table 4. Correlation coefficients matrix between seawater column parameters.

	T	Oxy	pH	S	NO_3	PO_4	Chl-a
T	1.00						
Oxy	-0.89	1.00					
pH	-0.90	0.92	1.00				
S	-0.42	0.32	0.45	1.00			
NO_3	-0.36	0.25	0.30	0.47	1.00		
PO_4	-0.61	0.52	0.57	0.66	0.89	1.00	
Chl-a	-0.65	0.67	0.68	0.58	0.53	0.78	1.00

3.2 Seawater column parameters measurements

Granulometry analysis shows that all sampling stations are characterised by high percentages of the Arenite "Sand" fraction (between 2 and 0.063 mm) with values ranging between 66% and 99% and very low values of Silts and Clays (<63 µm) ranging between 0.42% and 4.32% (Table 5). Moreover, the six sampling stations could be divided in two different groups, one assembling the three stations situated toward offshore (S-R, S-2 and S-3; designated as offshore stations) characterised by high gravels (and shells debris) contents, ranging between 17.35% and 32.34% and the other gathering the three others stations, situated toward the coastal shore, (S-1, S-4 and S-C; designated as coastal stations), characterised by high sand contents ranging between 90.75% and 99.09%.So, the offshore stations have a sandy shelly type sediment while the coastal stations have a sandy type sediment.

Table 5. Particles sizes distribution in the six sampling stations.

Stations	Gravels	Sand	Silts &Clays
	> 2 mm	2 mm - 63 µm	< 63 µm
S-R	32.34	66.50	1.16
S-1	0.16	99.09	0.69
S-2	23.88	75.35	0.76
S-3	17.35	81.82	0.83
S-4	0.34	95.33	4.32
S-C	8.82	90.75	0.42

3.3 Chemical sediment parameters

Results of chemical sediment parameters analysis are shown in Table 6. In 2013, the highest percentage of OM (2.58%) was recorded in station S-3 while in 2016, a slight increase of OM was observed in all stations with a higher percentage recorded in station 4 (3.2%).

Similar situation was recorded for TOC and Phosphorus and to a lesser extent for Nitrogen. In 2016, Station S-4 has relatively higher values in terms of OM, TOC, N and P comparing with the ones of the other five stations. However, statistical analysis, performed one-way ANOVA test (Table 7), showed that all parameters have a non-significate difference between the six stations (S-R, S-1, S-2, S-3, S-4 and S-C) and between the two study years (2013 and 2016).

Table 6. Analysis results of chemical sediment parameters obtained for 2013 and 2016 samples.

Parameters	Year	S-R	S-1	S-2	S-3	S-4	S-C
OM (%)	**2013**	0.90	1.00	2.50	2.58	0.96	1.20
	2016	1.40	1.50	2.90	3.15	3.20	1.70
TOC (%)	**2013**	0.52	0.58	1.45	1.5	0.56	0.7
	2016	0.84	0,86	1.75	1.85	1.87	1.00
Nitrogen (%)	**2013**	0.05	0.10	0.06	0.09	0.03	0.11
	2016	0.00	0.00	0.72	0.42	2.83	0.07
Phosphorus (%)	**2013**	4.0×10^{-3}	3.2×10^{-3}	4.5×10^{-3}	4.9×10^{-3}	6.0×10^{-3}	6.2×10^{-3}
	2016	0.04	0.00	0.06	0.04	0.29	0.5

Table 7. Values of one-way analysis of variance for of chemical sediment parameters.

	Stations			**Seasons**		
	d.f.	F	P Level	d.f.	F	P Level
OM	5	2.1416	ns	1	2.6961	ns
TOC	5	2.1254	ns	1	2.8812	ns
Azote	5	0.8216	ns	1	1.7986	ns
Phosphorus	5	0.7275	ns	1	3.3816	ns

3.4 Potential chemical parameters release

The studied farm is characterized by a monthly evolution in terms of cultured fish biomass, distributed amount of fish feed and site temperature as shown in Figure 3 related to the two sampling years (2013 and 2016).

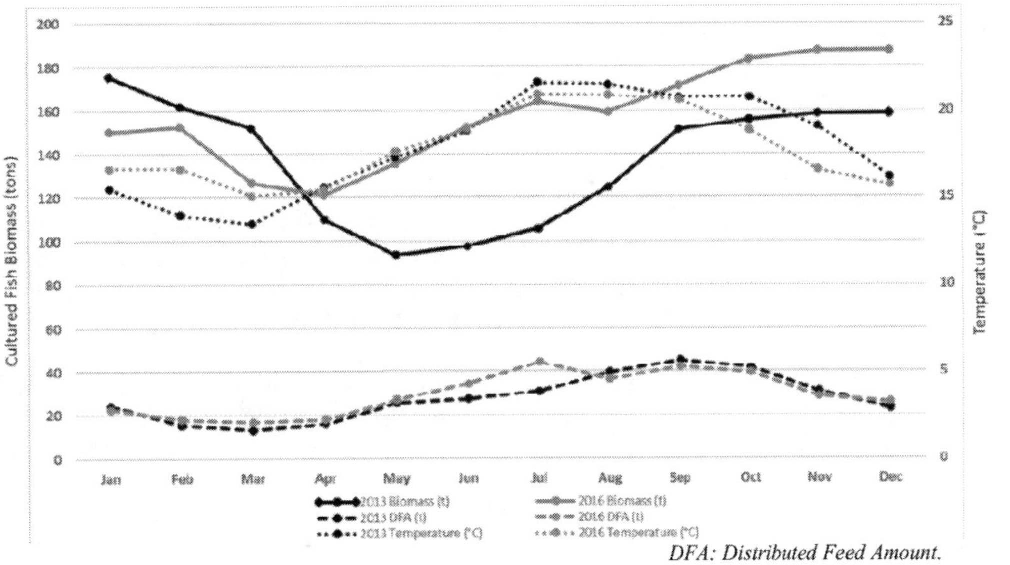

DFA: Distributed Feed Amount.

Fig. 3. Monthly evolution of cultured seabass biomass production, distributed fish feed and site temperature in the two sampling years (2013 and 2016).

The monthly cultured fish biomass differences between 2013 and 2016 resulted mainly from differences in fish sales amounts and in fish biomass growth increase. The decreases in biomass curve (Figure 3) resulted from the balance between the output (fish sales amounts) which was higher than the input (fish biomass growth increase). The annual produced seabass was 137.06 tonnes in 2013 and 157.58 tonnes in 2016 with an annual distributed amount of feed much higher in 2016 than 2013 of 20.64 tonnes. The annual mean temperature was 17.9°C in 2013 and 17.8°C in 2016.

Although these differences, statistical analysis (ANOVA test) showed that there is a non-significant difference in released carbon from faeces and wasted feed both in a cage surface ($P<0.05$, d.f.=1, F=0.1816) and in the concession surface ($P<0.05$; d.f.=1; F=0.2153). The same result is obtained for Nitrogen ($P<0.05$; d.f.=1; F=0.1391) and Phosphorus ($P<0.05$; d.f.=1; F=0.0091).

Table 8. Values of one-way analysis of variance for of chemical sediment parameters.

	Parameter	Year	Mean ($g.m^{-2}d^{-1}$)		Parameter	Year	Mean ($g.m^{-2}d^{-1}$)
Released as faeces in the exploited site area	C	2013	3.240	Released as faeces in the concession circumference surface	C	2013	0.349
		2016	3.442			2016	0.371
	N	2013	0.648		N	2013	0.070
		2016	0.688			2016	0.074
	P	2013	0.024		P	2013	0.003
		2016	0.026			2016	0.003
Released as wasted feed in the exploited site area	C	2013	23.005	Released as wasted food in the concession circumference surface	C	2013	2.477
		2016	24.438			2016	2.632
	N	2013	3.645		N	2013	0.393
		2016	3.872			2016	0.417
	P	2013	0.324		P	2013	0.035
		2016	0.344			2016	0.037
Total chemical elements released from faeces and wasted feed	C	2013	26.245	The sum of the chemical parameters Released from faeces and wasted food	C	2013	2.826
		2016	27.880			2016	3.002
	N	2013	4.293		N	2013	0.462
		2016	4.561			2016	0.491
	P	2013	0.348		P	2013	0.038
		2016	0.370			2016	0.040

3.5 Macro-benthic results

A total of six macro-benthic groups were found (Gastropods, Bivalves, Scaphopoda Amphipods, Nematoda and Polychaeta) in almost all stations. In both sampling years (2013 and 2016), gastropod and polychaeta groups were generally more abundant than the other groups. Among the six macro-benthic groups, Gastropods were more abundant in Station 1 and station 4 while Polychaeta were having a slightly higher in abundance in S-R. In generally speaking, marco-benthic abundance in the different groups ranged between 64 and 350 individuals per m^2 (Ind./m^2) in 2013 and between 32 and 382 ind./m^2 in 2016, with the exception of some macro-benthic groups (Scaphopoda and Bivalves) which are absent in some stations (S-2, S-3 and S-C) (Figure 4). Scaphopoda were absent in Station S-2 in 2013 while Bivalves were absent in S-3 in 2016. They were both absent in S-C as well as in 2013 than in 2016. The total number all groups individuals is ranged between 510 ind./m2 in S-C and 1,019 ind./m^2 in S-4 for 2013 and from 446 ind./m^2 in S-C to 1,019 ind./m^2 in S-

4 for 2016. The specific proportion in macro-benthic groups ranged from 7% to 37% in 2013 while it ranged 4% and 43% in 2016.

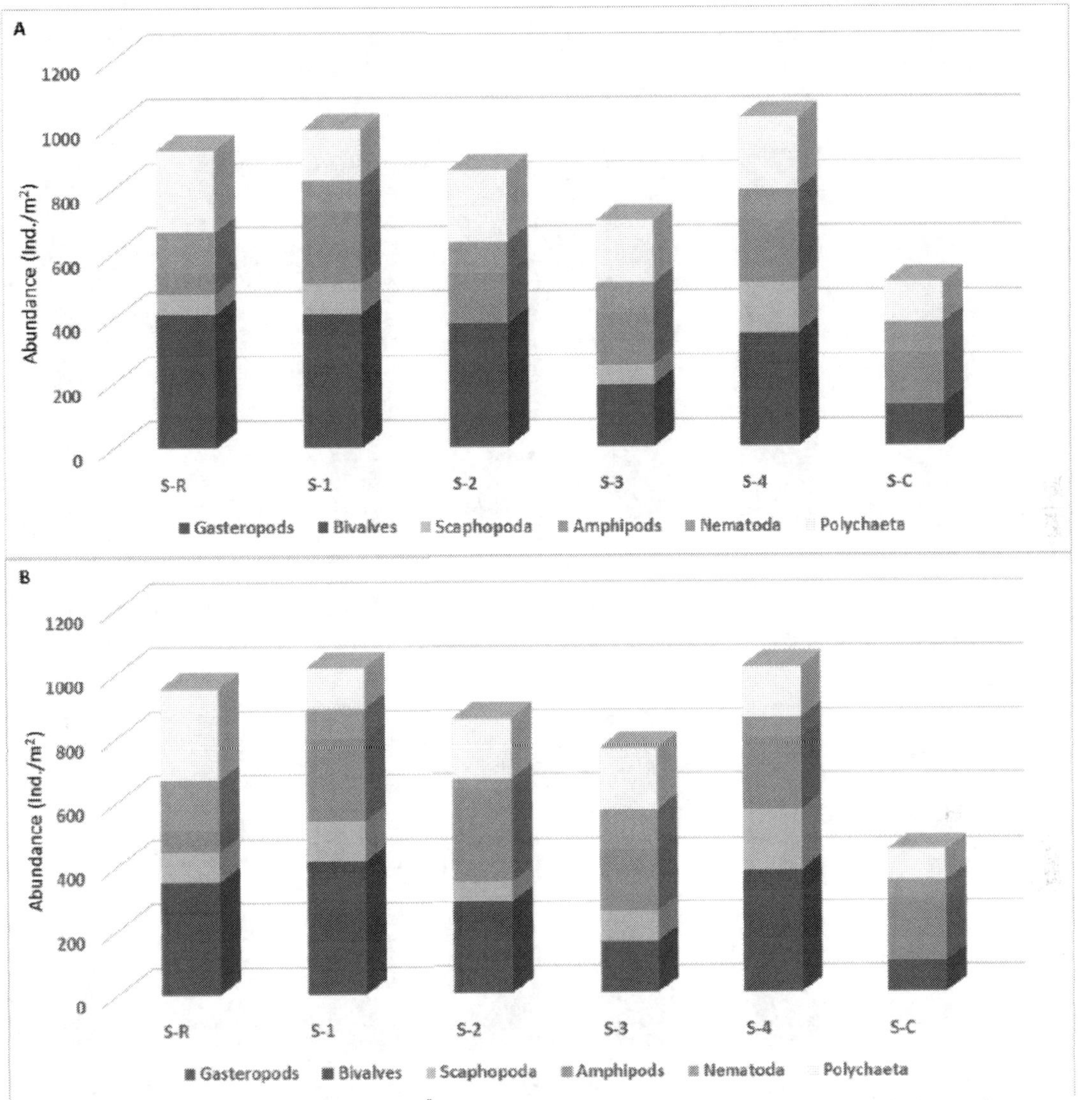

Fig. 4. Abundance values (individuals/m^2) of macro-benthic in sampling stations (with A graph showing 2013 results and B graph showing 2016 results).

Statistical analysis performed with ANOVA test did not shown a significant difference in terms of groups individuals between sampling stations both in 2013 and in 2016; Gastropods ($P<0.05$; d.f.=1; F=0.0085), Bivalves ($P<0.05$; d.f.=1; F=1.2816), Amphipods ($P<0.05$; d.f.=1; F=1.2439), Nematoda ($P<0.05$; d.f.=1; F=0.3014), Polychaeta ($P<0.05$; d.f.=1; F=0.4079). The number of organisms were not significantly different between stations in both years 2013 ($P<0.05$; d.f.=5; F = 0.8366) and 2016 ($P<0.05$; d.f.=5; F=0.8602).

4 Discussion

According to the recorded data of the present study, most recorded results are non-significantly different from reference station in terms of physicochemical parameters and macro-benthic organisms. Moreover, these results seemed to be within the acceptable environment limits and could allow to say that a fish farm located in opened bay characterised by high hydrodynamic is well integrated in its local environmental with no significant impacts.

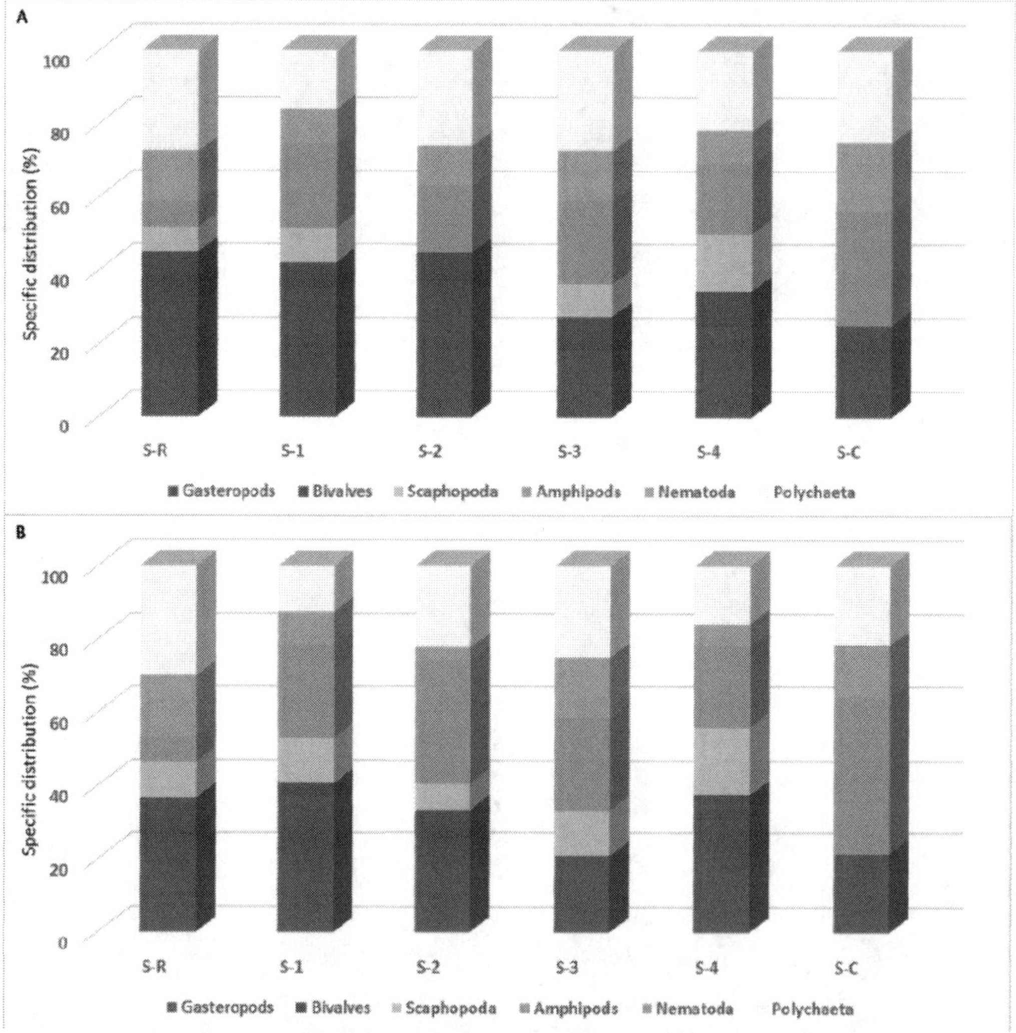

Fig. 5. Specific distribution proportion (%) of macro-benthic in sampling stations (with A graph showing 2013 results and B graph showing 2016 results).

Actually, concerning TOC values, they did not exceed 1.87% in all sampling stations, comparing with results reported by [28], based on data collected from 951 stations in 7 coastal regions in 3 different continents. They reported there was no stress for communities associated with TOC concentrations <1%, while stress was severe at TOC concentrations >3.5%. In our study, all sampling stations, including reference station located at a distance of 500m from farm limit, showed TOC values lower than 2%. These results are also lower than those obtained by some authors for fish farms located in some Mediterranean areas, such as [29] whose obtained TOC values of 4.7% at a distance of 0 m from seabass and seabream breeding cages, which decreased to 3.14% and 3.01% moving away from the farm respectively at 5m and 25m.

Furthermore, in Jinhae Bay (in Korea), [30] reported that, for a mono-aquaculture of Pacific oysters, TOC percentages were ranged between 1.97% and 4.15%, while [31] reported that, in six rearing seabream and seabass sites in the Aegean and Ionian seas (eastern Mediterranean Sea), the TOC percentage varied between 1% and 5% and remained included between 1% and 3% 50m far from cages. In the present study, TOC values in sampling stations are not significantly different from the reference station's ones and did not exceed 1.87%; it is lower than TOC values reported by [29]. It does not affect benthic community, while a medium TOC values have a visible change on the total abundance of the macro-benthic fauna.

However, in the other side, [32] recorded, a TOC percentage ranged between 0.35% and 0.68% in an integrated multi-trophic aquaculture farm in Sanggou Bay (China). They concluded that these low percentages of TOC and also other nutrient fluxes were impacted by the oxidation degree at the different stations.

Concerning OM, all sampling stations, including reference station located at a distance of 500m from farm limit, showed values ranging between 0.96% and 3.2%. [33] have studied the background effects of organic waste from seven Norwegian fish farms and they have obtained values ranged from 15 to 47% while in our case OM values are lower than 3.5%. These results could be considered as a good indicator of the normal level of OM in fish farm in the local opened bay.

In case of Nitrogen, [34] found 0.28% before farming activity in sea trout cages (in Faroe Islands) and afterwards 0.57% during the farming period. Comparing to these results, Nitrogen percentages obtained in our case study appear to be in the same levels, except for station 4 where the recorded percentage is a little bit high (2.83%).

Regarding Carbon, results reported by [27, 35] showed a potential Carbon releasing values which were higher than those obtained in the present case study. These results could be explained taking into consideration the M'diq fish farm production capacity which does not exceed 200 tons per year. This farm is smaller than the ones studied by [27, 35]. Comparatively, this resulted in low nutrients discharges. In addition, the current state in the present studied area is very important and disperse the released nutrients far away from the local fish farm.

In [35] study, the founding fluxes of Carbon were ranging from 8.90 to 54.24 g dry weight $m^{-2}d^{-1}$; however, the determined deposition fluxes were ranging between 3.44 and 20.96 g of carbon per m^2 per day, taking in consideration that carbon contents in feed and faeces was respectively 49.6% and 27.7% [36]. It is assumed that the deposition factor may be 30% of the founding fluxes. In our case, measured data of C, N and P in sediment, obtained in 2013 and 2016, had no significant difference between these two years and were very low comparing to others Mediterranean fish farm. The estimated potential released values did not affect C, N and P content in local sediment because of the importance of the local tidal current in the dispersion of dissolved discharges and also the importance of the population of wild fish populated around the cages in the consumption of uneaten feed.

According to [37], excess feed requires a different probability distribution for settling velocity, generally with much higher mean value compared to the faecal pellets. For example, [38] examined organic wastes around their studied fish farm in the Slovenian part of the Adriatic and have found mean settling velocity of faecal pellets and excess feed of 0.64±0.05 cm/s and 5.7±0.15 cm/s, respectively. Similarly, [39] consider that appropriate range for settling velocity of focal pellets and excess feed is 2–4 and 6–12 cm/s, respectively.

The environmental impact on a given aquaculture site depends mainly upon the farm's size and its production and the local hydrodynamic conditions and the site's topography. Results showed there is a maintain of satisfactory environmental conditions in and around the studied fish farms.

The effects of organic matter and nutrients enrichment on benthic communities are depending on each site location and its specific conditions. For example, in salmon farming area, studied by [33, 40, 41, 42] the recorded data gave similar results regarding macro-faunal succession but revealed large differences regarding spatial extent of the impacts.

In general speaking, little information was known on fish farming impacts in the Mediterranean [43], where marine fish farming, particularly sea bream (*Sparus aurata*) and sea bass (*Dicentrarchus labrax*) farming, has been grown exponentially during the last period of more 30 years. Mediterranean marine environment has typical characteristics that might result in considerable differences when compared with patterns induced by salmon industry [44]. However, it worth a lot to say that some previous studies curried out in the Mediterranean have addressed some fish farming impacts on different local environment components, such as impact on water column chemistry and parasites [45], effect on nutrients and plankton (46), dynamics of sediment accumulation beneath cages [47] and recovery process of benthos after cessation [48]. The low culture biomass and the high benthic abundance combined with the important marine biodiversity, are key factors in reducing organic content in sediment. Tidal regime and current state could affect the dispersion of the organic and inorganic material and may reduce values of the studied parameters near fish farm cages such as the present studied case.

The results of the present study showed important findings highlighted in two sampling periods spaced with three years each other. The studied parameters, such as TOC, OM and macro-benthic abundance have then been important parameters to evaluate the effect of fish farm on its exploited ecosystem, using some available environment parameters knowledge (such as current velocity, current direction and frequency, distance from the rivers, adaptive fish density, good fish farming husbandry). Even if the studied parameters have shown a non-significant difference from the reference station, which is almost 500m far from fish farm site, it seems necessary to recommend the extension the study duration and increase number of sampling stations and also sampling frequency (dates) according to the seasons. This would allow to refine more the present study and to seek to for understanding the mechanism that governs the interaction of this activity and its local conditions environment and to provide relevant information that can be useful for the local fish farm which has a wish to increase its production capacity in the future.

5 Conclusion

The main problem in fish cages culture, particularly the intensive mono or polyculture ones using carnivorous species, is generally considered to be related to the amount of released waste from the farm, mainly as uneaten feeds and faeces.

The present study has shown that M'diq fish farm has a very weak effect on local marine environment, particularly in the vicinity and surrounding area of its cages. Taking into consideration local conditions and the effects of different feeding practices, these finding could support a sustainable management of this farm and even its extending strategy (if needed). Actually, this study has allowed to gather preliminary information on how the local farm site interacts with its ecosystem with no significant negative impacts and could further be extended both in time and in space, mainly in terms of production capacity increase, farming space enlargement and cultured species diversification. It is obvious that the environmental impacts depend on the amount of food given to fishes, the mode of feeding, the fish density in cages, the annual production and the years of farming operation. The local hydrology and geomorphology are suitable factors for environmental quality preservation. Then, the location of the studied fish farm in opened bay is the most important aspect and no eutrophication phenomenon could be happened. Sediment beneath fish cages and in surrounding area was not significantly enriched in nutrients nor in organic carbon. However, simulated organic matter fluxes could be used in the framework of a suitable model of potential changes in sediment, in order to predict the evolution of the biogeochemical variables of sediment under different scenarios of organic matter deposition, depending on the development strategy of the local aquaculture activity and the evolution of environment conditions.

The sustainability of this fish-farming in operation rely then on the quality of its local environmental. This study has provided an important information about the accurate state of this local fish farm and could then give a reflection basis for a suitable extending production strategy. It is obvious that investment decision for harmonious development of fish farming should be decided taking into consideration an integrated management, using methods such as environmental impact assessment and risk assessment, in addition to economic evaluation, vulnerability assessments, resource accounting, cost-benefit analysis and outcome-based monitoring.

References

1. FAO, *The State of World Fisheries and Aquaculture 2016: contributing to food security and nutrition for all*. Rome (2016)
2. H.E. Froehlich, R.R. Gentry, M.B. Rust, D. Grimm, B.S. Halpern, PLoS ONE **12**(1): e0169281(2017)
3. D. Whitmarsh, P. Wattage, European Environment. **16**: 108-121 (2006)
4. R.H. Findlay, L. Watling, L.M. Mayer, Estuaries, **18**: 145-179 (1995)
5. L. Guillou, C. Alves-de-Souza, R., Siano, H., González, Microbiology Today **37**: 92-95 (2010)
6. MAPA (Ministerio de Agricultura, Pesca y Alimentación), *La Gestión Medioambiental en la Acuicultura Española*. Ediciones Mundi-Prensa, (2002)
7. F. Aguado, *Impacto ambiental de los sistemas piscícolas marinos: la acuicultura en jaulas flotantes*. Series De Química Oceanográfica. Serie I, Monografías **1**, 35-83 (2001)
8. N. Preston, P. Rothlisberg, *Aquaculture environmental impacts*, in Proceedings of the National Outlook Conference 29th February-2nd March Volume 1, Natural Resources, ABARE (Canberra), (2000)
9. A. Dosdat, *Environmental impact of aquaculture in the Mediterranean: Nutritional and feeding aspects*, in Uriarte A. (ed.), Basurco B. (ed.), Environmental impact assessment of Mediterranean aquaculture farms, Cahiers Options Méditerranéennes, CIHEAM Zaragoza, **55**, 23-36 (2001)
10. U.C. Barg, D.M. Bartley, A.G.J.Tacon, R.L. Welcomme, *Aquaculture and its environment: a case for collaboration*, in Hancock, D.A., Smith, D.C., Grant, A., Beumer, J.P. (Eds.), Developing and Sustaining World Fisheries Resources, CSIRO Publishing, Australia, 462-470 (1997)
11. R. Boaventura, A.M. Pedro, J. Coimbra, E. Lencastre, Environmental Pollution **95** (3), 379-387 (1997)
12. B.A. Costa-Pierce, *Environmental impact of nutrients from aquaculture: towards the evolution of sustainable aquaculture systems*, in Baird, D.J., Beveridge, M.C.M., Kelly, L.A., Muir, J.F. (Eds.), Aquaculture and Water Resource Management, Blackwell Scientific, (1996)

13. M.C.M. Beveridge, M.J. Phillips, R.M. Clarke, *A quantitative and qualitative assessment of wastes from aquatic animal production*, in Burne, D.E., Tomasso, J.R. (Eds.), Aquaculture and Water Quality, The World Aquaculture Society, (1991)
14. H., Nhhala, B., Abdellaoui, E, Talbaoui, *Morocco: country report*, in S., Cataudella, D., Crosetti, F., Massa (eds), Mediterranean coastal lagoons: sustainable management and interactions among aquaculture, capture fisheries and the environment, Studies and Reviews, General Fisheries Commission for the Mediterranean, FAO, Rome, (2015)
15. S. Merinero, S. Martínez, A. Tomás-Vidal, M. Jover, AquaTIC: revista electrónica de acuicultura, **23**, 1-19 (2005)
16. J. Lakhdar-Idrissi, A. Orbi, F. Zidane, K. Hilmi, F. Sarf, *Caractérisation physico-chimique de la Baie de M'diq en relation avec l'activité aquacole*. Bulletin Institut Scientifique Rabat, Section Sciences de la Vie, **23**, 65-70 (2001)
17. D. Nachite, *Le développement touristique du littoral de la région Tanger-Tétouan : une évolution vers des scenarios non désirables*, in S. DominguezBella, A. Maate, (Eds.), Geología y geotourismo en la orilla sur delEstrecho de Gibraltar, MCN-UCA, Cadiz, Spain (2009)
18. J.C. Bertheses, N. Tesson, *Observations hydrologiques sur la Sebkha Bou Areg (lagune de Nador),* Travaux et documents, Bulletin de l'Instit Scientifique des Pêches Maritimes, **24**, Morocco(1978)
19. A. Orbi, J. Lakhdar-Idirissi, K. Hilmi, F. Sarf, *Etude hydrodynamique (courantologique) dans la baie de M'Diq*, Travaux et Documents, INRH, **104** : 80 Morocco(1997)
20. A. Aminot, and M. Chaussepied, *Manuel des analyses chimiques en milieu marin*, CNEXO, Éditions Jouve, 395 pages, Paris, France (1983)
21. G. Yucel-Gier, F. Kucuksezgin, F, Kocak, Aquaculture Research, **38,** 256-267 (2007)
22. N. Neofitou, D. Vafidis, S. Klaoudatos, Aquaculture Environment Interactions, **1**, 95-105 (2010)
23. P. Di Marco, T. Petochi, G. Marino, A. Priori, M.G. Finoia, P. Tomassetti, B.M. Poli, Aquaculture, **471**, 92-105 (2017)
24. P. Bellair, C., Pomerol. *Éléments de géologie*, in A. Colin (Ed.), Géologie, 526 pp., Paris, France (1977)
25. A. Walkley, I.A. Black, Soil Science, **37**, 29-37 (1934)
26. B., Dabin, *Application des dosages automatiques à l'analyse des sols*, CahierOrstom, série Pédologie, **III** (4), 335-366 (1965)
27. D. Brigolin, V. Meccia, C. Venier, P. Tomassetti, S. Porrello, R. Pastres, Aquaculture Environment Interactions **5**, 71–88 (2014)
28. J. Hyland, L. Balthis, I. Karakassis, P. Magni, A. Petrov, J. Shine, R. Warwick, Marine Ecology Progress Series, **295**, 91-103 (2005)
29. D. Brigolin, F. Pranovi, S. Kholeif, K. Abdelsalam, R. Pastres, Regional Studies Marine Science, **7**, 129-135 (2016)
30. J.H. Hyun, S.H. Kim, J.S. Mok, J.S. Lee, S.U. An, W.C. Lee, R.H. Jung, Marine Pollution Bulletin, **74**(1), 187-198 (2013)
31. C.J. Cromey, H. Thetmeyer, N. Lampadariou, K. Black, J. Kögeler, I. Karakassis, *MERAMOD:* Aquaculture Environment Interactions, **2**(2), 157-176 (2012).
32. Z. Ning, S. Liu, G. Zhang, X. Ning, R. Li, Z. Jiang, J. Zhang, Aquaculture Environment Interactions, **8**, 221-232 (2016).

33. P. Kupka-Hansen, K. Pittman, A. Ervlk, *Organic waste from marine fish farms-effects on the seabed*, in T. Makinen, (ed), Marine aquaculture and environment, Nordic Councilof Ministers, Copenhagen, Nord 105-119 (1991)
34. G. á Norði, R. Glud, E.Gaard, K. Simonsen, Marine Ecology Progress Series, **431**, 223-241 (2011)
35. M. Jusup, J. Klanjšček, D. Petricioli, T. Legović. Ecological Modelling, **220**(19): 2407-2414 (2009)
36. D. Brigolin, R. Pastres, T.D. Nickell, C.J. Cromey, D.R. Aguilera, P. Regnier, Marine Ecology Progress Series, **388**, 63-80 (2009)
37. M. Jusup, S. Geček, T. Legović, Ecological Modelling, **200**, 459-466 (2007)
38. V. Malačič, J. Forte, Annales Histoire Sciences Sociales, **13**, 3–8(2003)
39. A.M. Doglioli, M.G. Magaldi, L. Vezzulli, S. Tucci, Aquaculture, **231**, 215-235 (2004)
40. J.R. Brown, R.J. Gowen, D.M. McLusky, Journal of Experimental Marine Biology and Ecology, **109**(1), 39-51 (1987)
41. B.D.S. O'Connor, J. Costelloe, B.F. Keegan, D.C. Rhoads. Marine Pollution Bulletin, **20**(8), 384-390 (1989)
42. D.P., Weston. Marine Ecology Progress Series,**61**, 233-244 (1990)
43. B. Munday, A. Eleftheriou, M. Kentouri, P. Divanach. *The interactions of aquaculture and the environment: A bibliographical review.* Commission of the European Communities. Dir. Gen. for Fisheries, Brussels, 325 p. (1992)
44. I. Karakassis, E. Hatziyanni, Marine Ecology Progress Series,**203**, 247-253 (2000)
45. S. Papoutsoglou, M.J. Costello, E. Stamou, G. Tziha. Aquaculture Research, **27**, 25-34 (1996)
46. P. Pitta, I. Karakassis, M. Tsapakis, S. Zivanovic, Hydrobiologia 391: 181–194 (1999).
47. I. Karakassis, M. Tsapakis, E. Hatziyanni. Marine Ecology Progress Series,**162**, 243-252 (1998)
48. I. Karakassis, E. Hatziyanni, M. Tsapakis, W., Plaiti. Marine Ecology Progress Series,**184**, 205-218 (1999)

Towards sustainable food productions in Morocco: Aquaponics

Haytam Rharrhour [1]*, *Fatima* Wariaghli [1], *Simon* Goddek [2], *Mohamed* Sadik [3], *Aziz* El Moujtahid [4], *Hassan* Nhhala [5], and *Ahmed* Yahyaoui [1]

[1]BIOECOGEN Laboratory, Faculty of Sciences of Rabat, Mohammed V University, 4 Avenue Ibn Battouta B.P. 1014 RP, Rabat;
[2]Mathematical and Statistical Methods (Biometris), Wageningen University, P.O. Box 16, 6700 AA Wageningen, The Netherlands;
[3]Water and Forests Department, National Hydrobiology and Fish-farming Center, Azrou 53100, Morocco;
[4]Sciences and Technologies Faculty, Soultan Moulay Slimane University, Beni-Mellal, Morocco;
[5]Faculty of Science and Technics, University of Abdelmalek Essaadi, Ancienne Route de l'Aéroport, Km 10, Ziaten. BP : 416. Tangier – Morroco.

Abstract. This article reviews the development potential of Aquaponics in Morocco, an emergent approach which integrates aquaculture with Hydroponics, whereby aquaculture effluent is used to fertilise plants in the hydroponic system. Aquaponics is the subject of an alternative solution to earth modern problems, linked to climate change, such as drought, soil degradation, food and water availabilities, negative impact of aquaculture on the environment, the use of chemicals and pharmaceuticals. This food production system has gained interest in the past five years, however, there is yet to be successful experiences or scientific publications in Morocco, although the economy of this country is based on two main sectors: agriculture and fisheries. The current paper was developed using a systematic, strategic and comprehensive literature review on peer reviewed publications on several themes, including aquaponics, aquaculture, hydroponics and horticulture, from 1960 to 2020. Over 59publications were reviewed and analysed. The review found that aquaponics has gained interest in the past five years, over 500 scientific articles have been published, since 2015, with the term "Aquaponics" as title, abstract or keyword, highlighting its many advantages and major drawbacks, but unfortunately none of these publications were of Moroccan origin, which consolidates the need for a review which will scale up the importance of this technique and its development potential in Morocco.

1 Introduction

The economy of Morocco is principally based on fisheries and agriculture, that contribute by 2.3% and 15% to the GDP respectively [1-2]. However, these two sectors suffer from several problems, such as over exploitation of natural resources, pollution, drought and

*Corresponding author: haytam.rharrhour@um5r.ac.ma

climate change. Also, demand for fish and aquaculture products in Morocco are increasing due to population growth (36 million in 2017, 39 million forecast for 2025).

The Moroccan Agricultural sector is at the centre of the national economy, having an added value of around 13.92billion USD for 2017, with an average GDP (between 2008 and 2017) exceeding 11.80 billion USD/ year. The agriculture sector remains the primary provider of jobs [2]. In rural areas, nearly three-quarters of the working population gets their income from the food industry (i.e. agriculture, forestry, and fishing). 38% of total employment, as a minimum, are offered by agricultural activities. The Moroccan agro-industry produced a total of 335 thousand tonnes (equivalent to 711Million USD) of diverse products such as fruits and vegetables, cereals, seeds, spices and more in 2017 [2]. The Moroccan fishery sector is considered an essential industry of the country, its annual production is around 1,4 million tons of fish, divided into 1.3 million tons of Coastal and artisanal fishing, 64 thousand tons of deep-sea fishing, and about twenty thousand tons of other activities such as aquaculture. The Moroccan fishery sector employs in total about120,000 fishers and about50,000 people who are working in the processing, transport, and fish trade[1]. While most commercial fisheries can be found in the south of the country.

Integrated Agriculture Aquaculture Systems (IAAS) were practised to treat aquaculture effluents. These systems were adopted to develop an emergent approach, in the past fiveyears, scientific publications about aquaponics exponentially increases since 2015. Aquaponics is the amalgamation of two major food production sectors: aquaculture and agriculture. It is the combination of the RAS (Recirculation Aquaculture System) with hydroponics [3-4-5-6] from where name "Aquaponics" comes. It has gained interest due to its potential for supplying food security and providing local food self-sufficiency in arid areas.

Aquaponics technology is based on the coupling of RAS with hydroponics (soil-less crop culture), in which nutrient rich aquaculture effluent is used to grow plants in hydroponic system [3-4-7]. Recirculation aquaculture system "RAS" is a closed tank based fish farming system, considered as an alternative solution for water consumption by pond aquaculture [8], due to its land and water low requirements [9]. In aquaponics, RAS fish tanks are linked to hydroponic production beds, whereby plants use nutritional elements dissolved in water [3], therefore, hydroponic unit serves as abiological filter in order to purify water before sending it back to the fish tanks in a closed system context. Aquaponics is a quasi-closed food production system that produces both fish and plants without the need of chemicals and pharmaceuticals, using 90% less water than Aquaculture and hydroponics. However, this technique has shown some drawbacks, especially coupled aquaponics systems that have a low economic profitability and inefficient nutrients availability for plants.

This paper aims to give an overview of aquaponics potential for responding to food productions environmental impact, food insecurity and inequality in its availability, poverty and water scarcity in developing countries like Morocco.

2 Method

The present study was developed using a systematic, strategic and comprehensive literature review on peer reviewed publications on several themes, including aquaponics, aquaculture, hydroponics and horticulture, from 1960 to 2020 on both Scopus and Google Scholar. More than 100 publications were reviewed, analysed, collected in the reference software Zotero, and classified depending on key words and year of publishing, among which we choose to work on 59 papers that were considered valuable.

3 Aquaponics

3.1 Definition

Aquaponics is a closed-loop recirculated food production system, that uses nutrient-rich aquaculture effluent to grow plants in hydroponic systems. Aquaponics is considered an alternative sustainable solution for both aquacultural and agricultural productions, addressing water scarcity, pollution, soil erosion and degradation [4-6-7-10].Food production industries have an effect on air and water quality, ecosystem functions, and energy consumption [11]. Aquaponics is considered a model of sustainable agriculture, a newly emerging approach that combines intensive food production with waste-stream recycling and water conservation [4-7].

Considering population growth and environmental issues linked to climate change, aquaponics can cover the local demand for fish and vegetables if its basic principles are respected.

3.2 History

Early Asian and American civilisations used fish wastes to fertilise crops [12]; Aztecs used surrounding cities waste materials to fertilise cultivated agricultural islands known as *Chinampas*, while in south China, Thailand and Indonesia, rice was cultivated in combination with fish, in a sort of early aquaponics.

One of the pioneering works of aquaponics returns to The New Alchemy Institute, in East Falmouth, Massachusetts. In the late 1970s, the institute provided several publications on aquaponics [3-12], which were the seed of modern aquaponics. In 1977, [13] validated the feasibility of combining fish and plants productions in a recirculating water system, by producing ice-lettuce and tomatoes and highlighting the nitrification process. The evolution of the "flood and drain systems" adopted in backyard aquaponics comes back to the pioneering work of Mark McMurtry [3-14], in the 80s, inventor of Integrated Aqua-Vegeculture System (IAVS). Tom and Paula Speraneo adopted, in the early 1990s, the North Carolina State University system to invent a full-size commercial-scale solar greenhouse [3].

Over the past five years, aquaponics has gained a growing interest considering its key benefits which respond to the disadvantages of aquaculture and the unsustainable use of water by agriculture.

3.3 Aquaponics principles

A question that needs to be asked is whether plants can assimilate fish waste? In general, fish excreta are mainly composed of ammonia [9]. Vines and wedding (1960) [15] suggested that ammonia is toxic for both plants and animals, mentioning that ammonia's toxicity is represented by its impact on the electron transport system in plants mitochondria, while [16] claims that ammonium is necessary for plants, NH_4^+ ions stimulates the uptake of nitrate NO_3^-, suggesting a ratio of 75% $NO_3^-$to 25% NH_4^+.

The nitrifying bacteria, plays a vital role in aquaponic systems [3-6-7], by transforming ammonia into nitrates, the form of nitrogen used by plants [17]. This shows that aquaponics is a mini-ecosystem wherein fish, plants, and bacteria coexists in a symbiotic way.

3.4 Key elements

3.4.1 Nutrient balance

In aquaponic systems, uneaten feed and fish excreta are the primary source of plants nutrients, which fall into two categories, based on required quantities, macronutrients and micronutrients; Carbon, hydrogen, oxygen, nitrogen, phosphorus, potassium, calcium, Sulphur and magnesium are required in large quantities, also called macronutrients, while iron, copper, zinc, manganese, molybdenum, boron, and chlorine are considered as micronutrients [18-19-20-21].

Fish tanks effluent provides both macronutrients such as ammonia, nitrate, nitrite, phosphorus, potassium, and micronutrients to produce hydroponic plants, however, one of the limitations with aquaponics is iron deficiency, which is needed for both plants and fish growth [10]. Aquaponic systems are often nutrient deficient especially Potassium (K^{+}), Calcium (Ca^{++}) and Magnesium (Mg^{++}) as macronutrients, and iron (Fe^{++}) as a micronutrient [7], therefore, chelated iron is added to the system in order to compensate for the lack [3]. Nevertheless, some farmers use some supplements to regulate both nutrient deficiencies and pH such as calcium and potassium hydroxides for increasing pH and phosphoric acid to reduce it.

Given the fact that fish feed is the main source of nutrients, we can consider it as a key element for aquaponics. However, one main challenge of aquaponics is to manage nutrient concentrations to promote optimal plant growth without causing any adverse effects on fish and bacteria.

3.4.2 Water quality

Water isthe matrix through which plants receive nutrients, and the medium in which both fish and bacteria live. In RAS, good water conditions are necessary for plant development [3]; since aquaponics is a branch of RAS, these systems bound to the same water quality parameters of recirculation aquaculture systems [22].

Aquaponics water quality requires balancing tolerance ranges of three groups of organisms for each water quality parameter; in general, the tolerance ranges of the three organisms are similar [12], however, optimising one unit comes at the expense of the other.

pH has a significant role in aquaponic systems functioning. It is an important water parameter that can affect either positively or negatively plant, fish and bacteria well-being. These three living organisms have different optimal pH ranges [7]. [23] recommend an acid pH for an optimum plant growth (5.5–5.8) [23], while nitrifying bacteria requires a neutral pH range (7.5–8) according to [19]. Fish optimal pH range varies depending on fish species, in general, RAS systems maintain pH between 7.0 and 8.0 [24-25]. Acidic pH yields to an optimal plant growth [26], on the other hand, it inhibits bacterial functioning and increase fish mortality [7-27], therefore, a neutral pH range of 7-9 is recommended for aquaponic systems [3-6-7-19]. The figure below (Fig.1) represents the operational pH for aquaponics systems according to [25].

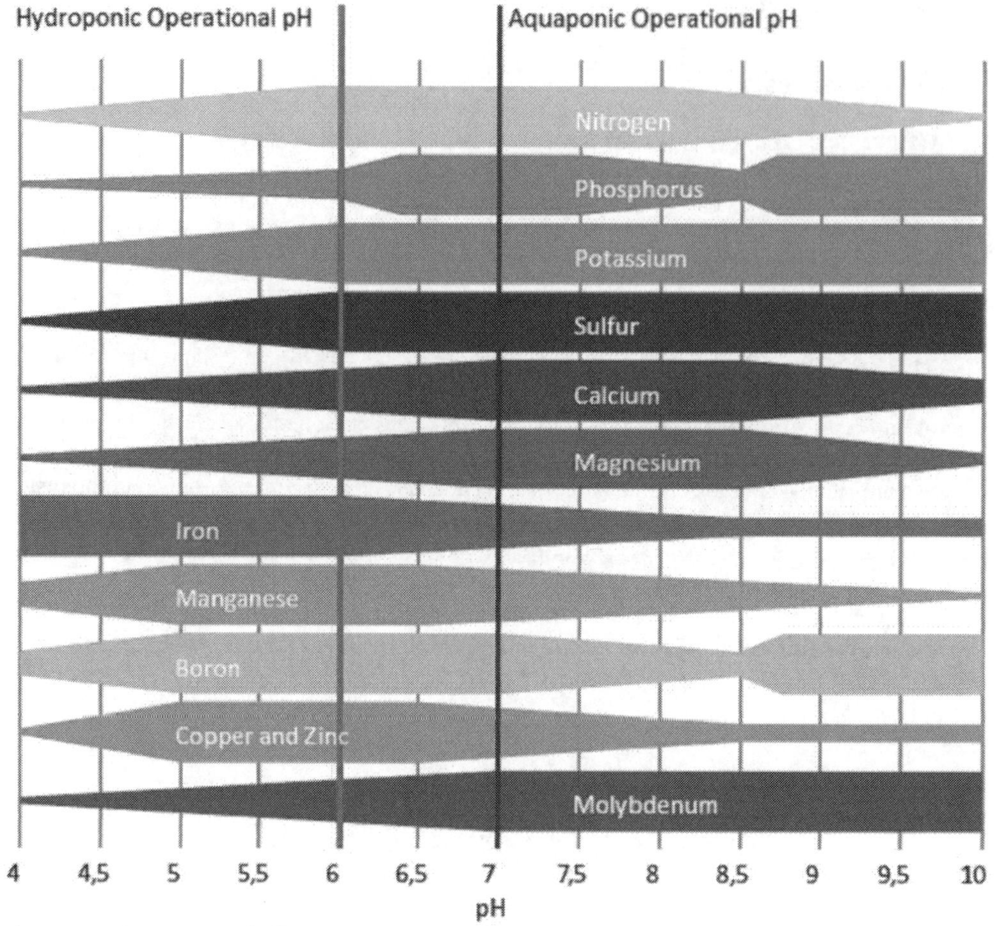

Fig. 1. Aquaponic operational pH and nutrient availability [25].

Dissolved Oxygen (DO) is another key water quality parameter for aquaponics. DO must be maintained in aquaponic systems [21-25]; considering that plant and nitrifying bacteria have lower DO requirements than fish [25], DO must be maintained depending on fish species. In general, warm water fish species can tolerate lower DO concentrations than cool water species [25], thus, DO concentration should be maintained at 5 ppm (parts per million) or higher. There is a negative correlation between water temperature and oxygen solubility, therefore, water temperature can affect aquaponics productivity [12]; however, water temperature is maintained in aquaponics depending on fish species requirements. [25] correctly argues that water-carbon dioxide (CO_2) is another important water quality parameter that is often ignored in aquaponic systems. Because the fish tolerance is generally lower than plants and bacteria [25], carbon dioxide concentration should be set according to fish requirements;[9] claims that carbon dioxide concentrations should be kept under 20mg/l in RAS systems, which is, in general, the same case for aquaponics.

Total nitrogen (Ammonia, Nitrite, Nitrate) is a significant water quality parameter [12]. Nitrogen is essential for plants at all stages of vegetation (growth, reproduction, and storage), primordial for plant growth and well-being, the basis of all proteins. It is involved in proteins, nucleic acids and chlorophyll production [20-28]. In aquaponics, fish waste (gill excretion, urine and faeces) and fish food residue are the main source of Total Ammonia Nitrogen "TAN" [19]. Fish excreta (urine and faeces) contain 10 to 40% of nitrogen as ammonia [29].Nitric nitrogen, also called Nitrate [NO_3^-], is the most soluble form and mainly used by plants. Ammonium [NH_4^+] is also used but in small quantities, most of it is oxidised into nitrate by nitrifying bacteria. Organic nitrogen is indirectly usable; it undergoes mineralisation which will first produce ammoniacal nitrogen and then Nitric nitrogen. The main source of nitrate in aquaponic systems is nitrification process [3]; Nitrifying bacteria oxidise toxic forms of nitrogen excreted by fish, which is ammonia, into a non-toxic absorbable form by plants, which is nitrate [30]. The nitrification process is consisted of two dependent steps; first, ammonia oxidising bacteria convert ammonia[NH_3/NH_4^+] into nitrites [NO_2^-] which in turn is used by nitrite oxidising bacteria to produce nitrates[NO_3^-] [21]. Build-up solids provide although another part of nitrate through uneaten food mineralisation [7]. Nitrate is harmless for Aquaponic living components at concentrations of 150-300 mg/l [31-32]. Nitrite is more toxic than ammonia, above 1 mg/l, it can be harmful for both plants and fish species and can cause the lethal Brown-blood disease for aquatic organisms [33]. Therefore, it is not recommended to add fish and plant during the cycling process.

Other water quality parameters that should be maintained are water clarity and water alkalinity. Suspended solids can affect both fish and plants well-being [21-24-25]; thus, suspended solids loads should be maintained below 30mg/l [25]. Water alkalinity, also called carbonate hardness (KH) is a key parameter due to its unique relationship with pH [12]; it is the total amount of carbonates (CO_3^{2-}) and bicarbonates (HCO_3^-) dissolved in water. Carbonate Hardness buffer the pH (create resistance to pH change), therefore, the optimum level of alkalinity for aquaponics is about 60–140 mg/l [12].

Fish unit is the most sensitive among the three units. Fortunately, nitrifying process conditions matches perfectly fish species, while plant can tolerate it, therefore, aquaponics water quality parameters should be adapted to fish requirements.

3.5 Aquaponic systems

In general, an aquaponic system is consisted of a fish farming unit, a solids removal unit, a biofilter, a hydroponic unit and a sump [21]; in some systems, the hydroponic component, the biofilter and the solids removal component can be combined, such as media bed and deep water culture techniques.

There are generally three types of hydroponic subsystems: media bed culture, Nutrient Film Technique "NFT", and Deep Water Culture technique "DWC" [3-7-19-21]. Many experiments carried out by several researchers were able to test the three systems; as result, most of them agreed that NFT systems are the least productive among the three systems. [34] tells us cucumber culture yields in media bed are better than NFT's [34]. Another study by [35] tested the difference between the three hydroponic subsystems using an aquaponic system coupling Murray Cod with green oak lettuce; this study shows that NFT is less efficient in terms of removing nutrients from fish tanks effluent water and producing plant biomass or yield than media bed culture or DWC in an Aquaponic context. This approach was explained by the difference between plant roots contact with water in NFT (only half submerged in water) which makes the contact surface smaller comparing to DWC technique which provides more contact surface between plant roots and water.

Media bed culture technique is used mostly in small scale aquaponics dedicated to research experiments or small operations [7-12-21]. Media bed culture advantages include its simplicity and low initial costs [12]; this hydroponic system combines biological and mechanical filters with hydroponics which eliminates the expense of separate filters [21]. Another advantage of this system is that media act as a substrate for nitrifying bacteria [35] providing them more surface material for growth and nitrifying process [7-36]. This technique uses inert solids as a substrate such as gravel, hydroton, perlite, cocopeat, and sand [7-21] which provide a fixation substrate for plants root, offering a chance to grow larger plants. On the other hand, this technique has drawbacks such as high coasts when it comes to a larger scale [12], water channelling, inefficient biofiltration, inefficient nutrient delivery to plants and clogging of the media [7-12-35-37-38] if aquaculture unit effluent exceeds hydroponics' carrying capacity.

Deep water culture "DWC" also called floating or raft hydroponics is the best system design for commercial scale. It consists of suspending plants in polystyrene sheets, with their roots fully submerged in water [12]. Some of DWCs' benefits are their low maintenance, maximised nutrient uptake by plants and ability to support a large number of plants with minimal materials [7-39]. That's not all, DWC plants can support power outage up to two weeks [38], use only 1% of total water per day [40] and have lower environmental impact than media bed culture [7-41]. However, this technique requires a good oxygenation [12]and roots exposure to harmful organisms associated with aquaculture systems [21] such as fish fry and zooplankton.

There are two types of aquaponics systems: Coupled and decoupled systems. Coupled systems (Fig.2) are the classic systems of aquaponics, largely practiced in the USA, based on combining three classes of living organisms: Aquatic organisms, plants and bacteria taking advantage of each other. These systems consist of one recirculated closed loop, consisted of one or more fish tanks, a clarifier, a biofilter and a hydroponic unit [42], in which water flows over one outlet in each unit [7].

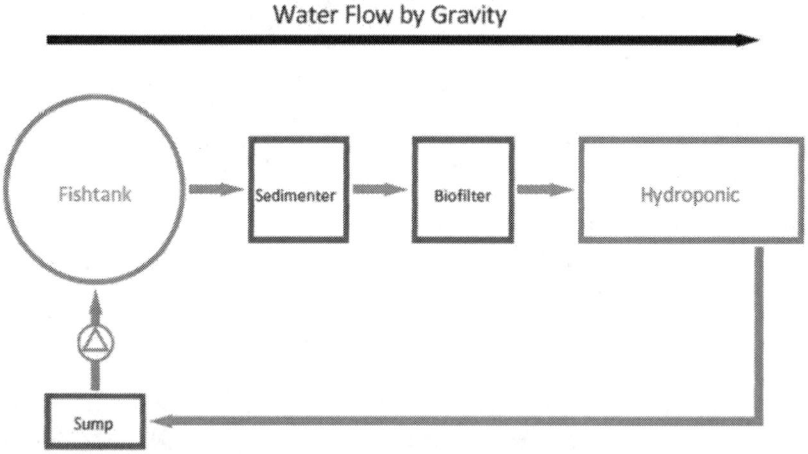

Fig. 2. Basic technical system design of a coupled aquaponic system [42].

Water flows from the fish tank by gravity to a water treatment unit consisted of a sedimenter or a clarifier and a biofilter, then to a hydroponic unit. A sump is used in order to pump the water back to the fish tanks. Some designs of hobby scale aquaponics dispense the use of a water treatment unit and a sump such as media bed culture techniques, in which the media gravel plays a dual role, of both mechanical and biological filters, and the water is pumped from the fish tanks directly to the hydroponic unit and flows back to the fish tanks by gravity.

Modern aquaponics systems are adapting Decoupled system designs that are more complex but also more efficient (in terms of harvest yield and water and nutrient use efficiency) than coupled aquaponics systems. These designs separate the aquaculture unit from the rest of the system [43], in which water have more than one direction in some units [7]. Decoupled systems offer an efficient filtration and better control of nutrient concentrations and water pH [44-45-46]. Furthermore, these designs extract nutrients from the sludge and provide it to the plants in a soluble mineral form [43]. However, decoupled aquaponics systems remain complex and not are not practised enough yet, in addition to their high initial cost comparing to coupled designs.

3.6 Aquaponics profitability

Until now, scientific publications on aquaponics generally dealt with the technical aspect, there are very few researchers who have studied the economic feasibility of aquaponics projects especially on a commercial scale. Aquaponics is generally profitable [37-47-48-49-50], it is a viable and sustainable food production technology. It is complex to calculate an aquaponics systems cost or to compare it to another system, this is due to the differences between specific site conditions, climatic factors, market prices, etc. These factors differ from region to region and from country to country. Energy cost is one of the highest costs in aquaponics systems [19-51], this parameter varies between countries, which makes the very valid point that studying aquaponics profitability is complex. [52] claims that aquaponics systems reach financial equilibrium after 2 yearswhile [19] correctly argues that aquaponics systems profitability should be calculated per square meter.

A financial feasibility study by [50] in a Mediterranean urban context found that aquaponics is profitable after an annual operating positive income, illustrating that labour is the main cost item of their system (50%), however, this study was limited in a small scale system (a five square meter unit of one cubic meter fish tank). Given the fact that nutrients main source in aquaponics systems is fish feed, it seems to be one of the highest cost in this technology. Another study by [51] chose to focus on a small scale aquaponics system, to produce tilapia and a variety of crops products by adopting FAO model in order to evaluate its economic sustainability. This study tells us that greenhouse construction cost, fingerlings, and energy cost are the main costs of aquaponics systems, however, they correctly argue that horticultural profitability would be higher if they focused on crops with short production cycle such as lettuce. In a point of view economic profit, this system was non-profitable, the authors suggest that small scale aquaponics should be utilised for autonomy and self-consumption.

One of the limitations with aquaponics development is the lack of commercial scale financial feasibility, considering that private companies do not share their studies with the public, while those scientific experiences are limited to small scale systems.

4 The contribution of Aquaponics to Moroccan development

4.1 Moroccan aquaculture

In nearly 100 years, Moroccan aquaculture started. The first freshwater fish farming activity has begun with the creation of the first fish hatchery in Azrou, which aimed to promote recreational fishing through restocking lakes in the Middle Atlas [53]. Freshwater aquaculture has turned, since the 80s, from a restocking activity into a food production aquaculture, through the creation of multiple private units practicing eel, carp, tilapia and trout farming, and exploiting the dams for public extensive aquaculture activities [54]. Freshwater aquaculture production is estimated at 15,000 tonnes in 2015, divided into 13,000 tonnes of restocking and 2,000 tonnes of intensive production. The 90s were the most productive years of Moroccan aquaculture. During this period, Morocco has benefited from FAO assistance within the framework of the Technical Cooperation Programme (TCP). In the late 90s, multiple shellfish farming companies were installed on the Atlantic coast practicing St. Jacques shells, mussels, oyster and clam farming [53-55]. Therefore, Moroccan aquaculture recorded its peak production in 1999, with 2793 tonnes [53]. On the other hand, the 2000s were critical for the Moroccan aquaculture sector. Multiple problems accumulation led to a significant decline in national production. Most of mariculture companies have marketed their products in the European market, however, European health measures were the origin of the quantity limitation produced of bivalves [55], likewise, the

European market knew a price drop of aquaculture products explained by public aid from the European Union to European aquaculture farms. On the other hand, some companies closed due to financial issues, while MAROST was forced to close due to European competition. Marine aquaculture has suffered from environmental problems such as the appearance of phytotoxins in the Mediterranean, and the degradation of the Atlantic lagoons [55].

Since 2009, the kingdom has decided to move towards the blue economy and this by adopting a strategy which encourages development and enables sustainable and reasonable competitiveness in the national fishing sector, baptised Halieutis. Actually, marine aquaculture produces 600 tonnes according to [1], which is considered negligible compared to national fishery production (1.37 million tonnes).

Moroccan Aquaculture is still between hypes and hopes, suffering from several problems such as inexistence of hatcheries of marine products and good quality fish feed producers, high competition from European and Asian productions, lack of product diversification, especially marine products, which remain limited to Sea bass, Sea bream, Oysters and Clams, and the absence of a common administrative framework which inhibits the sustainable development of the sector.

4.1.1 Framework of inland aquaculture

The management and promotion of freshwater aquaculture is attributed to the Department of Water and Forests (DEF) of the Ministry of Agriculture, Fisheries, Rural Development, Water and Forests. Inland aquaculture is governed by the Dahir of April 11, 1922 and its texts taken for its application. The Department has succeeded until today in developing inland aquaculture by implementing a development strategy (2005-2014) for fishing and fish farming in continental waters [53], which helped inland production to jump from 2500 tonnes in 2005 to 15000 tonnes in 2014. National Center for Fish and Hydrology in Azrou (CNHP) of the High Commission for Water, Forests and Desertification Control is invested in the Research and Development mission; its role is restocking dams essentially. The mission of this center is to generate knowledge and appropriate technologies to enhance the aquatic environment, create jobs and contribute to the development and promotion of continental aquaculture. In 2015, DEF has set another pivotal strategy (2015-2024) who aims to enhance inland production to reach 50000 tonnes by 2024 and to create 10000 aquaculture jobs [54], through the adoption of advantageous fiscal measures, such as reduction of the import tax on feed for aquaculture.

Actually, there are five hatcheries (three public and two private) and seven breeding stations including two Eel culture farms, four Tilapia farms and only one Trout farm. Compared to marine aquaculture, inland farming is playing a significant role in the development of the aquaculture sector, and that's by mastering freshwater species production steps, however, the marketing of its products remains modest given the weak demand of the national market. The favourable climate of the kingdom, its 3,500 km long coastline and 120,000 ha underexploited massive freshwater resources, are potentials to increase national aquaculture production levels in both marine and freshwater.

4.1.2 Fish species with aquaponics potential

There are six major families reared in Moroccan inland aquaculture: *Anguillidae*, *Centrarchidae*, *Cichlidae*, *Cyprinidae*, *Esocidae*, and *Salmonidae*, both native and non-native fresh water species are reared.

- *Anguillidae* family is represented by the European eel *Anguilla Anguilla* (Linnaeus, 1758), an endemic species with a high commercial value, purely economic breeding intended for export to Asian and European countries.
- *Centrarchidae* family is represented by the largemouth bass *Micropterus salmoides,* (Lacépède, 1802), a non-native species also known as The black bass, of North American origin, which found its ideal in the Moroccan climate.
- *Cichlidae* family is represented by the Nile tilapia *Oreochromis niloticus,* (Linnaeus, 1758), of North African origin, introduced from Egypt, known for its resistance and rapid growth, widely practiced in breeding.
- The common carp or European carp *Cyprinus carpio,* (Linnaeus, 1758), is a non-native species of the *Cyprinidae* family, widely introduced to lakes for sport fishing. The grass carp *Ctenopharyngodon idella* and the silver carp *Hypophthalmichthys molitrix* are another species of *Cyprinidae* family, of East-Asian origin, largely introduced to Moroccan dams for aquatic weed control.
- The northern pike *Esox Lucius,* (Linnaeus, 1758) is the *Esocidae* family representative, a brackish carnivorous species introduced to Morocco for sport and commercial fishing.
- Salmonidae family is represented by two important species, a native species which is the brown trout *Salmo trutta macrostigma*, and the rainbow trout *Oncorhynchus mykiss* which was introduced in Morocco since 1924.

So far, Moroccan native species are rarely reared despite having good economic potential. Taking for example some species of trout: *Salmo akairos* which is a trout species that inhabits Ifni lake of the Atlas Mountains, and *Salmo multipunctata* a new species of Atlantic trout identified in 2015 [56] that inhabits Draa basin in southern Morocco. These species inhabit cold freshwater water streams, these habitats suffer catastrophic seasonal floods that affect the demography of the populations; The deforestation of the Atlas Mountains and overfishing are the main problems that affect trout species in Atlas Mountains, therefore, these species, that are considered endangered, need to be reared for economic and environmental reasons, otherwise, *Salmo pallaryi*, another trout species of Atlas Mountains which populated Sidi Ali Lake, is probably extinct.

The blue barbell *Pterocapoeta maroccana* (Günther, 1902) is a native species of *Cyprinidae* family, only found in Morocco, included in the IUCN category of endangered species since 2006. It's an omnivorous species that populate fast flowing streams [57]. *Pterocapoeta maroccana* is the only representative of the genus, threatened by barant net fishing, climate change, habitats destruction by pollution from the use of pesticides and fertilizers and household garbage. This species needs to be reared in order to repopulate its habitat, as a freshwater species, the blue barbell is a native species with aquaponics and aquaculture potentials.

4.2 Socio-economic impact of Aquaponics in Morocco

The Moroccan agricultural sector is one of the main sectors of activity at the national level, it is one of the pillars of the Moroccan economy. The share of agricultural upstream in the national economy is considerable with 74 billion dirhams, which corresponds to 14% of the gross domestic product [58-59].

Food insecurity in Morocco is a real threat to the nation and this is due to multiple causes; drought, water scarcity and other problems linked to climate change affect food production, hailstorms is one of them, this phenomenon threatens Moroccan agricultural production; A hailstorm hit the region of Fes-Meknes and affects crops on the previous sixth of June, the affected area could reach 9,100 hectares in 27 rural municipalities in the region (Ministry of Agriculture and Fisheries). These problems can lead to food insecurity in rural areas, it also can increase agricultural products prices, for this reason, rational food production technologies need to be practiced. Fish consumption increased up to 16 kg/year per capita in 2020 and demand is likely to increase further, and this is due to Moroccan population growth, furthermore, poverty is one of the main causes of inequality in food availability, especially in rural areas.

Local food production should cover local food demand in order to enable inhabitants of poor and / or arid regions, where water supply is relatively limited, to have access to a source of animal and vegetable protein; this is the major advantage of this activity and it is a support for food security. Aquaponics is a sustainable food production approach that uses 90% less water compared to aquaculture and agriculture [12], due to the minimal water loss represented by evaporation and transpiration from plants. Its strengths are represented by the fact that this activity eliminates the use of chemicals or pharmaceuticals which allows to produce more food with less resources, reduces negative impacts of Agriculture and Aquaculture on the environment (Organic and chemical pollutions), and according to some scientific publications aquaponics increases productivity compared to Hydroponics supporting food security. Furthermore, this activity fights against soil degradation, allowing environment conservation and restoration, which indirectly preserve Biodiversity.

5 Conclusion

The Moroccan economy is mainly based on two major food production sectors: Agriculture and sea fishing. Unfortunately, Moroccan agriculture is threatened by rising temperatures and lower rainfall, as well as fish products are overexploited. For this reason, sustainable agriculture and aquaculture are alternative solutions. Aquaculture should be practiced not only for its economic benefits, but also for its positive environmental impact which is biodiversity conservation. Moroccan native species are endangered and threatened, due to climate change, habitats degradation and overfishing; controlling reproduction and breeding of these species can be an effective solution for this threat to Moroccan biodiversity, and that's by repopulating natural habitats while exploiting these species rationally in aquaponics to grow vegetables. Today, the principal concern of Moroccan aquaculture should be rearing endemic species in order to repopulate natural habitats, while using aquaculture wastes for growing plants on aquaponics.

Acknowledgements

This study was supported by the MENFPESRS and CNRST from Morocco under grant N° PPR/2015/1 for the project "Impact des changements climatiques sur la diversité génétique des poissons des eaux douces du Maroc".

References

[1]. Ministère de l'Agriculture, de la Pêche Maritime, du Développement Rural et des Eaux et Forêts, Département de la Pêche Maritime, *La mer en chiffres* (2019)

[2]. Ministère de l'Agriculture, de la Pêche Maritime, du Développement Rural et des Eaux et Forêts, Département de l'agriculture, *Filières végétales*, (2020). [Online]. Available: https://www.agriculture.gov.ma.

[3]. S. Diver, Natl. Sustain. Agric. Inf. Serv. 28:1–28 (2006)

[4]. D. Klinger and N. Rosamond, ARER**37**, pp. 247-76, (2012)

[5]. S. Goddek, C. R. Espinal, B. Delaide, M. H. Jijakli, Z. Schmautz, S. Wuertz and K. J. Keesman, Water**8**, pp. 303, (2016)

[6]. Z. M. Gichana, D. Liti, H. Waidbacher, W. Zollitsch, S. Drexler and Waikibia Joseph, AI**26**, pp. 1541-1572, (2018)
[7]. B. Yep and Y. Zheng, JCP**228**, p. 1586–99., (2019)
[8]. T. Losordo , M. Masser and J. Rakocy, SRAC, (1998)
[9]. M. P. Masser, J. Rakocy and T. M. Losordo, SRAC, (1999)
[10]. N. Kasozi, R. Tandlich, M. Fick, H. Kaiser and B. Wilhelmi, AR**15**, (2019)
[11]. Z. Wu, S. Chunhua and v. d. H. Anton, PR**90**, p. 119–33, (2019)
[12]. C. Sommervile, M. Cohen, E. Pantanella , A. Stankus and A. Lovatelli, *Small-Scale Aquaponic Food Production : Integrated Fish and Plant Farming*, in : Fao Fisheries and Aquaculture Technical Paper **589,** Rome, (2014)
[13]. L. C. Naegel, Aquaculture**10**, pp. 17-24, (1977)
[14]. B. König, J. Janker, T. Reinhardt, M. Villarroel and R. Junge, JCP**180**, pp. 232-243, (2018)
[15]. H. M. Vines and R. T. Wedding, PP*35*, p. 820–825, (1960)
[16]. J. B. Jones, *Hydroponics : A Practical Guide for the Soilless Grower*, 2nd ed., Boca Raton: CRC Press, (2005)
[17]. S. J. Leghari, N. A. Wahocho , G. M. Laghari, A. H. Laghari, G. M. Bhabhan, K. H. Talpur, T. Ahmed, S. A. Wahocho and A. A. Lashari, AEB**10**, pp. 209-219, (2016)
[18]. J. W. Salisbury and D. M. D'Aria, RSE**46**, pp. 83-106, (1992)
[19]. S. Goddek, B. Delaide, U. Mankasingh, K. V. Ragnarsdottir, H. Jijakli and R. Thorarinsdottir, Sustainability**7**, pp. 4199-4224, (2015)
[20]. A. McCauley , C. Jones and J. Jacobsen, NMM, p. 11–12, (2009)
[21]. J. Rakocy, M. P. Masser and T. Losordo, SRAC, (2006)
[22]. C. Espinal and D. Matulic, *Recirculating Aquaculture Technologies*, in : S.Goddek, A.Joyce, B.Kotzen, G.M.Burnell, Aquaponics Food Production Systems: Combined Aquaculture and Hydroponic Production Technologies for the Future, pp. 35-76, (2019).
[23]. B. Bugbee, AH , pp. 99-112, (2004)
[24]. M. B. Timmons, J. M. Ebeling, F. Wheaton, S. Summerfelt and B. Vinci, Cayuga Aqua Ventures, LLC., (2002)
[25]. W. Lennard and S. Goddek, *Aquaponics : The basics,* in : S.Goddek, A.Joyce, B.Kotzen, G.M.Burnell, Aquaponics Food Production Systems: Combined Aquaculture and Hydroponic Production Technologies for the Future, pp. 113-143, (2019)
[26]. Y. Zou, Z. Hu, J. Zhang, H. Xie, C. Guimbaud and Y. Fang, BT **210** , p. 81–87, (2016)
[27]. R. V. Tyson, E. H. Simonne, D. D. Treadwell, J. M. White and A. Simonne, HS **43**, pp. 719-724, (2008)
[28]. R. Novoa and R. Loomis, PS **58**, pp. 177-204, (2005)
[29]. S. Wongkiew, Z. Hu, K. Chandran, J. Woo Lee and S. K. Khanal, AE **76**, pp. 9-19, (2017)
[30]. D. E. Canfield , A. N. Glazer and P. G. Falkowski, Science, **330**, pp. 192-196, (2010)
[31]. A. Graber and R. Junge, Desalination**246**, pp. 147-156, (2009)
[32]. W.-H. Hu, L. C. Cajas-Monson, S. Eisenstein, L. Parry, B. Cosman and S. Ramamoorthy, NJ **14**, p. 91, (2015)
[33]. M. B. Timmons and J. M. Ebeling, Cayuga Aqua Ventures, LLC 2nd edition, (2010)
[34]. S. Wren, *Comparison of hydroponic crop production techniques in a recirculating fish culture system*, Texas A&M University, College Texas, Texas, (1984)
[35]. W. A. Lennard and B. V. Leonard, AI **14**, pp. 539-550, (2006)

[36]. C. Maucieri , A. A. Forchino, C. Nicoletto, R. Junge, R. Pastres , P. Sambo and M. Borin, JCP **172**, pp. 3119-3127, (2018)
[37]. J. Rakocy, D. Bailey, K. Shultz and W. Cole, *Evaluation of a Commercial-Scale Aquaponic Unit for the Production of Tilapia and Lettuce*, in : Tilapia In Aquaculture**106**, 357-372(1997)
[38]. J. Rakocy, *Aquaponics Q and A (The Answers to Your Questions About Aquaponics)*, Nelson and Pade, Inc. (2003)
[39]. D. Patillo, NRAC *Technical Bulletin Series* **19**, pp. 1-10, (2019)
[40]. D. C. Love, M. S. Uhl and L. Genello, AE **68**, pp. 19-27, (2015)
[41]. Forchino, H. Lourguioui, D. Brigolin and R. Pastres, AE **77**, pp. 80-88, (2017)
[42]. H. W. Palm, U. Knaus, S. Appelbaum, S. M. Strautch and B. Kotzen, *Coupled Aquaponics Systems*, in : S.Goddek, A.Joyce, B.Kotzen, G.M.Burnell, Aquaponics Food Production Systems: Combined Aquaculture and Hydroponic Production Technologies for the Future, pp. 163-199, (2019)
[43]. S. Goddek and O. Körner, AS **171**, pp. 143-154, (2019)
[44]. W. Kloas, R. Gross, D. Baganz, J. Graupner, H. Monsees, U. Schmidt, G. B. Staaks, J. Suhl , M. Tschirner, B. Wittstock, S. Wuertz, A. Zikova and B. Rennert , AEI**7**, pp. 179-192, (2015)
[45]. S. Goddek and K. J. Keesman, Desalination **428**, pp. 76-85, (2018)
[46]. S. Goddek and K. J. Keesman, AI **28**, pp. 2481-2490, (2020)
[47]. J. Rupasinghe and J. Kennedy, AEM **14**, pp. 81-96, (2010)
[48]. K. Tokunaga, C. Tamaru, H. Ako and P. Leung, JWAS**46**, pp : 20-32(2015)
[49]. K. K. Quagrainie, R. Valladão Flores, H.-J. Kim and V. McClain, *Economic Analysis of Aquaponics and Hydroponics Production in the US Midwest*, JAA **30**, pp : 1-14(2017)
[50]. A. Ascuito , E. Schimmenti, C. Cottone and V. Borsellino, UFUG **38**, pp. 397-402, (2019)
[51]. J. Lobillo-Eguíbar, V. M. Fernández-Cabanás , L. A. Bermejo and L. Pérez-Urrestarazu, Agronomy**10**, p. 1468, (2020)
[52]. R. Laviale, (2015)
[53]. FAO, *Vue Générale Du Secteur Aquacole National - Maroc*, (2015). [Online]. Available: https://www.fao.org/fishery/countrysector/naso_morocco/fr.
[54]. REAPC, *Etude préliminaire sur l'aquaculture au Maroc*, **72**, (2016).
[55]. FENIP, *Etude de l'état des lieux de l'aquaculture au Maroc et identification des marchés aquacoles cibles et de leurs conditions d'accés*, Juillet , (2010)
[56]. Doadrio, s. Perea and A. Yahyaoui, Graellsia**71**, pp. 1-26, (2015)
[57]. A. Yahyaoui, S. Perea, P. Garzon and I. Doadrio, *Atlas des poissons des eaux continentales du Maroc*, (2019)
[58]. The World Bank, *Rapport de suivi de la situation économique au Maroc*, (2020)
[59]. HCP, *Prospective Maroc 2030 _ Agriculture 2030, quels avenirs pour le Maroc*, (2011)

Investigation of the Response of onion (Allium Cepa L.) to continuous deficit irrigation as smart approaches to crop irrigation under Mediterranean conditions

Aziz Abouabdillah[1*], *Omnia* El Bergui[1,2], *Rachid* Bouabid[1], *Mohamed* Bourioug[1], *Youssef* Brouziyne[3], *Nabil* El Jaouhari[4] and *Ahmed* Bouaziz[2]

[1]Ecole Nationale d'Agriculture de Meknès, Meknès, Morocco
[2]Institut Agronomique et vétérinaire Hassan II, Rabat, Morocco
[3]Mohammed VI Polytechnic University, International Water Research Institute, Benguerir, Morocco
[4]Moulay Ismail University, Faculty of Science, Meknès, Morocco

Abstract. Water is a crucial resource for food production and its scarcity associated with frequent droughts has increased the need for a more efficient use of it along with new irrigation management technologies. This study addressed onion crop responses to continuous deficit irrigation with trigging thresholds of readily available water content. The experiment was conducted on an experimental plot in open field in Morocco. Three water regimes were applied T1 control (100%), T2 (75%) and T3 (50%) of crop evapotranspiration ETc combined with two trigging thresholds (10% and 5%). This is a complete random block device with four repetitions. The measurements concerned the monitoring of vegetative, Eco physiological and yield parameters. The results obtained show that: (i) 100% ETc irrigation at a threshold of 5% of RAW recorded the maximum bulb diameter and weight, thus achieving the best marketable bulb yields. However, in terms of yields, this treatment is not significantly different from the other irrigation regimes with the exception of the irrigated treatment at 50% daily ETc and at a threshold of 10% RAW. The latter recorded the lowest values in terms of production parameters. (ii) For the eco-physiological parameters, significant effects of irrigation dose were observed for proline content, stomatal conductance and leaf temperature, and the effect of the trigging threshold was clearly observed for the moisture content of the leaves. (iii) Water restrictions have minimized the rate of premature run and population density of Thrips tabaci in the onion. (iv) Finally, the best agronomic efficiencies in the use of irrigation water were recorded in treatments with a water restriction of 50%.

Key word: Continuous deficit irrigation, onion, Crop coefficient, available water content

*Corresponding author: aziz.abouabdillah@gmail.com

Introduction

Decrease in water resources has become a worldwide problem and there may be insufficient water for food production by 2025 (Girona et al., 2010), especially in Mediterranean region which is a climate change "hot spot" (Abouabdillah et al., 2010). As a consequence, saving water in irrigation is the most critical issue to be considered (Ayas., 2019). Irrigation water is considered as acrucial element for crop production (Howell, 2001; Steduto et al., 2012; Abdelkhalik et al., 2019). In Morocco, about 75% of water is allocated to agriculture (Abdelkhalik et al., 2019). Therefore, agriculture is a key sector in the Moroccan economy, as it represents approximately 15% of the Gross Domestic Product (Brouzyine et al., 2018). It should be pointed out that 43 % of Moroccan population makes a living from agricultural activity (Bzioui, 2004). Actually, major challenges exist in maximizing water use efficiency (WUE) and increasing crop productivity per unit of water applied. Within this context, deficit irrigation (DI) is used as a practice that provides water below optimal crop water requirements leading to increased WUE (Pereira et al., 2002; Costa et al., 2007; Fereres and Soriano, 2007; Capra et al., 2008; Evans and Sadler, 2008; Sharma et al., 2014 : Chai et al., 2016). However, DI can save a significant amount of irrigation water with the risk of reducing the yield of some crops. Generally, a distinction is made between two types of deficit irrigation that can be applied: Sustainable Deficit Irrigation (SDI) which means the application of continuous water restriction throughout the entire crop cycle (Goldhamer et al., 2006), or a water stress during particular crop developmental periods (RDI: Regulated Deficit Irrigation) (English et al., 1990; Chalmers et al., 1981; Hueso and Cuevas, 2004).
Among vegetable crops, onion is one of the main foods widely used around the world. According to the most recent data of the United Nations Food and Agriculture Organization (FAO), worldwide onion production was approximately 96.7 million tons from 5.03 million hectares (FAO, 2018). Average world yields increased from 18.4 t/ha in 2004 to 19.2 t/ha in 2018. The leading onion producing countries are China, India and the Unites States of America accounting for 20% of the world production (FAO, 2018). Based on FAO data, Morocco produces 954801 tons of onions annually, which is approximately 1% of the world's onion production. This study was carried out in the Fes-Meknes region of Morocco. This region is the country's leading onion producer, accounting 54 % of the total production.
Water requirements for onion varies depending on location and irrigation system (López-Urrea, 2009). The maximum root penetration of onion is 76 cm, but most of itis in the first 18 cm of soil limiting the amount of soil water available for roots (Hanci and Cebeci, 2014). As a result, rainfall is insufficient to sustain its production, making irrigation crucial.
Some previous experiments on onion field grown in different climates were studied. Deficit irrigation (50 % ETc) saved 50 % of irrigation water with a 25 % reduction in bulb yield while at 75 % of ETc, the reduction in bulb yield is not significant in comparison with the control (Igbadun et al., 2012). Deficit irrigation increased the water use efficiency of the onion (Santa Olalla et al., 2004; Singh et al., 2017; Igbadun et al., 2012). In addition to water savings, deficit irrigation may also have positive effects on fruit quality (Sharma et al., 2014). Therefore, the objectives of this study were to determine optimal irrigation parameters for yield optimization and increased water use efficiency, as well as to study the combined effect of deficit irrigation and irrigation frequencies on the onion crop.

1 Materials and methods

1.1 Experimental site and climatic conditions

The field experiment was conducted during the growing season of 2018 from April to august at the National School of Agriculture of Meknes, situated in Fès-Meknès region, Morocco(Fig. 1). Its geographical coordinates are: latitude 33°50'36'' N, longitude 5°28'39'' W and its altitude is 546 m above sea level. The soil texture at the site is loamy-clayey. The volumetric water content for the first 0.3 m was 38 % at field capacity and 21.3 % at wilting point. Soil characteristics are detailed in Table 1.

The climate of the study area is Mediterranean with an average annual rainfall of approximately 450 mm. Maximum and minimum temperature values at the experimental site during the experiment were 40°C and 4°C, respectively (Fig. 2a). The highest and lowest relative humidity values were found as 100% and 16%, respectively. As shown in Fig. 2(b), daily reference evapotranspiration, values vary between 0.5 and 7mm.Rainfall during the irrigation season was 55 mm, which was taken inconsideration in the irrigation water allocation for each treatment.

Table 1. Soilphysico-chemical characterization at the experimental site.

Parameters (unit)	Soil property
Clay (%)	37.3
Silt (%)	56.5
Sand (%)	16.2
Electrical conductivity (dS m-1)	0.102
pH	8.41
Total limestone (mg kg-1)	24.4
Organic matter (%)	2.78
Mineral nitrogen (mg kg-1)	25.9
Olsen P2O5 (mg kg-1)	9.0
Extractable K2O (mg kg-1)	317.0

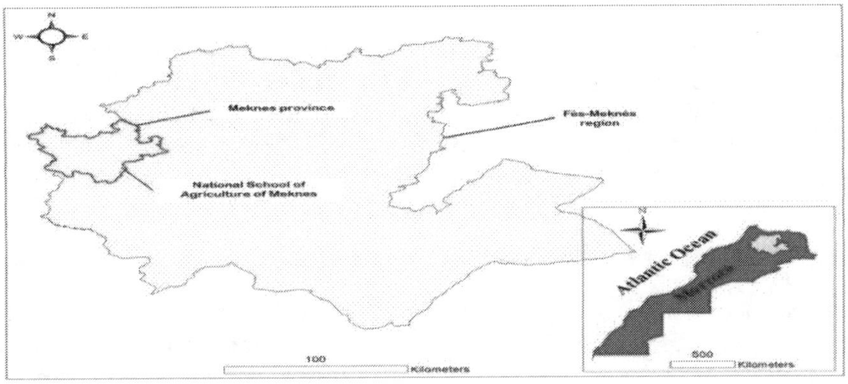

Fig. 1. Geographic location of the experimental site (prepared using ArcGIS software 10.3).

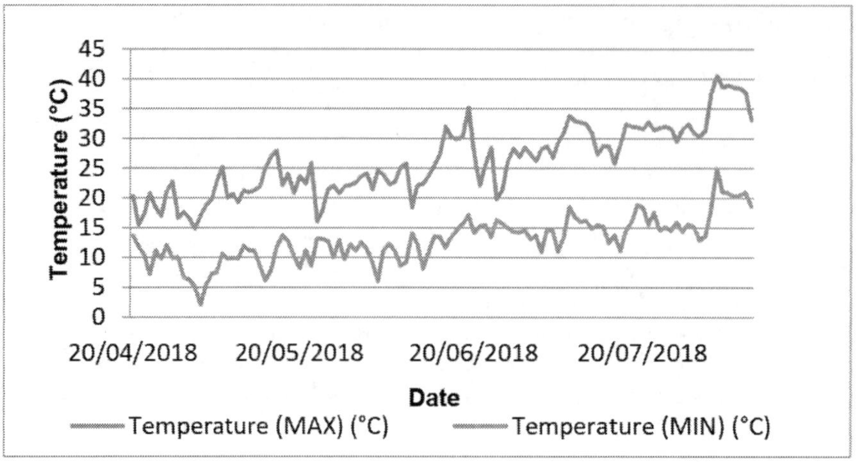

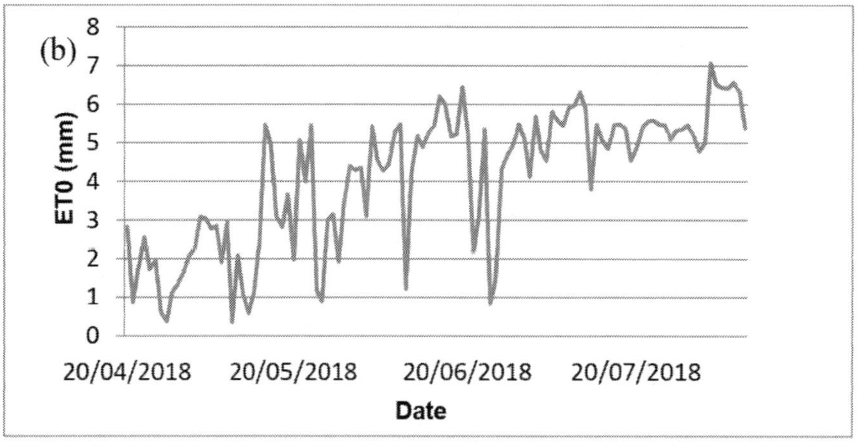

Fig. 2. Daily Average (a) temperature (°C) (b) reference evapotranspiration (ET0) during the growing season

1.2 Crop Management

The onion plant (Allium cepa L.) was transplanted on April 29, 2018 at a spacing of 0.1 m by 0.8 m at a depth of 1 cm. After transplanting, all treatments were fully irrigated to restore the soil moisture content to the field's capacity. The type and rate of application was based on the results of the soil analysis. Fertilizer units applied during the crop cycle were 73 kg/ha N, 44 kg/ha P2O5 and 60 kg/ha K2O. The crop was harvested 99 days after transplanting, when the bulb onions were mature. Mature bulb onions were hand pulled from the ground and sun-dried for one month and then stored in silos before measuring yield and quality parameters.

1.3 Irrigation treatments and experimental design

The plot is provided with a drip irrigation system. Each planting line has a single polyethylene drip line, with self-regulating drippers, uniformly spaced 40 cm apart on the drip line and delivering 1.6 l/hour. Crop evapotranspiration (ETc) was calculated using the following formula(Allen at al., 1998) :ETc (mm) = (ET0 x Kc) – effective rainfall. ET0 is the reference daily evapotranspiration calculated by the modified Penman-Montheith formula based on climatic parameters collected continuously from a weather station located approximately 30 m from the experimental site. The Kc used in this study is that published by López-Urrea (2009)(Table 2). Effective rainfall is the fraction of precipitation that directly meets the crop water requirements, it corresponds to 80 % of total rainfall, taking in consideration the amount of rain lost through surface runoff, deep percolation or evaporation (El Jaouhari et al., 2018). In this experiment there were six treatments combining three doses of irrigation (100%, 75% and 50% of daily ETc) and two thresholds of the Readily available water RAW (10% and 5%). On the one hand, a threshold of 10 % with 100 % (T1), 75 % (T2) and 50 % (T3) of ETc, on the other hand a threshold of 5 % with 100 % (T4), 75 % (T5) and 50 % (T6) of ETc throughout the cycle.

In order to determine the water amount to be applied for each irrigation, it was necessary to calculate the Readily Available Water (RAW) equivalent to the Net Maximal Amount of irrigation, which corresponds to the amount of water that a soil can hold in the effective root depth, and which is calculated by using the formula below:

$RAW = f \times (Fc - PWP) \times Rd \times PHS$

With

f: Irrigation triggering threshold (5%) and 10% in this study

Fcc: Humidity at field capacity

Hpfp: Humidity at permanent wilting point

Rd: effective root depth

PHS: Percentage of humidified soil using the installed drip irrigation system

All the key parameters of the RAW were calculated at the beginning of the experiment. The factor f is an important indicator for judging the work quality in irrigation management. The values of f chosen for this test were 5% and 10%.

The basic meaning of f = 5% or 10% is that irrigation is triggered when 5% or 10% of the available soil water storage is consumed.

A randomized complete block design was used with four block replicates. Each replicated block contained all six treatments. Each elementary plot consisted of six planting line of 20 m long each.

Table 2. Proposed Kc model for onion growing by López-Urrea (2009).

Stage	Kc	Duration
1 leaf	0.70	52
2 leaf	0.80	
3 to 4 leaf	0.90	62
Bulb growth	1.05	30
Bulb thickening	1.20	
Ripening	0.7	24

1.4 Data analysis

Statistical analysis methods were used to analyze the data obtained from the experiment for the effect of water stress and irrigation frequency on yield and quality components of onions. Analysis tools used comprised Analysis of Variance (ANOVA) and Student-Newman-Keuls

(SNK) multiple comparison test using SPSS software (version 20). Treatments effects were considered significant at $P < 0.05$.

1.5 Plant measurements

1.5.1 Growth parameters

For growth monitoring, observations and measurements included leaf number and plant height. They were carried out every 15 days during the trial period on 48 selected randomly (8 plants per treatment). Plant height (cm) was measured using a double decameter.
The Relative Growth Rate (RGR in cm.cm-1.j-1) is an indicator of plant growth and was calculated in relation to plant height measurements according to the following equation (James and Richards, 2006) :

$$\mathrm{RGR\ (cm.cm-1.day-1)} = \frac{\mathrm{Ln\ (Height)t2\ -\ Ln\ (Height)t1}}{\mathrm{t2-t1}}$$

Where : (Height) t1 = plant height measured at time t1 ; (Height) t2 = plant height measured at time t2 ; t2-t1= the number of days between (Height) t1 and (Height) t2.

1.5.2 Eco-physiological parameters

In this study, the relative chlorophyll content was non-destructively measured with the use of chlorophyll content meter model CCM-200 (Opti-Sciences, Inc., Hudson, NH). Three measurements were taken every 20 days on a random sample of 24 plants per treatment. The stomatal conductance was measured on mature leaves using a porometer (Leaf Porometer Decagon Devices, INC). Three leaf temperature measurements were taken using an infrared thermometer on two plants per experimental unit.
The leaf relative water content (LRWC ; %) was determined in fresh leaf discs. The discs were weighed (FW), and immediately floated on water to saturate them with for 6h in darkness. The adhering water of the discs was blotted, and turgor weight (TW) was recorded. After dehydration at 70°C for 48h, the dry weight (DW) of the disc was noted. LRWC was calculated using the equation developed by Turner (1981) :

$$\mathrm{LRWC\ (\%)} = \frac{\mathrm{FW-DW}}{\mathrm{TW-DW}} \mathrm{x\ 100}$$

Proline was extracted from a sample of 100 mg fresh leaves and roots crushed for each treatment in 10 ml of 3% aqueous sulpho-salycylic acid and estimated using the ninhydrin reagent according to the method described by Bates et al. (1973). The absorbance of fraction was read at a wavelength of 520 nm. Concentration of proline (expressed as µmol g-1 of fresh weight) was determined by referring to a calibration curve prepared from a standard range of proline based on the following concentrations: 5, 10, 15, 20 and 25 µg/ml proline per tube.

1.5.3 Fruit yield, quality and water use efficiency

The average bulb weight per treatment was determined for two plants from each experimental unit were randomly picked. The marketable bulb yield (MBY)per treatment was calculated using the bulb weight and the number of plants per line. The equatorial diameter (mm) of

https://doi.org/10.1051/e3sconf/202233704001

onion bulbs were measured using a digital caliper as one of the parameters of crop quality (Rop, 2007).
Water use efficiency (WUE) is the ratio between the fruit yield (kg.ha-1) obtained and the amount of water applied (m3) (Bouaziz et al., 2002). The WUE was calculated as follows:

$$\text{WUE (kg/m3)} = \frac{\text{Yield (kg/ha)}}{\text{total of water received (mm)}}$$

2 Results and discussion

2.1 Growth parameters

The evolution of the growth parameters (Leaf number, plant height, root mass and volume) is presented in Fig. 3. During the experiment, the six applied irrigation regimes did not show any significant effect. Fig. 3 (a) indicated that the highest average number of leaves per plant (8.29 leaves) was recorded within T2, while T6 recorded the lowest leaves number (7.54 leaves). These results are in accordance with Enchalew et al. (2016), who reported that the highest onion leaves number was found in the treatment receiving 90% ETc and up to 20% water restriction was tolerable to obtain at least seven leaves per plant. Karim (1981) worked on the effect of irrigation frequencies on the onion crop and reported that as soil water supply increased, plant growth parameters increased. Fig.3 (b) shows that plant growth was influenced with the amount of water applied between the treatment. The evolution of onion plant height followed the same pattern, with the exception of the T6 treatment where the decrease in height started as early as the 68th day after transplanting. The average height of plants varies most of the time between 39 and 48 cm. Our results do not match with those of Bagali (2010) who noted that the plant height show a significant effect of irrigation frequency as the highest heights were observed at 100% of ETc. In the same study by Enchalew et al. (2016), the highest and lowest plant heights were obtained within the treatments receiving 90% and 50% ETc, respectively. However, another study, published by Kanton et al. (2003), who examined the effect of irrigation frequency on onion crop, indicates that as irrigation intervals increased, plant height decreased significantly. The effect of SDI and irrigation frequencies on relative growth rate (RGR) are shown in Fig. 3(c). The RGR peaks are reached between days 68 and 82 after transplanting for the T1 treatment, and between days 56 and 68 after transplanting for the rest of treatments. However, significant difference was found between T1 and the others treatments. The highest RGR was recorded by T1 during the 14–26 day interval after transplanting. Our results showed that the highest relative growth rate is obtained in the non stressed treatment, which is consistent with the results of Galmes et al. (2005). It was found by Hegde (1986) that irrigation had no significant effect on onion RGR except between 21 and 35 days after transplanting.

https://doi.org/10.1051/e3sconf/202233704001

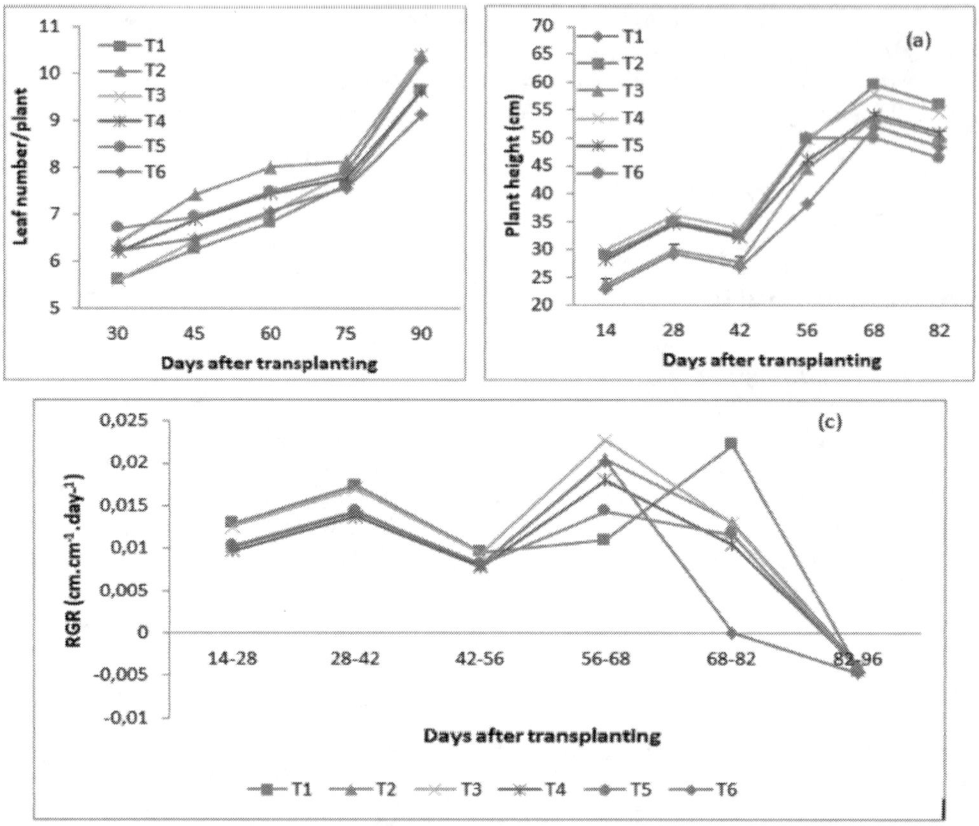

Fig. 3. Plant growth (a) number of leaves per plant, (b) plant height, (c) evolution of the relative growth rate after transplanting

2.2 Eco-physiological parameters

The relative chlorophyll content was unaffected significantly by the irrigation doses and frequencies. Indeed, T4 treatment had the highest value and the lowest was recorded by T3, T5 and T6 treatments (Fig. 3A). Pirzad et al. (2011) concluded that a water stress (excessive or deficit water) decreases significantly leaf chlorophyll concentrations (chlorophyll-a, chlorophyll-b and total chlorophyll). It was found that water deficit can destroy or prevent the production of chlorophyll (Montagu and Woo, 1999). For the stomatal conductance, significant effect of SDI treatments (T3 and T6) was observed (Fig. 3B). The results of this study are not in accordance with those obtained by Cornic and Briantais (1991), who noted that the stomatal conductance increases for treatments receiving more water. In this experiment, it may be due to the fact that T1, T2, T4 and T5 treatments plants were not stressed during the measurement. A significant difference was discerned between the treatments under SDI for leaf temperature (Fig. 3C). The lowest values were 27.75 and 27.25 °C recorded in the T1 and T4 treatments, respectively. According to Hsiao (1973), the increase in leaf temperature depends strongly on environmental factors.

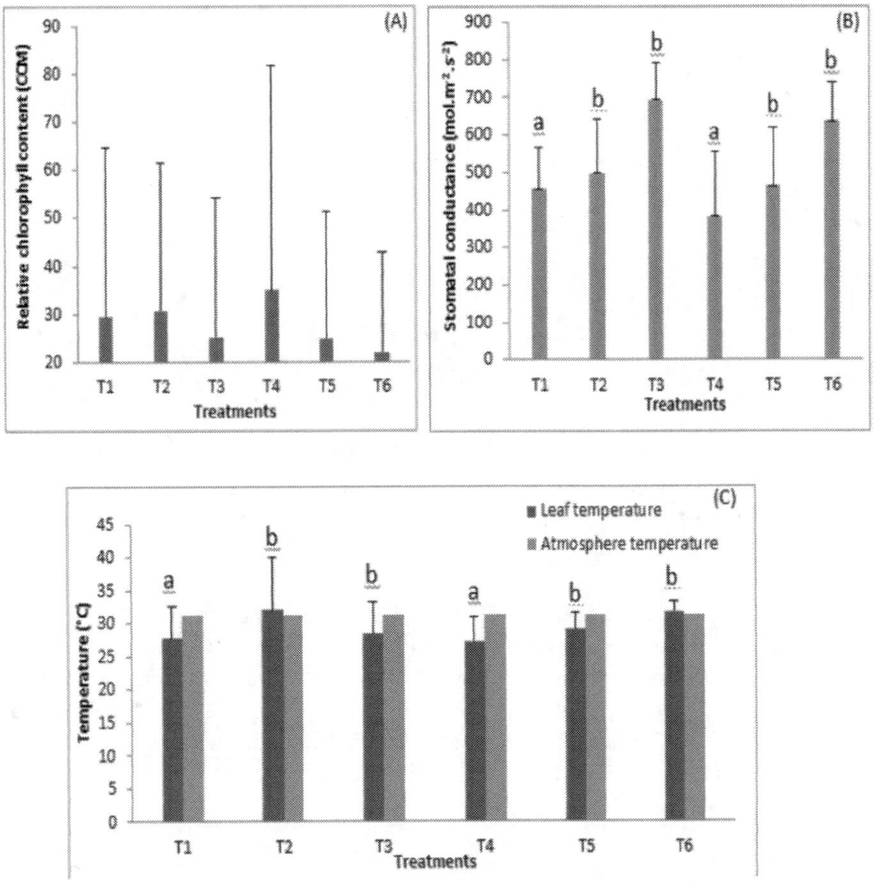

Fig.4. Eco-physiological parameters (A) relative chlorophyll content, (B) stomatal conductance, (B) leaf temperature under different irrigation treatments. Different letters indicate significant differences according to the SNK test at $p < 0.05$. The error bars show the standard deviation of measurements between replications.

Deficit irrigation at 50 % of ETc had significant effect on proline content (Fig. 5). Proline value increased significantly under drought stress regardless of frequency. The highest concentration of proline was found in plants that were under maximum stress (50% of the ETc throughout the cycle) within the T3 and T6 treatments, in comparison with other treatments. Plants generally accumulate free proline in response to many environmental stresses (Hare and Cress, 1997). The results obtained in this study are considered similar to those obtained by Pirzad et al. (2011) and those found by Delauney and Verma (1993), in which maintaining high-water levels prevents the accumulation of osmotic elements, such as proline and total soluble sugars.

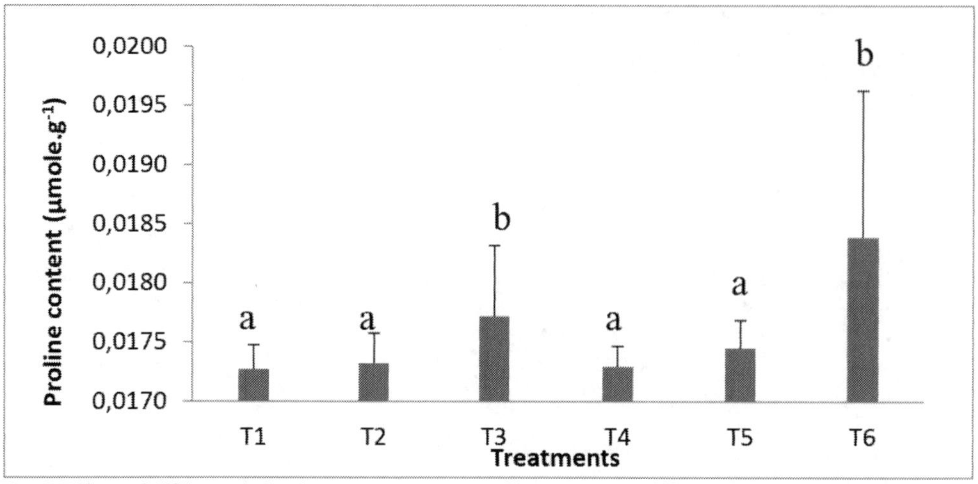

Fig. 5. Effect of different irrigation treatments on proline content. Different letters indicate significant differences according to the SNK test at $p < 0.05$. The error bars show the standard deviation of measurements between replications.

The results of variance analysis highlighted the presence of significant effect of the irrigation frequency factor on the leaf relative water content during the experiment, the irrigation regime did not have an effect on this parameter. The LRWC values for T1, T2, T3, T4, T5 and T6 were 92.74, 93.98, 91.65, 95.47, 95.31 and 95.66 %, respectively. Treatments with 5 % threshold (T4, T5 and T6) had the highest LRWC values (Fig. 6). It appears that LRWC increases with decreasing the threshold of the RAW. Surendar et al. (2013) reported that a reduction of net photosynthesis when the LRWC was reduced.

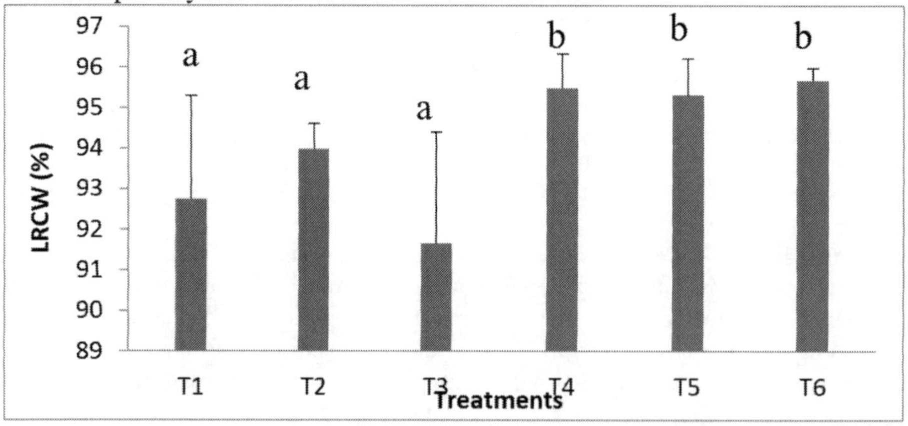

Fig. 6. Effect of different irrigation treatments on leaf relative water content (LRWC) of onion plants. Different letters indicate significant differences according to the SNK test at $p < 0.05$. The error bars show the standard deviation of measurements between replications.

2.3 Fruit yield, quality and water use efficiency

In the present study, no significant differences were observed between the six treatments concerning the marketable bulb yield (MBY) (Table 3). The T4 treatment reached the highest MBY (36.9 t/ha), followed by the T2 treatment (34.9 t/ha). It should be noted that in our study, as a result of experimental constraints, we planted at a density of 134,663 plants/ha, whereas the known planting density in the region varies between 200,000 and 250,000 plants/ha. This is in agreement with the findings of Santa Ollala et al. (2004), who did not obtain in their experiment a significant difference in dry matter yied between treatments that were water restricted at different levels and those that received 100 % of their water requirements. Other studies (Hegde, 1986; Al-Jamal et al., 2000) showed that onions can reach maturity with different levels of water restriction. According to Pelter et al. (2004), water stress can reduce the yield of yellow onion and that the greatest reduction could be observed when irrigation is withheld at the 5-leaf, 7-leaf stage. The highest onion yields have been recorded among low RAW thresholds (Mermoud et al., 2005). Irrigation at 100 % of the field capacity every 2 days recorded the maximum yield (Mbagwu et al. (1985). Statistical analysis showed a highly significant effect of the interaction between dose and frequency on bulb weight (Table 3). Bulb weight varied significantly between the T4 and the rest of treatments. Irrigation at 100 % ETc with 5 % RAW threshold produced the maximum bulb weight. These results are in line with Al-Moshileh (2007), which indicates that a high application of soil moisture leads to significant photosynthesis. Onion bulb diameter was found not being significantly influenced by the irrigation treatments applied. The largest diameter (70 mm) was from T4 which received maximum amount of water with 5 % RAW threshold while T5 treatment gave the smallest diameter (Table 3). Santa Ollala et al. (2004) found the similar results, considering that the increase in bulb was attributed to the enhancement in the water quantity applied. The highest onion bulb diameter was recorded by the irrigation at 100 % ETc (Enchalew et al., 2016).

The reduction of 50 % ETc in irrigation water amount had a significant increase in the WUE, reaching 0.83 kg/m3 and 0.93 kg/m3 for T3 and T6, respectively (Table 3). The lowest WUE was 0.52 kg/m3 obtained with the irrigation level equivalent to 100 % ETc and 10 % RAW threshold. The maximum WUE for onion was obtained in treatment with 50 % restriction in irrigation water quantity compared to control (Shock et al., 2000; Kebede, 2003). This result is supported by the results of Singh et al. (2017). With deficit irrigation, the reduction in yield is compensated by increased production from the additional irrigated area with the water saved (Ali et al., 2007).

Table 3. Effect of the irrigation dose and frequency on marketable bulb yield (MBY), bulb weight (BW), bulb diameter (BD) and water use efficiency (WUE). Mean values followed by different letters in each column indicate significant differences at $P \leq 0.05$ according to the SNK test.

Treatment	MBY (t/ha)	BW(g)	BD(mm)	IWUE (kg/m^3)
T1	34.2 a	292.1 a	65.6 a	0.52 a
T2	34.9 a	286.1 a	67.5 a	0.71 a
T3	27.0 a	318.9 a	60.6 a	0.83 b
T4	36.9 a	381.1 b	70.0 a	0.57 a
T5	28.2 a	302.3 a	60.1 a	0.58 a
T6	30.1 a	300.0 a	64.6 a	0.94 b

3 Conclusion

In this study, the effect of sustainable deficiency irrigation and irrigation frequencies are being tested on the growth, yield and quality of the onion crop. Our results showed no significant effect on growth parameters and relative chlorophyll content. The sustainable water restriction to 75% of the crop water requirement increased proline content and water use efficiency compared to the 100% ETc control irrigated treatment throughout the cycle.. However, this treatment recorded the highest marketable bulb yield with a RAW threshold of 10%. Severe deficit irrigation at 50% of the water requirement could significantly influence stomatal conductance and leaf temperature, and also recorded the lowest values in terms of yield and quality parameters. Concerning the eco-physiological parameters; total chlorophyll, no significant difference was generated except for the leaf relative water content, which was significantly affected by the irrigation frequency. According to this study, it is advisable to apply 100 % of water requirements with a trigging threshold of the readily available water content of 5%, if water is not a limiting factor. But, if water is rare, it could be better to apply 75 % of water requirements, saving 25% of irrigation water without having a significant effect on production and quality.

Acknowledgements

This work has been carried out under the auspices of the MCRDV research project MCRDV "COMPETITIVE MECHANISM OF DEVELOPMENT RESEARCH ANDVULGARISATION" and we gratefully acknowledge the financial support provided by the
Ministry of Agriculture, Fisheries, rural development, water and forest. We are also grateful to the National School of Agriculture of Meknes (ENAM) for their support andcooperation during the project which made this work possible.The authors also wish to thank students Lamrahli Douae, and Dahane Mahdi for their valuable contribution during their final engineering projects.

References

1. A. Abdelkhalik, N. Pascual-Seva, I. Nájera, A. Giner, C. Baixauli, B. Pascual, 2019. Yield response of seedless watermelon to different drip irrigation strategies under Mediterranean conditions. Agric. Water Manage. 212, 99–110.
2. A. Abouabdillah, O. Oueslati, A.M. De Girolamo, A. Lo Porto, 2010. Modeling the impact of climate change in a Mediterranean catchment (Merguellil-Tunisia). Fresenius Environ. Bull. 19, 2334–2347.
3. A.Al- Moshileh, 2007. Effects of planting date and irrigation water level on onion (Allium cepa L.) production under central Saudi Arabian conditions. J Basic Appl Sci. 8, 14-28.
4. A. Bouaziz, K. Belabbes, 2002. Efficience productive de l'eau en irrigué au Maroc. Hommes, Terre & Eaux, 32, 4p.
5. A. Capra, S. Consoli, B. Scicolone, 2008. Water management strategies under deficit irrigation. J. Agric. Eng. 39, 27.
6. A. J. Delauney, D. P. S. Verma, 1993. Proline biosynthesis and osmoregulation in plants. PLANT J, 4, 215-223.
7. A. Mermoud, T. D. Tamini, H. Yacouba, 2005. Impacts of different irrigation schedules on the water balance components of an onion crop in a semi-arid zone. Agric. Water Manage. 77, 282-295.
8. A. N. Bagali, 2010. Studies on Integrated Nutrient Management and Scheduling of Drip Irrigation in Onion (Allium cepa L.) cv. Telagi Red (Doctoral dissertation, UAS Dharwad).
9. A. Pirzad, M. R. Shakiba, S. Zehtab-Salmasi, S. A. Mohammadi, R. Darvishzadeh, A., Samadi, 2011. Effect of water stress on leaf relative water content, chlorophyll, proline and soluble carbohydrates in Matricaria chamomilla L. J. Med. Plant Res. 5, 2483-2488.

10. A. Singh, S. K. Srivastava, K. J. Kumar, D. M. Denis, A.Thomas, 2017.Yield, irrigation production efficiency and economic return of onion under variable irrigation methods. The allahabad farmer, 73(1).
11. Al-Jamal, M. S., Sammis, T. W., Ball, S., Smeal, D., 2000. Computing the crop water production function for onion. . Agric. Water Manage. 46, 29-41.
12. B. Enchalew, S. L. Gebre, M. Rabo, B. Hindaye, M. Kedir, 2016. Effect of Deficit Irrigation on Water Productivity of Onion (*Allium cepal.*) under Drip Irrigation. Irrigat Drainage Sys Eng. 5 :172. doi : 10.4172/2168-9768.1000172
13. D. K. Rop, E. C. Kipkorir, J. K. Taragon, 2016. Effects of Deficit Irrigation on Yield and Quality of Onion Crop. J. Agric. Sci. 8, 112.
14. D. M. Hegde, 1986.Growth analysis of onion (*Allium cepa* L.) in relation to irrigation and nitrogen fertilization. J AGRON CROP SCI. 157, 227-234.
15. D.J. Chalmers, P.D. Mitchell, L. van Heek, 1981. Control of peach tree growth and productivity by regulated water supply, tree density, and summer pruning. J AM SOC HORTIC SCI. 106, 307-312.
16. E. Fereres, M.A. Soriano, 2007. Deficit irrigation for reducing agricultural water use. J. Exp. Bot. 58, 147–159.
17. F. Hanci, E. Cebeci, 2014.Investigation of Proline, Chlorophyll and Carotenoids Changes Under Drought Stress in Some Onion (*Allium Cepa* L.) Cultivars. TURKJANS. 2, 1499-1504.
18. F. M. Santa Olalla, A. Domınguez-Padilla, R. Lopez, 2004. Production and quality of the onion crop (Allium cepa L.) cultivated under controlled deficit irrigation conditions in a semi-arid climate. Agric. Water Manage. 68, 77-89.
19. Food and Agriculture Organization of the United Nations (FAO), 2018. FAOSTAT Website. Accesed on 12 May 2020. http://www.fao.org/faostat/en/#data.
20. G. Cornic, J. M. Briantais, 1991. Partitioning of photosynthetic electron flow between CO2 and O2 reduction in a C3 leaf (*Phaseolus vulgaris L.*) at different CO2 concentrations and during drought stress. Planta. 183, 178-184.
21. G. Q. Pelter, R. Mittelstadt, B. G. Leib, C. A. Redulla, 2004. Effects of water stress at specific growth stages on onion bulb yield and quality. . Agric. Water Manage. 68, 107-115.
22. Goldhamer, DA., Viveros, M., Salinas, M., 2006. Regulated deficit irrigation in almonds : effects of variations in applied water and stress timing on yield and yield components. Irrig. Sci. 24, 101-114.
23. H.E. Igbadun, A.A. Ramalan, E. Oiganji, 2012.Effects of regulated deficit irrigation and mulch on yield, water use and crop water productivity of onion in Samaru, Nigeria. Agric. Water Manage. 109, 162-169
24. J. Galmes, J. Cifre, H. Medrano, J. Flexas, 2005. Modulation of relative growth rate and its components by water stress in Mediterranean species with different growth forms. Oecologia. 145, 21-31.
25. J. Hueso, J. Cuevas, 2004. Deficit irrigation effects on flowering of loquat. Options Méditerranéennes Serie A, 58, 105-108.
26. J. S. C. Mbagwu, J. O. Osuigwe, 1985. Effects of varying levels and frequencies of irrigation on growth, yield, nutrient uptake and water use efficiency of maize and cowpeas on a sandy loam ultisol. PLANT SOIL. 84, 181-192.
27. J.Girona, M. H. Behboudian, M. Mata, J. Del Campo, J. Marsal, 2010. Exploring six reduced irrigation options under water shortage for "Golden Smoothee" apple : Responses of yield components over three years. Agric. Water Manage. 98, 370–375.
28. J.J. James, J.H. Richards, 2006.Plant nitrogen capture in pulse-driven systems: interactions between root responses and soil processes. J. Ecol. 94, 765–777.

29. J.M. Costa, M.F. Ortuño, M.M. Chaves, 2007. Deficit irrigation as a strategy to save water : physiology and potential application to horticulture. J. Integr. Plant Biol. 49, 1421–1434.
30. K. D. Montagu, K. C. Woo, 1999. Recovery of tree photosynthetic capacity from seasonal drought in the wet–dry tropics : the role of phyllode and canopy processes in Acacia auriculiformis. FUNCT PLANT BIOL. 26, 135-145.
31. K. Surendar, D. Devi, I. Ravi, Jeyakumar, P., Velayudham, K., 2013. Effect of water deficit on relationship between yield and physiological attributes of banana cultivars and hybrids. African J. Plant Sci. 7, 374–383.
32. Kebede Woldetsadik., 2003. Shallot (*Allium cepa var, ascolonicum*) responses to plant nutrients and soil moisture in a sub-humid tropical climate. PhD. Thesis. Swedish University of Agricultural Sciences, Alnarp, Sweden.
33. L. S. Bates, R. P. Waldren, I. D. Teare, 1973. Rapid determination of free proline for water-stress studies. Plant and soil, 39, 205-207.
34. L.S. Pereira, T. Oweis, A. Zairi, 2002. Irrigation management under water scarcity. Agric. Water Manage. 57, 175–206
35. L'Economiste, 2018. Culture de l'oignon : Comment Fès-Meknès est devenue championne. Accesed on 12 May 2020. https://www.leconomiste.com/article/1032222-culture-de-l-oignon comment-fes-meknes-est-devenue-championne.
36. M. Bzioui, 2004.Rapport national 2004 sur les ressources en eau au Maroc. UN Water-Africa, Nov.2004.
37. M. H. Ali, M. R. Hoque, A. A. Hassan, A. Khair, 2007. Effects of deficit irrigation on yield, water productivity, and economic returns of wheat. Agric. Water Manage. 92, 151-161.
38. N. El Jaouhari, A. Abouabdillah, R. Bouabid, M. Bourioug, L. Aleya, M. Chaoui, 2018. Assessment of sustainable deficit irrigation in a Moroccan apple orchard as a climate change adaptation strategy. Sci. Total Environ. 642, 574–581.
39. N.C. Turner, 1981. Techniques and experimental approaches for the measurement of plant water status. Plant Soil. 58, 339–366.
40. P. D. Hare, W. A. Cress, 1997. Metabolic implications of stress-induced proline accumulation in plants. Plant Growth Regul., 21, 79-102.
41. P. Steduto, T.C. Hsiao, E. Fereres, D. Raes, 2012. Crop Yield Response to Water, FAO Irrigation and Drainage Paper No. 66. Rome, Italy.
42. Q. Chai, Y. Gan, C. Zhao, H.L. Xu, R.M. Waskom, Y. Niu, K.H.M. Siddique, 2016. Regulated deficit irrigation for crop production under drought stress. A review. Agron. Sustain. Dev. 36, 1–21.
43. R. A. L. Kanton, L. Abbey, R. H. Gbene, 2003. Irrigation schedule affects onion (*Allium cepa L.*) growth, development, and yield. J. Veg. Crop Prod. 9, 3-11.
44. R. López-Urrea, F. M. de Santa Olalla, A. Montoro, P. López-Fuster, 2009. Single and dual crop coefficients and water requirements for onion (Allium cepa L.) under semiarid conditions. Agric. Water Manage. 96, 1031-1036.
45. R.G. Allen, L.S.Pereira, D. Raes, M. Smith and others, 1998. Crop Evapotranspiration Guidelines for Computing Crop Water Requirements—FAO Irrigation and Drainage Paper 56. FAO, Rome vol. 300, D05109.
46. R.G. Evans, E.J. Sadler, 2008. Methods and technologies to improve efficiency of water use. Water Resour. Res. 44, 1–15.
47. S. Ayas, 2019. Water-Yield Relationships in Deficit Irrigated Onion. TURJAF. 7, 1310-1320.
48. S. P. Sharma, D. I. Leskovar, K. M. Crosby, A. Volder, A. M. H., Ibrahim, (2014). Root growth, yield, and fruit quality responses of reticulatus and inodorus melons (Cucumis

melo L.) to deficit subsurface drip irrigation. Agricultural Water Management, 136, 75–85.
49. Shock, C.C., E.B.G. Feibert L.D. Saunders., 2000.Irrigation criteria for drip-irrigated onion. Horticultural Science. 35, 63-66
50. T. C. Hsiao, (1973). Plant responses to water stress. Annu. Rev. Plant Physiol. 24, 519-570.
51. T.A. Howell, 2001. Enhancing water use efficiency in irrigated agriculture. Agron. J. 93, 281–28.
52. Y. Brouzyine, A. Abouabdillah, A. Hirich, R. Bouabid, R. Zaaboulc, L. Benabdidate, 2018. Modeling sustainable adaptation strategies toward a climate-smart agriculture in a Mediterranean watershed under projected climate change scenarios. Agric. Syst. 162, 154–163.
53. Z. Karim, (1981). Water use by onion in grey terrace soil. Bangladesh Horticulture. 9, 61-64.

Advancing knowledge about restricted irrigation strategies on commercial peach plantation under Mediterranean condition

Aziz Abouabdillah[1]*, *Soukaina* Radi[1], *Adil* Asfers[2], *Youssef* Brouziyne[3], *Omnia* El Bergui[1,4], *Mohamed* Bourioug[1], *Fouad* Amlal[5], and *Rachid* Bouabid[1]

[1]Ecole Nationale d'Agriculture de Meknès, Meknès, Morocco
[2]Agricultural Training and Research Center, Providence Vertecompany,LOUATA farm, Séfrou, Morocco
[3]Mohammed VI Polytechnic University, International Water Research Institute, Benguerir, Morocco
[4]Institut Agronomique et vétérinaire Hassan II, Rabat, Morocco
[5]Université Chouaib Doukkali Avenue Jabran Khalil Jabran, B.P 299-24000, El jadida Grand-Casablanca, Maroc

Abstract. This study was conducted to investigate the effects of different restricted irrigation strategies on peach trees. Sustainable and regulated deficit irrigation treatments were applied during the 2018/2019 production season in a commercial plot of "Bénédicte" variety located at the Atlas Mountain's region of Morocco. Five different treatments were tested: T1, T2 and T3 with an application of respectively 125% (over-irrigation), 100% (control) and 75% (deficit-irrigation) of Crop Water Requirement (CWR) throughout the whole crop cycle; These treatments were classified under Sustainable Deficit Irrigation treatments (SDI). Regulated Deficit Irrigation treatments (RDI) comprising of treatments T4, and T5 which correspond respectively to applications of 75% ETc and 50% ETc during the pit hardening stage (PH), and 100% ETc during the rest of the cycle. The results showed that deficient irrigation treatments had no effect on vegetative growth parameters.A downward trend in average fruit weight and size at harvest was observed in the T3 treatment. A significant increase in sugar content was observed in T3 and T5 compared to the control T2. With regard to biochemical parameters, the deficient treatment (T5) recorded the highest proline content in response to water stress, followed by T3 and T4. Therewas no significant difference between the crop yields under SDI and RDI treatments, but T4 produced arelatively higher yield of 47 T/Haamong the treatments.Thisled to the water use efficiency (WUE) performance ranking: T3 with 10.63 kg/m^3, T4 (75% PH) was in second place with a WUE of 9.6 kg/m^3, finally T5 with an efficiency of 9.35 kg/m^3.

---

*Corresponding author:aziz.abouabdillah@gmail.com

1 Introduction

In many parts of the Mediterranean region, water is a major constraint for crop production, given the dramatic increase in demand for this element on one hand and the consequences of climate change on the other hand. Indeed, by 2050, rainfall amounts in North African countries would be reduced by 20 to 50% compared to current average annual values [1].Globally, irrigation consumes around 85% of available freshwater[2]; and this ratio goes up to 93% in Morocco[3]. Due to various agricultural development plans, irrigated fruit trees in Morocco have been developed considerably in recent years and becameone of the largest consumers of irrigation water;this raises questions about the sustainability of fruit tree crops in such water stressed country. Therefore, and under this context, the improvement of the water use efficiency is crucial.Thismight be achievedthrough the application of adequate irrigation strategies, such asdeficit irrigation [4].

Despite of the large number of studies reported on deficit irrigation strategies, there is still significant knowledge to gain on the response of tree crops (especiallypeachestrees) to restricted irrigation scheme under the Mediterranean climate. Among the various attempts to explore this field, there have been some works on the improvement of water use efficiency in peaches trees using deficit irrigation strategies in Lleida area in Northern Spain[5], The exploration of root dynamics of peach trees under partial rootzone drying and continuous deficit irrigation in Murcia region (Spain)[6]; other researchers focused more on physiological response of peaches tree submitted to deficit irrigation strategies in Tunisia[7]and in France[8]. As per the very few works on the consequences of restrictive irrigation management on both productivity and fruit quality of peaches trees in the Mediterranean area, Faci et al., [9], tried to monitor a set of yield and fruit quality parameters while applying different irrigation schemes in Ebro Valley (Spain), and they proved that saving irrigation water is possible without affecting productivity and fruit quality under the experiment conditions.

In any regulated deficit irrigation management strategy, the selection of the growth stage (s) where water restriction will be applied is of crucial importance.Most specialists agree that that pit hardening (PH) stage (Stage II) is an adequate period for the application of deficit irrigation (DI)on peaches trees [10]. In addition to water saving, DI at stage II might improve fruit quality too [11, 12, 13]. The improvement in fruit quality was mainly expressed as increase of fruit soluble solids (FSS) content. As a result, a consumer panel indicates an increased preference for fruits grown under restricted irrigation [14]. Moreover, applying deficit Irrigation at stage II reduces fruit drop phenomenon at maturity[11, 13] and could decrease shoot growth[11, 12, 13]. As peryield, the application of regulated deficit irrigation (RDI) at the pit hardening (PH) stage does not affect the final yield according to some authors[11, 13, 15]. However, in other studies, a significant reduction in yield and fruit size at harvesting stage has been recorded[12, 16]. In addition, for sustainable deficit irrigation (SDI) and since water stress affects all stages of fruit development, the effect on production is most apparent through the reduction in size, fruit weight and yield[11, 17, 18, 19]. This paper discusses the effect of sustainable deficit irrigation (SDI) and regulated deficit irrigation (RDI) on peach trees; The specificobjective of this study is to refine the crop coefficient Kc model of peaches published in FAO-Bulletin 56 to the real conditions of Atlas Mountainsarea in Morocco in order to achieveoptimal production in terms of both quantity and quality standards.

2 Materials and methods

2.1 Experimental trial

This research was carried out during the 2018-2019 growing season on a 17 years old peach cultivar 'Benedicte'. The trial was set up at the orchard: "*LOUATA*", located in the North-East of the Middle Atlas Mountains chain, inMorocco; the growing area is surrounded bythe two cites Fez and Séfrou, at an altitude varying from 400 to 770 m above meansea level. This region has a warm Mediterranean climate with dry summers according to the koppen-Geiger classification. The mean annual rainfall and temperature were respectively 502 mm and 15.9 °C based on 1982 to 2012 climate dataset. The trees were planted in 2003 at 5 m row spacing and 3 m tree spacing within the rows. The trees were grafted onto GF-667 rootstock and conducted on goblet. The soil has the following textural composition: 50.0% clay, 32.3% silt and 18.1% sand over a depth of 40 cm. Thewater content at the field capacity (FC) was estimated to 42.1% and a wilting point moisture content (WP) of 2.96%,therefore water retention capacity is around125mm per 1m depth. During the experiment trial, the sum of the reference evapotranspiration ET0 during the irrigation season (from fruit set to harvest) was about 520 mmwhile the cumulative precipitation (mm) was 36.2 mm. Precipitation was taken into account in the estimation of irrigation water for each treatment. Apart from the irrigation protocols described below, the experimental plot was managed in accordance with the standard farm practices including fertilization, pest management and weed control. This means that all the studied treatments havereceived the same amounts ofinputs except for irrigation. Thinning and winter pruning were also carried out in the same manner for all treatments. The experimental plot was equipped with a localized drip irrigation system. Each planting line had two polyethylene lateral pipes, equipped with integrated drippers evenly spaced by 60 cm on the lateral and delivering 2.2 liters per hour of water.

2.2 Irrigation treatments

The table 1 represents the different irrigation treatments subject to investigation in this study. The tested five treatments are:T1 the over-estimate of water crop requirement (125% ETc), T2 representing the control (100% Crop evapotranspiration ETc), T3 the sustainable deficit irrigation (SDI) treatment (75% ETc) that was applied over the whole fruit development stage from early March to late July, and the two regulated deficit irrigation (RDI) treatments T4 and T5 where respectively 75% and 50% ETc have been applied during the phase II of fruit development (pit hardening PH) and 100% ETc was applied during the rest of the cycle.

The calculation of the irrigation water requirements is based on a water balance, which expresses the difference between the crop's water requirements and the water input from natural sources, given by the formula:

$$\boldsymbol{GWR} = \left(\frac{\boldsymbol{Kc \times ET_0}}{\boldsymbol{IE \times UC}}\right) - \boldsymbol{E_r} \qquad (1)$$

With, GWR: Gross Irrigation water requirement (mm); Er: Effective rainfall (mm), also called useful rainfall, calculated by removing at total rainfall, the losses by runoff or deep percolation (70%); Kc: Crop coefficient of the peach proposed in FAO Bulletin 56[20] (Kc in = 0.55 during flowering and fruiting; Kc mid = 0.9 at development and Kc end = 0.65 during ripening); IE is the irrigation efficiency (0.9 since the irrigation network was new), and UC, uniformity coefficient was calculated 0.93 ; and ET0 (mm) is the reference daily evapotranspiration calculated by the modified Penman-Montheinth formula[21]. The estimation of ET0 was based on climate parameters obtained from the weather station installed within the trial orchard.

Table 1. Description of the different irrigation treatments subject to investigation in the study.

Treatment type	Treatment name	Description
Over-irrigated treatment	T1 (125%)	supply of 125% of the crop's water requirements throughout the growth cycle
Control treatment	T2 (100%)	supply of the exact needs of the crop (100% ETc) during the whole growth cycle
Sustainable Deficit Irrigation treatment	T3 (75%)	supply of 75% of the crop's water requirements throughout the growth cycle
Regulated Deficit Irrigation treatment	T4 (75% PH)	supply of 75% of the crop's water needs during the stage of pit hardening and 100% of the needs during the rest of the growth cycle.
Regulated Deficit Irrigation treatment	T5 (50%PH)	supply of 50% of the crop's water needs during the stage of pit hardening and 100% of the needs during the rest of the growth cycle.

2.3 Experimental design

A randomized full block design was used in this trial with four block repetitions. Each replicated block hostedall the five treatments. Each treatment grouped 4 trees in the same block; three of them with the same trunk diameter were selected to measure parameters while the others were considered as guard trees.

2.4 Measured parameters

Three trees per replicate (12 trees per treatment) were selected for monitoring growth, yield and quality parameters. Trees were selected based on trunk diameter to reducevariability amongtrees.

2.4.1 Plant growth parameters

The growth of mixed twigs was monitoredby measuring the length and basal diameter of the twigs. On four twigs per tree, distributed over the four cardinal points, three measurements per parameter were performedfrom the cell multiplication stage to harvest. Measurements were carried outusing a decameter for the length (cm) and a precision caliper for the basal diameter (mm) of the twigs. Measurements were made once every 30 days. To assess the effect of the water regime on the growth of fruiting structures in the following year, the evolution of the length of young shoots was monitored on four shoots per tree at a rate of one shoot per orientation. This parameter was measured once every 15 days.

2.4.2. Production Parameters

Considering the feature of at maturity' fruit fall of the studied variety, the number of fruits was calculated on fourbranches/tree after thinning till the last day before harvesting. In order to determine fruit growth rate for each treatment, weekly monitoring of equatorial and longitudinal diameters was carried out on fourfruits per tree marked at fruit set stage on all the 60 trees. Fruits were equally distributed over the four orientations of the tree. The final fruit diameter was also measured after harvesting on a sample of 72 fruits/block using a precision caliper. On the same sample, fruitweight was measured using a precision balance. Harvest was carried out over 4 passes. In addition, weighing of the whole production was carried out at harvest for the 60 trees of the trial, and the average yield was then calculated for each treatment. The yield evaluation was based on the sum of the productions of the fourharvesting passes.

2.4.3. Fruit quality parameters

Three key fruit quality parameters were selected to be monitored in this experiment:
-Sugar content (°Brix): on 72 fruits/block, the soluble solids content (°Brix) was measured using a digital refractometer.
-Firmness: the firmness (kg/cm^2) of the fruit was measured by a penetrometer on both sides of the fruit. The measurements were performedon the same sample used for the sugar content measurements.
-Acidity of the fruit: to measure the acidity of the fruit (g of malic acid/L), 10 ml of peach juice is mixed with 50 ml of distilled water and3 drops of phenolphthalein. The titration was then carried out by slowly pouring in the 0.1 N NaOH solution until the juice/water mixture change from colorless to pink. The titratable acidity is then determined by multiplying the volume of added sodium hydroxide (0.1N NaOH) by the coefficient 0.67 (factor corresponding to the dominant acid in peach juice which is malic acid).

2.4.4. Biochemical parameters

The extraction and determination of proline was carried out according to the method described by [22]. 500 mg of leaf samples are ground in 10 ml of 3% Sulfosalicylic acid solution and the extract is then filtered through Whatman filter paper. Then 1 ml extract is taken to which 1 ml acetic acid (CH3COOH), 25 mg ninhydrin and 1 ml mixture (12 ml distilled water, 30 ml acetic acid and 8 ml Ortho-phosphoric acid) are added. The tubes are incubated for 30 min at 100 degrees in a water bath. They gradually turn red. After cooling, 4 ml Toluene is added to each tube. After stirring two phases are formed: the upper one containing proline is recovered and finally the optical density is determined by a spectrophotometer at wavelength 528 nm. The concentration of proline in the plant leaves is determined by referring to a standard straight line prepared from a standard range of proline.

2.4.5. Water productivity

Agronomic water use efficiency (or water productivity) is an important index that measures quantity produced from each cubic meter of consumed water for each treatment. The calculation formula is as follows: WUE = Production (kg)/ quantity of water applied (m3).

2.5 Statistical analysis

The results were analyzed in SPSS and Microsoft Office Excel programs using an analysis of unidirectional variance (ANOVA) followed by the Student Newman and Keuls (SNK) test used to compare means at the 5% threshold.

3 Results

3.1 Vegetative growth parameters

3.1.1. Mixed twig evolution

In order to assess the effect of RDI and SDI on the growth of mixed twigs, the development of these structures was monitored on the basis of one observation per month. The measurements involved the length and basal diameter evolution of the mixed twigs.

Concerning the SDI (fig 1 & 2), the first measurement of twig elongation carried out in 55 days after flowering (cell division stage) and after 13 days of irrigation according to the experimental protocol, showed that T1 treatment (125%) in the first place with 62.72 cm in length, followed by T2 (100%) with 61.01 cm and finally T3 with an average length of 59.67 cm. Two months later, T1 showed a length increment of 9.75 cm, T2 showed 11.5 cm longer and T3 has increased its average length by 9.53 cm. However, these elongation differences are not statistically significant, which makes it possible to say that a 25% excess or restriction in water supply compared to requirements does not affect the growth of mixed twigs in terms of length.Similarly, for basal diameter, the analysis of variance did not show any significant difference between T1 (125%ETc); T2 (100%ETc) and T3 (75%ETc). Furthermore, between 55 and 115 days after flowering, the basal diameter of mixed twigs of T1 (125%) revealed an increase of 1.30 mm, followed by T2 (100%) with 1.08 mm longer and finally T3 (75%) with 0.93 mm increase.

As for the RDI (fig 3 and 4), no significant difference between the T2 control and the RDI treatments T4 (75% PH) and T5 (50% PH) was revealed, neither for the length of the mixed twigs nor for their basal diameter evolution. Furthermore, the mixed twig growth difference between T5 and the control was about 1.09 cm less than T2 in terms of length and 0.07 mm in terms of basal diameter. On the other hand, T4 showed higher growth deviations from the control T2, 2.16 cm in lengthand 0.24 mm in basal diameter.

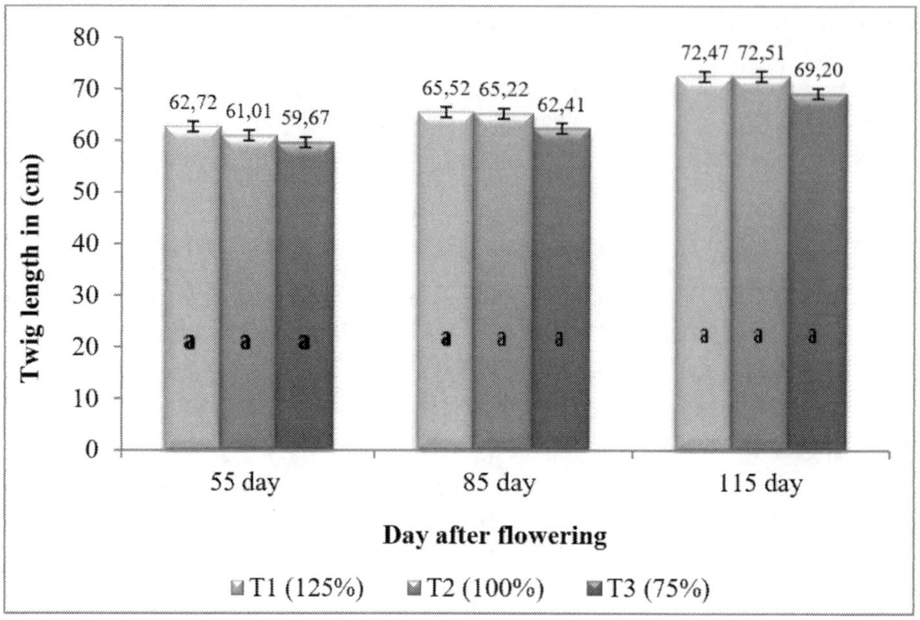

Fig. 1. Effect of Sustainable Deficit Irrigation (SDI) on the mixed twigs length during threemeasurements at 55 day; 85 day & 115 day after flowering. The lines represent the standard deviation. The treatments are described in section 2.2. The different letters indicate significant differences according to the SNK test at P ≤ 0.05

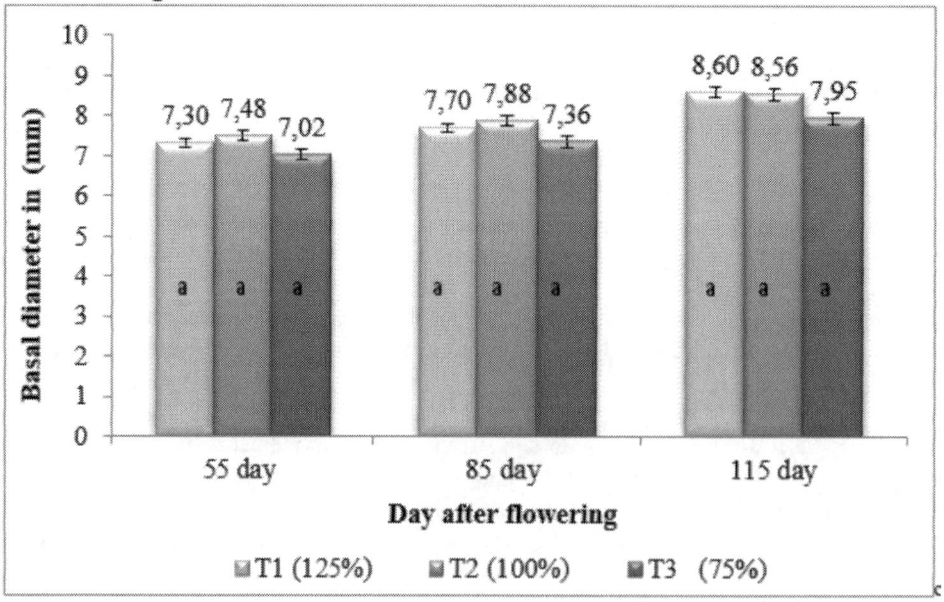

Fig. 2. Effect of Sustainable Deficit Irrigation (SDI) on the evolution of the basal diameter of mixed twigs during threemeasurements at 55 day; 85 day & 115 day after flowering. The lines represent the standard deviation. The treatments are described in section 2.2. The different letters indicate significant differences according to the SNK test at P ≤ 0.05

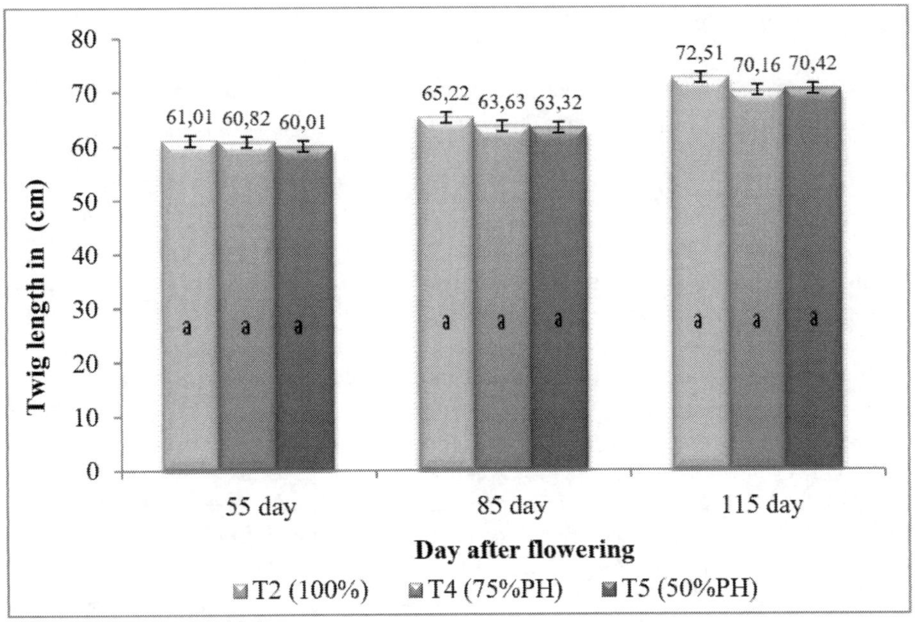

Fig. 3. Effect of Regulated Deficit Irrigation (RDI) on the mixed twigs length during threemeasurements at 55 day; 85 day & 115 day after flowering. The lines represent the standard deviation. The treatments are described in section 2.2. The different letters indicate significant differences according to the SNK test at $P \leq 0.05$.

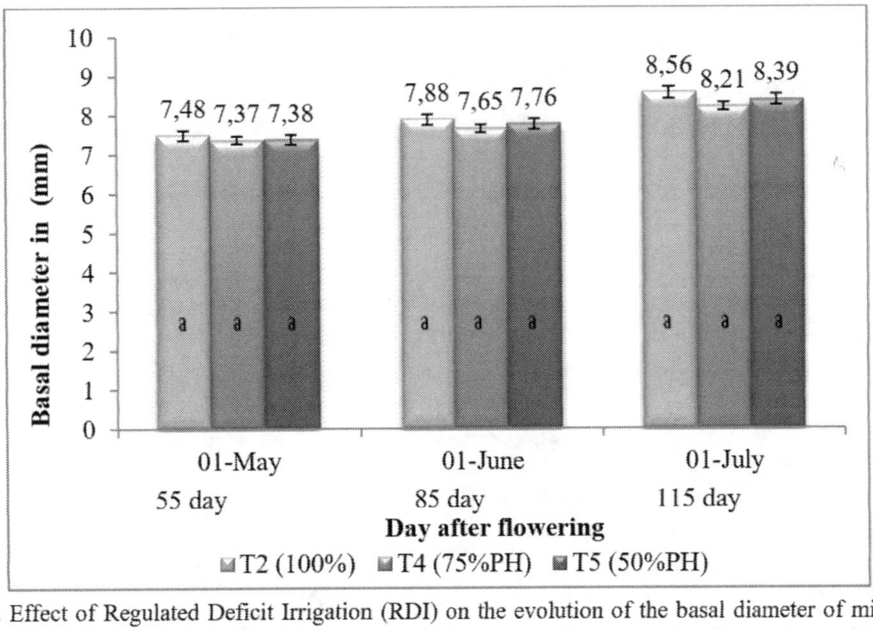

Fig. 4. Effect of Regulated Deficit Irrigation (RDI) on the evolution of the basal diameter of mixed twigs during threemeasurements at 55 day; 85 day & 115 day after flowering. The lines represent the standard deviation. The treatments are described in section 2.2. The different letters indicate significant differences according to the SNK test at $P \leq 0.05$.

3.1.2. Shoot elongation

For peaches, the young shoots are the structures that bear the fruit for the following year; hence the importance of monitoring their evolution and highlighting the effect of the water regime on the elongation of these structures. Indeed, 4 measurements were taken every 15 days to monitorthe length of these structures. Regarding the SDI, statistical analysis did not show any significant effect of irrigation dose on the elongation of the young shoots (fig 5). Furthermore, starting from 0 cm length in the flowering stage for all treatments (T1; T2 and T3), the T2 control was able to reach an average length of 19.11 cm by 115 days after flowering (fruit swelling stage), followed by T1 with 17.01 cm as average length and finally T3 with 14.58 cm as average length. A presentation of the growth rate under the different treatments during this periodmakes it possible to separate between two phases, a first one going from cell multiplication to mid-swelling, and a second one lying between late swelling and maturity. As for the first period, the three treatments T1; T2 and T3 did not show any significant difference between each other in terms of elongation rate of the young shoots;it is only from mid-swelling (second period) that the difference between the growth rate of shoots in the three treatments begins to appear, putting on the one hand T2 treatment with the highest elongation rate 0.42 cm/day and on the other hand T1 and T3 with lower growth rates 0.14 and 0.10 cm/day respectively. No significant difference was observed between T1 and T3 in growth rate. However, these differences in elongation rate did not show any significant difference between the three treatments T1, T2 and T3 in final shoot length.

For RDI, there was no significant difference in shoot length between the T2 control; T4 (75% PH) and T5 (50% PH) (fig 6). Indeed, T2 recorded a shoot length of 19.11 cm on 01-July, which is 3.11 cm longer than T4 and 4.4 cm longer than T5. For the elongation rate, it should be noted that no significant difference between the T2; T4 and T5 was recorded, except for the late growth period when shoot growth in T2 was faster than in T4 and T5.

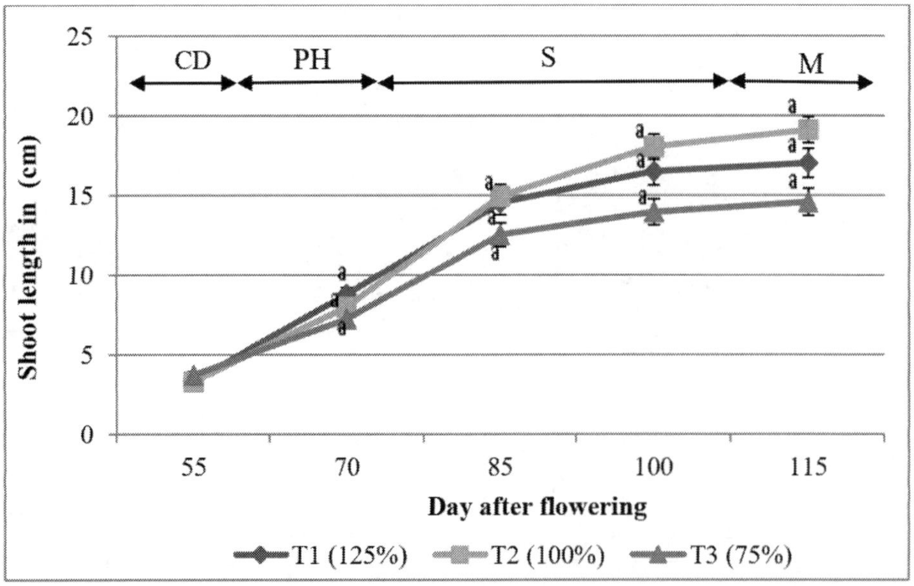

Fig. 5. Effect of Sustainable Deficit Irrigation (SDI) on the shoot length during fivemeasurements at 55; 70; 85; 100 & 115 days after flowering. The lines represent the standard deviation. The treatments are described in section 2.2.CD: cell division stage; PH: pit hardening stage; S: swelling fruit stage;

M: maturity. The different letters indicate significant differences in the corresponding date, according to the SNK test at $P \leq 0.05$.

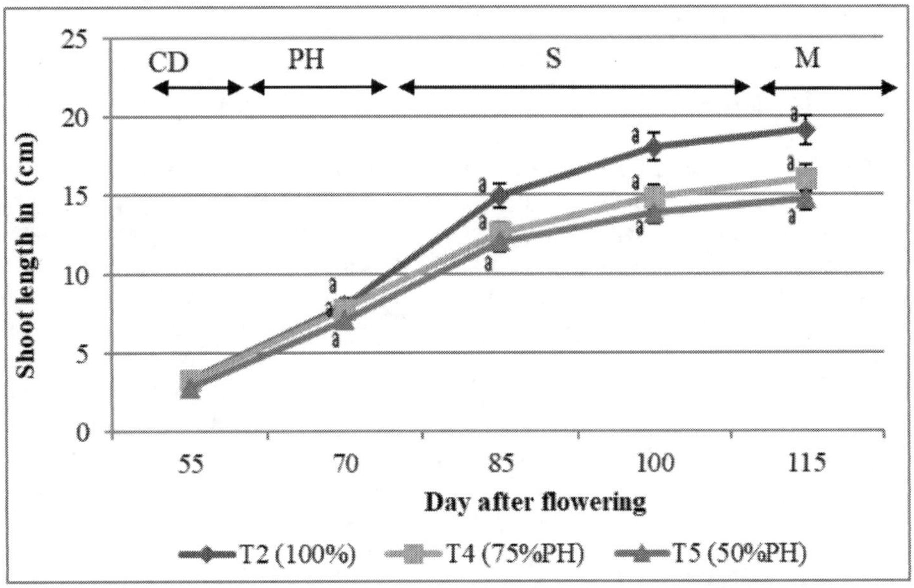

Fig. 6. Effect of Regulated Deficit Irrigation (RDI) on the shoot length during fivemeasurements at 55; 70; 85; 100 & 115 days after flowering. The lines represent the standard deviation. The treatments are described in section 2.2.CD: cell division stage; PH: pit hardening stage; S: swelling fruit stage; M: maturity. The different letters indicate significant differences in the corresponding date, according to the SNK test at $P \leq 0.05$.

3.2 Production parameters

3.2.1 Fruit number/tree

Our starting point was a number of around 320 fruits per tree after thinning operation, with no significant difference between T1, T2 and T3 on the one hand (Table 2), and between T2, T4 and T5 on the other (Table 3). It is noteworthy that T2 recorded a deviation of 18 fruits/tree from the 320 fruits intended. Whereas T1 and T3 recorded, respectively, only 1 and 4 additional fruit per tree. For RDI treatments, T4 was found to increase by 5 fruits/tree compared to T2, while T5 recorded 6 fewer fruits/tree compared to T2.

3.2.2 Average yield

In their 17th year, the trees in the experimental plot achieved a maximum yield of 46.43 T/ha recorded for the T2 control. For the SDI, statistical analysis revealed that irrigation rate has no effect on average yield; and therefore, no significant difference between the T1 treatment resulting in 45.50 T/ha and T2 on the one hand; and T3 reaching -2.72 T/ha lesser than T2 on the other hand. This makes it possible to say that a 25% increase or decrease of the irrigation dose compared to the crop water requirements (100% ETc) does not affect yields. Similarly, for RDI, the analysis of variance revealed no significant difference between the T2 control and the regulated deficit irrigation treatments T4 (75% PH) and T5 (50% PH). In fact, T4 recorded +0.93 T/ha more than the control treatment, a difference that could be due to the number of fruits/tree which was higher for T4 than it was for T2. A non-significant decrease in yield was observed in T5 treatment compared to T2, with almost -3% less (1.45T/ha).

3.2.3. Fruit drop rate

The Benedicte cultivar, subject of this study, is among the varieties that are characterized by fruits dropping off at maturity. To investigate the effect of the irrigation dose on the intensity of this feature, the variation in the number of fruits between thinning and harvesting was monitored. This allowed us to show a clearly significant difference between T1 and T3 in the fruit drop rate, in fact, T1 showed a high trend of fruit falling at maturity, with 13.25% of fruit loss at maturity, compared to T3, which contributed to losses of around 6.32%. So a shift from 125% ETc to 75% ETc might allow the preservation of almost 7% of the production from this phenomenon. On the other hand, no significant difference has been revealed between T1 and T2 or between T2 and T3 in terms of fruit loss at maturity. As for RDI, the analysis of variance revealed no significant difference between T2; T4 and T5. However, the T2 control recorded losses of 9.39%, which is 0.77% and 3.85% higher than T4 and T5 respectively.

3.2.4 Fruit weight

Fruit weighting results in the different treatment are presented in table 2 and 3. Compared to the control one, the T3 treatment (SDI) showed no significant difference in terms of fruit weight, T1 also did not show a significant difference from the control. Furthermore, a significant difference was noticed between T1 and T3, with 8.7% additional weight for T1 compared to T3. As for the RDI, it was found that the fruit weight is insensitive to regulated deficit irrigation at 75% and 50% of the requirements in the pit hardening phase.

3.2.5 Fruit size

Figure 7 shows the fruit diameter development speed for the three irrigation doses T1(125%), T2(100%) and T3(75%) over fruits development phase; The reference evapotranspiration values are also represented in the same figure. An initial interpretation of the figure allows us to say that the evolution of the fruit diameter in the three treatments T1, T2 and T3 followed the same trend during the three stages of fruit development. It is only at the end of the fruit development phase that a difference in growth speed begins to appear between the treatments, putting T1(125%) fruits in the first place with the highest growth speed, at around 1 mm/day, followed by T2(100%) with almost 0.93mm/day and lastly T3 at 0.74mm/day rate. This difference in growth speed could have resulted in a significant difference in the final size of the fruit. Indeed, the fruit size at harvest of T1 came in the first position with an average of 76.17mm, followed by T2 with 75.17mm and T3 was the last with 73.03mm. The difference was not significant enough between T1 and T2 which did not allow to have a significant difference after the statistical analysis, but the difference was clearly significant between T3 and T2 on the one hand and T3 and T1 on the other hand. As regards RDI, it was found that the application of a water restriction by 25% and 50% in the pit hardening phase did not have a significant effect on the fruit's growth rate or on its final size. It should be noted that the difference in size between T2 and T4 was around 0.38 mm and between T2 and T5 was 0.82 mm (fig 8).

Table 2. Effects of Sustainable Deficit Irrigation (SDI) on yield parameters (average yield, fruit drop rate, fruit weight, fruit size and fruit number/tree) of peach trees (var. Benedict). The treatments are described in Section 2.2. Data represent means ± standard deviation. The different letters indicate significant differences according to the SNK test at P ≤ 0.05.

Parameter	**Treatment**		
Fruit number/tree	**T1 (125%)**	**T2 (100%)**	**T3 (75%)**
	321 ± 40,96 a	**338 ± 74,54 a**	**324 ± 61,79 a**
Average yield (T/ha)	**T1 (125%)**	**T2 (100%)**	**T3 (75%)**
	45,50±5,97 a	**46,43±10,49 a**	**43,71±8,42 a**
Fruit drop rate (%)	**T1 (125%)**	**T2 (100%)**	**T3 (75%)**
	13,28±6,13 a	**9,39±3,91 ab**	**6,32±3,89 b**
Fruit weight (mg)	**T1 (125%)**	**T2 (100%)**	**T3 (75%)**
	238,80± 34,20a	**234,95 ± 31,07ab**	**219,82± 27,76b**
Fruit size (mm)	**T1 (125%)**	**T2 (100%)**	**T3 (75%)**
	76,17± 3,57a	**75,17± 3,63 a**	**73,03± 3,50b**

Table 3. Effects of Regulated Deficit Irrigation (RDI) on yield parameters (average yield, fruit drop rate, fruit weight, fruit size and fruit number/tree) of peach trees (var. Benedict). The treatments are described in Section 2.2. Data represent means ± standard deviation. The different letters indicate significant differences according to the SNK test at P ≤ 0.05

Parameter	**Treatment**		
Fruit number/tree	**T2 (100%)**	**T4 (75% PH)**	**T5 (50% PH)**
	338 ± 74,54 a	343 ± 59,34 a	332 ± 53,85 a
Average yield (T/ha)	**T2 (100%)**	**T4 (75% PH)**	**T5 (50% PH)**
	46,43 ± 10,49 a	47,36 ± 8,19 a	44,98 ± 7,66 a
Fruit drop rate (%)	**T2 (100%)**	**T4 (75% PH)**	**T5 (50% PH)**
	9,39 ± 3,91 a	8,62 ± 3,85 a	5,54 ± 4,20 a
Fruit weight (mg)	**T2 (100%)**	**T4 (75% PH)**	**T5 (50% PH)**
	234,95 ± 31,07 a	231,27 ± 31,64 a	224,55 ± 31,99 a
Fruit size (mm)	**T2 (100%)**	**T4 (75% PH)**	**T5 (50% PH)**
	75,17 ± 3,63 a	74,79 ± 3,51 a	74,35 ± 3,98 a

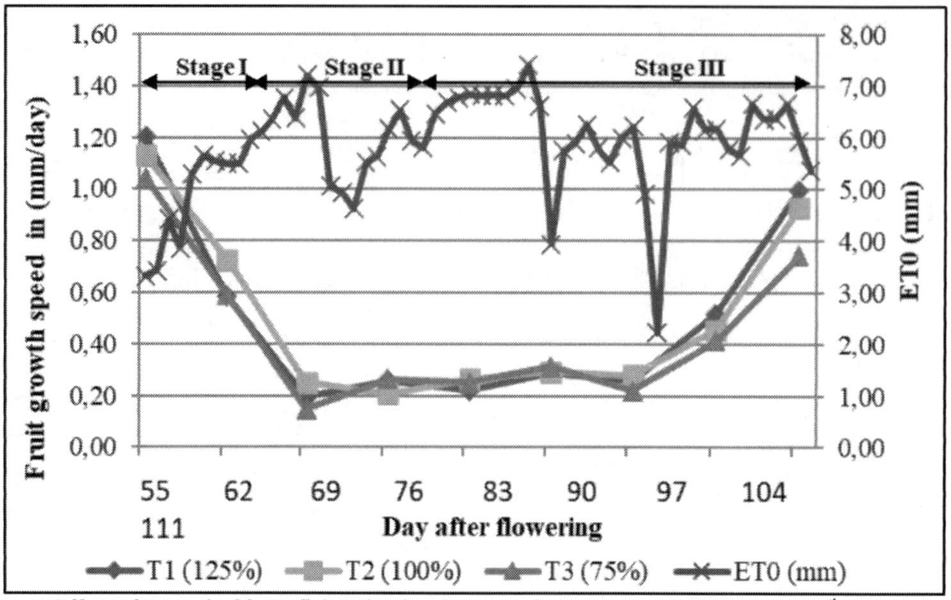

Fig. 7. Effect of Ssustainable Deficit Irrigation (SDI) on fruit growth speed from the 55th day until the 111th day after flowering. Stage I: celle division; Stage II: pit hardening; Stage III: fruit swelling stage. The reference evapotranspiration (ET0). The treatments are described in section 2.2.

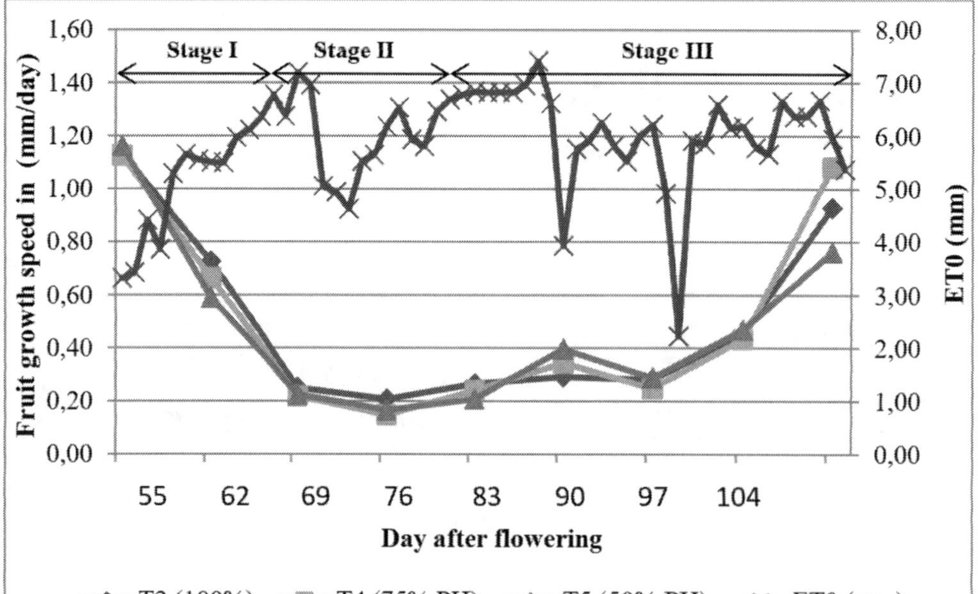

Fig. 8. Effect of Regulated Deficit Irrigation (RDI) on fruit growth speed the 55th day until the 111th day after flowering. Stage I: celle division; Stage II: pit hardening; Stage III: fruit swelling stage. The reference evapotranspiration (ET0). The treatments are described in section 2.2.

3.3 Quality parameters

3.3.1 Sugar content

Concerning the sugar content of peach fruits, Table 4 shows the effect of the irrigation dose on the °Brix content. For the SDI, the results show that reducing the irrigation dose tends to increase the sugar concentration in the fruit at harvest stage. This finding was confirmed by the results of the statistical analysis which showed the presence of a significant difference in the sugar level of the fruit by identifying two homogeneous groups using the Student-Newman and Kleus test at the 5% threshold. The first group included the T3 treatment (75% ETc) with an average of 13.42 °Brix, the second group included the other two treatments T2 (100% ETc) with an average of 12.08 °Brix and T1 (125% ETc) with an average of 11.62 °Brix. As for the RDI, the application of a 50% water restriction during the pit hardening stage led to a significant sugar content increase in the fruit compared to the other treatments. This was confirmed through the results of the statistical analysis which showed the presence of a significant effect of the T5 treatment (50% PH) on the fruit sugar content compared to the T2 treatment (100%). Whereas the 25% water restriction, T4 (75% PH), did not significantly affect the sugar content of the fruit compared to the T2 control (100%).

3.3.2 Acidity

Tables4 & 5 summarize the effect of water restriction regime on the fruit's acid content at harvest. The results show that reducing the irrigation amount tends to significantly reduce fruit acidity. In fact, a transition from 125% ETc to 75% ETc leads to a 13% reduction in fruit acidity. However, statistical analysis did not reveal any significant difference between the T3 deficiency treatment (75% ETc) and the T2 control treatment (100% ETc). As for the controlled deficit irrigation, the statistical analysis showed no significant effect of water restrictions (during the pit hardening stage) on fruit acidity.

3.3.3. Firmness

In Peaches industry, firmness is a key factor of external quality; and plays an important role in the commercial valuation of the final production. In our study, firmness monitoring results at harvest showed that there is no significant effect of SDI treatments on this parameter. However, treatment T3 (75%) recorded the highest value (6.05 kg/cm²), followed by treatments T2 (100% ETc) and T1 (125% ETc) with 5.95 kg/cm² and 5.93 kg/cm² respectively. As for RDI, statistical analysis reported no significant difference between the two water deficient treatments T4 and T5 and the T2 comfort treatment. In fact, the average fruits firmness of the treatment T5 (50% PH) was the highest with a value of 6.04 kg/cm².

Table 4. Effects of Sustainable Deficit Irrigation (SDI) on Quality parameters (Sugar content, acidity and firmness) and Biochemical parameters (Proline content). The treatments are described in Section 2.2. Data represent means ± standard deviation. The different letters indicate significant differences according to the SNK test at $P \leq 0.05$.

Parameter	Treatment		
Sugar content (°Brix)	T1 (125%)	T2 (100%)	T3 (75%)
	11,62± 1,02 b	12,08± 1,09 b	13,42± 1,35 a
Acidity (g of Malic acid/l)	T1 (125%)	T2 (100%)	T3 (75%)

	9,25± 0,78**a**	**8,70**± 0,65**ab**	**8,01**± 1,10**b**
Firmness (kg/cm²)	**T1 (125%)**	**T2 (100%)**	**T3 (75%)**
	5,93± 0,48 **a**	**5,95**± 0,56 **a**	**6,05**± 0,49 **a**
Proline content (mmol/L)	**T1 (125%)**	**T2 (100%)**	**T3 (75%)**
	0,065± 0,014**b**	**0,094**± 0,018**b**	**0,148**± 0,022**a**

Table 5. Effects of Regulated Deficit Irrigation (RDI) on Quality parameters (Sugar content, acidity and firmness) and Biochemical parameters (Proline content). The treatments are described in Section 2.2. Data represent means ± standard deviation. The different letters indicate significant differences according to the SNK test at P ≤ 0.05.

Parameter	**Treatment**		
Sugar content (°Brix)	**T2 (100%)**	**T4 (75% PH)**	**T5 (50% PH)**
	12,08± 1,09 **b**	**12,14**± 0,82 **b**	**12,70**± 2,14 **a**
Acidity (g of Malic acid/l)	**T2 (100%)**	**T4 (75% PH)**	**T5 (50% PH)**
	8,70± 0,65**a**	**8,56**± 0,79**a**	**8,47**± 0,78**a**
Firmness (kg/cm²)	**T2 (100%)**	**T4 (75% PH)**	**T5 (50% PH)**
	5,95± 0,56 **a**	**5,96**± 0,60 **a**	**6,04**± 0,55 **a**
Proline content (mmol/L)	**T2 (100%)**	**T4 (75% PH)**	**T5 (50% PH)**
	0,094± 0,018**b**	**0,116**± 0,025**b**	**0,187**± 0,016**a**

3.4 Biochemical parameters

3.4.1 Proline content

Proline accumulation is often reported as a consequence of water and/or salt stress. The high accumulation of proline in the leaves could be the result of a decrease in its oxidation and/or a reduction in its use in protein synthesis. Similarly, hydrolysis of proline-rich proteins and/or activated synthesis of this amino acid would lead to its accumulation in cells[23]. The results illustrated in tables 4 and 5 shows that water stress induced a remarkable accumulation of proline in peach leaves. Indeed, between T1, T2 and T3, proline synthesis was negatively correlated with irrigation dose. Similarly, for RDI, T5 was found to be more stressed than T4 and T2.

3.5 Water use efficiency

Agronomic water use efficiency (WUE) is an important index that measures the productivity of provided water amount from the perspective of irrigation efficiency. The calculation formula is as follows: WUE = Yield in kg/ amount of water applied in m3.

In this study, WUE was improved by applying treatment deficits T3, T4, and T5 leading to respectively:946 m3/ha, 124 m3/ha, and 248 m3/ha water savings (table 6). Compared to the control treatment, the T3 treatment (75%), which had no effect on fruit yield, improved the WUE by 16%; and the Regulated Deficit Irrigation (RDI) treatments increased the WUE by 6% and 2% respectively for T4 (75%) and T5 (50%). It should be noted that the increase in T4 in terms of yield allowed it to have higher water use efficiency than T5 where 50% less water was supplied compared to the control T2 (100%).

Table 6. Water use efficiency (WUE) of T1 (125%), T2 (control, 100%), T3 (75%, SDI), T4 (75%, RDI) and T5 (50%, RDI).

Traitement	Average yeild (Kg/ha)	Amount of water applied (m^3/ha)	Water use efficiency (Kg/m^3)
T1 (125%)	45502	6002	7,58
T2 (100%)	46433	5057	9,18
T3 (75%)	43712	4111	10,63
T4 (75% PH)	47356	4933	9,60
T5 (50% PH)	44984	4809	9,35

4 Discussion

4.1 Vegetative growth

The applied water deficit , whether SDI (T3; 75%ETc) or RDI (T4 andT5; 75% and 50%ETc) in pit hardening phase did not affect neither the vegetative growth of the mixed shoots (the organs carrying fruit during the season), nor the young shoots (the origin of future mixed shoots). This allows us to say that for peaches, the development of these organsis insensitive to a 25% reduction in water supply during the entire growth cycle (T3); or a water restriction in the stone hardening phase equivalent to 25% or 50% of the crop water requirements(T4 and T5 respectively). Furthermore, it was remarkable that among the three deficit treatments (T3, T4 and T5), the RDI treatments (T4 and T5) were the closest in terms of vegetative growth behavior to the T2 control. On the other hand, it was found that despite an increase in intake to 125% (T1) there was no increase or acceleration of the growth of mixed twigs and shoots compared to the control (100% ETc).

These conclusions are in contradiction with other results found by several authors in the same climate or in different climates. Indeed,Razouk et al., [12]adopted similar water restriction level as in this study (75% and 50% in pit hardening stage) and in almost equivalent climate conditions (Fez-Meknes region; Morocco) and found a significant decrease in shoot growth compared to the control. This can be explained by the climate effect throughout the growing season which intervenes in terms of evapotranspirationdemand ET0 and rainfall (the two experiments were not carried out in the same year). In different climates (France and Spain) and under severe water restriction,(Li, et al., and Sotiropoulos et al., [11, 13] irrigated respectively 33% ETc and 35% ETc during the pit hardening and 100% during the rest of the cycle and both reported that the applied water deficit has significantly reduced the growth of mixed shoots and the development of shoots. This difference between the results of present study and those of Li et al [11]and Sotiropoulos et al.,[13]may be related to the stress threshold applied during the stone hardening. We believe that the selected stress threshold in this study was not as severe as in those of Li, et al and Sotiropoulos et al [11, 13]; which probably allowed to highlight the growth difference between these deficient treatments and the control. This leads to the conclusion that the more severe the water stress in the pit hardening phase, the more apparent the decrease in vegetative growth is when compared to a 100% ETc treatment.

As for SDI, Abrisqueta et al., in Spain[17],, and by providing 50% of the water needs throughout the fruit growth cycle, found a significant effect of deficit irrigation on stem growth and pruning weight. Whereas in our experience, SDI (T3; 75%) had no effect on tree vegetative growth.

4.2 Production

It was concluded in this study that the yield was insensitive to water regime changes. In other terms, the water deficit induced by the three treatments T3 (75%; SDI); T4 (75% PH; RDI) and T5 (50% PH; RDI) did not significantly affect the yield. Similarly, the increase in water intake to 125% of the evapotranspiration needs did not increase the yield.

Furthermore, a significant difference was concluded for the fruit average weight and size at harvest between SDI treatment (T3; 75%) on the one hand and both T1 and T2 on the other hand. Applying water restriction has been revealed as tending to decrease both fruits weight and size; This result may not be encouraging for farmers interested in managing fruit size to comply with market specifications. However, it should be noted that the reduction in fruit size in T3 compared to T1 and T2 only began to appear at the end of the swelling phase. This leads us to consider a reduction in the water restriction level applied towards the end of this phase, or even to reduce this level throughout the cycle and examine whether the difference in fruit size compared to the control is still persisting. On the other hand, it should be noted that the RDI treatments T4 and T5 did not show any significant difference with T2 in terms of yield, size or average fruit weight.

In addition, and for Regulated Deficit Irrigation (RDI), many studies have also shown that deficit irrigation during Phase II does not affect production. Among others, a 3years study about the "Andross " variety has shown that deficit irrigation by 35% ETc during the pit hardening phase has no effect on fruit production[15]. This finding was again confirmed by Sotiropoulos et al. [13], in a study carried out in Greece on the variety "A-37" with a water deficit of 35% ETc. In this study, authors reported that the tested irrigation regime had no effect on productivity, and on both fresh and dried fruit weight. Moreover, the size of "Merrill Sundance" variety' fruits at harvest was not affected by a water deficit of 66% ETc in France [11].

Nevertheless, other authors have found different responses of peaches to irrigation restrictions during fruit growth phase II.In a study carried out in Morocco revealed a significant reduction in yield and fruit size at harvest for an irrigated treatment at 50% of evapotranspiration requirements[12]. Similarly,a stady in spane showed that restriction of irrigation (by 60% ETc) during pits hardening of untinned peaches of the "Andross" variety intended for the processing industry affected stem growth, total yield and fruit size[16]. These different behaviors could be due to other factors such as fruit load, variety, climatic conditions and soil water retention properties.

Concerning sustainable deficit irrigation (SDI), and in contradiction with our results, all the studies we have reviewed agree that for peaches, sustainable deficit irrigation (SDI) significantly reduces the yield. Indeed, in Tunisia, yield of the late production variety "Carnival" seems to be linked to irrigation regimes, and thus sustainable deficit irrigation (SDI) at 33% ETc leads to significant water savings (33%) but a yield reduction of 14% [19].Once again in Spain and on an early production variety, a water restriction of half of the needs considerably penalizes peach yields[17]. The discrepancy between our results from Abrisqueta and Ghrab can be explained by the fact that the water restriction in our study was the lowest compared to those of these studies (25% vs. 67%; 50%); which did not allow, probably, a difference in yield to occur. Another explication could be linked to the irrigation approach we used in this study. The high frequency in irrigation application per day could increase the water use efficiency as well as the nutrient use efficiency. It is well known that irrigation water management and humidity in soil are key factors to succeed in irrigation management. In our case, to satisfy the crop water requirements, the number of irrigations per day was between one to six irrigations according to the Readily Available Water (RAW) which corresponds to the amount of water that a soil can hold in the effective root depth. Moreover, the considered irrigation triggering percentage was (5%) leading to high frequency in irrigation.

As for fruit weight, Crisosto et al., [18]restricted the irrigation supply by 50-75% ETc to 'O Henry' variety plantations in California and highlighted a reduction in fruit weight; confirming our results on the effect of T3 on fruit weight.

For the rate of fruit falling at maturity, which is typical for the "Benedicte" variety, no significant difference of this percentage was recorded among the deficit irrigated treatments T3, T4 and T5 compared to T2, but the difference is clearly noticeable between (T2; T3; T4; T5) on the one hand and T1 on the other hand. However, Sotiropoulos et al, [13], found, under severe water stress (35% ETc) during stone hardening , a significant decrease in this percentage compared to a treatment irrigated at 100% crop evapotranspiration.

4.3 Fruit quality

Several studiesagreed with our findings and confirm that water stress during the second phase of fruit development induces an increase in soluble sugar levels TSS[11, 12, 13]. However, it has no effect on firmness according to Sotiropoulos et al., [13] works. In addition, it reduces acidity according to a Moroccan study[12].

Similarly, for sustainable deficit irrigation (SDI), and in harmony with our results, the majority of studies agree that the level of total sugars increases in the fruits of trees exposed to SDI whereas titratable acidity decreases[11, 18, 24].

4.4 Agronomic Water use efficiency

It is clear that to achieve the same performance as the control treatment, the more water is saved compared to the control treatment, the more efficient the water use. In our experience and due to non-significant performance differences between treatments, the water use efficiency of the different treatments did not follow the same ranking as the water saved. In fact, the SDI T3 treatment (75%) resulted in a water saving of 18.70% compared to the amount of water supplied for the control, which is a saving of 946 m^3/ha compared to the control, and with an average efficiency of 43.7 t/ha, the water use efficiency (WUE) was 10.63 (kg/m^3). As per RDI, and in terms of water saving, we found T5 (50% PH) in first place with 4.90% water saved compared to the control T2 (100%) (248 m3/ha less than the control), followed by T4 (75% PH) with 2.45% (124 m3/ha) compared to the control. However, in terms of yield, T4 came first with 47.35 t/ha compared to T5, which could only reach 44.98 t/ha. The fact of having a low water saving, but a high potential yield, allowed T4 to be in first place in terms of water use efficiency with 9.60 (kg/m^3) compared to T5, which recorded 9.35 (kg/m3) of water productivity.

On the other hand, T1 is found to be the treatment with the maximum amount of water applied among all the treatments, but without any advantage in terms of efficiency compared to the control. This classifies it as the most inefficient treatment in terms of water use efficiency, with only 7.58 (kg/m^3).

In an experiment carried out in Spain[25], Tapia Vargas et al., tested "Flordestar" variety behavior under SDI (50% ETc) and achieved showed a water use efficiency of around 2.7 kg/m^3 compared to the control (120% ETc) 2.1 kg/m^3.The difference between the WUE in our experience and theirs is due to the difference between the cultivars ("Flordestar" is a late production variety while "Benedicte" is a seasonal variety);as well as a difference in rainfall during the growing cycle and in yields achieved; which will result in a different water use efficiency.

5 Conclusion

This study had as an objective to assess the response of commercial plantation of peaches under Mediterranean climate conditions to various irrigation restrictions. Based on the evolution of the parameters monitored under the water restrictions conditions, it is concluded that a restriction of 25% of the crops water requirement treatment gavea good compromisein peach tree irrigation, allowing a net improvement in water use efficiency, fruit quality without any significant reduction in yield or vegetative growth. In addition, this water regime has the disadvantage of reducing fruit size, making it unattractive to farmers targeting markets of large size fruits. On the other hand, the regulated deficit irrigation treatmentsT4 and T5 showed similar performance to the control treatment in terms of yield and size. Therefore, it is likely that the negative effects of the T3 treatment on the final fruit size can be reduced by combining sustainable deficit irrigation with regulated deficit irrigation during pit hardening;therefore, it is recommended to integrate this irrigation program as a treatment in future studies to confirm this hypothesis.

Funding

This work was supported by Moroccan Ministry of Agriculture, Maritime Fisheries, Rural Development and Water and Forestry. It's a part of the research project Competitive Research Development and Extension Mechanism (MCRDV). The project consists in the installation of a modern field platform dedicated to research and development of smart irrigation management in the region of Fez-Meknes.

Acknowledgements

This study was conducted by the research unit SPEC-trum (Sol-Plante-Eau-Climat Continuum) within the National School of Agriculture of Meknes (ENAM) in Morocco. The authors gratefully acknowledge the Louata administration staff and in particular to Pr. Ait Houssa Abdelhadi and Mr. Hadri karim for their significant interest and technical support, and for their efforts to ensure the projects progression in the best conditions. The authors also wish to thank the student Ait Ijja Sara for her valuable contribution during her final engineering project.

References

1. Ragab, prudhomme. *Impact du changement climatique sur les régimes hydrologiques des fleuves côtiers Libanais*, 3ème conférence internationale sur les "Ressources en Eau dans le Bassin Méditerranéen", WATMED3, Tripoli, Liban, Novembre, (2002).
2. J.Van Schilfgaarde. Agric. Water Manag. **25**, 203–219 (1994).
3. M.M .Jellali, *Développement des ressources en eau au Maroc. Direction Général de l'Hydraulique, Rabat, Maroc.*(1997)
4. A. Bouaziz, K. Belabbes. H T E, **32**, 57–72. (2002)
5. M. Pascual, J.M. Villar, J. Rufat, Agric. Water Manag. **164**, 253–266.(2016).
6. J.M. Abrisqueta, O. Mounzer, S. Álvarez, W. Conejero, Y. García-Orellana, L.M. Tapia, J. Vera, I. Abrisqueta, M.C. Ruiz-Sánchez, Agric. Water Manag. **95**, 959–967. (2008).
7. M. Guizani, S. Dabbou, S. Maatallah, G. Montevecchi, H. Hajlaoui, M. Rezig, A.N. Helal, S. Kilani-Jaziri. Agric. Water Manag. **217**, 81–97. (2019).
8. O. Forey, A. Metay, J. Wery. Eur. J. Agron. **81**, 106–116. (2016).
9. J.M. Faci, E.T. Medina, A. Martínez-Cob, J.M. Alonso, Agric. Water Manag. **143**, 102–112. (2014).
10. M.H. Behboudian, John Wiley, Researchgate, **38**, 149 – 189. (2011).
11. S.H. Li, J.G. Huguet, Schoch, P.G., P. Orlando,H S, **64**, 5. (1989).
12. R. Razouk, J. Ibijbijen, A. Kajji, M. Karrou, A.J.P.S, **4**, 561-570.(2013).
13. T. Sotiropoulos, D. Kalfountzos, I. Aleksiou, S. Kotsopoulos, N. Koutinas, Sci. Agric. **67**, 164–169. (2010).
14. X. Vallverdú, J. Girona, G. Echeverría, J. Marsal, M.H. Behboudian, G. Lopez, Sci. HORTSCI **47**, 656–659. (2012).
15. J. Girona, M. Gelly, M. Mata, A. Arbonès, J. Rufat, J. Marsal, Agric. Water Manag. **72**, 97–108. (2005).
16. X. Domingo, A. Arbonés, J. Rufat, M. Pascual, J.M. Villar, Acta Hortic. 213–220. (2011).
17. I. Abrisqueta, L. Tapia, W. Conejero, M. Sanchez-Toribio, J. Abrisqueta, J. Vera, M.C. Ruiz-Sánchez, J. Agric. Res. 8. (2010).
18. C.H. Crisosto, R.S. Johnson, J.G. Luza, G.M. Crisosto, HortScience,**29**, 1169–1171.(1994).
19. M. Ghrab, R. Zitouna, M. Mimoun, M. Masmoudi, N. Ben Mechlia, Biol. Agric. Hortic. **29**, 29–37. (2013).
20. FAO-56, *Chapter 6 - ETc - Single crop coefficient (Kc) [WWW Document]. URL http://www.fao.org/3/X0490E/x0490e0b.htm (accessed 4.9.20).*
21. Penman. *Estimation de l'évaporation - Penman - 1956 - Eos, Transactions American Geophysical Union . Wiley Online Libr.*
22. L.S. Bates, R.P. Waldren, I.D. Teare, P .S.**39**, 205–207. (1973).
23. L. Ben Khaled, A. Gõmez, M. Honrubia, A. Oihabi, Agro**23**, 553–560. (2003).
24. H. Engin, F. Şen, G. Pamuk, Z. Gökbayrak, Crop Hortic. Sci. **38**, 39–46. (2010).
25. M. Tapia Vargas, A. Larios Guzmán, I. Abrisqueta Villena, O. Mounzer, J. Vera Muñoz, J.M. Abrisqueta García, M.C. Ruiz Sánchez, Rev. Fitotec. Mex. **33**, 89–93.(2010).

Carcass and meat quality characteristics of Beni-Guil autochthonous ovine breed: effect of live weight at slaughter

Kamal Belhaj[1,2*], *Farid* Mansouri[2], *Youssef* Rbah[2], *Aymane* Allay[2], *reda* Melhaoui[1], *Aziz* Tikent[2,3], *Yassine* Taaifi[2,3], *Abdessamad* Benmoumen[1], *Mohamed* Boukharta[4], *Marianne* Sindic[5], *Hana* Serghini-Caid [2], *Ahmed* Elamrani[2]

[1] Laboratoire de Management de l'Agriculture Durable, EST Sidi Bennour, Université Chouaib Doukkali, El Jadida
[2] Laboratoire d'Amélioration de la Production Agricole, Biotechnologie et Environnement (LAPABE), F Université Mohamed Ier, Oujda; Maroc. 717, 60000 Oujda, Maroc.
[3] Service Vétérinaire, Office National de Sécurité Sanitaire des produits Alimentaires, Oujda, Maroc.
[4] Institut Supérieur Industriel agronomique, Huy, Belgique.
[5] Laboratoire Qualité et Sécurité des Produits Alimentaires, Gembloux Agro-Bio Tech, Université de Liège ; Belgique.

Abstract. Ninety female lambs of the Beni-Guil sheep breed were used to study the effect of slaughter live weight (SLW) on carcass characteristics and meat quality. The animals were raised in Morocco's eastern area under a pastoral-transhumant system. The lambs were slaughtered at three weight classes (SLW1 =20-25 kg, SLW2 = 26-30 kg, SLW3 = 31-35 kg). Ultimate pH value (pHu), fat and meat color, zoometric measurement, conformation score, fatness state, and fat consistency were evaluated 24 h post-slaughter. The carcass measures and compactness indexes increase with higher SLW ($p<0.001$). However, the dressing percentage and shrinkage were not affected. The *longissimus lumborum* muscle was used for meat quality determination. The pHu was not affected ($p>0.05$) by SLW effect. With respect to meat color, the lightness (L*) decreased with increasing SLW, while the redness increased ($p<0.05$). The female lambs with a carcass weight of 20-25 kg have a higher L* value (41.89), and lambs with a carcass weight between 31 and 35 kg have a higher redness value (21.73). In relation to the c¹arcass fatness state and carcass conformation, the SLW had a significant effect ($p<0.001$). The conformation and the fatness increase in the logical sequence of SLW.

1 Introduction

The total ruminant population in Morocco currently numbers about 29 million heads divided between cattle, sheep, goats and camels. The sheep sector ranks first with a workforce of around 20 million heads, contributing 30% to national meat production [1]. The Moroccan sheep population is composed by more than 95 % of autochthon breeds [2]. The Beni-Guil, Sardi, Timahdite, D'man, Boujâad, Beni-Hsen, Black Siroua Mountain White, and Saghro are the main sheep breeds having major socio-economic concerns [3]. The sheep production is mainly based on pastoral-transhumant systems exploiting local breeds. To have sustainable use, create added value to products, and provide more opportunities for breeders, the Moroccan government focuses on developing local breeds in their cradles through the preservation of local breeds in their origin area, average daily gain improvement and product certifications. Since 2011, the Moroccan Institute for Standardization has granted the Beni-Guil sheep meat as a Protected geographical indication (PGI) the from eastern region of Morocco, (ministerial order n° 1164-11 du joumada II1432 (5 may 2011)). This PGI breed is among the main local breeds exploited in Morocco with a workforce of around 2 million heads. This sheep breed is born, reared, and slaughtered in the eastern Moroccan region, fed a particular diet and reared under specific pastoral-transhumant system of this region. It is well suited to the harsh climate and poor pasture conditions, and is classified as a meat production breed. It is a breed characterized by fire brown head, ears, prothorax, and legs, without white spots, white fleece, and no horns in the female (Figure 1). The rams usually have coarse and long horns that are pale and spiral (Figure 2). The Beni-Guil sheep meat has a unique sensory and nutritional quality linked to this distinct environment and natural grazing system. The heavy lamb's meat is less tender with a strong flavour and high fat content [4]. The consumers in Moroccan rural areas prefer heavy carcasses from adult animals; while in urban areas, the meat from light animals

* Corresponding author: belhaj.kamal90@gmail.com ; k.belhaj@ump.ac.m

is favoured (generally meat of young animals). The Beni-Guil sheep meat label only takes into account the meat from animals with milk teeth, slaughtered at 120-180 days of age. However, the meat from adult animals also needs to be considered. Various factors affect the sheep meat quality, such as feed [5], live weight at slaughter [6], genotype [7], and sex [8]. There are no previous studies on carcass and meat quality characteristics of Beni-Guil sheep breed slaughtered at different weight. Hence, this study aims to determine the effect of live weight at slaughter (20-25 kg; 26-30 kg; 31-35 kg) on carcass and meat quality of Beni-Guil-PGI sheep, and therefore, to contribute to the characterization of this Moroccan label.

2 Materials and methods

2.1 Animals

Animals were reared under transhumant-local production systems in the pasture of eastern Morocco highlands. A total of 90 single-born female lambs reared extensively were slaughtered during spring 2018 at three slaughter weights classes; **SLW 1**: 20-25 kg, **SLW 2**: 26-30 kg, and **SLW 3**: 31-35 kg (n = 30 for each group) for determining slaughter, carcass, and meat quality characteristics traits. The studied animals were weaned at the age of 3 months and were not shorn. The selected animals were weighed and slaughtered after a 12-h fasting period with free access to water. Female lambs were slaughtered in their areas of origin in a slaughterhouse. Further details about rearing, diet composition and slaughter conditions are described by Belhaj, Mansouri [9]. The study period was from February to July 2018. The *longissimus lumborum* samples were taken 24 hours *post-mortem* with the assistance of a slaughterhouse veterinarian for color measurement.
The study was conducted on two slaughterhouses located in the eastern region of Morocco that of Oujda and Jerrada. The climate of this area is semi-arid. The annual rainfall varies between 100 and 250 mm/year and the average annual temperature is around 19 °C with lows that can reach -4 °C in winter and average maximums of around 42.5 °C in summer season [10].

Fig 1. Female lambs of Beni-Guil sheep breed

Fig 2. Male lambs from the super category of Beni-Guil sheep breed

2.2 Carcass and meat measurements

The details of the methods used to assess the Beni-Guil sheep breed's carcass and meat quality characteristics carcass and meat quality characteristics of Beni-Guil sheep breed are described by Belhaj, Mansouri [11].

2.3 Statistical analysis

Statistical analyses were conducted using Statistical Package for the Social Sciences (IBM SPSS. 20). The normal distribution was verified according to the Shapiro Wilk test for quantitative variables. A two-way analysis of variance (ANOVA) was carried out for the live weight at slaughter. Tukey's post-hoc test was used for means comparison. The difference was considered significant at $p<0.05$. Principal component analysis (PCA) was performed on the data set to differentiate the samples according to their breed and age at slaughter and to obtain more information on the variables that mainly influence the carcass of our samples.

3 Results and discussion

3.1 Carcass measurements

The obtained values for the effect of slaughter live weight on carcass characteristics of Beni-Guil female lambs are given in table 1. The weight measurement particularly the carcass weight and the yield are important production and slaughter criteria for breeders, butchers, and retailers. These slaughter parameters are affected by several factors such as alimentary tract size and fill, slaughtering processes, fleece or skin mass, fatness state, and sexual organs. The mean hot and cold carcass weight ranged between 10.72-15.96 kg and 10.55-15.67 kg respectively for hot and cold carcass weight. These slaughter parameters varied according to the slaughter live weight: carcasses of SW-class 3 were heavy than other groups. However, the slaughter data highlighted the no significant effect of live weight on economic and commercial yield. The animal's age significantly influenced ($p<0.001$) the yield at slaughter (data not shown). Authors reported similar results for other sheep breeds [12-14]. Contrary to this finding, Liu, Guo [6] obtained a significant effect of SLW on carcass yield for Oula lambs slaughtered at five live weight. These contradictory results could be attributed to the difference in SLW, in particular the difference between studied weight classes (Liu, Guo [6] : 25–45 kg present study : 20-35 kg), animals' age at slaughter, and sex. The recorded carcass yield values are higher than those reported by Blasco, Campo [15] for Texel breed and by Yaranoğlu and Özbeyaz [16] for three Turkish breeds. However, these values are lower than those reported by Santos, Cabo [14], and Santos, Silva [13] for Churra Galega Mirandesa breed (50-57 kg) and Churra da Terra Quente breed (48-50 kg). Firstly, this difference could be explained by alimentary tract size and fill, because the animals studied by Santos, Cabo [14], and Santos, Silva [13] are not functional as ruminants, and therefore have a lower digestive tract content and size (lower weight of offal). Secondly the fleece size because the female lambs have a fleece with long hair in the present study.

Table 1. Effect of live weight at slaughter on carcass traits of Beni-Guil sheep breed from eastern Morocco (means ± SD).

Parameters	Live weight classes			P value
	20-25 kg	26-30 kg	31-35 kg	
Weight measurements				
Hot Carcass Weight (Kg)	10.72 ± 0.98^a	14.07 ± 0.78^b	15.96 ± 0.89^b	<0.001
Cold Carcass Weight (Kg)	10.51 ± 0.98^a	13.80 ± 0.78^b	15.67 ± 0.87^c	<0.001
Economic yield (%)	48.08 ± 1.47^a	48.16 ± 2.48^a	47.37 ± 1.49^a	NS
Commercial yield (%)	47.11 ± 1.47^a	47.24 ± 2.45^a	46.49 ± 1.46^a	NS
Shrinkage loss (%)	2.02±0.40	1.90±0.42	1.85±0.26	NS
linear measurements				
G (cm)	14.17 ± 3.59^a	16.60 ± 0.52^b	16.81 ± 0.68^b	<0.01
K (cm)	58.05 ± 1.08^a	61.43 ± 0.99^b	61.22 ± 0.35^b	<0.001
F (cm)	27.68 ± 0.90^a	31.35 ± 1.60^b	31.36 ± 1.09^b	<0.001
LCI = G/F	0.54±0.04	0.53±0.03	0.54±0.03	NS
CCI1 = G/K	0.26 ± 0.02^a	0.27 ± 0.01^b	0.27 ± 0.01^b	<0.05
CCI2=CCW/K(g/cm)	181.02 ± 16.70^a	224.76 ± 12.4^b	255.95 ± 14.80^c	<0.001
Subjective characteristics				
EUROP-Conformation	2.00 ± 0.00^a	2.33 ± 0.49^a	2.85 ± 0.36^c	<0.001
EUROP-Fatness	2.95 ± 0.22^a	3.16 ± 0.38^a	3.58 ± 0.51^b	<0.001
Fat color	1.00±0.00	1.00±0.00	1.00±0.00	NS
Fat firmness	1.00±0.00	1.00±0.00	1.00±0.00	NS

Values with different superscript letters are significantly ($p<0.05$) different within row.
G: Basin width; **K**: Carcass length; **F**: Leg length; NS: not significant.
LCI: Leg compactness index; **CCI1**: Carcass compactness index1; **CCI2**: Carcass compactness index2; **Conformation** (1 = Poor to 5 = Excellent); **Fatness** state (1=Very low to Very high); **Fat color scale** (1 = very white to 4 = strongly colored); **Fat firmness scale** (1 = hard to 4 = very soft and oily).

All-the linear measurements were significantly affected by live weight at slaughter (Table 1). The results showed that the carcass linear measures and carcass compactness indices (CCI 1 and CCI 2) increase gradually in the logical sequence of their SLW ($p<0.05$), showing higher values for animals of SLW-class 3 ($p<0.001$). As slaughter weight increases, the carcass compactness increases, as shown by Liu, Guo [6], and Santos, Cabo [14]

who compared carcass and meat quality traits of Oula and Churra Galega Mirandesa sheep breed slaughtered at different weights. The obtained carcass compactness values based on carcass weight are higher than those found in the carcass of the Churra Galega Mirandesa breed (23.1 g/cm) slaughtered at two different slaughter weights [14]. Moreover, these carcass indices are comparable to those reported by Blasco, Campo [15] for the carcass of Texel sheep breed (220-230 g/cm) in the case of weight-class 2 and 3 (224-250 g/cm).
The meat market wants well-muscled carcasses with a high percentage of lean meat and a particular degree of fatness, to minimise weight loss during cold storage [12]. As shown in Table 1, carcass of SLW 1 group expelled more water than other groups, showing higher shrinkage loss value, but this difference was not significant. This water loss was correlated negatively with fatness state, since the fat layer represents an obstacle that prevents water loss during storage. A Similar correlation was reported by D'Alessandro, Maiorano [17] in Kivircik's sheep breed meat. The estimated shrinkage loss values are lower than those reported by Ekiz, Demirel [18], Carrasco, Ripoll [19], and D'Alessandro, Maiorano [17] for Kivircik (2.38-3.36%), Chura (2.3-2.9%) and Leccese (2.42-3.32%) sheep breed, respectively. Concerning the carcass subjective quality, the results showed a significant effect of SLW on conformation and fatness state (p<0.001). According to the community scale for classification of ovine carcasses (EUROP), all assessed carcasses were classified within fair (O) and good (R) score. As expected, the SLW 3 animal carcasses showed a higher conformation and fatness score. In line with previous studies [20, 21], increased slaughter weight increased the carcass conformity and fatness state. This could be explained in part by the tissue development at young age. The conformation of exanimated carcasses was comparable to that reported by Carrasco, Ripoll [19] for Churra Tensina lambs and by Blasco, Campo [15] for Texel breed and its crossbreeding (Texel × Segureña). In respect to fatness state, the obtained scores are slightly higher than those observed by Mazzone, Giammarco [22] in Apenine lamb carcasses (2.18-2.26) and are comparable to those estimated by Blasco, Campo [15] for Texel breed (2.25-3.45). Regarding the carcass fat and consistency, no significant difference was observed. All the assessed carcasses have a white and hard carcass fat. A similar result was reported by Carrasco, Ripoll [19] for Chura sheep breed.

3.2 Carcass colour

The colour is the principal quality for meat before purchase. It is the meat commercial quality. In term of consumer preference, the muscle colour must be bright red, indicates meat freshness. The results of colour coordinates lightness (L*), redness index (a*), yellowness index (b*), Chroma (C*), and hue angle (H*) are presented in table 2. All obtained L* (41.19-41.89) values are in the bright red meat colour range [6, 23]. The L* values decrease with increasing SLW, while the a* increase (p<0.05). These findings are in agreement with those reported by Beriain, Bas [24], Santos, Cabo [14] and Budimir, Trombetta [25] for other sheep breeds. This result can be attributed to difference in slaughter age, in particular the myoglobin content. This muscle pigment is linked with meat redness (higher a* and lower L*). The studied carcasses were darker in comparison with several lambs' carcasses [6, 14, 20, 25]. This difference could be ascribed to the intensive physical exercise of studied animals, reared under the typical traditional-transhumant production system of eastern Morocco (long-distance travel by animals to graze).

Table 2. Effect of live weight at slaughter on ultimate pH and meat color of Beni-Guil sheep breed from eastern Morocco (means ± SD).

Parameters	20-25 kg	26-30 kg	31-35 kg	P value
pH ultimate	5.89±0.08	5.77±0.09	5.71±0.07	NS
L* (lightness)	41.89±0.2^{b}	41.59±0.42^{ab}	41.19±0.64^{a}	<0.05
a* (redness)	20.95±0.93^{a}	21.22±0.59^{ab}	21.73±1.04^{b}	<0.05
b* (yellowness)	7.74±0.61	7.37±0.46	7.57±0.93	NS
Chromaticity	22.35±0.9	22.47±0.53	23.02±1.2	NS
Hue angle°	29.29±1.75	19.19±1.57	19.54±1.16	NS
a*/b*	2.72±0.24	2.88±0.21	2.90±0.32	NS

Values with different superscript letters are significantly (p<0.05) different within row.
Scales of L*, a*, b*, chroma and Hue: a* and b*: −60 to 60; L*: 0 to 100; chromaticity: 0–60; hue: 0–360°
NS: not significant

3.3 Ultimate pH

The meat pH is a vital chemical parameter because it influences practically all qualities of meat. The effect of slaughter live weight on ultimate pH is listed in Tables 2. The recorded values ranged from 5.71 to 5.89, indicating the normal post-mortem pH evolution (animals slaughtered under stress-free conditions). According to Tukey's test, there is no significant difference in the obtained pH values between the studied groups. These results a agree with those reported by several authors, who don't find differences in meat pHu of lambs slaughtered at the different live weight. These values are within the normal pH range of commercial meat [26], and are comparable to those previously reported by several authors for sheep meat [4, 6, 17]. In the current study, the pHu value showed a decreasing trend with increasing slaughter weight, but these differences were not statistically significant (5.89 vs 5.77 vs 5.71; $p>0.05$). This negative correlation could be explained by sensitivity to pre-slaughter manipulation. The heavy animals are more stressed at slaughter and consequently exhibit a lower ultimate pH [26]. In addition this correlation could be linked to the muscle glycogen reserve because the animal of SLW 3 is older than those of other weight classes and has received more cereals promoting propionic acid production, which is considered as a muscular glycogen precursor [27]. This correlation was also observed by numerous researchers in several lambs meat [6, 28].

3.4. Chemometric analysis of the carcass and meat quality characteristics

A principal component analysis (PCA) was carried out to identify and determine the correlations between the studied animals and measured parameters. This analysis was performed using 14 variables, including weight measurements, carcass measures, and carcass subjective characteristics. PCA results indicate that the first three principal component (PC) account for more than 80% of the total variation (Table 3). The first PC explained 52.62% and the second explained 19.63%. The studied lambs' projections on the factorial map reveal discriminating between the studied carcasses, allowing for a simplified summary of the interpretations already indicated. Figure 3 represents the factorial plane defined by the first and second PCs. The results of PCA showed that the most important variables, which explained a large part of the total variance, were the weight and carcass measurement. Two separated groups can be observed according to PC 1. Animals with SLW of 20-25 kg are located on the left side of the factorial map and differentiated from other groups (SLW 1 and 2). These latter are correlated positively with the weight and carcass measurements indicating that these animals have a heavy carcass (HCW, CCW and CCI 2) with greater carcass zoometric measurements (K, F, G, H and Th). These correlations between carcass measurements and slaughter live weight agree Cañeque, Pérez [29] who found that while carcass weight was increased, carcass weight, CCI and zoometric measurements improved. Regarding the second PC, the relationship between studied animals is not clear. The most important variables, which are regrouped into the PC 2, were the carcass yield and leg compactness.

The PCA exposed the slaughter live weight effect on the carcass and meat quality characteristics of Beni-Guil-PGI sheep breed. Researchers reported similar discrimination for carcasses from animals slaughtered at different weight classes [29-31].

Table 3. Three main components explaining more than 80% of the total information on the carcass and meat quality traits of Beni-Guil sheep breeds.

Variables	Principal component			Variables	Principal component		
	1	2	3		1	2	3
HCW	0,967	-0,104	-0,004	F	0,762	-0,333	0,175
CCW	0,966	-0,133	-0,009	EUROP-Fatness	0,631	0,073	0,180
CCI 2	0,957	0,008	0,032	EUROP-Conformation	0,383	-0,311	-0,212
H	0,915	-0,070	0,050	EY	0,160	0,853	0,460
Th	0,878	-0,019	0,087	CY	0,212	0,850	0,433
G	0,862	0,084	-0,208	LCI	0,229	0,790	-0,521
K	0,849	-0,283	0,121	CCI 1	0,511	0,583	-0,539

HCW: Hot carcass weight; **CCW**: Cold carcass weight; **EY**: Economic yield; **CY**: Commercial yield; **SL**: Shrinkage loss; G: Basin width; **K**: Carcass length; **F**: Leg length; **LCI**: Leg compactness index; **CCI 1**: Carcass compactness index1; **CCI 2**: Carcass compactness index 2.

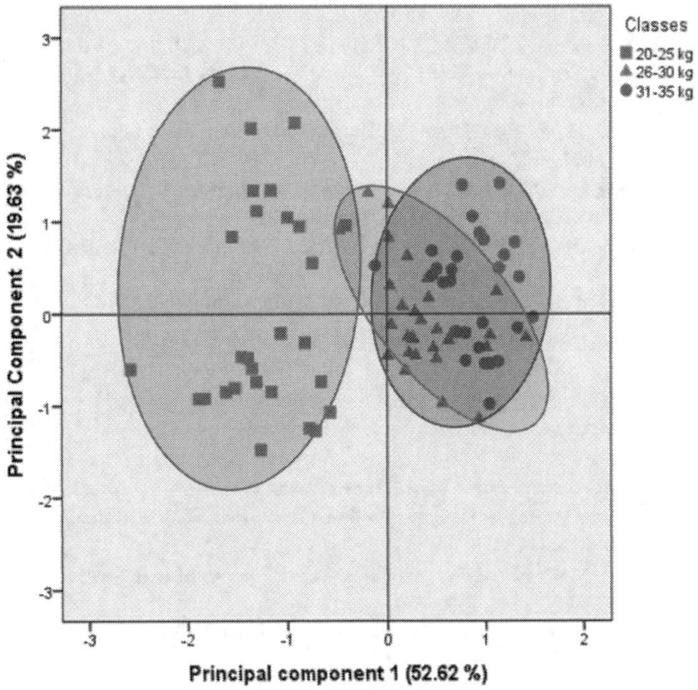

Fig. 3. Projection of the studied female lambs of Beni-Guil breed in the plane defined by the two principal components

4 Conclusion

The effect of live weight at slaughter on the carcass and meat quality traits has been studied. The results showed that this endogenous factor had a significant effect on the lamb's carcass and meat quality and should be considered in the PGI label to offer consumers a carcass with homogeneous traits from one purchase to another. This could increase consumer fidelity. In particular, the meat of female lambs slaughtered at 30-35 kg have a heavy carcass with better compactness, better conformation, and a high fatness state. On the other hand, those slaughtered at 20-25 kg have the optimal red coloration of the carcass (bright-red colour) with an average fatness state.

Acknowledgments

This paper is supported by Wallonie-Bruxelles International (WBI) as part of the project 1–6. In addition, the authors are indebted to the ANOC "Association Nationale Ovine et Caprine" for its collaboration, particularly to Mr. A. Kacem, B. Belhaj, and A. Elmejdoubi.

References

1. HCP. *Morocco in figures*. 2019 [cited 13/06/2019; Available from: https://www.hcp.ma/downloads/Maroc-en-chiffres_t13053.html.
2. A. Elbeltagy, *Characterization and value addition to local breeds and their products in the Near East and North Africa*, in *Regional Workshop Rabat, Morocco*. 2012, FAO: Rabat, Morocco. p. 19-21.
3. I. Boujenane, *Small ruminant breeds of Morocco*, in *Characterization of small ruminant breeds in West Asia and North Africa*, L. Iniguez, Editor. 2005, FAO: Rome. p. 5-54.
4. J.F. Tejeda, R.E. Peña, and A.I. Andrés, Meat science, **80**, 1061-1067 (2008)
5. B. Ekiz, A. Yilmaz, H. Yalcintan, O. Kocak, and M. Ozcan, Large Animal Review, **26**, 67-72 (2020)
6. J.-b. Liu, J. Guo, F. Wang, Y.-j. Yue, W.-l. Zhang, R.-l. Feng, T.-t. Guo, B.-h. Yang, and X.-p. Sun, Small Ruminant Research, **123**, 251-259 (2015)

7. A. Sabbioni, V. Beretti, E.M. Zambini, and P. Superchi, Italian Journal of Animal Science, **15**, 2-9 (2016)
8. E. Miguel, B. Blázquez, and F. Ruiz de Huidobro, *Liveweight and Sex Effects on Instrumental Meat Quality of Rubia de El Molar Autochthonous Ovine Breed. Animals 2021, 11, 1323*. 2021, s Note: MDPI stays neu-tral with regard to jurisdictional claims in ....
9. K. Belhaj, F. Mansouri, M. Sindic, M.-L. Fauconnier, M. Boukharta, H. Serghini Caid, and A. Elamrani, Journal of Food Quality, **2021**, (2021)
10. A. Bechchari, A. El Aich, H. Mahyou, B. Baghdad, and M. Bendaou, Journal of Materials and Environmental Science, **5**, 2572-2583 (2014)
11. K. Belhaj, F. Mansouri, A. Tikent, A. Ouchatbi, M. Boukharta, C.H. Serghini, and A. Elamrani, Revue d'élevage et de médecine vétérinaire des pays tropicaux, **73**, 21-26 (2020)
12. F. Pena, J. Perea, A. García, and R. Acero, Meat Science, **75**, 543-550 (2007)
13. V. Santos, S. Silva, E. Mena, and J.M.T.d. Azevedo, Meat Science, **77**, 654-661 (2007)
14. V. Santos, A. Cabo, P. Raposo, J. Silva, J. Azevedo, and S. Silva, Small Ruminant Research, **130**, 136-140 (2015)
15. M. Blasco, M.M. Campo, J. Balado, and C. Sañudo, Journal of the Science of Food and Agriculture, **99**, 3335-3342 (2019)
16. B. Yaranoğlu and C. Özbeyaz, Eurasian Journal of Veterinary Sciences, **35**, 15-23 (2019)
17. A.G. D'Alessandro, G. Maiorano, M. Ragni, D. Casamassima, G. Marsico, and G. Martemucci, Small Ruminant Research, **114**, 97-104 (2013)
18. B. Ekiz, G. Demirel, A. Yilmaz, M. Ozcan, H. Yalcintan, O. Kocak, and A. Altinel, Small Ruminant Research, **114**, 26-34 (2013)
19. S. Carrasco, G. Ripoll, A. Sanz, J. Álvarez-Rodríguez, B. Panea, R. Revilla, and M. Joy, Livestock Science, **121**, 56-63 (2009)
20. S. Martínez-Cerezo, C. Sañudo, B. Panea, I. Medel, R. Delfa, I. Sierra, J. Beltrán, R. Cepero, and J. Olleta, Meat Science, **69**, 325-333 (2005)
21. M. Juárez, A. Horcada, M. Alcalde, M. Valera, O. Polvillo, and A. Molina, Meat Science, **83**, 308-313 (2009)
22. G. Mazzone, M. Giammarco, G. Vignola, L. Sardi, and L. Lambertini, Meat science, **86**, 474-478 (2010)
23. C. Salifou, A. Youssao, G. Ahounou, P. Tougan, S. Farougou, G. Mensah, and A. Clinquart, Annales de médecine vétérinaire, **157**, 27-42 (2013)
24. M. Beriain, P. Bas, A. Purroy, and T. Treacher, Sheep and goat nutrition: Intake, digestion, quality of products and rangelands, 75-86 (2000)
25. K. Budimir, M.F. Trombetta, M. Francioni, M. Toderi, and P. D'Ottavio, Small Ruminant Research, **164**, 1-7 (2018)
26. C. Devine, A. Graafhuis, P. Muir, and B. Chrystall, Meat science, **35**, 63-77 (1993)
27. R. Vernon, *Lipid metabolism in the adipose tissue of ruminant animals*, in *Lipid metabolism in ruminant animals*, W. William and Christie, Editors. 1981, Elsevier. p. 279-362.
28. B. Ekiz, A. Yilmaz, M. Ozcan, C. Kaptan, H. Hanoglu, I. Erdogan, and H. Yalcintan, Meat science, **82**, 64-70 (2009)
29. V. Cañeque, C. Pérez, S. Velasco, M. Dıaz, S. Lauzurica, I. Álvarez, F.R. De Huidobro, E. Onega, and J. De la Fuente, Meat science, **67**, 595-605 (2004)
30. A. Karlsson, Meat science, **31**, 423-433 (1992)
31. K. Belhaj, F. Mansouri, A. Tikent, Y. Taaifi, M. Boukharta, H.C. Serghini, and A. Elamrani, The Scientific World Journal, **2021**, (2021)

 https://doi.org/10.1051/e3sconf/202233704004

Chemical characteristics of seed oil from wild prickly pear (*Opuntiaficus - indica*) in eastern Morocco

Ahmed Marhri[1,*], *Kamal* Belhaj[1,3], *Reda* Melhaoui[1], *Aziz* Tikent[1], *Abdessamad* Benmoumen[1], *Aatika* Mihamou[1], *Hana* Serghini-Caid[1], *Ahmed* Elamrani[1], *Christophe* Hano[2], *Malika* Abid[1]and *Mohamed* Addi[1]

[1]Laboratory for Agricultural Productions Improvement, Biotechnology and Environment (LAPABE), Faculty of Sciences, University Mohammed First, BP-717, 60000 Oujda, Morocco.
[2]ligneous and Field Crops Biology Laboratory, INRA USC1328, Orleans University, CEDEX 2, 45067 Orléans, France.
[3]Laboratory of Sustainable Agriculture Management, Higher School of Technology Sidi Bennour, University Chouaib Doukkali, Street Jabran Khalil Jabran BP 299-24000 El Jadida

Abstract. Prickly pear cactus (*Opuntia ficus indica*), is a Cactaceae plant with a great economic relevance in the world. Nowadays, nopal cactus industry is active and rapidly expanding.In fact, the seeds oil sector is the best exploitation due to the high price of this oil. Moisture content, oxidative stability, quality index, total phenol, flavonoids, chlorophyll and carotenoid contentof prickly pear seed oil were analyzed.The seeds used in this study were provided by a cooperative DAR ATABIAA located in Chouihia (Eastern region of Morocco) during the harvesting season 2020. The samples were obtained from wild trees grown in this region. Results showed that moisture content of prickly pear seeds was 9.12%, whereas the acidity and peroxide index value was 1.11% and 4.44 meq O2/kg respectively, andan oxidative stabilityof 17.31 hours. The total phenolic and the flavonoid content values for theseseed oilsare 260.07 mg/kg and 65.99 mg/kg, respectively. Furthermore, the results show a carotenoid contentof 0.61 mg/kgand total chlorophyll of 1.52 mg/kg.

1 Introduction

Prickly pear (*Opuntiaficus indica*),a member of the Cactaceae family, originally fromAmerican continent more exactly from the south of this continent.Grows abundantly in arid and semi-aridregions around the world, including Mediterranean countries, South Africa, Middle East, India, Australia and other areas[1] and grows everywhere in Morocco[2].Prickly pear speciesare well-adapted to arid lands and to a variety of climates, theygrow wild in areas with restrictedaccess to water, limiting the development of a lot of succulent food plants[2].On one side, this plant arecultivated in dry regions as an important nutrient and food source, anditcan alsobe used as natural wind break barrier, helping to protect the soil from the spread of the desert and controlling erosion[2].Moreover, Prickly pears are considered to be a versatile plant thatcan be cultivated and grown for usein several field, including food, cosmetic and pharmaceutical industries[2]. On the other side, the cultivation ofthis plant requires little investment. All these characteristics have contributed toan unceasing extension of this culture in several areas all over the world, and the plant is becoming increasingly economically important[1]. The fruit is the most consumed part of the plant, it contains around 300 seeds (2% to 10% of the fruit)[1],while the pericarp and the pulp of the fruits accounts for 33 to 55% and 45 to 67% respectively[3].However,until now, this part of the fruit(seeds) remains poorly exploited. In fact, Millions of pounds of seeds are thrown away each year as waste[4], although proper utilization of these by-products leads to a production of a very important and a costly oil. Prickly pear seed oils (PPSO) are of great interest because they are edible oils with high level of unsaturationand antioxidant radical scavenging properties[5].This oil has been shown to have considerable amounts of antimicrobial, biological activity, cardio protective, anti-thrombotic, anti-inflammatory, anti-arrhythmic, hypolipidemic and anti-hyperglycemic[4]. The seeds are composed mainly by 45.1 wt% of cellulose, followed bylipids (23%)[6]. Thus,they are rich in albumin protein with 6.5 kDa molecularweight, and proanthocyanidins withstrong antioxidant activity [7], they are composed by49.6% fiber, 17 fat, 16.6% protein and 3.0% ash.Aspartic acid, glutamic acid, arginine and glycine make up the most abundant part

*Email: ahmed20maghri@gmail.com

of the amino acids [8].According to literature data ,*Opuntia ficus indica*seed oil is characterized by a fairly low acidity (1.27 %) [9] andPeroxide values ranged between 9.50 and 33.67 meq O2/kg,another study conducted by De Wit, Hugo [10]revealed that these oils have a relatively low owidative stability ranging between 2.16 and 4.15 h.Thus, this previous research has shown that chemical properties such as peroxide number and oil content correlate with oxidative stability.According to Khemiri and Bitri [11], PPSOhas a totalphenolic and the flavonoidcontentof 26.5 Gallic acid eq/g and 3.1 mg Quercetineq/g respectively,carotenoid content (10.52mg/kg)oil and a total chlorophyll content of 4.57 mg/kg.
The oil content and composition are influenced by the environmental and genetic factors such as soil, cultivar and geographical area[12].Moisture content, oxidative stability, quality index, and pigments contentof *Opuntia ficus-indica*seed oil from Chouihia region of eastern Morocco were investigated in this study. These chemical properties will provide further information on the valorization of seeds in various fields, such as the food, cosmetic, and pharmaceutical industries. The results were compared with other studies to evaluate the quality of this vegetable oil from Eastern Morocco.

2 Material and Methods

2.1 Samples and plant material

Prickly pear seeds used in this study were extracted from the fruits provided by the cooperative DAR ATABIAAobtained from wild trees locatedin rural zone of Chouihia (Berkane Province) eastern Morocco.The fruits were peeled by hand, and the pulps were scraped from the seeds with multiple washes of water. The seeds were then dried externally in the sun. The oil was extracted via mechanical pressing which is the most prevalent method for obtaining oils in the world[13].

2.2 Moisture content

The determination of the moisture percentage of the seeds was carried out according to AOAC [14], 2 g of prickly pear seeds are dried in stove at 103 ° C for 3 hours until weight stability, are then placed in the desiccator for one hour before being weighed.
The loss of mass found after steaming is then equated with the mass of water contained in the product. The moisture content is equal to the average of the results of the grading tests, expressed in g of water / 100 g of seed.

$$Humiditylevel(\%) = [(P2 - P3) \div (P2 - P1)] \times 100$$

P1 : mass in grams of the tare vessel
P2 : mass in g of (tare vessel + sample) before drying
P3 : mass in g of (tare vessel + sample) after drying

2.3 Free acidity

The acidity was determined according to the official method of the European Commission (EEC, 2003).Acidity value determines the percentage of free fatty acid expressed by the major fatty acid (linoleic) present in the oil resulting from the hydrolysis of triglycerides.Using the official method of the European Commission: 0.5 g of *Opuntia ficus indica* oil was dissolved in 5 ml of a mixture of absolute ethanol and diethyl ether(2.5ml, 2.5 ml;v/v),to which eight drops of phenolphthalein (1%) were added. The free fatty acids are titrated under agitation by an alcoholic solution of 0.1 N potassium hydroxide (the pink color of the phenolphthalein persists for at least 10 seconds). The free acidity is calculated according to the following formula[15]:

$$\%\ \text{of linoleic acid} = \frac{C\ (KOH) \times V\ (KOH) \times M\ (C18H32O2)}{P.E}$$

C (KOH) : KOH concentration expressed in mol/l
V (KOH) : Poured volume of KOH expressed in ml
P.E : test sample in gram

2.4 Peroxide index

The peroxide indice is used to estimate the content of the primary oxidation products present in the oil (hydroperoxides), which is expressed in meq O2/kg of oil.This index is determined by the so-called acetic acid and chloroform method according to the AOAC [14]standard.Acetic acid-chloroform mixture (15:20 v/v) was used to dissolve 1 g of tested oil. To this solution, 0.5 ml of a saturated solution of potassium iodide (KI)is added. Finally, 30 ml of distilled water is added to the mixture after one minute of stirring. This solution titrated with 0.01 N sodium thiosulfate using starch as a color indicator (colored indicator starch paste). The Peroxide values calculated according to the following equations.

$$\text{Peroxide index (meq O2 / Kg)} = \frac{(Vt - Vw) \times 0.01 \times 1000}{P.E}$$

Vt :Volume in ml of sodium thiosulphate added to the titration
Vw : Volume in ml of sodium thiosulphate used for the white.
P.E : test sample in grams

2.5 Oxidative stability

The stability oxidative was estimated viathe measure of induction time (oil resistance to thermo-oxidative stress) using the Rancimat 743(Metrom Co., Basel, Switzerland). In order to achieve this, 3g of oil was weighedand placed in a heating block at 100°C. The principle of this method is to accelerate the aging of the oil by thermo-oxidative deterioration. The degradation products resultsare transferred to the measuring cell containing distilled water, and a conductivity meter was used in measuring. The operating conditions are as follows: an air flow rate of 20 L/h and a temperature of 100 ° C (Delta T: 1.6° C)[16]. The result was expressed in hours.

2.6 Determination of chlorophyll and carotenoids content

The analysis of the chlorophyll and carotenoid content (expressed in mg/kg) was determined by the colorimetric method described byIsabel Minguez-Mosquera, Rejano-Navarro [17]. 1.5g of oil was dissolved in 5 mL cyclohexane, and the absorption was measured with a uvspectrophotometer (RAYLEIGH UV1800; UV-Visible)at 670nm for chlorophylls and 470 nmfor carotenoids. The content of these pigments is calculated by the following equations:

$$\text{Chlorophyll mg/kg} = \frac{A670 \times 106}{613 \times 100 \times d}$$

$$\text{Carotenoid mg/kg} = \frac{A470 \times 106}{2000 \times 100 \times d}$$

A470 :Absorbance at wavelength 470 nm
A670 : Absorbance at wavelength 670 nm
L : Tank thickness in cm

2.7 Determination of total phenolic content

The extraction of phenolic compounds from cactus seed oil done using liquid–liquidextractionmethod described by Khemiri and Bitri [11]. In a centrifuge tube, 2 ml of methanol/water mixture (80/20, V/V) are addedto 2 g of cactus oil. After stirring for 10 minutes, the tubes are centrifuged for 15 min at 3000 rpm. The hydro-methanolic phase was recovered by a Pasteur pipette.This extract is then used for the colorimetric determination by the Folin-Ciocalteu reagent in order to determine the total content of phenolic compounds[18].2 ml of each polyphenol extract are added to 1 ml of Folin-Ciocalteu reagent diluted 10 times in distilled water, and 1 ml of an aqueous solution of 10% Na2CO3 (P/V). After stirring, this solutionwas put in the obscurity for 90 min at room temperature, the absorbance measurements are carried out using a UV-visible spectrometer (RAYLEIGH UV1800) at 760 nm.Total phenolic were expressed as mg of Gallic acid equivalents per g of oil (G.A. eq/g oil) [1].

Expression of results:

$$T = \frac{C \times V}{M}$$

T : Total polyphenol content
C : Concentration of extract equivalent to Gallic acid (mg/ml)
V : Volume of the extract (mL)
M : Mass of the dry matter (g)

2.8 Total Flavonoid Content

The Bahorun method[19] was employed to determine the total flavonoids in extracts, and aluminumtrichloride (AlCl3) was utilized to quantify the flavonoids in the various extracts. 1.5 mL of each extract is added to 1.5 ml of an AlCl3 solution (2% in methanol),after 30 min of incubation in the dark, the absorbance is measured at 430nm[20]using a UV-visible spectrometer (RAYLEIGH UV1800). The flavonoids concentration is deduced from a calibration range established with quercetin at different concentrations.

Expression of results:

$$T = \frac{C \times V}{M}$$

T : Flavonoid content
C : Equivalent extract concentration in quercetin (mg/ml)
V : Volume of the extract (ml)
M : Mass of the dry matter (g)

3 Results and Discussion

3.1 Moisture Content of prickly pear seeds

The amount of water detected in prickly pear seeds in the current study was (9.12 g/100 g), which is generally a low value compared to other species[21, 22].In fact, this reduction of water content is one of the mechanisms adopted by cactus to reduce respiration, which causes oxidation and stockdepletion. Therefore, the seeds of prickly pear resist well to dehydration, In other words, water reduction is a strategy to increase the viability of the seeds, which is why they are classified orthodox seeds. In comparison with the results found in the literature, our finding found slightly higher compared to that of Bahorun, Gressier [19], Karabagias, Karabagias [4], BENATTIA [20], and El Mannoubi, Barrek [9], which are 6.0, 6.0, 6.43, and 6.9 g/100 g respectively.Our study shows a slightly high humidity, which may be due to the climatic conditions of the eastern region, which are more favourable in comparison with the arid climate.Moreover, the high moisture content probably gives information about the water absorption capacity of the wild cultivar and the soil conditions.

The obtainedresults of acidity, peroxide index, total phenol, flavonoids, chlorophyll and carotenoid content of prickly pear seed oil are presented in the table 1.

Table 1: Chemical characteristics of wild prickly pear (*Opuntia ficus indica*) seed oil in eastern Morocco

Chemical quality parameters	Mean	SD	Min	Max
Acidity %	1.11	0.03	1.09	1.15
Peroxide (meqO_2/kg)	4.44	0.44	3.99	4.88
Stability oxidative (h)	17.31	1.79	16.33	20.94
Phenol (mg/kg)	260.07	30.1	201.19	314.9
Flavonoid (mg/kg)	65.99	12.77	46.33	84.95
Chlorophyll (mg/kg)	1.52	0.19	1.12	1.69
Carotenoid (mg/kg)	0.61	0.03	0.55	0.63

SD: standard deviation

3.2 Quality index

The measurementof free acidity is the main parameters used to determine oil quality and therefore provides information on its aging. Over time, the triacylglycerolshydrolyze to free fatty acids and glycerol,the degraded oil contains more free acids, consequently, higher acidity index.The recordedvalue of this parameter was about1.11% of linoleic acid (table 1). This result is lowercomparable to that obtained by El Mannoubi, Barrek [9] whofound a free acidity value of 1.27 %. In addition our outcome is higher thanthatrecorded byKarouiet al[23](0.27%),and isslightly lower compared tothe value reported byOzcanet al[24]and by Khemiri and Bitri [11] whofound a free acidity about 1.41 % and 1,95% respectively.

According to the joint Food and Agriculture Organization/World Health Organizationfood standards program codex alimentarius commission 2019, the examination of hydroperoxides allows for the determination of the oxidation stage of fatty acid. The acceptable peroxide value must be less than 20meq O2/kg for Virgin oils.The obtained value (4.44 meq O2/kg) is below the standard for virgin oil. This result allows us to concludethat the studied oil has a less oxidized.In comparison with other studies, this value was slightly higher compared to the results reported byKaroui, Ayari [23] (3.71 meq O2/kg)and much higherthan the finding ofOzcan and Al Juhaimi [24],andKhemiri and Bitri [11]who found (1.63 meq O2/kg) and (2.230 meq O2/kg), respectively.Low acidity and peroxide values indicate a low degree of deterioration of the tested oil, these quality indices are mainly affected by several factors such as improper harvesting systems, transportation, storage of the fruit, and in particularoil extraction method.In fact, the oil acidity increased during storage of fruit. It was highest when fruit were stored at 7.5°C and decreased with lower storage temperature.Oil stored in a controlled atmosphere had lower acidity than oil stored in air. The increase in acidity was probably the result of fungal lipase activity[25]. Regarding the peroxide value of oil obtained from fruit stored at various temperatures and controlled atmospheres increased during storage, compared to oil from freshly harvested fruit. The increase was significant only for fruit stored at 7.5°C[25].

3.3 Oxidative stability

The induction time in the Rancimat test is used to assess oxidation sensitivity to thermo-oxidative stress.According to the literature, cactus oils have a relatively low oxidative stability index ranging from 1.79 to 4.15 hours[10, 26]. This higher sensitivityis due mainly to its richness in polyunsaturated fatty acids. However, the studied seed oil obtained from wild cultivar of Chouihia region shows higher stability oxidative index (17.31 hours), which is confirm the lower recorded value of acidity and peroxide index.This higher SOI recorded could be linked to thephenol and chlorophyll content.Similar results have been demonstrated by Salama, El Harkaoui [16]who found that natural antioxidant richness increases the SOI for vegetable oil.

3.4 Total phenolic and flavonoid content

The phenolic and flavonoids compounds have significant antioxidant activities and are classified among the most main antioxidants in the oils. This secondarymetabolites are considered as a determining factors in term of organoleptic quality[27]. The flavonoids are the most important polyphenolic class. This bioactive molecules are to be able to liberate the hydrogen that allows oxidative stability improvement [26].The colorimetric evaluation of the total phenolic and flavonoid content show that the studied seed oil of prickly pear has a value of 260.07 mg/kg and 6.599 mg/kg (table 1), respectively. Concerning the total phenol content, our results arelower than those found by Karabagias, Karabagias [4], which reported a total phenol concentration of 551 mg/kg.The process of oil extraction affects the amount of polyphenols, and therefore decrease the content of this natural antioxidants resulting inlow oxidation resistance.Indeed, the total phenolic content were rised with increasing temperature[28].Regarding the flavonoid, the estimated value is lower than that reported by Chavez-Santoscoy, Gutierrez-Uribe [29]who found a values ranging between 95 and 374 mg/kg respectively.This difference could be explained by the difference in the method utilized and the conditions extraction, which have an impact on the total concentration of phenols and flavonoids, and hence on antioxidant activity[30].The richness in polyphenol increases the interest of the oil,since itreduces glucose level in the blood and lowered level of bad cholesterol. Polyphenols arenatural calming and anti-fatigue compounds, and have a sedative action in conditions such as angina, heart spasms, headaches and stomach aches[20].

3.5 Carotenoidand chlorophyll content

These two natural components are the majorpigments of vegetable oil, ensuring an orange, red, or yellow color [31]. Green oils are rich in chlorophyll pigments, whilst oils with high levels of carotenoids are more yellowish [32].Because of their antioxidant nature in the dark and pro-oxidant nature in the light[28-29], chlorophyll play an important role in the oxidative activity of the oil during storage and in the preservation of its quality.On the other hand, carotenoids are the photo-oxidation inhibitors, and are able to neutralize active-singlet oxygen or triplet activator [33].In the present study,*Opuntiaficusindica*seed oil hasthe lowest carotenoid and chlorophyll contents with respectively0.61 and1.52mg/kg. In comparison with theresults ofEl Mannoubiet al[9](8.01and 2.403 mg/kg)the studied oil is lower in terms of pigments.The presence of these antioxidants has contributed in the remarkable improvement of the oxidative stability of this oil[9].

4 Conclusion

The chemical composition of PPSO has recently attracted the interest of scientists and experts. According to the results of this study, Chouihia PPSO is characterized by better oxidative stability on the one hand, richness in antioxidants on the other hand and could be considered as a good source of bioactive compounds (polyphenols, carotenoids and chlorophyll). This result shows that this Prickly Pear is a versatile plant and a source of bioactive components for industrial application for food and pharmaceutical purposes. Further studies are needed to better characterize this seed oil produced in eastern Morocco by evaluating fatty acid profile, triacylglycerol composition, tocopherols and phytosterols content.

References

1. M. Chbani, B. Matthaus, Z. Charrouf, H. El Monfalouti, B. Kartah, S. Gharby, and I. Willenberg, Foods, **9**, (2020)
2. Z. Ghazi, M. Ramdani, M.-L. Fauconnier, B. El Mahi, R.J.J.o.M. Cheikh, and E. Science, **5**, 967-972 (2013)
3. J. Arrizon, C. Calderón, G.J.J.o.I.M. Sandoval, and Biotechnology, **33**, 921-928 (2006)
4. V.K. Karabagias, I.K. Karabagias, I. Gatzias, and A.V. Badeka, Processes, **8**, (2020)
5. E. Ramírez-Moreno, R. Cariño-Cortés, N.d.S. Cruz-Cansino, L. Delgado-Olivares, J.A. Ariza-Ortega, V.Y. Montañez-Izquierdo, M.M. Hernández-Herrero, and T. Filardo-Kerstupp, Journal of Food Quality, **2017**, 1-8 (2017)
6. R. Ciriminna, R. Delisi, L. Albanese, F. Meneguzzo, M.J.E.J.o.L.S. Pagliaro, and Technology, **119**, 1700013 (2017)
7. V. Bittrich, E.A.J.B.s. Maria Do Carmo, and ecology, **19**, 319-321 (1991)
8. W. Sawaya, J. Khalil, and M.J.P.F.f.H.N. Al-Mohammad, **33**, 91-97 (1983)
9. I. El Mannoubi, S. Barrek, T. Skanji, H. Casabianca, and H.J.C.o.N.C. Zarrouk, **45**, 616-620 (2009)
10. M. De Wit, A. Hugo, N.J.J.o.f.p. Shongwe, and preservation, **41**, e12898 (2017)
11. I. Khemiri and L. Bitri, Oxid Med Cell Longev, **2019**, 1568720 (2019)
12. F. El Hachimi, A. El Antari, M. Boujnah, A. Bendrisse, and C. Alfaiz, Journal of Materials and Environmental Science, **6**, 1488-1502 (2015)
13. R. Savoire, J.-L. Lanoisellé, E.J.F. Vorobiev, and B. Technology, **6**, 1-16 (2013)
14. AOAC, *Official Methods of Analysis of the Association of Official Analytical Chemists*. 15 Edi. ed. Vol. 17. 1990, Washington: Association of Official Analytical Chemists.
15. F. Mansouri, A. Ben Moumen, G. Richard, M.-L. Fauconnier, M. Sindic, H. Serghini-Caid, et al. **24**: 9 (2017).
16. M.A. Salama, S. El Harkaoui, I. Nounah, H. Sakr, M. Abdin, M. Owon, M. Osman, A. Ibrahim, Z. Charrouf, and B. Matthäus, Ocl, **27**, (2020)
17. M. Isabel Minguez-Mosquera, L. Rejano-Navarro, B. Gandul-Rojas, A.H. SanchezGomez, and J.J.J.o.t.A.O.C.S. Garrido-Fernandez, **68**, 332-336 (1991)
18. C. El Kar, A. Ferchichi, F. Attia, and J.J.J.o.f.S. Bouajila, **76**, C795-C800 (2011)
19. T. Bahorun, B. Gressier, F. Trotin, C. Brunet, T. Dine, M. Luyckx, J. Vasseur, M. Cazin, J. Cazin, and M.J.A.-f. Pinkas, **46**, 1086-1089 (1996)
20. F.K. BENATTIA, *Analyse et Application des Extraits de pépains de Figues de Barbarie*. 2018, 08-01-2018.
21. M. Amin, M. Hossain, and K. Roy, Journal of Food Engineering, **65**, 83-87 (2004)
22. R. Ellis, T. Hong, and E. Roberts, Annals of Botany, **61**, 405-408 (1988)

23. I.J. Karoui, J. Ayari, N. Ghazouani, and M. Abderrabba, Ocl, **27**, (2020)
24. M.M. Ozcan and F.Y. Al Juhaimi, Int J Food Sci Nutr, **62**, 533-6 (2011)
25. A. Kiritsakis, G. Nanos, Z. Polymenopulos, T. Thomai, and E. Sfakiotakis, Journal of the American Oil Chemists' Society, **75**, 721-724 (1998)
26. E.J.O.J.o.t.E.C. Regulation, **295**, 57-66 (2003)
27. P. Andrewes, J.L. Busch, T. de Joode, A. Groenewegen, H.J.J.o.a. Alexandre, and f. chemistry, **51**, 1415-1420 (2003)
28. H. Özdemir, E. Bakkalbaşı, and I. Javidipour, Journal of Food Science and Technology, **58**, 2606-2616 (2020)
29. R. Chavez-Santoscoy, J. Gutierrez-Uribe, and S.J.P.F.f.H.N. Serna-Saldívar, **64**, 146-152 (2009)
30. K.W. Lee, Y.J. Kim, H.J. Lee, C.Y.J.J.o.a. Lee, and f. chemistry, **51**, 7292-7295 (2003)
31. L. Cerretani, M.-J. Motilva, M.-P. Romero, A. Bendini, G.J.E.F.R. Lercker, and Technology, **226**, 1251-1258 (2008)
32. O. Baccouri, A. Bendini, L. Cerretani, M. Guerfel, B. Baccouri, G. Lercker, M. Zarrouk, and D.D.B.J.F.C. Miled, **111**, 322-328 (2008)
33. K. Viljanen, S. Sundberg, T. Ohshima, M.J.E.J.o.L.S. Heinonen, and Technology, **104**, 353-359 (2002)

Chemical characterization of local "Beldi"almond oil for food and cosmetic purposes in northeastern of Morocco

Reda Melhaoui[*1], *Souhayla* Kodad[1], *Nadia* Houmy[1], *Kamal* Belhaj[2], *Farid Mansouri*[1], *Malika* Abid[1], *Mohamed*Addi[1], *Aatika* Mihamou[1] ,*Hana* Serghini Caid[1], *Ahmed* Elamrani[1]

[1]Laboratoire d'Amélioration des Productions Agricoles, Biotechnologie et Environnement, (LAPABE), Faculté des sciences, Université Mohammed Premier, Oujda, Maroc.
[2]Laboratoire de Management de l'Agriculture Durable, EST Sidi Bennour, Université Chouaib Doukkali, El Jadida, Maroc

Abstract. The new trend of almond planting intensification aims for better profitability at the expense of biodiversity, which will lead to forsaking of local Beldiecotypes. Indeed, European late-flowering cultivars, especially *Ferragnes* and *Ferraduel*, were recently largely planted in northeastern Morocco. In parallel with our studies on these new almond crops, which are profitable in favorable areas, this research aims to conserve local almond biodiversity. Thus, it seeks added value for products from Beldialmonds, particularly for defective kernels (Doubles, malformed, split, and broken almonds). Therefore, these defective kernels could be used as raw materials for almond oil extraction to be used in cosmetology or even in food industry. in addition, this study is the first that has been conducted to determine the characterization of the Beldi ecotype has been carried out in this region. Low acidity and peroxide indexes values, which are respectively 0.44% and 2.67 meq O2/Kg, and a medium oxidative stability value (19 hours)were recorded for the studied almond oils obtained from seeds by mechanical pressing extraction. Total tocopherols content showed significant values 528.28 mg/Kg, with α-tocopherol as dominant. The total phenolic content of the analyzed Beldi almond oil is 147.59 mg/kg,The analyzed fatty acid profiles show that their composition is mainly constituted by oleic acid (64.68%). Finally, it should be noted that despite its low production, the Beldiecotypes will continue to spread in the marginal agro-ecological zones due to their remarkable resistance to drought and their almond oils highly valued in cosmetic, especially for skin and hair care products.

1 Introduction

[*] Corresponding author:reda_melhaoui@hotmail.fr

 https://doi.org/10.1051/e3sconf/202233704005

Sweet almonds are edible nuts of *Prunus amygdalus*, (var. *dulcis),* belonging to the Rosaceae family [1], approximately 80% of the global almond cultivated area is concentrated in California and the Mediterranean region[2]. The majority of almond cultivars are self-incompatible and must be cross-pollinated, at least two independent genes expressed in the pistil or pollen control the self-incompatibility that explains the high polymorphism of almond cultivars[3,4], and the necessity to safeguard this biodiversity. The almond tree is a typically Mediterranean crop, with large economic importance. It is able to withstand the high temperatures during the summer and the intense winter cold. It can survive long periods of drought and be cultivated on very poor land and marginal areas [5].The Almonds have numerous nutritional advantages and potential health benefits, they are considered excellent sources of energy, fats, proteins, minerals, carbohydrates, unsaturated fatty acids, phenoliccompounds and tocopherols[6–9]. In addition, the oil that is the main constituent of almond kernels, which is largely used in cosmetic especially for skinand hair care.

Currently, Morocco is the sixth-largest producer of almond nutsin the world, with an average production of 102.185 tons in 2019[2].In marginal areas, in addition to the varieties of almonds commonly cultivated, the Beldi almond trees represent a set of local almond ecotypes reflecting the richness of biodiversity and showing high adaptation to diverse edaphoclimatic conditions of the different agro-ecological areas of Morocco. However, the early pre-flowering and a relative sensitivity to late frosts, which sometimes occur in February, could be limiting factors that could affect the flowering of these almond ecotypes and therefore affect their fruit production. [10].

The resulting variability from seedling leads to high levels of heterozygosity, and therefore to a strong heterogeneity of the fruits showingpoor pomological quality (kernels of different shapes and sizes, high rate of doubles and defective kernels. However, based on the chemical analysis, Beldialmonds show excellent quality in terms of nutritional values and health benefits [11].

In Morocco, more than 50% of almond plantations are located in the regions of Taza-Taounate-Alhouceima and Sous-Massa-Daraa that represent 30% of national production. However, the Saïss region produces as much with only 6% of plantations; this is due to the more intensive nature of its plantations, the availability of water, use of high-yield varieties. Indeed, the production is highly variable due to drought and the incidence of frost, which affect flowering and subsequent fruit set. Therefore, in eastern Morocco, the latest almond planting program aimed at increasingfarmer's profitability at the expense of biodiversity. This results in the introduction of new late-flowering European varieties, mainly *Ferragnes* and *Ferraduel*, which in the short term seem better adapted and more profitable but in the long term this could lead to the abandonment of the Beldi ecotypes and a loss of biodiversity.

Beldialmond kernels, showing defects in shape, size, and kernel development, unfit for human consumption,are used as raw material for almond oil extraction. This is a valuation operation for creating adding value to farm products made fromBeldialmonds and so for sustainable agriculture. This is a real opportunity for the small farmers from these marginal zones which constitutes a financial resource at the service of rural development and therefore to promote the cultivation of biodiversity preservation of theBeldialmond. Thus, our study concerns the chemical analysis of the oil extracted from Beldialmonds as a high benefit product for food and cosmetic uses.In Morocco, some studies have been conducted to determine the characterization of oil content and quality indices for almond cultivars [12–14]. however, no specific study is done on the Beldi ecotype except for preliminary work byNadia et al and Kodad et al [11,15]

2. Material and methods

2.1. Almonds' harvesting and oil extraction

Harvesting of Beldialmonds was carried in crop year 2018 out on the southern side of the Tafoghaltmountains in northeastern of Moroccoknown as ancestral areafor traditional cultivation of Beldialmond trees. This area shows great biodiversity within the same field, where plantations were done by seed germination. Thus, Almonds nuts of Beldiecotype were harvested in August at the last phase of maturity in the rural zone of Sidi Bouhria (34°44'13.6'' N, 002°20'15.0'' W). Six kilograms of shelled almonds were harvested from three almond orchards, which are mainly rain-fed but receive additional irrigation during periods of drought. Almond oil is extracted usinga mechanical oil screw press, at the speed of 70 RPM, oil's fine particles were removed by centrifugation at 3000 RPM for 15min. Finally, the obtained almond oil samples were stored for further analysis in the dark at a temperature of 4°C.

2.2. Analytical methods

The chemical quality parameters of Beldi almond oil were analyzed, its oxidation stability was tested, and its fatty acid and triacylglycerol profiles were determined. Each parameter is analyzed in triplicate

2.2.1 Chemical parameters and oxidative stability of Beldialmond oil

Almond oil acidity (% oleic acid) and peroxide value (meq O_2/Kg),
The official European methods[16] used to determine the oil acidity (% oleic acid) and peroxide index (meq O_2/Kg) of olive oil were maintained for the evaluation of the studied almond oil (AO) quality. 3g of almond oil heated at 100°C with an airflow of 15L/h, was used to determine the oxidative stability index (OSI) using a MetrohmRancimat 743 apparatus[17].OSI value is expressed as the oxidation induction time (hours).

2.2.2 Phenol analysis

TheFolin-Ciocalteu method described by Ollivier et al, (2004) was used for analyzing the total phenolic contents of almond oil, were the absorbance was made at 750 nm by using caffeic acid as a standard [18].

2.2.3 Tocopherols analysis

HPLC-FLD equipped with an automatic injector was used for the tocopherols profile analysis, according to the AOCS method Ce 8-89 AOCS, [16] where α, β, and γ tocopherols which are the main tocopherol isomers were separated by using an Uptisphere 120A° NH2 column (150 mm *3 mm, 3 μm, Interchim, Montluçon, France), with an injection volume of 10 μL and a maintained temperature at 30 °C. Hexane/2-propanol (99:1, v/v) constitute the mobile phase which was eluted in isocratic conditions at a flow rate of 1mL/min. The tocopherols were identified and quantified by external standardization (Mixture of tocopherols: α-tocopherol, β-tocopherols, γ-tocopherols, δ-tocopherols) obtained from Sigma-Aldrich (Steinheim, Germany).

2.2.4 Fatty acids analysis

Fatty acid profile of almond oils was identified by a standard composed of 37 methyl esters of fatty acids. Before analysis, fatty acids were converted into fatty acid methyl esters; after adding 8 mL of hexane, the mixture was subjected to gas-chromatograph analysis using an HP 6890 series gas-chromatography system equipped with an FID detector and a capillary column (Supelco Omega wax: 30 m × 0.25 mm × 0.25 µm). The carrier gas was nitrogen, supplied at a flow rate of 1.7 ml min-1. The initial temperature of the injector was 50°C, it was gradually increased to 150°C at a rate of 30°C min-1 and then at a rate of 4°C min-1 up to 250 °C. The injection volume was 1 µl in the split-less mode. This method of analysis was described in by Mansouri & al[19].

2.2.5 Triacylglycerols analysis.

Analysis of molecular species of Triacylglycerols (TAGs) was carried out using 2 ml ofBeldialmond oil dissolved in acetone (9%) and filtered through 0.45µm membranes. TAGs profile was determined by high-performance liquid chromatography (HPLC) coupled with a refractive index detector. The column used is type C18 reversed-phase column (ODS C18: 250× 5mm, 5µm), and the mobile phase consisted of acetone/acetonitrile (60% - 40%, v/v). Elution was conducted in isocratic conditions out at 1ml/min. TAGs peaks were identified by comparison with standards triglyceride (Sigma chemical products).

3. Results and discussion

3.1. Quality indexes, phenol and tocopherol of*Beldi*almond oil

Lipolysis and lipid oxidation,which causes undesired flavors and smells, are the main reasonsfor virgin almondoils spoilage,hence the importance of measuring the acidity, peroxide index, and oxidative stability index (OSI) values[20]. Those parameters are essential factors to determine the shelf-life and acceptability of almond oil via the determination of enzymatic or chemical hydrolysis of acylglycerols, as well as the secondary oxidation products such as ketones, aldehydes [21]. In this regard, the analysis resultsof physico-chemical characteristics shows a low acidity (0.43%), a low peroxide value (2.67 meq O_2/kg) and a medium OSI value (19 hours)expressed as the oxidation induction time (Table1)

Like all vegetable oils, the presence of natural antioxidants (phenols and tocopherols) in almond oil improves its stability, and therefore its storability. Thus, phenolic compounds present in almond oil are important secondary metabolites with great antioxidant capacity. In fact, Beldialmond oil's richness of phenols delaying the oil oxidation speed by different actions and it is closely associated with the nutritional quality and health benefits.These natural antioxidantsplay a prominent role in the prevention of various diseases[22]. The total phenolic contentof the analyzed Beldialmond oil is 147.59 mg/kg, (Table 1). This observed value is higher than that recorded for oils extracted from almond varieties (*Ferragnes, Ferraduel,* and*Marcona*) commonly cultivated in Morocco, which varied from 18.53 mg/kg to 237 mg/kg[23,24]. This variability iscertainly attributed to the genotype and the agro-climatic conditions but also to the oil extraction methods.

In addition to their richness of phenols, almond oil is a good dietary source of tocopherols, which constitute a large class of fat-soluble compounds with vitamin E activities. In fact, vitamin E is a mixture of different forms of tocopherols and tocotrienols, with α-tocopherol being the most effective form of vitamin E[25]. Due to its high content of antioxidants, almond oil may help reduce the risk of cardiovascular disease and plays an important role in preventing diseases associated with aging and protecting cell membranes from oxidative damages[26,27]. Therefore, almond oil is a natural cosmetic product that helps soothe the skin from irritation, protects from UV damage, and restores the skin's moisture barrier[28]

Beldialmond oil was analyzed for their quantitative and qualitative tocopherol contents by HPLC-FLD. Total tocopherols content and proportion of each homolog (α-, β-, and γ tocopherols) are summarized in Table 1. Beldialmond oil has a high content of tocopherols estimated at 528.28 mg/kg, with α tocopherol as the major homolog (516.33 mg/kg) followed by γ-tocopherol (9 mg/kg) and β-tocopherol (2.95 mg/kg). This oil is very rich in tocopherols (528.28 mg/kg)compared to that from *Ferragnes* and *Ferraduel* cultivarswhere tocopherol contents range between (385-433 mg/kg)[13,29].This high content of tocopherols a (particularly α-homologue, the most effective form of vitamin E), make this oil a product of nutritional interest and beneficial for human health. Actually, this oilis considered an excellent source of natural antioxidants for food and cosmetic purposes.Therefore, these benefits will encourage the extension of Beldialmond cultivation for oil production. Besides this will help farmers' improving incomes, contribute to rural development, and led to protect the biodiversity of almond trees in marginal areas.

Table 1:Quality indexes, phenols and tocopherol contents in analyzedBeldialmond oil

Physicochemical parameters	values
Free acidity (% C18:1)	0.43± 0.1
peroxide value (meqO_2/kg)	2.67± 0.05
Total phenols (mg/Kg)	147.59± 14.50
α-tocopherol (mg/Kg)	516.33± 14.01
β-tocopherol (mg/Kg)	2.95± 0.25
γ-tocopherol (mg/Kg)	9.00± 1.50
total tocopherol (mg/Kg)	528.28± 15.26
oxidative stability index (hours)	19.08± 0.1

The oxidative stability of almond oil should be understood as the oil's sensitivity to lipid oxidation, which leads to unwanted flavors and odors. It could be affected by the storage conditions of the oil, namely light, temperature, and exposure to oxygen, but also by its content of natural antioxidants. It is, therefore, necessary to measure the OSI of Beldialmond oil, which is a key parameter that refers to the susceptibility of the oil to lipid oxidation and which provides information about oil quality and its aptitude for storage [33].The result of the analyzed Beldialmond oil showed an OSI value of 19.08 h, which's comparable to those of literature that range from 10.2 to 24.2 h[23,34]. The richness in antioxidants of Beldialmond oil would explain its resistance to oxidation and therefore its good stability during storage

by comparing these results with other type of vegetable oil such as olive oil, the latter presents high phenol values with 316.59 mg / kg and SOI with 44.55 h for the picholine ecotype. For tocopherol olive oil has less tocopherol concentration with 254.13 mg/Kg [30]. regarding of oil characteristics of non-industrial hemp (Cannabis sativa L.), the total tocopherol content was between 376.46 and 796.06 mg kg^{-1} of oil, with a dominance of γ-tocopherol [31], which is not the case for almond oil with a dominance of α- tocopherol. for argan oil, the total phenolic content was lower than 10 mg/Kg. the total tocopherol content range between (383-485 mg/Kg), the oils from unroasted seeds presented same stability of almond oil with value ranged between 16h and 32h [32]

3.2 Fatty acids and triacylglycerols profiles

The fatty acidsare classified into three groups, including saturated fatty acid (SFA), monounsaturated fatty acid (MUFA), and polyunsaturated fatty acid (PUFA).They plays multiple roles in human's health and other organisms[35].The content of saturated fatty acids in Beldialmonds oil is very low (10.03%) while the proportion of unsaturated FA (UFA= MUFA+ PUFA) is very important (89.89%). The most important unsaturated fatty acids are oleic acid and linoleic acid, the contents of which are estimated at 64.68% and 24.37% respectively. Table 2 presents the Beldialmond oil fatty acids profile, the mean values of SFA, MUFA, PUFA, and the ratios UFA/SFA and O/L. Among the eleven fatty acids identified, four of them are major and which are respectively in order of quantitative importanceC18:1 (64.68%), C18:2 (24.37%), and C16:0 (7.22%),C18:0 (2.62%). The GC-FIDanalysis allows the detection and identification of seven other fatty acidsin the form of traces, As stated by several studies that refer to the benefits of unsaturated FA intake on risk factors for cardiovascular disease, the consumption of Beldialmond oil will provide nutritional and health benefits. in addition, its O/L ratio (2.65) reflects its richness oleic acid (C18: 1), considered to be the healthiest FA in the diet and equally as an indicator of the good nutritional value of this oil [36].the fatty acid profile of almond oil is comparable to that of olive oil and argan oil with higher values of oleic acid 75.68% and 48.1% respectively[19,37],but it is different compared to non-industrial hemp oil (Cannabis sativa L.) with a predominance of linoleic acid (48.26%–55.39%)[31].

The HPLC analysis of triacylglycerols (TAGs) of Beldialmond oil allows the identification of eleven molecular species; four of them are major ones (OOO, OOL, LLO and POO) covering together more than 83% of total TAGs. The other eight TAGS are minors triacylglycerols (LLL, LPL, SOO, LLS+ POL, POP, SLO, and, PPL, together they represent less than 20% of the total TAGs, Where P is palmitic acid, O is oleic acid, S is stearic acid and L is linoleic acid (Figure 1).

Table2: Fatty acids (FA)profile of Beldialmond oil (SFA: Saturated FA; UFA Unsaturated FA: MUFA: Monounsaturated FA; PUFA: Polyunsaturated FA, O/L=Oleic/ Linoleic ratio

Fatty acids (%)	Mean± Standard deviation
C14:0	0.04 ± 0.01
C16:0	7.23 ± 1.00
C16:1	0.65 ± 0.20
C17:0	0.06 ± 0.02
C17:1	0.09 ± 0.01
C18:0	2.62 ± 0.59
C18:1	64.68 ± 4.00
C18:2	24.37 ± 3.00
C18:3	0.04 ± 0.01
C20:0	0.08 ± 0.02
C20:1	0.06 ± 0.02

SFA	10.03 ±1.63
MUFA	65.48 ± 4.23
PUFA	24.41 ± 3.02
UFA/SFA	8.96 ± 0.64
O/L	2.65 ± 0.13

The analyzed Beldialmond oil presents two main predominant TAGs, OOO (37.87%) and OOL (23.47%) which in total represent more than 60% of total TAGs. This is in perfect agreement with its richness in oleic acid (64.68%).

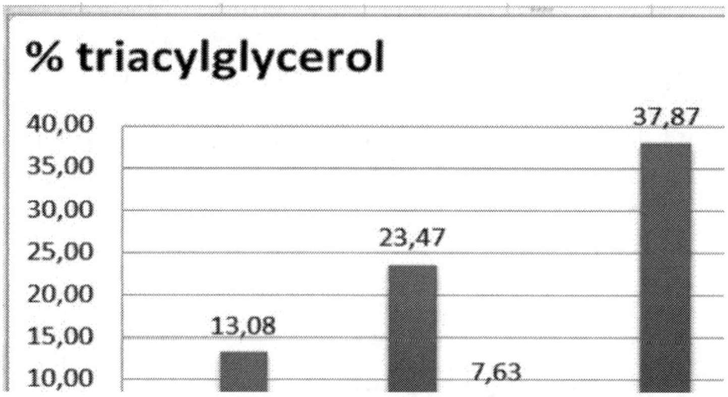

Figure 1: Profile of triacylglycerols (TAGs) of almond oil extracted from defective kernels of almonds Beldiecotype. (Eleven TAGs species were identified, OOO, OOL, LLO and POO as majors ones, LLL, LPL, SOO, LLS+ POL, POP, SLO, PPL as minors and where O = oleic acid, L = linoleic acid, P = palmitic acid and S =stearic acid.

4. Conclusion

This work is the first assessment of the quality of Beldialmond oil produced in a rural cooperative located in eastern Morocco. The study concerns the use of defective Beldialmonds (small, split and broken, and malformed kernels) as raw material for the extraction of almond oil by mechanical pressing of defective kernels unfit for human consumption. As shown by chemical analyzes, Beldialmond oil has the required biochemical qualities which make it a product of interesting nutritional quality as edible oil. In addition, its richness in tocopherols and other bioactive molecules gives it other beneficial properties for its cosmetic use. Thus, both for food use but especially for cosmetic applications Beldialmond oil extracted from defective kernels is profitable. Finalybeldi almond oil generates benefits that could directly improve the small farmers' income, which therefore contributes to the preservation of this culture and its biodiversity as well. in addition to the physico-chemical characterization of almond oil, it is necessary to develop an effective protocol for clonal propagation of interesting "Beldi" almond genotypes carrying the desired agronomic traits

Acknowledgments
The Moroccan Ministry of High Education and Research supported this work. The authors are indebted to the "Centre National de Recherche Scientifique et Technique (CNRST), they also gratefully acknowledge the financial participation from the "programme de recherché appliquée(PARAI)

References

1. Chin S-W, Shaw J, Haberle R, Wen J and Potter D. *Mol Phylogenet Evol* **76**: 34-48 (2014)
2. FAO-Stat. FAO-Stat. 2020. Crops. Items: Almond with Shell. Year: (2019)
3. Gómez EM, Dicenta F, Batlle I, Romero A and Ortega E. *Scientia Horticulturae* **245**: 218-223 (2019)
4. Tao R and Iezzoni AF. *Scientia Horticulturae* **124**: 423-433 (2010)
5. Abdini A, Imani A, Rasouli M, Zinalabdini M and Abdoosi V. *Journal of Nuts* **11**: 73-90 (2020)
6. Altamimi M, Zidan S and Badrasawi M. *Nutr. Metab. Insights* **13**, (2020)
7. Liu Y, Hwang H-J, Ryu H, Lee Y-S, Kim H-S and Park H. *Nutr. Res. Pract.*; **11**: 479-486 (2017)
8. García MC, González-García E, Vásquez-Villanueva R and Marina ML. *Food Funct.* **7**: 4693-4701 (2016)
9. Ojeda-Amador RM, Fregapane G and Salvador MD. *European Journal of Lipid Science and Technology* **121**: 1900114 (2019)
10. Parker L, Pathak T and Ostoja S. *Science of The Total Environment* **762**: 14 (2021)
11. Kodad O, Estopañán G, Juan T and Socias i Company R. *Journal of Food Composition and Analysis* **33**: 161-165 (2014)
12. Sakar EH, El Yamani M and Rharrabti Y. *J Mater Environ Sci* **8**: 2679-2686 (2017)
13. Melhaoui R, Fauconnier M-L, Sindic M, Addi M, Abid M, Mihamou A, Serghini-Caid H and Elamrani A. *Comm. Appl. Biol. Sci, Ghent University, 83/1*(2018)
14. Melhaoui R, Houmy N, Kodad S, Belhaj K, Mansouri F, Addi M, Abid M, Mihamou A, Hano C, Sindic M, C HS and Elamrani A. *E3S Web Conf.* **240**: 04004 (2021)
15. Houmy N, Mansouri F, Benmoumen A, Elmouden S, Boujnah M, Sindic M, Fauconnier M-L, Serghini-Caid H and Elamrani A. *Cahiers Options Méditerranéennes*(2015)
16. AOCS. Determination of Tocopherols and Tocotrienols in Vegetable Oils and Fats by HPLC. Offic. Method (Ce 8–89).; (1989)
17. Mansouri F, Ben Moumen A, Lopez G, Fauconnier M-L, Sindic M, Serghini-Caid H and Elamrani A. In Book of Proceedings InsideFood Symposium 6 (2013)
18. Ollivier D, Boubault E, Pinatel C, Souillol S, Guérère M and Artaud J. In Annal Expert Forum Chem Toxicol. Vol 965169-196 (2004)
19. Mansouri F, Benmoumen A, Richard G, Fauconnier M-L, Sindic M, Serghini-Caid H and Elamrani A. *Rivista Italiana Sostanze Grasse* **93**: 21-30 (2016)
20. Roncero JM, Álvarez-Ortí M, Pardo-Giménez A, Gómez R, Rabadán A and Pardo JE. *Rivista Italiana delle Sostanze Grasse* **93**: 237-243 (2016)
21. Baltazar P, Hernández-Sánchez N, Diezma B and Lleó L. *Agronomy* 2020; **10**: 41.
22. Alu'datt MH, Rababah T, Alhamad MN, Al-Mahasneh MA, Almajwal A, Gammoh S, Ereifej K, Johargy A and Alli I. *Food Chemistry* **218**: 99-106 (2017)
23. Rabadán A, Álvarez-Ortí M, Gómez R, Miguel C de and Pardo JE. *Journal of the Science of Food and Agriculture* **98**: 2402-2410 (2018)
24. Ozcan MM, Matthaeus B, Aljuhaimi F, Ahmed IAM, Ghafoor K, Babiker EE, Osman MA, Gassem MA and Alqah HAS. *J. Food Sci. Technol.-Mysore* **57**: 4182-4192 (2020)
25. Combs Jr GF and McClung JP. The Vitamins: Fundamental Aspects in Nutrition and Health. Academic press; (2016)
26. Tungmunnithum D, Elamrani A, Abid M, Drouet S, Kiani R, Garros L, Kabra A, Addi M and Hano C. *Applied Sciences* **10**: 3313 (2020)
27. Chen B, McClements DJ and Decker EA. *Critical Reviews in Food Science and Nutrition* **51**: 901-916 (2011)
28. Lin T-K, Zhong L and Santiago JL. *International journal of molecular sciences* **19**: 70

(2018)
29. Kodad O, Estopanan, G, Teresa Juan, Ali Mamouni and Rafel Socias i Company. *J Agric Food Chem*.6137-6141(2011)
30. Mansouri F, Benmoumen A, Richard G, Fauconnier M-L, Sindic M, Serghini-Caid H and Elamrani A. *Rivista Italiana Sostanze Grasse***93**: 21-30 (2016)
31. Taaifi Y, Benmoumen A, Belhaj K, Aazza S, Abid M, Azeroual E, Elamrani A, Mansouri F and Serghini-Caid H. *International Journal of Food Science & Technology***65**-11 (2021)
32. Cayuela JA, Rada M, Pérez-Camino M del C, Benaissa M, Abdelaziz E and Guinda Á. *European Journal of Lipid Science and Technology***110**: 1159-1166 (2008)
33. Rabadán A, Álvarez-Ortí M, Pardo JE and Alvarruiz A. *Food Chemistry***259**: 31-35 (2018)
34. Kochhar SP and Henry CJK. *International Journal of Food Sciences and Nutrition***60**: 289-296 (2009)
35. Sales-Campos H, Reis de Souza P, Crema Peghini B, Santana da Silva J and Ribeiro Cardoso C. *Mini reviews in medicinal chemistry***13**: 201-210 (2013)
36. Čolić SD, Fotirić Akšić MM, Lazarević KB, Zec GN, Gašić UM, Dabić Zagorac DČ and Natić MM. *Food Chemistry***234**: 455-463 (2017)
37. Rueda A, Seiquer I, Olalla M, Giménez R, Lara L and Cabrera-Vique C. *Journal of Chemistry*(2014)

E3S Web of Conferences **337**, 04006 (2022) https://doi.org/10.1051/e3sconf/202233704006
I2CNP 2021

Effect of pre-harvest practices on physical and colorimetric characteristics of date fruit (cv. 'Aziza Bouzid') in Figuig oasis

Kawtar Jdaini [1*], *Fouzia* Alla [1], *Kamal* Guerrouj [2], *and Mohamed Aziz* Elhoumaizi[1]

[1]Laboratory for Agricultural Productions Improvement, Biotechnology and Environment (LAPABE), Faculty of Sciences, University Mohammed First, BP-717, 60000 Oujda, Morocco
[2] Higher Institute of Health Professions and Nursing Skills, 60000 Oujda, Morocco

Abstract. Dates represent a very important energy source that contains a wide range of functional, nutritional, bioactive, and aromatic components. Dates characteristics have been widely studied but none has been conducted on the pre-harvest factors responsible for the quality variation. The aim of this study is to evaluate the effect of pre-harvest practices on physical and colorimetric characteristics. The analyses were conducted on twelve samples of 'Aziza Bouzid' cultivar and collected at Tmar stage from Figuig oasis. Results obtained for physical parameters show a significant variation between different samples. Weight of 'Aziza Bouzid' varied from 5.44 g to 9.04 g. Geometric mean diameter and Porosity varied from 13.08 mm to 17.01 mm and 31.88 % to 42.3 % respectively for 'Aziza Bouzid'. Highest Chroma and hue values registered for 'Aziza Bouzid' were 38.14 and 83.22 and the lowest were 21.1 and 67.15. This study contributes to develop good agricultural practices, help improve the quality and increase the quantity of Moroccan date production.

1 Introduction

Date palm (Phoenix dactylifera L.) is one of the oldest fruit trees grown in arid regions of Arabian Peninsula, North Africa and the Middle East [1]. It has been cultivated for ecological, economical and social purposes. Dates fruit are considered as a good source of sugars (fructose, glucose and sucrose) and fibers, important vitamins and minerals [2].

Morocco production areas were estimated to 50.000 ha concentrated mainly in the oases of Draâ, Ziz, Tata, Tiznit, Guelmim and Figuig [3-4]. Dates production in the country has reached an average of 129,562 tons in 2017 [5] composed of more than 220 cultivars including high proportion of Khalts (originated from seed) and other high value cultivars such as 'Boufeggous', 'Medjool', 'Bouskri', and 'Aziza bouzid' [6].

Total date production in the Figuig oasis for the 2012-2013 campaign is estimated at 3384 tones. It is dominated by the Assiane cultivar, followed by Boufeggous and Khalts. However 'Aziza Bouzid', which is the flagship profile of this region is only 39 T and therefore only represents 3.2% [7].

* Corresponding author: jdaini91@gmail.com

E3S Web of Conferences **337**, 04006 (2022) https://doi.org/10.1051/e3sconf/202233704006
I2CNP 2021

Fruits quality may be affected by different factors, they are classified in three groups; pre-harvest, harvest and post-harvest processes .Pre-harvest including firstly Climatic condition like temperature light rains winds and humidity [8].Secondly Tree condition as variety tree age, fruit positioning [9] and finally Cultural practices (Irrigation , use of fertilizers etc). Harvest factors relating at maturity at harvest and method and harvest time [8] .As for postharvest factors, Washing, Pre-cooling, Grading, Packaging, Transportation, Cold storage, Marketing can also affect fruit quality by modifying physiology, chemical composition, and morphology of fruits [9].

Knowledge of physical properties of agricultural produce and the parameters, which affected these characteristics, is a critical aspect to formulate the proper design standards of harvesting, handling, processing, and storing equipment [10]. The mass, size and shape are important for sorting, sizing and various separation. Density, surface areas and porosity of fruit solids must be known for accurate modelling of heat and mass transfer during drying and Colling [11].

Numerous studies have been carried out on the physical parameters of dates fruits. But none have been studied for the design, development of machinery and equipment for handling, processing, transporting and storing Moroccan cultivars. [12] Reported that the length, width and thickness of 'Zahidi' cultivars were 33.65mm, 22.12mm, 20.02mm respectively. GMD and AMD were 23.51-25.96mm and 24.29-26.59mm, respectively. The Sphericity value ranged from 0.73 to 0.737. [13] Investigated the physical parameters of dates fruit, and discovered that the values of axial dimensions and weight of cultivar Medjool were significantly higher than those of variety 'Khadrawy'. According to [14] the linear measurements of cultivar 'Mazafati' ranged from 32.10 to 43.50 mm in length, 20.50 to 28.20 mm in width and the average weight was 8.39 g. The density of the fruits was 0.82 g / cm3 and the density of the pitted fruits was 1.25 g / cm3, respectively. Bulk density and porosity were 0.44 g / cm3 and 44.90%. GMD, Sphericity and area were calculated to be 27.87mm, 0.73mm^2 and 2446.70 mm^2 correspondingly.

The effect of pre-harvest practices on the physical characteristics of dates has been discussed by some authors. [15] found out that thinning the center of the bunch by removing 15-30% of the total number of sprigs improved various physical and chemical aspects of the fruits of the "Khadrawi" date palm. According to research by [16], Bagging fruit bunches increased the weight of date bunches, accelerated ripening, and improved fruit quality compared to unbagged bunches. [17] found that irrigation systems can also alter the physical qualities of fruits.

The aims of this study are to determine physical properties for ‘Aziza Bouzid’ cultivar and to evaluate interaction between pre-harvest factors and these properties .To facilitate the design of some machines for its processing.

2 Materiel and methods

2.1 Study area

Figuig oasis situated in southeast at the Oriental region of Morocco (Latitude 32° 18N, longitude 1° 15 W), at 900m altitude. It’s characterised by arid climate, in which the annual average temperature of the region is 20 °C. This average covers large seasonal variations and daily amplitudes. Daily amplitudes range from 3 °C to 16 °C in winter and 24 °C to 42 °C in summer (DPA, 2009). As for rainfall, annual average does not exceed 125mm (Tag, 2003).

Fig. 1. Figuig oasis (sources: google earth 2009. Traitement: ArcGis 10 ©. Production: G. Janty, 2014; UMR Ladyss, University Paris Diderot)

2.2 Fruit Materiel:

Fruit material is composed of the most appreciated cultivar in the oasis of Figuig. Twelve samples of 'Aziza Bouzid' cultivar were collected at Tmar stage from twelve plots chosen randomly in Figuig oasis. Each sample was carried out according to a factorial schemas 20 x 5 x 1 (dates x trees x plot). Date palm trees used in this study were identified as homogeneous (same age, height and intact). Dates samples were transported in isotherm bags to the laboratory, then stored at –20 °C until further analysis.

Table 1. Names and code plots

	Areas denomination	**Code**
Aziza bouzid samples	Aarja	Arj
	Cooperative Aziza	Cop.A
	Znaga 1	Zng.1
	Znaga 2	Zng.2
	Znaga 3	Zng.3
	Znaga4	Zng.4
	Hamam Tahtani	Ht
	Laabidate	Lb
	Oulad lmaiz 1	OM.1
	Oulad lmaiz 2	OM.2
	Oulad Slimane	Os
	Loudaghir	Lg

2.3 Field investigation

Field investigation on each plot was reposed mainly on tree characteristics (Age and height of tree), and also on cultural practices used such as (limitation, thinning and bagging of bunches, irrigation and fertilization technics, method of pollination, presence or not of intercropping)

E3S Web of Conferences **337**, 04006 (2022) https://doi.org/10.1051/e3sconf/202233704006
I2CNP 2021

2.4 Physical analysis

2.4.1 Size and weight measurements

Dimensions measurements were realized on 100 fruits per samples. Width (W), Length (L), and thickness (T) are carried out using a digital caliper. Dates, seed, and pulp weight (Wt) were determined by weighing samples with precision balance (RADWAG PS 360.R2). The percentage of seed and pulp on the whole fruit was calculated using formulas (1) and (2) respectively;

$$Ratio\ P/D\ (\%) = (P.Wt/\ F.Wt)*100 \quad (1)$$
$$Ratio\ S/D\ (\%) = (S.Wt/\ F.Wt)*100 \quad (2)$$

2.4.2 Geometric and arithmetic mean diameter, Sphericity, aspect ratio and surface area

Geometric and arithmetic mean diameter (GMD, AMD) obtained from the measured geometric dimensions of Width (W) , length (L) and thickness (T) as (3) and (4) formulas [17-18].

$$GMD = (LWT)^{1/3} \quad (3)$$
$$AMD = (L+W+T)/3 \quad (4)$$

Sphericity expresses solid object shape relative to that sphere of same volume [18]. It is a relevant property to calculate fluid flow, heat and mass transfer. Sphericity was calculated by using the equations (5)

$$\Phi = GMD/\ L \quad (5)$$

Surface area is an important physical parameter that is related to size and shape. It is found by using relation reported by [19];

$$S = \pi *GMD \quad (6)$$

Aspect ratio of fruit indicates its tendency to be an oblong shape during handling and processing. It is calculate by equation (7) described by [19]

$$R = W/L \times 100 \quad (7)$$

2.4.3 Bulk and true density and porosity

The true density (ρT) is determined by the water displacement method [20] which the volume of water displaced was estimated by immersing the weighed quantity of fruits in water. Whereas Bulk density ρ b is the mass (M) of a group of individual fruit occupied a box of known volume (V) [21]. Fruits were poured into a standard container without compaction and afterward the content was weighed. The bulk density is calculated as the ratio of fruits mass only to container volume.

Bulk and true density respectively were expressed on (g cm^{-3}) and calculated as (8) equation

E3S Web of Conferences **337**, 04006 (2022) https://doi.org/10.1051/e3sconf/202233704006
I2CNP 2021

Porosity is the percentage of air between the fruits compared to a unit volume of fruit [21]; it was calculated using equation (9)

$$P = ((\rho_T - \rho_b)/\rho_T) * 100 \quad (9)$$

2.5 Color measurement

Fruits color properties were measured using a Chroma Meter CR 400 and were expressed according to the CIELAB color space system defined by International Commission on Illumination (CIE) in 1976. Coordinates L* define to lightness a* correspond a red and green ratio and b* indicate ratio yellow and blue. The chroma C* and Hue H* were defined according to the mathematical functions (10, 11).

$$C^* = \sqrt{(a^{*2} + b^{*2})} \; (10) \quad ; \quad H^* = tg^{-1}(b^* / a^*) \; (11)$$

2.6 Statistical analysis

Data analysis was performed using SPSS Statistics 19.0 package (SPSS Inc., Chicago, USA). One-way analysis of variance (ANOVA) and Duncan's multiple-range test were applied to determine significant differences between means and to separate groups of means with a probability $p \leq 0.05$ according to their physical and colorimetric characteristics. Values were expressed as an average of four determinations ± standard deviation. Principal component analysis (PCA) was performed on the physical data set in order to verify whether it was possible to differentiate the samples according to their soil condition and pre-harvest practices and to obtain information of variable that influence physical parameters of 'Aziza Bouzid' and Assiane cultivars

3 Results and discussion

3.1 Physical analysis

Size measurements (width, length, weigh and thickness) are presented in Table2. Measurements of size revealed a significant variation between different samples of 'Aziza Bouzid' cultivar collected from different plot in Figuig oasis, except percentage of pulp per date that show no significant difference. For length and width, Hamam Tahtani sample has the largest size with a length of 36.77 mm and a width of 19.69 mm. The smaller length was recorded for Znaga 1 (Zng.1) (length=29.60 mm) and smaller width for Loudaghir (width = 16.77). Weight was fluctuating between 5.92 g for Loudaghir (Lg) and 17.52 g for Hamam Tahtani (Ht). Weight of pulp varied between 8.20g for Hamam Tahtani (Ht) and 4.71g for Loudaghir (Lg). Percentage of seed per date is ranging between 13.52% for 'Aziza Bouzid' of Zng.4 to 20.72% Lg.

E3S Web of Conferences **337**, 04006 (2022) https://doi.org/10.1051/e3sconf/202233704006
I2CNP 2021

Table 2. Size measurements of date fruit (cv. 'Aziza Bouzid')

	Fruits			Seeds			Pulps		Ratio	Ratio
	Length (mm)	Width (mm)	Weight (g)	Length (mm)	Width (mm)	Weight (g)	Weight (g)	Thickness (mm)	Seed/Date	Pulpe/Date
Arj	33.79 ± 0.63^{de}	18.95 ± 0.54^{a}	8.37 ± 0.47^{f}	23.55 ± 1.20^{ef}	9.08 ± 0.56^{bc}	1.18 ± 0.2^{c}	7.20 ± 0.41^{g}	3.85 ± 0.18^{a}	14.96 ± 1.93^{abc}	85.67 ± 2.29^{a}
Cop.A	34.58 ± 1.15^{e}	18.56 ± 0.62^{de}	$7.24\pm0.56d^{e}$	23.12 ± 0.94^{de}	8.46 ± 0.36^{a}	1.01 ± 0.11^{ab}	6.23 ± 0.50^{ef}	4.08 ± 0.50^{a}	13.90 ± 1.41^{a}	86.10 ± 1.41^{a}
Zng.1	29.60 ± 0.77^{abc}	17.96 ± 0.92^{cd}	6.99 ± 0.55^{abc}	18.72 ± 0.38^{a}	8.71 ± 0.25^{ab}	0.97 ± 0.06^{a}	6.02 ± 0.49^{abcd}	4.59 ± 0.60^{b}	13.83 ± 0.47^{abc}	86.17 ± 0.47^{a}
Zng.2	34.59 ± 0.89^{e}	18.57 ± 0.34^{de}	7.07 ± 0.39^{cde}	22.75 ± 0.71^{cde}	8.82 ± 0.19^{abc}	1.10 ± 0.06^{abc}	5.97 ± 0.33^{def}	3.80 ± 0.34^{a}	15.63 ± 0.63^{abc}	84.37 ± 0.63^{a}
Zng.3	30.13 ± 1.09^{a}	17.25 ± 0.71^{ab}	6.44 ± 0.38^{abc}	21.10 ± 1.15^{ab}	8.94 ± 0.35	1.15 ± 0.12^{bc}	5.29 ± 0.38^{abcd}	3.69 ± 0.25^{a}	18.21 ± 1.75^{d}	81.79 ± 1.75^{a}
Zng.4	31.10 ± 1.02^{abc}	17.99 ± 0.54^{bcd}	7.59 ± 0.73^{ef}	20.73 ± 1.16^{ab}	8.74 ± 0.20^{abc}	1.00 ± 0.06^{ab}	6.59 ± 0.73^{fg}	3.96 ± 0.43^{a}	13.5 ± 1.68^{a}	86.48 ± 1.68^{a}
Ht	36.77 ± 0.88^{f}	19.69 ± 0.37^{f}	9.58 ± 0.43^{g}	24.66 ± 0.52^{f}	9.41 ± 0.19^{cd}	1.38 ± 0.07^{d}	8.20 ± 0.42^{h}	4.54 ± 0.37^{ab}	14.59 ± 0.64^{abc}	85.41 ± 0.64^{a}
Lb	32.73 ± 1.07^{cd}	17.87 ± 0.47^{bcd}	6.67 ± 0.51^{abcd}	21.85 ± 1.03^{abc}	8.83 ± 0.27^{abc}	1.10 ± 0.09^{abc}	5.58 ± 0.46^{bcde}	3.58 ± 0.56^{a}	16.92 ± 1.24^{cd}	83.08 ± 1.24^{a}
OM.1	32.19 ± 0.74^{bcd}	17.55 ± 0.38^{bc}	5.99 ± 0.33^{a}	21.03 ± 0.52^{ab}	8.59 ± 0.41^{ab}	1.07 ± 0.07^{abc}	4.91 ± 0.34^{ab}	3.58 ± 0.35^{a}	18.37 ± 1.62^{d}	81.63 ± 1.62^{a}
OM.2	30.59 ± 0.74^{ab}	17.55 ± 0.38^{bc}	6.09 ± 0.33^{ab}	20.66 ± 0.52^{ab}	8.77 ± 0.41^{abc}	1.00 ± 0.07^{ab}	5.09 ± 0.34^{abc}	3.75 ± 0.35^{a}	16.49 ± 1.62^{bcd}	83.51 ± 1.62^{a}
Os	31.81 ± 0.96^{abc}	17.94 ± 0.42^{bcd}	6.77 ± 0.33^{bcd}	21.46 ± 0.73^{abc}	8.81 ± 0.25^{abc}	0.98 ± 0.11^{a}	5.80 ± 0.24^{cde}	4.08 ± 0.49^{b}	14.35 ± 1.31^{ab}	85.65 ± 1.31^{a}
Lg	30.25 ± 1.42^{a}	16.77 ± 0.55^{a}	5.92 ± 0.63^{a}	22.02 ± 0.68^{bcd}	8.95 ± 0.47^{bc}	1.21 ± 0.15^{c}	4.71 ± 0.63^{a}	3.98 ± 0.84^{a}	20.72 ± 3.15^{e}	79.28 ± 3.15^{a}

Values are average (n=100) ± standard deviation. Different litters (a-g) show significant differences in the same row

Table 3. Some physical properties of date fruit (cv. 'Aziza Bouzid')

	AMD	GMD	R	Φs	S	Bulk density	True density	Porosity
Arj	20.37 ± 1.50^{de}	14.07 ± 0.74^{abc}	52.46 ± 4.06^{abc}	0.39 ± 0.05^{abcd}	655.66 ± 67.08^{ab}	0.64 ± 0.03^{def}	1.01 ± 0.06^{a}	36.60 ± 3.68^{abc}
Cop.A	19.07 ± 0.88^{e}	14.58 ± 1.42^{bc}	53.72 ± 2.11^{a}	0.42 ± 0.04^{ab}	677.34 ± 137.87^{ab}	0.65 ± 0.05^{cd}	0.95 ± 0.06^{a}	31.22 ± 6.72^{a}
Zng.1	17.38 ± 0.69^{ab}	14.31 ± 1.14^{abc}	60.87 ± 5.19^{d}	0.48 ± 0.05^{e}	649.74 ± 105.67^{ab}	0.66 ± 0.02^{def}	1.07 ± 0.09^{a}	37.90 ± 6.19^{bc}
Zng.2	18.99 ± 0.71^{de}	14.29 ± 1.08^{abc}	53.94 ± 3.78^{a}	0.42 ± 0.04^{ab}	647.25 ± 97.73^{ab}	0.66 ± 0.02^{def}	0.96 ± 0.03^{a}	30.87 ± 3.57^{a}
Zng.3	17.02 ± 0.91^{a}	13.20 ± 0.63^{a}	57.47 ± 3.85^{bc}	0.44 ± 0.02^{abcd}	549.17 ± 53.45^{a}	0.68 ± 0.03^{def}	1.10 ± 0.11^{a}	37.56 ± 7.16^{bc}
Zng.4	17.68 ± 0.55^{abc}	14.48 ± 2.84^{abc}	57.93 ± 2.18^{c}	0.47 ± 0.05^{cde}	668.90 ± 151.76^{ab}	0.65 ± 0.04^{cde}	1.10 ± 0.07^{a}	40.37 ± 5.18^{c}
Ht	20.33 ± 1.35^{f}	16.44 ± 0.79^{d}	56.36 ± 3.21^{abc}	0.45 ± 0.05^{bcde}	852.81 ± 82.82^{c}	0.68 ± 0.04^{f}	1.17 ± 1.33^{b}	41.88 ± 3.77^{c}
Lb	18.06 ± 1.66^{bcd}	14.11 ± 2.71^{abc}	54.99 ± 3.69^{abc}	0.43 ± 0.04^{abc}	650.71 ± 240.39^{ab}	0.58 ± 0.02^{a}	0.96 ± 0.05^{a}	38.98 ± 4.55^{c}
OM.1	17.30 ± 0.54^{ab}	13.39 ± 0.83^{ab}	54.47 ± 2.25^{bc}	0.44 ± 0.02^{abcd}	566.86 ± 69.91^{a}	0.68 ± 0.03^{ef}	1.06 ± 0.05^{a}	35.52 ± 3.17^{abc}
OM.2	17.77 ± 0.82^{abc}	13.30 ± 0.72^{ab}	54.62 ± 2.42^{ab}	0.41 ± 0.02^{a}	$558.25\pm61.25a$	0.61 ± 0.03^{ab}	1.04 ± 0.11^{a}	40.99 ± 6.44^{c}
Os	18.61 ± 1.33^{cde}	14.81 ± 1.77^{c}	56.55 ± 3.03^{abc}	0.47 ± 0.06^{de}	727.33 ± 209.99^{b}	0.63 ± 0.04^{bc}	1.06 ± 0.09^{a}	40.44 ± 3.87^{c}
Lg	17.00 ± 0.63^{a}	13.35 ± 0.89^{ab}	55.57 ± 2.43^{abc}	0.44 ± 0.01^{abcd}	562.82 ± 74.44^{a}	0.69 ± 0.03^{f}	1.02 ± 0.07^{a}	31.88 ± 6.90^{ab}

Values are average (n=100) ± standard deviation. Different litters (a-g) show significant differences in the same row
AMD: Arithmetic mean diameter, GMD: geometric mean diameter, R: Aspect ratio, Φs: Sphericity, S: Surface area

E3S Web of Conferences **337**, 04006 (2022) https://doi.org/10.1051/e3sconf/202233704006
I2CNP 2021

Pulp thickness was ranging between 3.58 mm for Loudaghir (Lg), Oulad Lmaiz 2 (OM.2) and 4.59g, 4.54g respectively for Hamam Tahtani (Ht) and Znaga 1(Zng.1). Percentage of seed per date is ranging between 13.52% for 'Aziza bouzid' of Zng.4 to 20.72% of Lg. These findings corroborate with [22] results for Tunisian cultivar has indicate an important differences between dates of different varieties, but also within same varieties. [23] Also revealed that morphological analysis of cultivar grown in Saudi showed a high level of diversity in length-width ratio, color, shape of the fruit, fruit-base and in the percentage of area covered by the fruit cap. Great variability was also reported between Pakistani cultivars by [24].

Geometric mean diameter, Sphericity, Surface area, Bulk and true density (table 3) show a significant difference ($p < 0.05$) .Geometric mean diameter GMD of Znaga 3 (Zng.3) sample had the lowest value (13.20 mm) whereas Ht had the highest (20.72 mm). Sphericity varied from 38.70% for Arj to 48.5% for Znaga 1 (Zng.1). The surface area varied from 852, 81 mm^2 Hamam Tahtani (Ht) to 549, 17 mm2 Znaga 3 (Zng.3). Lowest data of bulk density was 0.58g cm-3 for Lb and the highest was 0.69 for Lg. The physical parameters variability was reported also by [15-25], for Mazafati and 'Dairi' cultivars from Tehran.

The study revealed an important variability from different 'Aziza Bouzid' samples. Highest quality of dates samples was from the Hamam Tahtani (Ht), Aarja (Arj) and cooperative Aziz (Cop.A) plot may be due to the good practices used. The lowest quality registered for Oulad lmaiz and Loudaghir (Lg) owed to the use of traditional cultural practices. The investigation revealed that Hamam Tahtani, Aarja and cooperative Aziza (Good 'Aziza Bouzid' fruits quality) are the recent date palm plantations that adapt good cultural practices, such as respect of regular plantation and use of drips irrigation, use organic and mineral fertilizer. In addition to developed pre-harvest practices as cleaning trees and limitation, thinning, protecting fruit bunches .However, Oulad lmaiz and Loudaghir (low 'Aziza Bouzid') fruits quality plots belong to the ancient palm grove which is characterized by irregular planting, heterogeneous varietal profile, and use of ancestral practices. In the fact, according to [26], thinning of bunches induces to increase in fruit volume, weight, diameter, and length; along with seed and flesh weight in the 'Succary' cultivar. An irrigation system can also affect the physical properties of fruits; Findings given from the North African and Middle East countries marked that irrigation systems have an important influence on dates quality [17].

Variation interval of physical parameters raised in this study for 'Aziza Bouzid' allows to elaborate standards for machines designs. Dimensions are important physical parameters for drying, storing and are also useful to determinate the shape of the fruits. Arithmetic mean diameter and geometric mean diameter values are used to design of separating, harvesting, sizing and grinding machine. Determined Sphericity values serving to concept handling equipment of fruits. The porosity of dates fruits is important because it shows fruits resistance to air flow during, drying process and storage process.

3.2 Color measurement

Color results are presented in table 4. No significant variation was unregistered for Fruits color properties (a*, b*, C* and H*) except luminosity L* that values varied from 36.99 for Laabidate (Lb) to 44.18 for Hamam Tahtani (Ht) .Light colored dates are considered better mentioned. 'Aziza Bouzid' from Hamam Tahtani plot showed a better color with higher values unlike Laabidate plot which is the lowest one. [9] Indicate that climatic condition is responsible for color variability as temperature, light and humidity. Cultural factors as thinning, grading, and Bunching can also affect color.

E3S Web of Conferences **337**, 04006 (2022) https://doi.org/10.1051/e3sconf/202233704006
I2CNP 2021

Table 4. Color results of 'Aziza Bouzid' samples

	L*	a*	b*	c*	h*
Arj	43.16 ± 2.75^{cd}	5.80 ± 1.83^{a}	28.82 ± 4.58^{abcd}	29.40 ± 4.11^{abc}	78.62 ± 5.35^{c}
Cop.A	42.10 ± 1.43^{bcd}	9.86 ± 1.93^{c}	27.34 ± 2.27^{abcd}	29.06 ± 2.78^{abc}	70.17 ± 2.08^{ab}
Zng.1	39.08 ± 4.37^{ab}	5.91 ± 0.67^{ab}	22.82 ± 3.11^{abc}	23.57 ± 2.78^{ab}	75.48 ± 7.05^{bc}
Zng.2	$44,37\pm2.56^{de}$	$6,37\pm0.23^{ab}$	$30,11\pm3.57^{cd}$	$30,78\pm4.12^{bc}$	$78,05\pm2.85^{ab}$
Zng.3	40.38 ± 0.91^{abc}	9.08 ± 0.92^{bc}	21.85 ± 2.49^{ab}	23.66 ± 2.65^{ab}	67.45 ± 0.27^{c}
Zng.4	40.68 ± 0.26^{bcd}	7.72 ± 0.41^{abc}	26.18 ± 0.98^{d}	27.29 ± 1.07^{abc}	73.57 ± 1.41^{abc}
Ht	44.18 ± 0.18^{a}	7.77 ± 1.36^{ab}	29.77 ± 0.37^{a}	30.76 ± 1.40^{a}	75.38 ± 2.61^{bc}
Lb	36.99 ± 0.61^{de}	5.90 ± 0.10^{abc}	21.12 ± 0.73^{bcd}	21.92 ± 0.73^{bc}	74.40 ± 0.75^{bc}
OM.1	43.12 ± 0.29^{cd}	7.94 ± 0.25^{abc}	28.39 ± 0.83^{abcd}	29.48 ± 0.87^{abc}	74.37 ± 0.03^{bc}
OM.2	49.07 ± 1.51^{f}	8.33 ± 0.58^{abc}	31.70 ± 5.39^{abcd}	32.77 ± 5.42^{c}	75.27 ± 1.45^{bc}
Os	41.39 ± 0.22^{bcd}	6.05 ± 0.35^{ab}	26.79 ± 0.04^{abcd}	27.46 ± 0.35^{abc}	77.27 ± 0.73^{c}
Lg	47.33 ± 1.06^{ef}	6.84 ± 0.28^{abc}	32.77 ± 1.53^{abcd}	33.48 ± 1.56^{abc}	78.21 ± 2.18^{c}

Values are average (n=100) ± standard deviation. Different litters (a-e) show significant differences in the same row

L*: lightness, a*: red and green ratio, b*: ratio yellow and blue, C*: chroma, H*: Hue

3.3 Principal Component Analysis of physical quality characteristics of 'Aziza Bouzid' cultivar.

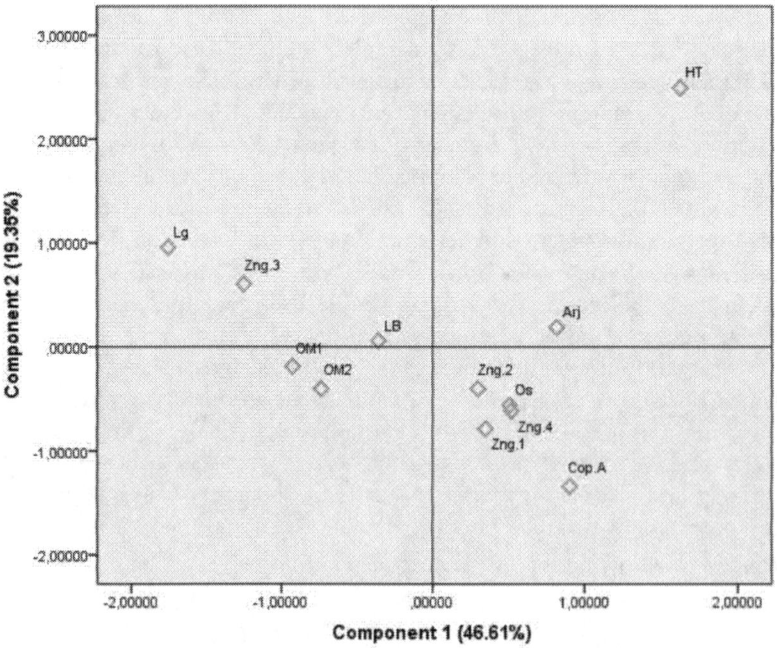

Fig. 2. Projection of the variables of the twenty studied dates fruits in the plane defined by two principal components; Arj:Aarja; Cop.A: Cooperative Aziza; Zng.1:Znaga 1; Zng.2:Znaga 2 , Zng.3 :Znaga 3;Zng.4 :Znaga4 ; Ht :Hamam Tahtani; Lb: Laabidate ;OM.1: Oulad lmaiz 1; OM.2:Oulad lmaiz 2; Os:Oulad Slimane; Lg :Loudaghir

E3S Web of Conferences **337**, 04006 (2022) https://doi.org/10.1051/e3sconf/202233704006
I2CNP 2021

Table 5. Two main components explaining more than 65% of the total information on the physical and colorimetric traits of 'Aziza Bouzid' dates

Variables	Principal Component		Variables	Principal Component	
	1	2		1	2
Width Fruit	0.924	0.242	**AMD**	0.812	0.271
Pulp Weight	0.900	0.365	**Length date**	0.696	0.331
GMD	0.899	0.359	**thickness**	0.586	0.207
Ratio Pulp/Date	0.899	-0.343	**Weight Seed**	0.095	0.941
Ratio Seed/Date	-0.891	0.353	**Width Seed**	0.236	0.903
Surface	0.878	0.341	**True density**	0.448	0.785
Weight Date	0.860	0.452			

Principal component analysis (PCA) was conducted to classify and determine the relationship between date samples and pre-harvest practices. Fig 2 and table 5 sum-up the results of PCA between 12 date samples. Two principal components were extracted from statistical analysis, which explained 65.96 % of the total variance in the data set. The first and second PCs (PC1 and PC2) accounted for 46.613 % and 19.943 %of the variation; respectively. Factorial map showed discrimination between the studied dates samples and also has approved the interpretations mentioned above in a very simplified way. Compared to PC1, the studied 'Aziza Bouzid' samples could be divided into two groups. The first consists of 'Aziza Bouzid' samples from Hamam Tahtani (Ht), Cooperative Aziza (Cop.A), Aarja (Arj) Oulad slimane (Os) and Znaga (Zng.2, Zng.4) plots. The second was composed of dates from loudaghir 'Lg', Oulad lmaiz 1 (OM.1), Oulad lmaiz 2 (OM.2) et Znaga 3 (Zng.3). Actually, the first group is correlated with Pulp weight, dates, weight, Geometric mean diameter, width of dates, Surface, Aritmetique mean diameter, dates length, Bulk density, true density, Ratio P/D and thickness. The second group is the opposite of the first one with respect to PC1. For principal component 2, the most distinct variables, were Aspect ratio R and Sphericity. According to the PC1, the effect of Pre-harvest practices is mainly manifested. In fact, Morphological 'Aziza Bouzid' samples from plot using good practices were opposed to those using traditional practices. 'Aziza Bouzid' sampled in Hamam Tahtani (Ht), Cooperative Aziza (Cop.A), Aarja (Arj) characterized by high morphological quality (high weight, high length and width dates, hand low seed percentage).

In opposition, samples collected from Loudaghir (Lg), present a low quality with low size and high portion of seed. In addition, the results of this statistical analysis allow us to highlight a clear variability between studied 'Aziza Bouzid' dates samples according to morphological quality characteristics. PCA analysis exposed clearly the effects of the pre-harvest practices on dates quality.

4 Conclusion

This study revealed the effect of pre-harvest factors on the physical and colorimetric characteristics of the cultivar 'Aziza Bouzid. The interaction between physical quality and pre-harvest practices showed that the highest physical quality was recorded in the plots for which farmers' adapted adequate pre-harvest practices. However, a low quality of the cultivar 'Aziza Bouzid' was observed especially in the plots where ancestral practices are still used.

The obtained results can be exploited for the raise yield of dates and also for the improvement of Moroccan dates quality. Firstly the study raised adequate pre-harvest

E3S Web of Conferences **337**, 04006 (2022) https://doi.org/10.1051/e3sconf/202233704006
I2CNP 2021

practices responsible for good dates quality (high weight and size of dates) secondly it presented very important values of the physical parameters (Sphericity, surface area, bulk, and true density), these results will be used for the design of harvesting, handling, processing and storing equipment that will keep the high quality of produced dates.

References

1. C.C.T. Chao, R.R Krueger, HortScience **42**, 1077- 1082 (2007)
2. A. Amanat, M. Waly, M.M. Essa, D. Sankar. Dates: Production, Processing, Food, and Medicinal Values (2012)
3. S.H. Larbi. Les zones phoenicicoles marocaines', in: Dollé V. (ed.), Toutain G. (ed.). Les systèmes agricoles oasiens. Montpellier: CIHEAM 12 (1990)
4. M. Haddouch. Situation actuelle et perspectives de développement du palmier dattier au Maroc Importance du secteur phoenicicole Importance socio-économique oasis', In : Ferry M. (ed.), Greiner D. (ed.). Le palmier dattier dans l'agriculture d'oasis des pays méditerranéens. Zaragoza : CIHEAM (1993)
5. FAOSTAT Agro-statistics, Database, Rome: Food & Agri. Organiz, United Nations (2017)
6. A. Hasnaoui, M.A. Elhoumaizi, M. Sindic, J. Sci. Res **3** ,139-149 (2010)
7. H.A. Chapron, M. Rahmani, A. Benmokhtar, R. Bouabid, Etude du positionnement commercial des variétés de dattes marocaines. Ministère de l'Agriculture et de la Pêche Maritime, O.R.M.V.A Ouarzazate. Maroc (2014)
8. M.D. Ramjan, M.T. Ansari, J. Med. Plant. Res **6**, 16-18 (2018)
9. M. Arshad, M. Shahnawaz ,S. Shahkeela ,H. Maqsood, A. Mukhtar ,K.S. Sher, Euro, J. Exp. Bio **4**, 82-89 (2014)
10. D. Bekele, J Plant Sci Res 5, 1-5 (2018)
11. V.G. Suchita, R. W. Vaishali, N.P. Bhagyashree, P.M. Nimkar, J. Appl. Hortic **17**, 225-229 (2015)
12. N.N. Desai, V.M. Modi, G.K. Saxena, R.H. Chaudhari, M.K. Jaipal, K.H. Gohil, A. Gora, Int J Curr Microbiol Appl Sci **8** , 1243-1249 (2019)
13. M.K Mahawar, K. Jalgaonkar, M. Kumar, V.S. Meena, B. Bhushan, ActaAgroph **24**, 271-277(2017)
14. M.K. Jahromi, S. S. Mohtasebi, A. Jafari, R. Mirasheh, S. Rafiee , J. Agric. Sci. Technol 2, 1-9 (2008)
15 A. R. Moustafa, N. Abdel-Hamid, A. Abd El-Hamid, M. R. El-Sonbaty and S. K. M. Abd El-Naby Bull Natl Res Cent **43**, 204 (2019).
16 R.A.A .Mostafa,A.M. El-Salhy, A.A. El-Banna and Y.M. Diab Middle East J. Agric. Res, **3**, 517-521, 2014
17 H. Darouich, M.R. Cameira, J.M. Gonçalves, P. Paredes, L.S. Pereira, MDPI Water **9**, 1-19 (2017)
18. O. Kabas, A. Ozmerzi, I. Akinci, J. Food Eng **73**, 198-202 (2006)
19. N.N. Mohsenin, Physical properties of plant and animal materials: Physical Characteristics and Mechanical Properties, New York: Gordon and Breach Science Publisher (1978).
20. W .Burubai, A.B. Burubai, J Food Process Technol **6**, 1-5. (2014)
21. C. Ozarslan, Biosyst. Eng, **83**, 169–174 (2002)
22. S. Balasubramanian, R. Viswanathan, J Food Sci Technol **47**, 279–284 (2014)
23. W. Herchi, H. Kallel, S. Boukhchina, Food Sci. Technol, Campinas **34**, 464-470(2014)
24. N.S. Al-Khalifah, E. Askari, A.E. Shanavas Khan, Emir. J. Food Agric **24**, 456-461 (2012)
25. R. W. K. Qadri, S. Waheed, M. S. Haider, I. Khan, S. A. Naqvi, M. Bashir, M. M. Khanthe, J. Anim.Plant. Sci **26**, 1268-1277 (2016)
26. M.K. Jahromi, S. Rafiee, A. Jafari, M.R. Ghasemi , R. Mirasheh , S.S. Mohtasebi, Int. Agrophysics **22**, 221-224 (2008)
27. M. Haddou, S. Babahani, A. Idder, RBRS 6, 46- 55 (2016)

The mealworm *Tenebrio molitor* (Coleoptera: Tenebrionidae) as a potential candidate to valorize crop residues

Jamaa Zim[1*], *Mohammed* Sarehane[1] and *Rachid* Bouharroud[2]

[1]Agronomic and Veterinary Institute Hassan II, Agadir, Morocco
[2]Integrated Crop Production Unit, Centre Régional de la Recherche Agronomique d'Agadir (INRA), Morocco

Abstract. Agricultural activities generate important amount of crop residues that need to be managed in appropriate way usingecofriendly and cost effective tools. The mealworm *Tenebrio molitor* (Coleoptera: Tenebrionidae) is a potential candidate to valorize crop residues thatcan produce atthe same time an alternative feed based on insect larvae. Four crop residues (citrus, tomato, pepper and potato) mixedwith wheat bran (ratio 50:50) were evaluated as diets for mass rearing of *T. molitor* under controlled environmental conditions (T= 28 ± 1°C; RH=70 ± 5 % and total darkness). Five zootechnical parameters (weight gain, growth rate, mortality rate, weight of consumed feed and conversion rate) were adopted to evaluate thegrowth and life developmental traits of larvae during 45 days. The effect of crop residues used as diet had a significant impact on bothweight gain and growth rate. Importantly, crop residues used alone has a negative effect on all zootechnical parameters evaluated. Indeed, the amendment of wheat bran has a significant positive impact on these parameters. Citrus and tomato were the most toxic diets showing a mortality rate of 92 and 98 %, respectively. The most efficient diets werethe mixture of citrus residues andwheat branfollowed by the mixture of potato residues and wheat bran and the mixture ofpepper residues and wheat bran. Fed substrate had no impact on conversion rate.

1 Introduction

Crop residues management and valorization raise more and more concern in environment departments in many countries. In the Souss valley (Southwest of Morocco), considered as the main region of production and export of vegetables and

*Corresponding author: j.zim@iav.ac.ma

fruits in Morocco, the annual crop residues was estimated at 3.4 tons in 2020 [1]. The residues and wastes from crops arenot usually easy to manage and a non-appropriate practice led to create a perfect environment in which varieties of pests and pathogens can develop and multiply [2]. Burning these residues has a negative impact on the emission of greenhouses gases [3]. As a result, poor waste management can pose a serious danger to surrounding crops and to the environment as they resist biodegradation, resulting in requirements of harsher pretreatment [4].
The yellow mealworms *Tenebrio molitor* Linnaeus (1758), has a four-stage life cycle: egg, larva, pupa and adult [5]. The larvae are raised as feed for birds, reptiles, amphibians, and fish using wheat bran (WB) as the primary feedstock, an agricultural by-product [6]. This insect is a considered as a more sustainable source of human edible protein compared tomilk, chicken, pork or beef [7].
Recent studies point to mealworms as an acceptable alternative source of protein for broiler chickens [8]. More than a million tons of mealworms are produced in China for domestic use and export as animal feedstock [9]. Mealworms are also considered as a valuable source of protein for insects in the USA and Europe [8]. Mealworms have also a biodegrading ability of plastic foams of polystyrene (PS) and polyethylene (PE) up to 50% after they have been absorbed into their guts [9, 10]; suggesting that mealworms have powerful digestive systems.
We hypothesized that the use of insects isa more economically feasible and environmentally friendly approach to convert crop residues into biomass and simple organic materials: a concept called biotransformation. This solution is sustainable, ecological and above all cost effictive. We plan to use a *T. molitor*, as a recycling engine for crop residues, producing a promising source of protein for animal feed. Researches have been done on different crop residues such as wheat straw, rice bran, maize straw and rice straw [9] but not on vegetables and fruits residues such as tomato, pepper and citrus.
This trial is a preliminary step carried out in order to evaluate the capacity of *T. molitor* larvae to transform vegetable and fruit wastes to resolve the issue of crop residue management and to reduce the production cost of these larvae mainly intended for animal feed.

2 Material and methods

2.1 Site of the experiment

The crop residues (citrus, tomato, pepper and potato) were collected from the experimental farm of the Agronomic and Veterinary Institute, Complexe Horticole-Agadir (30°21'12.0"N 9°28'38.8"W). Crop material was dried in a plastic greenhouse at ambient temperature for 8 days, then the dry residues were grinded with an electric grain mill (25 000 rpm). After weighing, the residues were put back in the greenhouse for 48 hours to be furtherdried. This operation has been repeated until the weight was stabilized. Twenty five mg of residue substrates were placed in cylindrical plastic boxes (7 x 5 cm, diameter and height) with smooth walls to prevent the larval escape. Table 1 shows the proportions of the 10 substrates used in this study.

2.2 Diets

The diet used for stock mass rearing and as control was prepared from wheat bran (60%) completed with animal feed (starter feed for broiler chicken, ALF ISSAN company) containing 20% of crude protein. Larva (22±3 days) were collected from a mass rearing carried out in the entomology laboratory of the Agronomic and Veterinary Institute, ComplexeHorticole-Agadir. In order to have homogeneous population of larvae, adults in peak of reproduction period were randomly selected and reared in large size boxes (L=41 , l=28 and H=7 cm) with a rearing density of 1 adult/cm², in the same diet (control). To serve as a water source and moisture, pieces of potato were added every 4 days. After 3 days the eggs were collected using a large-mesh sieve to separate the adults from the substrate containing the eggs, which will hatch after4 days (Cotton 1927). After 22 days the larvae were isolated from the substrate and put to fast for 24 hours before trials. Random samples of 100 larvae (average weight of a larva is 0.0011±0.0001 g) were added in the boxes containing substrate. The boxes were placed according to a completely randomized experimental design in a rearing chamber (2.8 x 2.9 x 2.5 m) under controlled environmental conditions T= 28 ± 1°C; RH=70 ± 5 % and total darkness.

In order to evaluate the effect of dry crop residues on the growth and biomass gain of *T. molitor*, we focused on four zootechnical parameters measured after 45 days of rearing which are the most important in mass rearing.

- $Weight\ gain = Final\ weight\ (after\ 45\ days) - initial\ weight$
- $Mortality\ rate\ = \frac{initial\ number\ of\ individuals - N \qquad of\ individuals\ after\ 45\ days}{Initial\ number\ of\ individuals}$
- $Weight\ of\ consumed\ feed =$
  $The\ initial\ weight\ of\ food - The\ final\ weight\ of\ food$
- $Conversion\ rate = Weight\ gain - weight\ consumed$

The crop residues used of tomato, pepper and citrus were the residues left from routine operations performed during the whole crop cycle (e.g. pruning).

2.3 Data analysis:

In this study, 9 substrates and control were studied. Each treatment was repeated 4 times. The recording of the results and the graphs were made with Excel software (version 2016). The statistical analysis of the results of this study was carried out using MINITAB software (version 2016) for the analysis of variance (ANOVA), using the general linear model procedure (α =0.05). The Tuckey test was used to compare the means between different treatments (significance level P ≤ 0.05).

Table 1. Composition of the substrates used in the experiment.

Substrate	Composition
S1	Tomato residue 50% + Wheat bran 50%.
S2	Tomato residue
S3	Potato residue 50% + Wheat bran 50%
S4	Potato residue
S5	Pepper residue 50 % + Wheat bran 50 %
S6	Pepper residue
S7	Citrus residue 50 % + Wheat bran 50 %
S8	Citrus residue
S9	25% of each of the 4 crop residues (pepper, tomato, potato and citrus)
Control	Wheat bran 60 % + Starter feed for broiler chicken 40 %

3 Results

3.1 Weight Gain and growth rate

The type of substrate used for mass rearing of *T. molitor* has a very significant effect on the weight gain and growth rate (Figure 1; F=113.21; P=0.000). The control showed the highest weight gain and growth rate (3.0024 g and 0,0751 g/day, respectively) for a means of samples of 150 larva over 45 days. The second diet with thehighest weight gain and growth rate was S8 containing 50% of citrus residues and 50% of wheat bran (2.1635 g). The other substrates also showed a moderate weight gain and growth rate. The main observation is that the use of dried crop residues-based diet alone caused a significant weight loss.

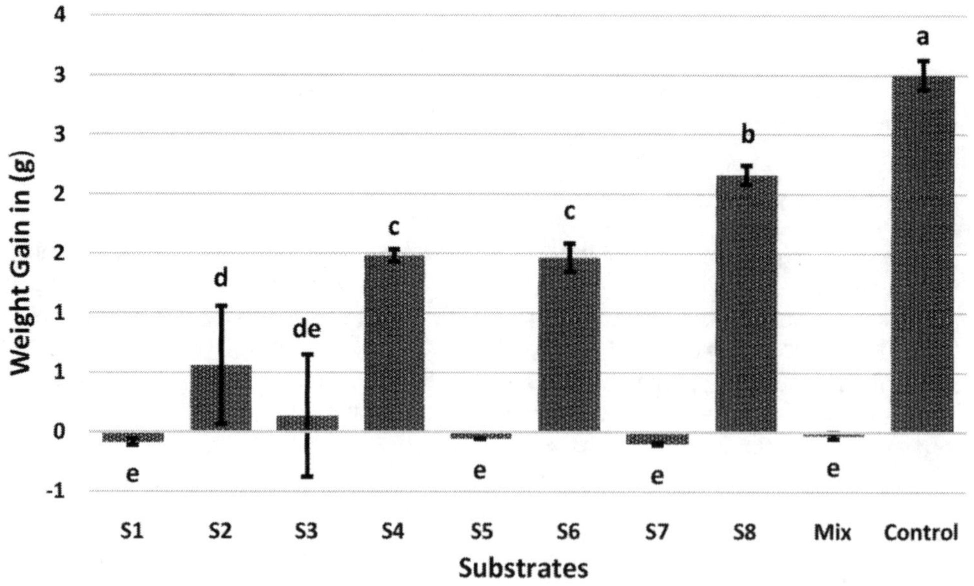

Fig.1. Effect of 10 substrates tested on the weight gain of *T. molitor*. The means followed by the same letter are not statistically different at P<5% according to the Tukey test.

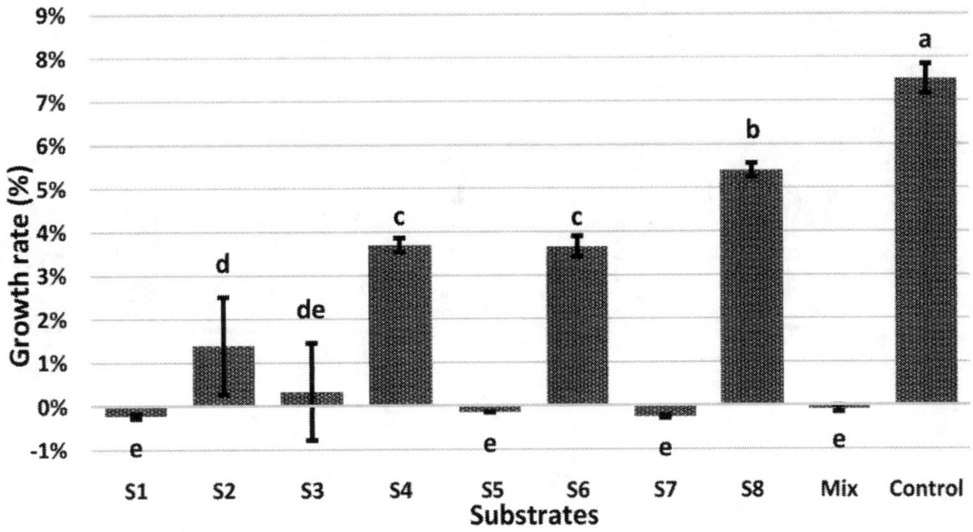

Fig.2. Effect of 10 substrates tested on the growth rate of *T. molitor*. The means followed by the same letter are not statistically different at P<5% according to the Tukey test.

3.2 Mortality rate

The analysis of variance for mortality rates showed a highly significant effect of substrate types on the mortality rates of *T. molitor* larvae (Figure 2; F=117,18; P=0.000). Based on the Tukey test for means comparison, the lowest mortality rates (less than 7%) were obtained in the control boxes and the 4 substrates containing 50% of wheat bran in addition to dried crop residues of citrus, pepper, tomato and potato. However, the substrates based on the dried citrus, pepper and tomato residues caused the most toxic effect to the larvae of *T. molitor* (80.75, 92.00 and 98.00 %, respectively). The substrates S5 (potato) and S9 (mixture of 25% of the 4 crop residues) caused a moderate mortality rate (48.25 and 32.00%, respectively).

https://doi.org/10.1051/e3sconf/202233704007

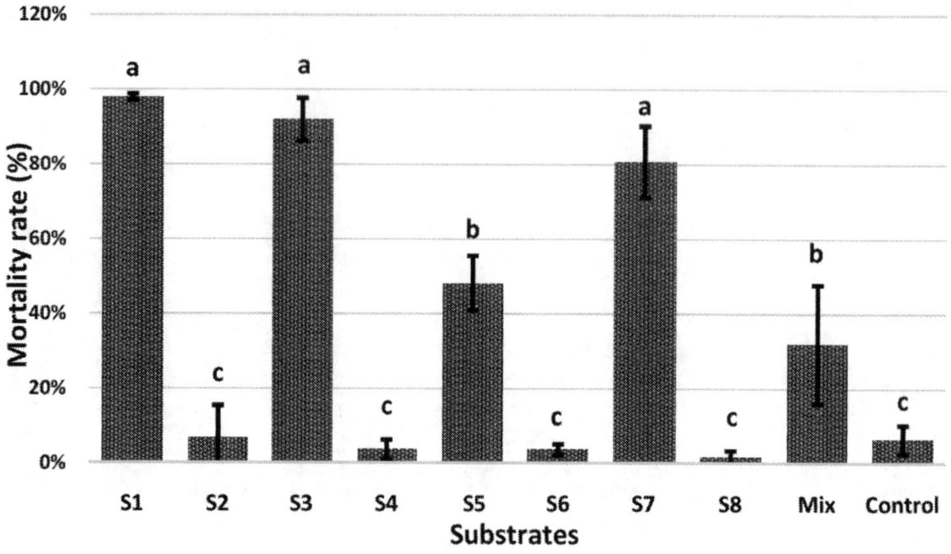

Fig.3. Effect of 10 substrates tested on the mortality rate of *T. molitor*. The means followed by the same letter are not statistically different at P<5% according to the Tukey test.

3.3 Weight of consumed substrate

The type of substrates used for rearing had a highly significant impact on the weight of consumed substrate (Figure 3; F=318.66 P=0.000). The mean comparisons test (Tukey) at a confidence limit of 95% showed that the control was the most consumed substrate (8.6238 g) followed by the substrate S8 based on 50% wheat bran and 50% dry citrus residue (5.0445 g). The third group was composed of 2 substrates S4 and S6 which are made by the mixture of 50% wheat bran and 50% dry potato or dry pepper waste. The highest mortality rates observed were in the substrates based only ofdried crop residues (without wheat bran), these residues also had a negative impact on the weight of consumed substrates.

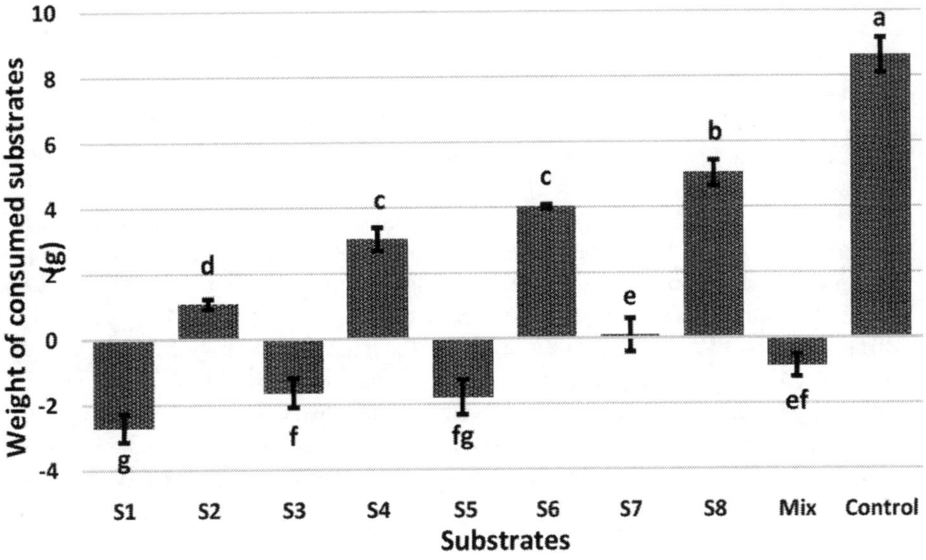

Fig.4. Effect of 10 substrates tested on the weight of consumed substrates of *T. molitor*. The means followed by the same letter are not statistically different at P<5% according to the Tukey test.

3.4 Conversion rate

Analysis of variance showed no significant effect of substrates on the conversion rate (Figure 4; P=0.631).

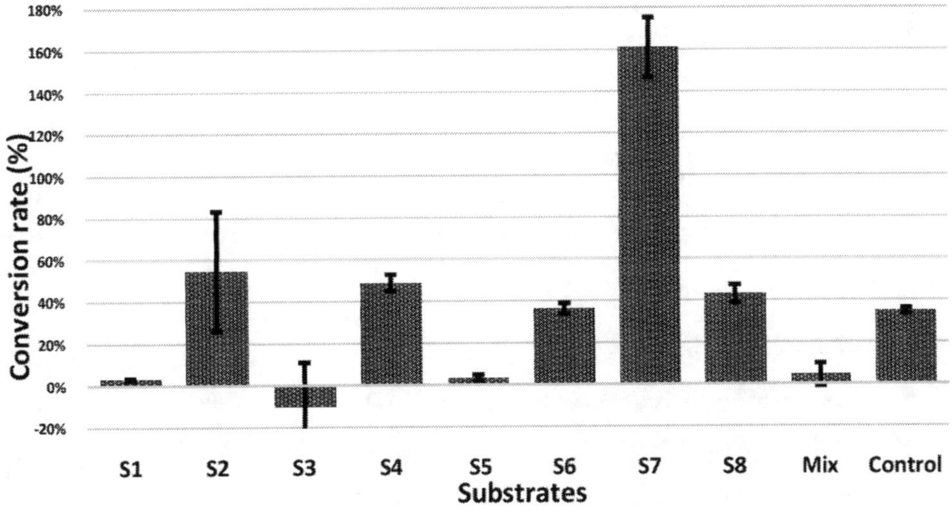

Fig.5. Effect of 10 substrates tested on the mortality rate of *T. molitor*. The means followed by the same letter are not statistically different at P<5% according to the Tukey test.

4 Discussion

The new concept of insect-based bioremediation using agricultural waste as feedstock promises an alternative to fossil resources to produce biodiesel and proteins, thereby mitigating climate change and improving energy, animal feed and food safety [11]. Sustainability is becoming increasingly important worldwide. This study has shown that the use of substrates based on crop residues: tomato, pepper, potato and citrus, as a single food for *T. molitor* larvae, blocks their growth and also causes a very high mortality rate. However, if the residues are mixed with wheat bran, the larvae gain some biomass, but it is still an unsatisfactory gain compared to our control or wheat bran alone. Our results are not far from those of [12] even though the wastes used are different from those used in our trial. In the study of Ramos-Elorduyet al. (2002), authorsused waste mixtures at variousproportions and showed that: a diet consisting of 80% dry organic waste mixture of fruit and vegetables, 5% yeast and 15% excretion of *T. molitor* showed a performance in weight gain of mealworms and a richness in proteins and energy. The mortality rate was also very low and did not exceed 7% except for the Tomato residues diet, which shows that the wastes, subjects of the test, are non-toxic, however, their nutrient intake remains low, which explains the drop in the zootechnical performance of the larvae of *T. molitor*. Citrus residues are much better than other residues in terms of substrate, substituting the ordinary control feed for raising the mealworm larvae. This low growth rate caused by the mixtures, compared to the wheat bran diet, indicates that the mealworms were not able to select the correct feed when mixed with other less favorable feeds. Probably the most effective way to correct nutritional imbalances is to seek nutritionally complementary foods [13, 14].

5 Conclusions andrecommendations

In light of these results, it can be concluded that it is possible to incorporate a small amount of crop residues, especially from citrus, into the diet of *Tenebrio molitor* larvae. The valorization of driedcrop wastes is influenced, in the first instance, by the growth of the larvae. Further trials on a wide range of locally produced wastes need to be carried out in order to find which wastes are more important to include in the diet of the mealworms. In countries where wheat bran is cheaper, its use as a main diet is recommended in industrial farms. The evaluation of the potential of the mealworms as a leader insects in waste recovery can solve the waste problem and give an ideal alternative to the classical method used in the management of organic waste in its broadest sense.

References

1. Agence pour le Développement Agricole (ADA), Evaluation environnementale stratégique du Plan Maroc Vert Rapport définitif - Pages 82-83 (2012).
2. Cruz, D.; Cisneros, R.; Benítez, Á.; Zúñiga-Sarango, W.; Peña, J.; Fernández, H.; Jaramillo, A. *Agronomy 11*, 2362,(2021).
3. Zhang, H.; Hu, D.; Chen, J.; Ye, X.; Wang, S. X.; Hao, J. M.; Wang, L.; Zhang, R.; An, Z. Environmental Science & Technology. 45 (13): 5477–82 (2011).
4. Mussoline W., Esposito G., Giordano A., Lens P.(2013).
5. Roberson, W.H. Cambridge University Press, Cambridge, UK, pp. 126e127 (2005).

6. Melis, R.; Braca, A.; Mulas, G.; Sanna, R.; Spada, S.; Serra, G.; Fadda, M.L.; Roggio, T.; Uzzau, S.; Anedda, R. Innov. Food Sci. Emerg. Technol. (2018).
7. Oonincx, D.G.A.B., De Boer, I.J.M. PLoS One 7, e51145 (2012).
8. Bovera, Fulvia&Loponte, Rosa &Marono, Stefania& Piccolo, Giovanni &Parisi, G. &Iaconisi, V. &Gasco, Laura &Nizza, Antonino. Journal of Animal Science (2015).
9. Shan-Shan Yang, Yi-di Chen, Ye Zhang, Hui-Min Zhou, Xin-Yu Ji, Lei He, De-Feng Xing, Nan-Qi Ren, Shih-Hsin Ho, Wei-Min Wu. Environmental Pollution, Volume 252, Part B, Pages 1142-1153 (2019).
10. Brandon, Anja Malawi & Gao, Shu-Hong & Tian, Renmao& Ning, Daliang& Yang, Shanshan& Zhou, Jizhong& Wu, Weimin&Criddle, Craig. Environmental Science & Technology (2018).
11. Wang, H., Rehman, K.u., Liu, X. et al.. Biotechnol Biofuels 10, 304 (2017).
12. Ramos-Elorduy, Julieta& González, Ernesto & Hernández, Alma &Pino, José. Journal of economic entomology (2002).
13. Waldbauer, G. P., & Friedman, S. *Annual Review of Entomology, 36(1), 43–63* (1991).
14. Behmer ST.. *Annu Rev Entomol.* (2009)

Phenotypic polymorphism, pomological and chemical characteristics of some local varieties of fig trees *(Ficus carica L.)* grown in Eastern Morocco

Aziz Tikent[1]*, *Ahmed* Marhri[1], *Atika* Mihamou[1], *Nargis* Sahib[1], *Hana* Serghini-Caid[1], *Ahmed* Elamrani[1], *Malika* Abid[1], and *Mohamed* Addi[1]

[1]Laboratoire d'Amélioration des Productions agricoles, Biotechnologie & Environnement (LAPABE), Faculté des Sciences, Université Mohammed Premier, BP : 717, 60000, Oujda, Morocco

Abstract. The fig tree (Ficus carica L.) is one of the earliest fruit trees cultivated in Morocco, especially in the north-eastern region; which is growing inextricably linked to our social and cultural heritage. This study concerns morpho-pomological and chemical parameters of five local fig varieties, namely: Aounk Hmam "AH", Chetoui "CH", Bounacer "BN", Ghoudane "GH", and Malha "MA". Significant differences were observed among the studied fig cultivars for some of the parameters investigated. In particular, the average fruit weight that ranged from 29.72 g to 58g for MA and AH respectively. At the ripening stage, fruits "MA" show the best skin cracks, while fruits "AH" are distinguished by their ease of peeling.In terms of titratable acidity, "AH" has the lowest value, while "CH" has the highest level. For the sweetest and tastiest variety of figs, Brix readings (expressed as total soluble solids content) registred range between 27 - 35% respectively for the MA and GH . According to the results of the parameters evaluated, particularly the acidity and the Brix index, the best varieties worthy of consideration from an agricultural and industrial point of view are "AH" and "CH" cultivars.

1 Introduction

The common fig (Ficus carica L., 2n = 26) belongs to the family Moraceae with more than 800 species[1], native to western Asia and the eastern part of the Mediterranean regions[2, 3]. It was one of the first Mediterranean fruit species to be domesticate[4, 5].The fig and olive trees are among the oldest domesticated trees in Morocco's northern region (Rif) [6]. Several studies on the fig tree's characterization have been conducted in Morocco. In the northern region of Morocco (Rif), Hmimsa et al [7] conducted a vernacular classification of fig varieties based on biological, agronomic, and use characteristics. The research revealed a high level of diversity among local fig varieties. The high level of intraspecific diversity, as well as the complexity of the naming system and classification system for fig varieties, is associated with high sociocultural and economic values. El Hajjami's [8] inventory of Moroccan local varieties of fig (Ficus carica L.) in four main production sites, (Chefchaouen, El Jadida, Ouezzane, and Taounate provinces) revealed the presence of 43 varieties. The principal component analysis (PCA) and agglomerative hierarchical clustering (AHC) have revealed similarities between some varieties. Hssaini et al [9] collected and analyzed phenotypic variation in fig accessions grown in three northern Moroccan regions (Taounate, Ouazzane, and Meknes). Based on pomological traits, they discovered three distinct groups with a total inertia of 90.92 %. In Morocco, Figs are also a valuable social resource. Figs are popular as gifts (fresh or dried) for family and friends, especially during religious events. The Ministry of Agriculture, Maritime Fisheries, Rural Development, Water and Forests has launched a project to plant 100 hectares of fig trees and build a drying unit in the rural municipality of Aghbal in the province of Berkane over the period 2019-2020 as part of the Green Morocco Plan. The project's objectives include the enhancement and development of the fig sector, as well as the improvement of the fig sector's productivity. The farmers involved in this project led by the cooperative "Gharmawne" will benefit from support equipment, training and supervision. In partnership with the municipality and other agricultural creators, this cooperative organizes each year the Aghbal festival of figs, a socio-economic event that aims to bring together consumers and producers of figs in eastern morocco. This event is held every year during the last weekend of August. The festival is an opportunity to promote the commercialization of figs and fig's derivative products, to encourage the innovation of small producers in this sector, and to show the natural and cultural assets of this region. The local fig

*Corresponding author:tikent.aziz@ump.ac.ma

https://doi.org/10.1051/e3sconf/202233704008

germplasm has not been subjected to intensive plant selection programs; as a result many fig populations and ecosystems had a great genetic diversity, which must to be fully exploited by its identification and classification[9].The fig has a wide geographical distribution in the country and is well-known for its high commercial value and best fruit flavor for fresh consumption. When compared to many other fig cultivars, the local fruits are known to be very high in sugars, the most important criterion of quality in figs. The purpose of this study is to assess the impact of different agro-ecological zones on the fruit quality of local fig varieties. Besides, this work aims to identify the morphological characteristics of local fig varieties and estimate the quality of their fruit, respectively, according to the morpho-pomological traits based on the IPGRI and CIHEAM descriptors [10, 11] and to the Aksoy's fig fruits quality descriptors[12], in order to conserve the fig germplasm and preserve this species from genetic erosion.

2 Materials and methods

2.1 Materials sampling

The sampling region (made up of 7 provinces and a prefecture) located in the northeast of the country (fig1), it covers an area of 90.130 km², accounting for or 12.7% of the country's total land area [13].

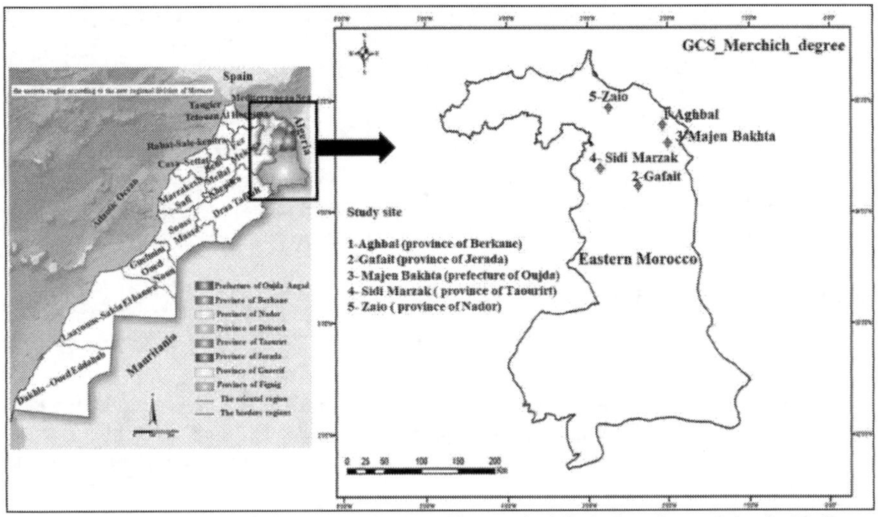

Fig1. Geographical location of the study in eastern Morocco. The area is delimited, to the north by the Mediterranean Sea, and to the south by the arid region of Daraa-Tafilalete. The climate of this region is Mediterranean in the north, with rare precipitation, often in the form of strong thunderstorms and continental in the south, sensitive to Saharan factors, with cold in winter and heat in summer, low rainfall generally poorly distributed in time and space[13].

Table1. Short name (Code) with important features of the five fig cultivars studied

N	Fig cultivars	Code	Figs category	Autumn fig skin color	utilisation
1	Aounk Hmam	AH	unifer figs	Purple spotted in green	Fresh
2	Bounacer	BN	bifer figs	Black more than purple	Fresh
3	Chetoui	CH	unifer figs	Green spotted in purple	Fresh/Dried
4	Ghoudane	GH	bifer figs	Black	Fresh/Dried
5	Malha	MA	unifer figs	Green	Fresh

The identification of trees (Table 1) was made with the local vernacular name given by farmer to crop and cultivar, and they were selected for their widespread distribution and their commercial value in this region. Prospection and sampling were carried out on traditional orchards of farmers (adult trees of roughly the same age " ten years") at different locations in eastern Morocco (fig 1), in the early of February for the apical buds and during the period (June, July ,August and September) for the fruits and leaves. At the level of three different zones in each site, samples of 150 fresh fruits (ten fruits per plant and three plants per cultivars in each zone for two successive years 2019-2020) ripe and disease free were chosen, selected (the most basal on shoot) and harvested. According to the description instructions, the same sampling approach was used for leaves and apical buds.

2.2 Phenotypic and pomological Study

A digital scale was used to take the fruit weight measurements. (KERN PCB 2000-1 Max 2000 g d = 0.1 g),the length and width of the fruit, the neck and the fruit stalk, the ostiole width ,the internal fruit pulp thickness carried out with a digital caliper PAQUIMETRO DIGITAL INOX 0-150MM),the index = fig width/fig length (Oblong < 0. 9; globose 0. 9–1.1; Oblate > 1. 1) and the ease of peeling (easy, medium, difficult) and lack of skin cracks (without cracks, longitudinal, Cracked skin) are determined. The leaf dimensions (length and width) were measured with a graduated ruler, while the thickness of the leaf stalk carried out with a digital caliper (in mm).As well as the terminal bud length and width were carried out with a digital caliper (in mm).

2.3 Chemical analysis

The chemical analyzes were carried out in triplicate and focused on the acidity and the Brix as a metric of fruit maturity parameters. The titratable acidity (expressed in % of citric acid) by titrating fig juice with 0.1 M NaOH to pH 8.1 in the presence of a colored indicator according to the AFNOR method (NF V O5−101, 1970). The Brix index indicates the total content of all sugars as the soluble solids in the fruit. Brix readings were done using digital refractometer (HANNA NUMERIQUE HI 96801 - ECHELLE 0-85% BRIX).

2.4 Statistical analysis

The data obtained was subjected to statistical analysis using IBM SPSS Statistics V21.0 software for descriptive statistics and also for analysis of variance by comparing average of characteristics studied using post hoc testing (Tukey's test) at the 5% threshold, graphs produced by the GraphPad Prism 8 software. The quality of the figs was assessed using the weighted ranked method by Aksoy [12], based on nine pomological characters.

3 Results and discussions

The morphological characterization of the figs was carried out on plant material available at the sampling site and therefore does not take into account the environmental component. Thus, the ease of peeling as well as the presence of skin cracks seems to be positive quality factors for figs intended for consumption in the fresh state. On the other hand, the absence of cracks would be a positive parameter allowing better preservation of dried figs. In addition, we note that this type of figs without skin cracks is intended both for consumption in the fresh state as for the production of dried figs. According to this study, AH had the best quality in terms of peelability while MA had the best quality in terms of skin cracking (Table 2).

Table2.Study results of two important parameter used to determine figs quality by Aksoy ranked method[12]

Cultivar	Ease of peeling	Skin cracks
Aounk Hmam (AH)	Easy	longitudinal
Bounacer (BN)	Medium	longitudinal
Chetoui (CH)	Medium	cracked skin
Ghoudane (GH)	Medium	longitudinal
Malha (MA)	Medium	without cracks

The susceptibility for skin cracks is favored by low temperatures associated with high humidity during the fig ripening and especially common in genotypes with a thin epidermis[14]. Varieties that have a too long maturity are more vulnerable to the problem of skin cracks; they should not be grown in early and rainy autumn areas. When the fruits are ripening, fig cracking begins at the ostiole and is exacerbated by low temperatures and/or an increase in relative humidity[15].

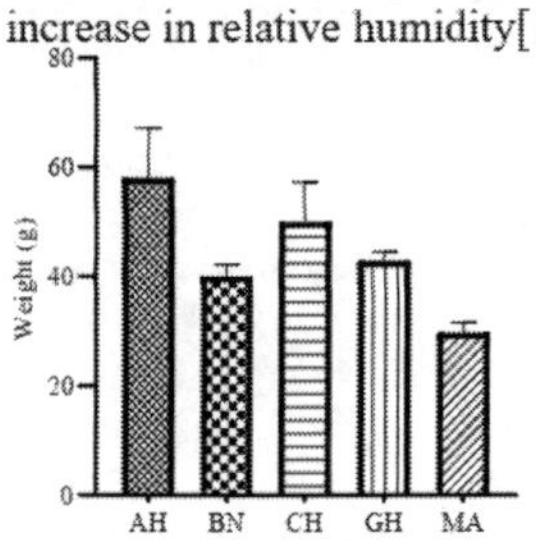

Fig2.Figs weight average

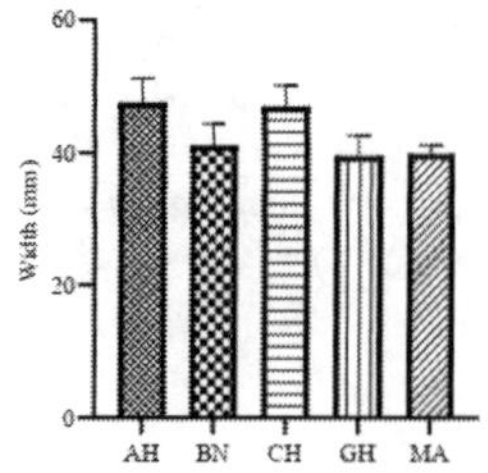

Fig3.Figs width average

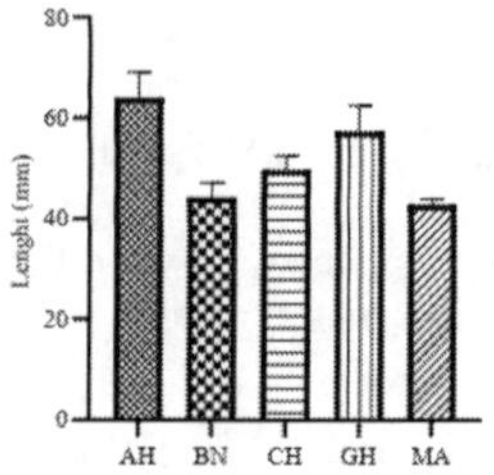

Fig4.Figs length average

The results show that fig weight (Fig 2) is high in AH (58. 09 ± 9.129 g) and CH (50.12 ±7.209 g) varieties and medium for GH (42.94 ±1.627 g) and BN (40.09 ± 2.182 g), while MA recorded the lowest weight (29.72 ± 1.767 g).The figs weight gives extremely attractive pricing for producers and traders in the market, for this purpose AH; CH and GH will certainly have the best chances of being appreciated by consumers. According to the results obtained concerning fig width (Fig 3), fig length (Fig 4) and their comparisons with those of the descriptor, we found that the width oscillates between small [28-38 mm]or medium [38-49 mm] at MA (39.88 ± 1.278 mm) , GH (39.56 ± 3.073 mm) and BN (41.16 ±3.202 mm) , for CH (47.03±3.140 mm) and AH (47.63 ±3.575 mm) it's between medium or large [50-60 mm]. The length is long [54-75 mm] for AH (64.07 ± 5.033 mm), and between medium [29-54 mm] and long for GH (57.44 ± 5.106 mm), and medium for CH (49.86 ± 2.728 mm), BN (44.32 ± 2.856 mm) and MA (42.88 ± 1.151 mm).

The Shape fruit changes depending to the stage of maturity and cultural maintenance [16], or because of environmental conditions and genotypic effect [17].In the fresh fig sector, shape or index is particularly important. According to the findings of this study (Fig 5), the oblong form (I < 0.9) is represented in GH (0.694± 0.080) and AH (0.745 ± 0.049), whereas the globose shape (I = 0.9-1.1) characterizes BN (0.928 ± 0.031) and MA (0.930±0.022) and CH (0.943± 0.044) varieties, which is suitable for commerce domain due to its appealing appearance and convenience of presentation.

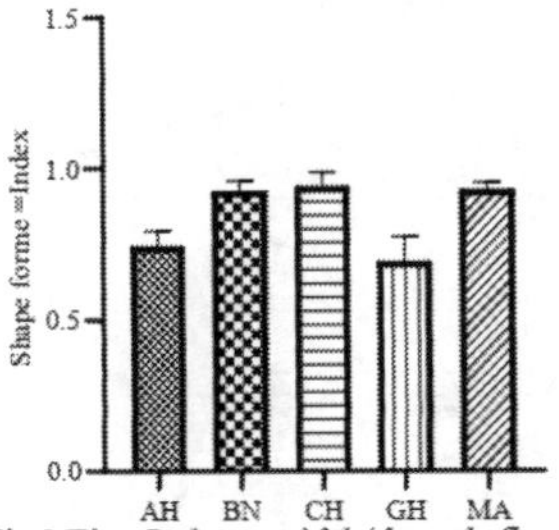

Fig5.Figs Index= width/ length fig average

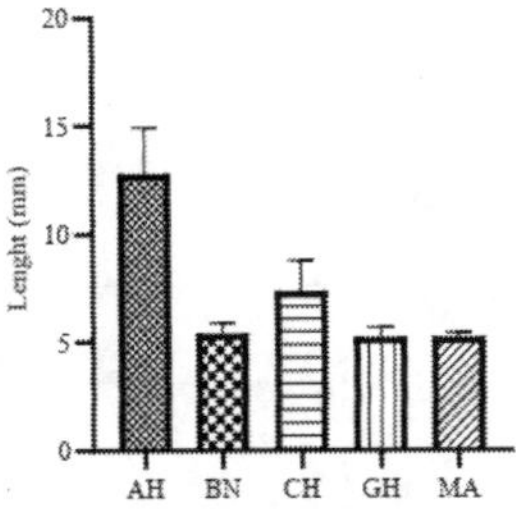

Fig6.Neck length average

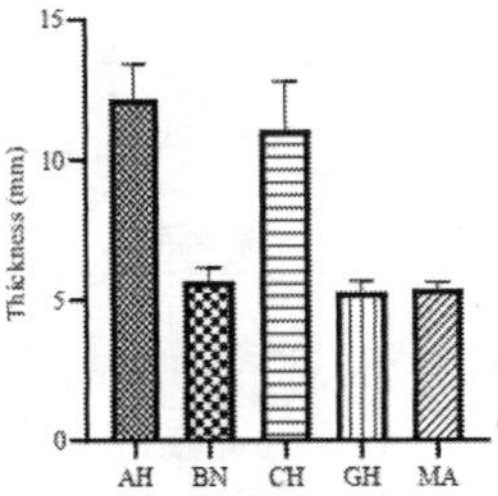

Fig7.Neck thickness average

Furthermore, AH (12. 81 ± 2.121 mm) variety has the longest neck followed by CH, BN, MA and GH (5.218 ± 0.456 mm) in that order (Fig 6). The same results have been observed in terms of fruit's neck thickness. The thickest neck) (Fig 7) belongs to the AH (12. 18 ± 1.244 mm) variety, followed by CH, BN, MA, and GH (5.408 ± 0.252 mm). It should be noted that for the neck length and the neck thickness BN, GH and MA are very similar. The fig farmers also considered that increasing these two characteristics would improve the freshness of the fig. Moreover, results show that AH (7.344 ± 3.287 mm), MA (8.407 ± 0.489 mm) and CH varieties have the longest fruit stalk (Fig 8) compared to BN and GH (4.905 ± 0.407 mm). In addition, AH (6.557 ± 0.728 mm), BN, and CH cultivars have thicker fruit stalks than GH and MA (5.106 ± 0.363 mm) varieties (Fig 9).

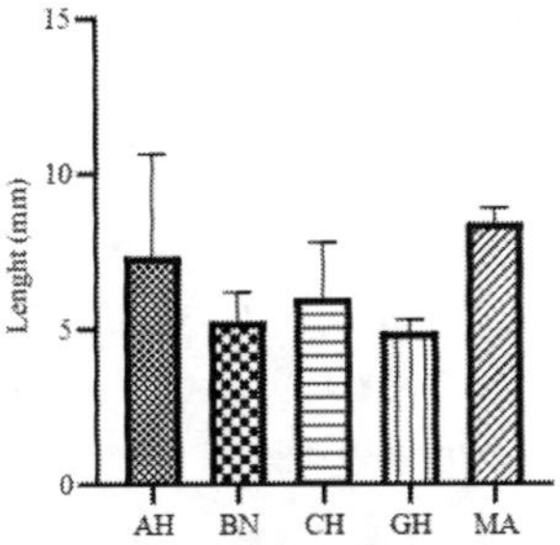

Fig8.Fruit stalk length average

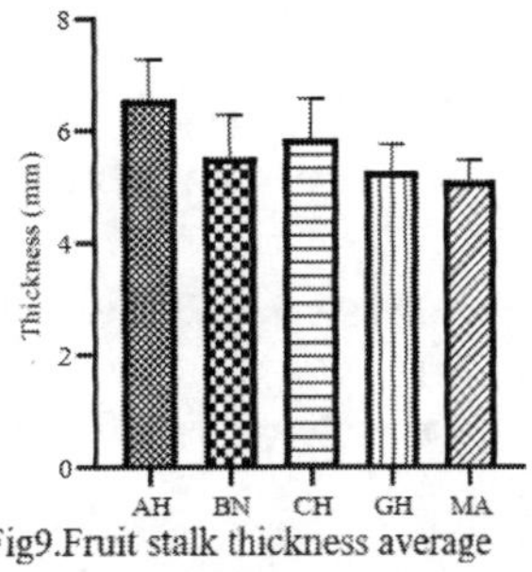

Fig9.Fruit stalk thickness average

Short necks, unlike long stalks, are considered as an undesirable characteristic by arborists because they make harvesting difficult and destructive to the fruit. Varieties with short necks, such as MA, may not be of interest to growers because they are difficult to harvest, whereas varieties with a long neck and stalk, such as AH, are preferable for their ease of harvest.

Results have shown that AH (6.887 ± 1.10 mm) figs have the greatest ostiole width (Fig 10) followed by CH, GH, BN and MA (3.788 ± 0.24 mm) in that order. The largest pulp width (Fig 11) is seen in AH figs (43. 59 ± 3.528 mm), followed by CH, GH, BN, and MA figs (36. 84 ± 1.293 mm). A large ostiole will facilitate the entry and flight of blastophagous as well as pesticides and pathogens, which make varieties with this character unsuitable for the fig industry and storage [18]. The thickness of the internal pulp is a determining pomological character in the fruit quality and their use.

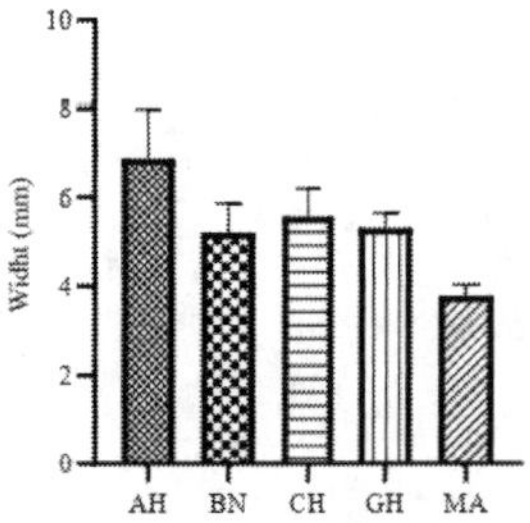

Fig10.Ostiole width average

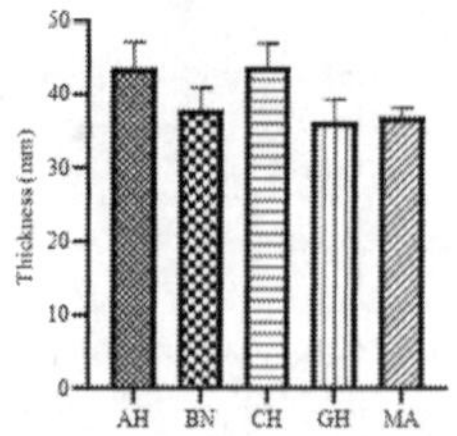

Fig11.Pulp thickness average

There was a lot of heterogeneity amongst the cultivars based on the fruit fig pomological qualities evaluation (Table3), with few exceptions in the following characteristics : width between AH/CH and GH/MA, Fig width/Fig length regarding BN;CH and MA between them, Ostiole width between BN/GH and CH/GH, Neck length regarding BN;GH and MA between them, Neck thickness regarding BN;GH and MA between them, Stalk length between BN/GH, Stalk thickness between GH/ MA, Pulp thickness between AH/ CH and BN/MA also between GH/ MA.

Table3. Ordinary one-way ANOVA of pomological fig characters (n=150) averages

Tukey's multiple comparisons test	F.W (g)	F.Wi (mm)	F.L (mm)	Wi/L Index	O.W (mm)	N.L (mm)	N.T (mm)	S.L (mm)	S.T (mm)	P.T (mm)
AH vs BN	****	****	****	****	****	****	****	****	****	****
Mean Diff	18.00	6.472	19.75	-0.1831	1.678	7.414	6.485	2.092	1.036	5.803
AH vs CH	****	Ns	****	****	****	****	****	****	****	Ns
Mean Diff	7.977	0.5983	14.21	-0.1985	1.314	5.426	1.097	1.341	0.7029	-0.178
AH vs GH	****	****	****	****	****	****	****	****	****	****
Mean Diff	15.15	8.067	6.639	0.05096	1.577	7.596	6.880	2.439	1.278	7.382
AH vs MA	****	****	****	****	****	****	****	***	****	****
Mean Diff	28.38	7.745	21.19	-0.1851	3.099	7.590	6.769	-1.063	1.451	6.751
BN vs CH	****	****	****	Ns	**	****	****	*	**	****
Mean Diff	-10.03	-5.873	-5.540	-0.01535	-0.363	-1.988	-5.388	-0.750	-0.332	-5.981
BN vs GH	**	**	****	****	Ns	Ns	Ns	Ns	Ns	**
Mean Diff	-2.849	1.595	-13.11	0.2341	-0.101	0.1818	0.3956	0.3476	0.2422	1.579
BN vs MA	****	*	Ns	Ns	****	Ns	Ns	****	***	Ns
Mean Diff	10.37	1.274	1.441	-0.001978	1.421	0.1756	0.2839	-3.155	0.4152	0.9481
CH vs GH	****	****	****	****	Ns	****	****	***	****	****
Mean Diff	7.176	7.468	-7.572	0.2494	0.2622	2.170	5.784	1,098	0.5751	7.560
CH vs MA	****	****	****	Ns	****	****	****	****	****	****
Mean Diff	20.40	7.147	6.981	0.01337	1.785	2.164	5.672	-2.404	0.7481	6.929
GH vs MA	****	Ns	****	****	****	Ns	Ns	****	Ns	Ns
Mean Diff	13.22	-0.3216	14.55	-0.2361	1.522	-0.006	-0.111	-3.502	0.1730	-0.631

* Weakly significant, ** Moderately significant, *** Highly significant, ****Very highly significant, Ns non-significant
F.W Fig weight, F.Wi Fig width, F.L Fig length, Wi/L Fig width /Fig length, O.W Ostiole width, N.L Neck length , N.T Neck thickness, S.L Stalk length , S.T Stalk thickness, P.T Pulp thickness.

This investigation has shown that the number of lobes per leaf (Fig 12) varies between 3 and 5 for AH (3.6 ± 0.9), MA (3.66± 0.94), CH (3.71 ± 0.96) and BN (4.2 ± 0.98) but for GH (5. 22 ± 1.38 mm) it is between 3 and 7. The great majority of the studied leaves have three lobes in AH, CH and MA in contrast to BN and GH they have five lobes.

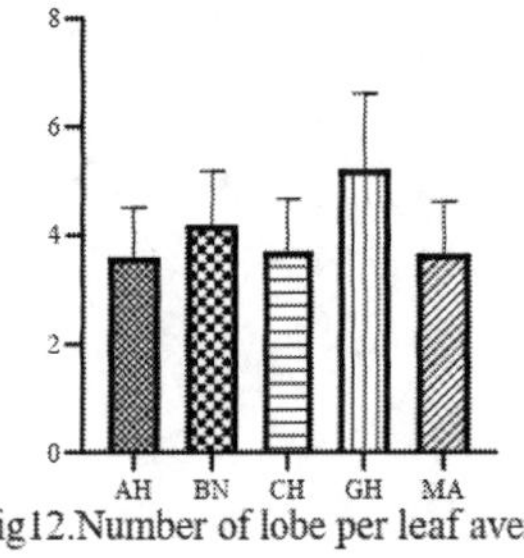

Fig12.Number of lobe per leaf average

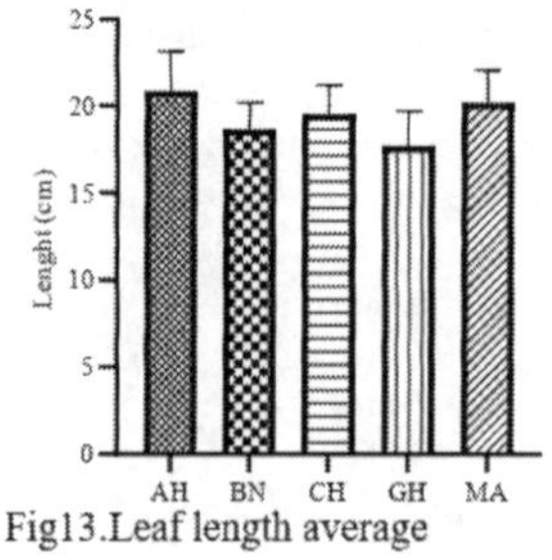

Fig13.Leaf length average

Moreover, AH (20.89± 2.26 cm) fig tree has the longest leaves followed by MA, CH, BN and GH (17.72 ± 2.009 cm) in that order (Fig13). AH (18.56 ± 2.75 cm) and CH fig trees has the greatest leaf width (Fig14) followed by GH, MA and BN (17.04 ±1.46 cm).The average leaf area of AH (392.9± 97.98 cm^2)followed by CH, MA, GH, and BN (319.7± 51.60 cm^2) in that order. These values show that their areas are Medium [250–400 cm^2], while the data demonstrate that all of these cultivars have large leaf areas [400-500 cm^2], with only AH and CH having a very large area [> 550 cm^2] and GH having a tiny area [250 cm^2] (Fig 15).

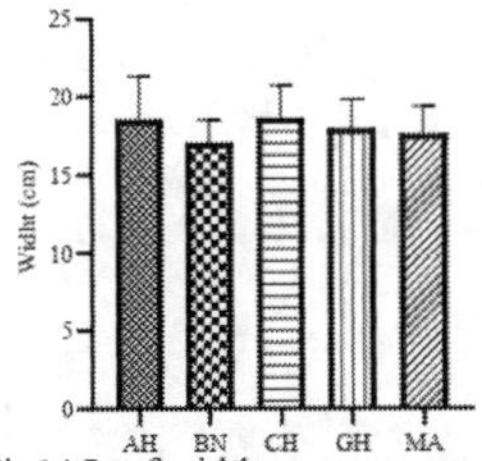
Fig14.Leaf width average

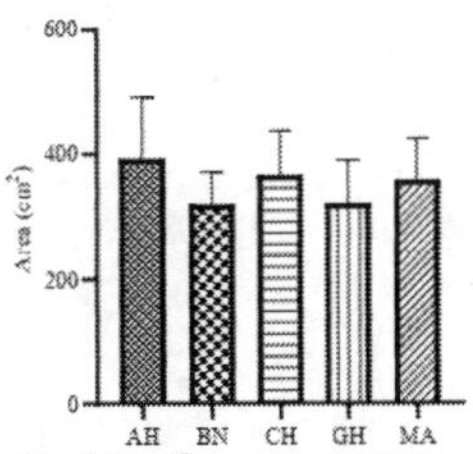
Fig15.Leaf area average

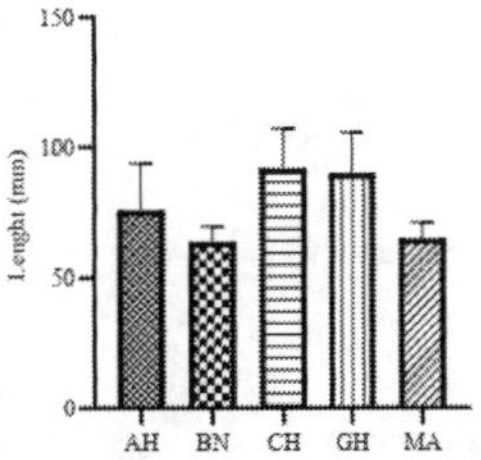
Fig16.Leaf stalk length average

Regarding leaf stalks length CH (91.92 ± 15.11mm) and GH (90.11 ± 15.42 mm) have a long leaf stalks [> 80 mm] while AH (75. 93 ± 17.92 mm); MA (65.05 ± 6.152 mm) and BN (63.90 ± 5.858 mm) have a medium leaf stalks [50-80 mm] (Fig 16). The highest average for leaf stalk length/leaf length ratio was recorded by GH (0.508 ± 0.072), followed by CH, AH, BN, and MA (0.322 ± 0.019) varieties (Fig17).

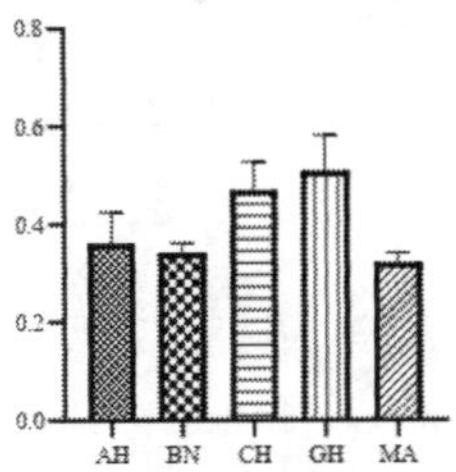
Fig17.Leaf stalk length/leaf length average

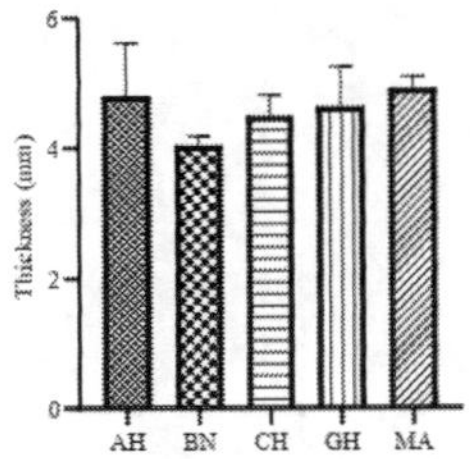
Fig18.Leaf stalk thickness average

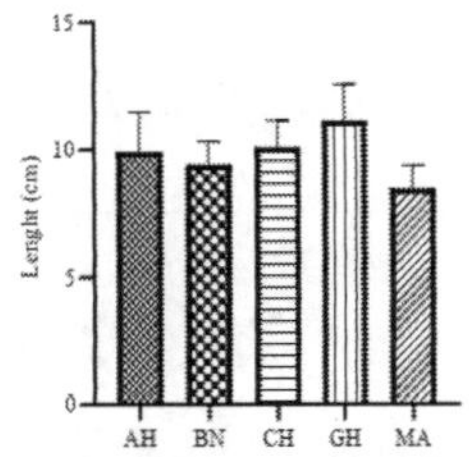
Fig19.Leaf lobe central length average

The leaf stalks length is not proportional to the leaves length. When it comes to leaf stalk thickness, MA (4.912 ± 0.177 mm) is the thickest, followed by AH, GH, CH, and BN (4.043 ± 0.141mm) in that order (Fig18). We noticed that the leaf stalks length is not proportional to the leaf stalks thickness.

Concerning the leaf central lobe length (Fig 19), GH (11.13 ± 1.415 cm) is the longest then CH followed by AH; BN and MA (8.472 ± 0.896 cm) in that order. According to the results of degree of leaf lobation [= Length of central lobe/ length of leaf] (Fig 20) GH (0.629 ± 0.055); CH (0.519 ± 0.058) and BN (0.504 ± 0.018) have a marked lobation [0.51 – 0.75], while AH (0.47 ± 0.045) and MA (0.41 ± 0.016) are on average lobation [0.26 – 0.50]. As for the number of leaves per shoot (Fig 21) BN (14.63 ± 1.55) and MA have the biggest number, then AH followed by CH and GH (6.467 ± 1.10) at last.

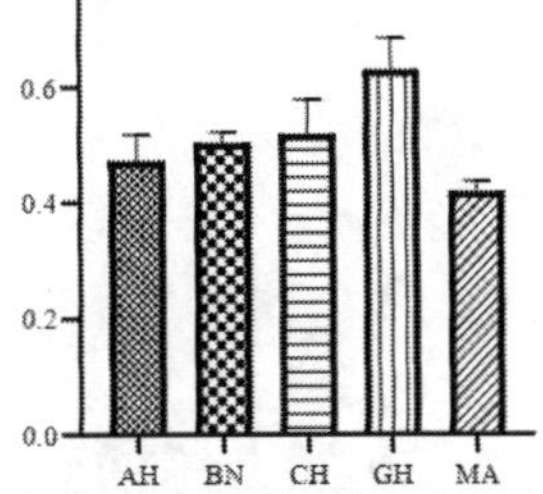
Fig20.Degree of leaf lobation average

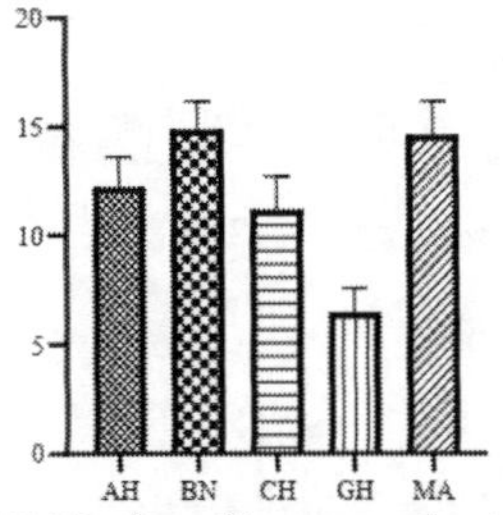
Fig21.Number of leaves per shoot average

Foliar analyzes show that the variability associated with the leaf area, leaf stalk size and the number of leaves per shoot, may be influenced by genetic characteristics. The leaf areas vary according to the varieties [19], this variation is independent of the number of leaves per shoot, of the number of lobes per leaf and their shape.

Morphological characteristics such as leaf size; number of leaves per shoot and pulp thickness; are important to selector and can be highly distinguishing variables. Nevertheless, differentiating fig tree varieties using their leaves is difficult because they are polymorphic within the same individual, particularly in terms of the number of lobes and the leaf stalk thickness[20].

The results of multiple comparisons of leaf character averages (Table 4) demonstrate a high degree of diversity among genotypes, with some exceptions in the characteristics listed below: Number of lobe regarding AH;CH and MA between them, Leaf length between AH/MA and CH/MA, Leaf width regarding AH;CH and GH between them also between BN/MA and GH/MA, Leaf area between AH/CH and CH/MA, leaf stalk length/Leaf length between AH/BN and BN/MA, leaf stalk length between BN/MA and CH/GH ,leaf stalk thickness regarding CH;GH and BN or MA between them also between AH/GH and AH/MA , Lobe central length between AH/CH, Degree of lobation leaf between BN/CH , Number of leaves per shoot between BN/MA.

Table4. Ordinary one-way ANOVA of leaves characters (n=150) averages

Tukey's multiple comparisons test	N.L	L.L (cm)	L.W (cm)	L.A (cm^2)	LS.L/L.L	LS.L (mm)	LS.T (mm)	LC.L (cm)	DL.L (mm)	N.L/S (mm)
AH vs BN	**	****	****	****	Ns	****	****	*	****	****
Mean Diff	-0.6000	2.225	1.515	73.21	0.01879	12.03	1.120	0.5027	-0.0302	-2.633
AH vs CH	Ns	****	Ns	Ns	****	****	*	Ns	****	*
Mean Diff	-0.1111	1.357	-0.07667	26.38	-0.1081	-15.99	0.6734	-0.1829	-0.0457	1.067
AH vs GH	****	****	Ns	****	****	****	Ns	****	****	****
Mean Diff	-1.622	3.168	0.6097	71.54	-0.1473	-14.17	0.5304	-1.212	-0.1555	5.767
AH vs MA	Ns	Ns	*	*	****	****	Ns	****	****	****
Mean Diff	-0.06667	0.6842	0.9566	34.48	0.03896	10.89	0.2516	1.445	0.05462	-2.392
BN vs CH	*	*	****	***	****	****	Ns	**	Ns	****
Mean Diff	0.4889	-0.8678	-1.592	-46.82	-0.1269	-28.02	-0.4466	-0.6856	-0.0154	3.700
BN vs GH	****	**	*	Ns	****	****	Ns	****	****	****
Mean Diff	-1.022	0.9428	-0.9053	-1.669	-0.1660	-26.20	-0.5896	-1.715	-0.1252	8.400
BN vs MA	**	****	Ns	**	Ns	Ns	**	****	****	Ns
Mean Diff	0.5333	-1.541	-0.5584	-38.73	0.02018	-1.147	-0.8684	0.9426	0.0848	0.2417
CH vs GH	****	****	Ns	***	****	Ns	Ns	****	****	****
Mean Diff	-1.511	1.811	0.6863	45.15	-0.03914	1.817	-0.1430	-1.029	-0.1098	4.700
CH vs MA	Ns	Ns	**	Ns	****	****	Ns	****	****	****
Mean Diff	0.04444	-0.6731	1.033	8.099	0.1471	26.87	-0.4219	1.628	0.1004	-3.458
GH vs MA	****	****	Ns	**	****	****	Ns	****	****	****
Mean Diff	1.556	-2.484	0.3469	-37.06	0.1862	25.06	-0.2789	2.657	0.2101	-8.158

* Weakly significant, ** Moderately significant, *** Highly significant, ****Very highly significant, Ns non-significant
N.L Number of lobe, L.L Leaf length, L.W Leaf width,L.A Leaf area , LS.L/L.L Leaf stalk length/Leaf length, LS.L Leaf stalk length , LS.T Leaf stalk thickness, LC.L Lobe central length , DL.L Degree of lobation leaf,N.L/S Number of leaves per shoot.

The tendency to form suckers is low [less than three] for AH; BN; CH and MA, medium [three to seven] for GH (Fig 22). The harvest period average is medium [21-40 days] for BN and GH; it coincides with that observed in Turkey varieties [21], Long [41-60 days] for AH and MA, Very long [>60 days] for CH (Fig 23).The beginning of fruit maturation for BN and GH is early [20-31 July], mid-season [1-15 August] for MA, and late [15-31 August] for AH and CH. The fig sector in eastern Morocco has the major advantages of fruit cultivation, early harvests (extended beginning of maturity from early to late) and great availability of fresh fruits on the market prolonged harvest period from medium to very long, allowing the market for fresh figs to be satisfied for an extended period of time.

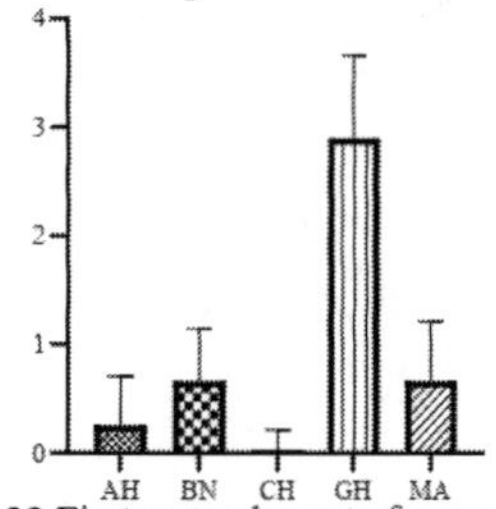

Fig22.Fig tree tendency to form suckers average

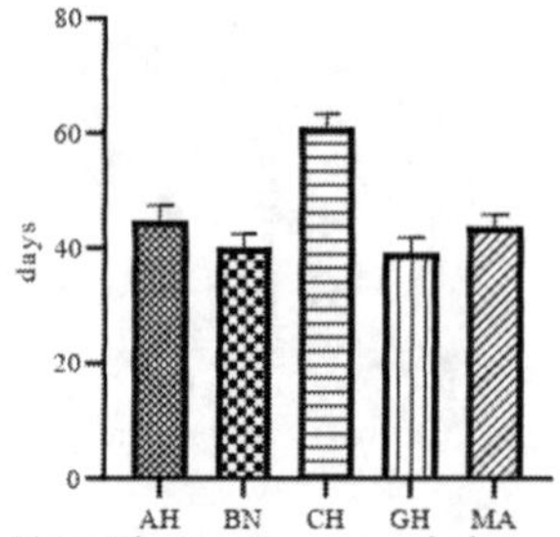

Fig23.Fig tree Harvest period average

https://doi.org/10.1051/e3sconf/202233704008

Furthermore, the terminals buds (Fig 24) of CH (11.53 ± 1.93 mm) and AH are the longest, followed by BN, GH, and MA (9.172 ± 0.96 mm) .In terms of terminal bud width (Fig 25), AH (7.196 ± 0.96mm) is the largest, followed by CH and BN, then GH and MA (5.319 ± 0.56 mm) . Regarding to the terminal bud length/width ratio (Fig 26) the biggest one is for CH (1.820 ± 0.20 mm) then BN and GH followed by MA and AH (1.524 ± 0.15 mm) in that order.

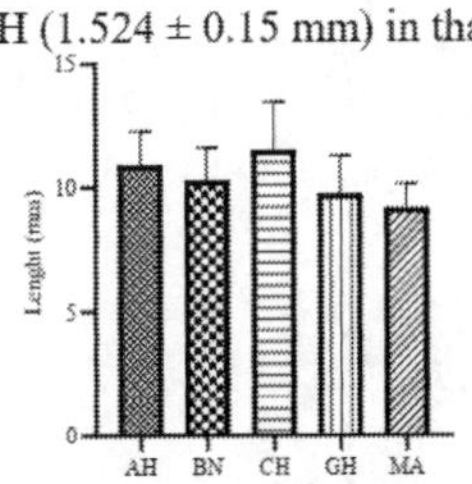

Fig24.Terminal bud length average

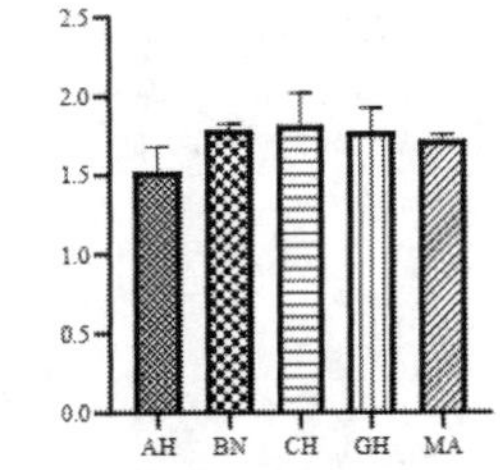

Fig25.Terminal bud width average

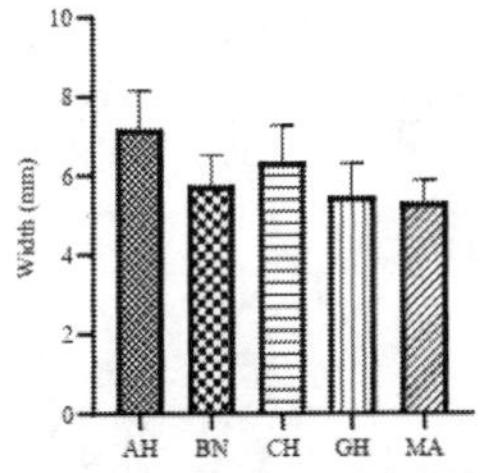

Fig26.Terminal bud length/width average

The terminal bud color is green for AH, BN, CH and MA, Pinkish brown for GH. The dominant terminal bud shape is conical for them at all (Fig27).

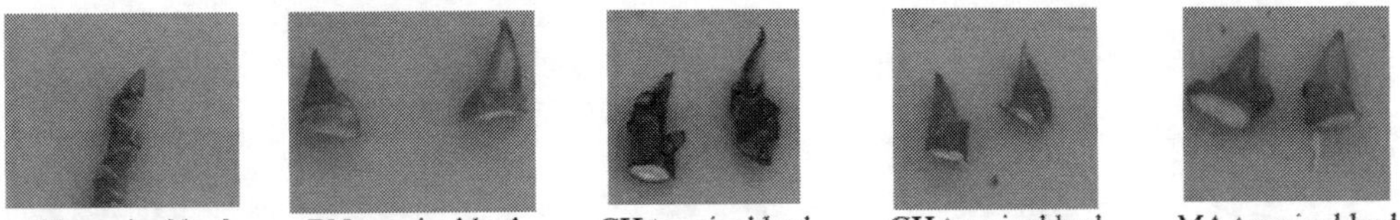

AH terminal bud BN terminal bud CH terminal bud GH terminal bud MA terminal bud

Fig27.Terminal bud shape and color of the studied fig cultivars

From the multiple comparisons (table 5), we notice a great variability in tree and terminal bud characters averages with some exceptions in the following variables: Bud terminal length between GH/MA, Bud terminal width between BN/GH and GH/MA, Bud terminal length/Bud terminal width regarding BN; CH and GH between them, tendency suckers between AH/CH and BN/MA, Harvest period between AH/MA and BN/GH.

Table5. Ordinary one-way ANOVA of tree characters (n=45); terminal bud (n=150); total soluble solids and titratable acidity averages

Tukey's multiple comparisons test	T.S	H.P (days)	Bt.L (mm)	Bt.W (mm)	L/W Bt	T.S.S %	T.A %
AH vs BN	*	****	*	****	****	****	****
Mean Diff	-0.407	4.667	0.6146	1.443	-0.266	1.575	-0.028
AH vs CH	Ns	****	*	****	****	****	****
Mean Diff	0.225	-16.13	-0.619	0.8487	-0.295	4.883	-0.040
AH vs GH	****	****	****	****	****	Ns	Ns
Mean Diff	-2.641	5.600	1.163	1.718	-0.255	-0.023	-0.004
AH vs MA	*	Ns	****	****	****	****	**
Mean Diff	-0.407	1.033	1.738	1.876	-0.200	8.19	-0.013
BN vs CH	****	****	****	****	Ns	****	**
Mean Diff	0.633	-20.80	-1.234	-0.594	-0.028	3.327	-0.011
BN vs GH	****	Ns	Ns	Ns	Ns	****	****
Mean Diff	-2.233	0.9333	0.5483	0.2751	0.0110	-1.580	0.024
BN vs MA	Ns	****	****	**	**	****	**
Mean Diff	0.000	-3.633	1.124	0.4332	0.0661	6.633	0.014
CH vs GH	****	****	****	****	Ns	****	****
Mean Diff	-2.867	21.73	1.783	0.8694	0.0399	-4.907	0.036
CH vs MA	****	****	****	****	****	****	****
Mean Diff	-0.633	17,17	2.358	1.028	0.0949	3.307	0.026
GH vs MA	****	****	Ns	Ns	*	****	*
Mean Diff	2.233	-4.567	0.5754	0.1581	0.0550	8.213	-0.009

* Weakly significant, ** Moderately significant, *** Highly significant, ****Very highly significant, Ns non-significant
T.S tendency suckers ,H.P harvest period , Bt.L Bud terminal lenght,Bt.W Bud terminal width, L/W Bud terminal length/Bud terminal width, T.S.S Total Soluble Solids, T.A Titratable Acidity

Regarding to the total soluble solids (Fig 28) GH (24.16 ± 0.04 %) and AH (24.13 ± 0.01 %) have the highest percentages, followed by BN; CH and MA (15.94 ± 0.08 %). As for titratable acidity (Fig 29) CH (0.272 ± 0.03 %C.A= Citric. Acid) has the highest percentage, followed by BN, MA; GH and AH (0.231 ± 0.004 % C.A) in that orders. The chemical fruits characteristics are important in the pomological quality determination and can also be used for varietal discrimination [22]. Fig fruit with a good quality should have a soluble solid content between 13.0% and 25.1%, and an acidity levels between 0.226 and 0.300 % [23]; these two criteria are met by our cultivars studied. Also significant differences of total soluble solids and titratable acidity averages in figs cultivar are noted except between AH and GH for these two traits (table5).

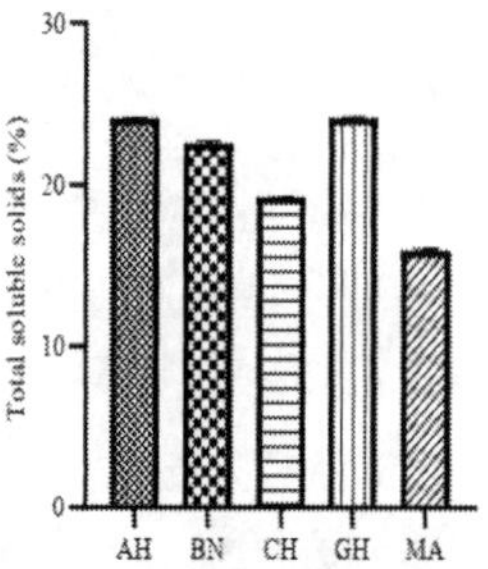

Fig28.Figs total soluble solids average

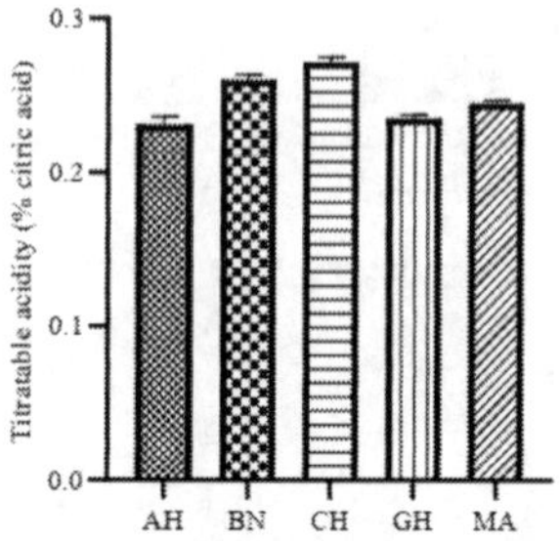

Fig29.Figs titratable acidity average

In order to determine the quality of figs; we use the weighted ranked method of Aksoy, based on nine characteristics which are with their weighting: the weight of the fruit (g) 20; beginning of maturation 20; shape of the fruit (index) 9; neck length (mm) 6; fruit skin cracks 10; ease of peeling 10; ostiole width (mm) 5; total soluble solids (%) 10 and titratable acidity (%) 10.The lowest score is obtained by MA (570),followed by BN (680) then GH (702), while the highest are recorded in particular at CH (720) and AH (738), theses scores reflect a very good pomological fig quality in the eastern Morocco (Fig 30). The fig quality is influenced by the genotypic effect, environmental conditions (temperature, humidity, wind...), the maturity stage and cultural maintenance[18].Certain pomological characteristics of apple, such as size and dry matter content, are more affected by the increase of the tree fruit load due to poor tree development. Thus, a weak tree development increases the effect of tree fruit load on fruit quality [24].

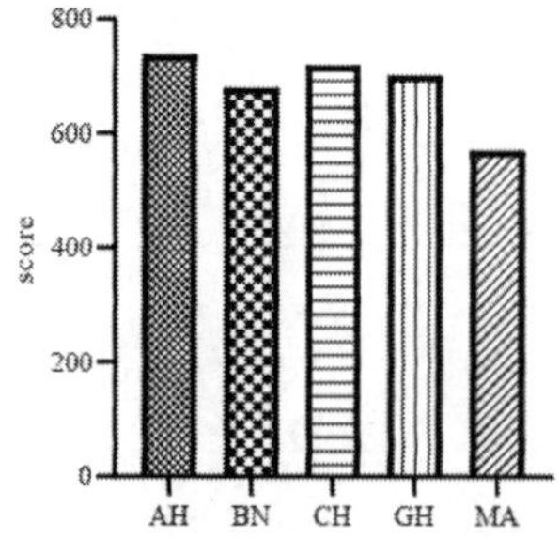

Fig30.Quality of figs by the weighted ranked method of Aksoy

4 Conclusion

The overall analysis of our results shows that morphological markers reveal phenotypic variability among the fig tree grown in eastern Morocco. In the studied area, this plant exhibits a wide range of morphological characteristics. Thus, the fig cultivars investigated appear to be well suited to the pedoclimatic conditions, requiring less water and phytosanitary treatment. In order to preserve and maintain the biological diversity of fig trees in this region, this research needs to be completed through the development of illustrated varietal sheets with useful recommendations. (e g A fig cultivars list with useful suggestions developed by nursery workers and arborists to help farmers choose varieties and use. furthermore, from an agricultural point of view, the studied cultivars are popular and recommended by local farmers because of their high fruit yields. Moreover, from an agri-food standpoint, as a summer fruit, these fresh figs produced in this region have an excellent taste, they are sought after and appreciated by consumers, in particular the fruits of the cultivar AH, CH. Finally, the cultivars AH, GH deserve to be candidates for IGP labeling as fresh fruits, not only because of their outstanding quality, but also because of their good pomological parameters and high Aksoy weighted method scores.

We are grateful to Gharmawne agricultural cooperative. Our thanks are also due at the Regional Directorate of Agriculture (DRA) in Oujda, at the Provincial Directorates of Agriculture (DPA) in Oujda, Berkane, Jerada, Nador and Taourirt for facilitating access for that work. Equal thanks go to the fig farmer producers in the collection sites and Aghbal municipality for the practical assistance and collaboration.

References

1. M,Pourghayoumi, et al., Phytochemical attributes of some dried fig (Ficus carica L.) fruit cultivars grown in Iran. Agriculturae Conspectus Scientificus, 81(3): p. 161-166 (2016)
2. A.H,Achtak, et al., Traditional agroecosystems as conservatories and incubators of cultivated plant varietal diversity: the case of fig (Ficus carica L.) in Morocco. BMC plant biology,. 10(1): p. 1-12 (2010).
3. Y,Hmimsa., Y. Aumeeruddy-Thomas, and M. Ater, Une forme spontanée de figuier (Ficus carica L.), le nābūt. Diversité de nomenclature, d'usage et de pratiques locales au Nord du Maroc. Revue d'ethnoecologie, (Supplément 1) (2017).
4. B,Khadari, et al., Contrasted genetic diversity and differentiation among Mediterranean populations of Ficus carica L.: a study using mtDNA RFLP. Genetic resources and crop evolution, 52(1): p. 97-109 (2005)
5. D,Zohary, and M. Hopf, Domestication of plants in the Old World: The origin and spread of cultivated plants in West Asia, Europe and the Nile Valley: Oxford University Press (2000).
6. Y,Aumeeruddy-Thomas, and Y. Hmimsa, Fig and olive domestication in the Rif, northern Morocco:Entangled human and tree lives and history, in Hybrid Communities, Routledge. p. 179-196 (2018).
7. Y,Hmimsa, Y. Aumeeruddy-Thomas, and M. Ater, Vernacular taxonomy, classification and varietal diversity of fig (Ficus carica L.) among Jbala cultivators in northern Morocco. Human Ecology, 40(2): p. 301-313(2012).
8. A,EL HAJJAM, and A. EZZAHOUANI, Conduite technique et inventaire des variétés locales marocaines de figuier (Ficus carica L.) dans quatre principaux sites de production, provinces de Chefchaouen, El Jadida, Ouezzane, et Taounate. Revue Marocaine des Sciences Agronomiques et Vétérinaires, 6(4): p. 494-504 (2018).
9. L,Hssaini, et al., Characterization of local fig clones (Ficus carica L.) collected in Northern Morocco. Fruits, 74(2): p. 55-64 (2019).
10. I IPGRI, C., Descriptors for fig. International Plant Genetic Resources Institute, Rome, Italy, and International Centre for Advanced Mediterranean Agronomic Studies, Paris, France, 52 (2003).
11. J,GIL,et al., Las higueras canarias y su diversidad: Bases orales y documentales(2011).
12. U,Aksoy,Descriptors for fig (Ficus carica L. and related Ficus sp.). Ege Uni. Faculty of Agriculture. Department of Horticulture, Izmir, Turkey, (1991).
13. H.C.P,Monographie de la région de l'oriental. [p 203](janvier 2021) Available from: https://www.hcp.ma/region-oriental/docs/Monographies/Monographie%20de%20la%20region%20-%20Janvier%202021.pdf.
14. A,Oukabli, S. Bartolini, and R. Viti, Anatomical and morphological study of apple (Malus× domestica Borkh.) flower buds growing under inadequate winter chilling. The Journal of Horticultural Science and Biotechnology, 78(4): p. 580-585 (2003).
15. A.K,Yadav, Phalsa: A potential new small fruit for Georgia. 1999, ASHS Press: Alexandria, VA, USA. p. 348-352(2009).
16. M,Simsek, Evaluation of selected fig genotypes from South east Turkey. African Journal of Biotechnology, 8(19).
17. O,Çalişkan,and A.A. Polat, Morphological diversity among fig (Ficus carica L.) accessions sampled from the Eastern Mediterranean Region of Turkey. Turkish Journal of Agriculture and Forestry, 36(2): p. 179-193(2012).
18. B,Gaaliche,et al, Pomological and biochemical characteristics of fig (Ficus carica L.) cv. Zidi in different agro-ecological zones of Tunisia. Pak. J. Agri. Sci, 49(4): p. 425-428 (2012).
19. A,Abo-El-Ez,R. Mostafa, and I.F. Badawy, Growth and productivity of three fig (Ficus carica L.) cultivars grown under upper Egypt conditions. Australian Journal of Basic and Applied Sciences, 7(2): p. 709-714(2013).
20. E,Khanfir, Identification of genetic diversity of Ficus carica: Morphological and molecular characterization of varieties from Kerkennah. Editions Universitaires Européennes. Saarbrücken, Allemagne, , 106 p(2015).
21. O,Çalişkan,and A.A. Polat, Fruit characteristics of fig cultivars and genotypes grown in Turkey. Scientia horticulturae, 115(4): p. 360-367(2008).
22. M.B,Miklavcic , B.B., Valencic V., Sugars and Organic Acids Content in Fig Crops, 90–100 pp 2006 and 2007: In The Common Fig (Ficus carica L.) in Istria.Morphological, Molecular and Some Chemical Characteristics. . University of Primorska, Science and Research Centre Koper, Publishing House Annales. Project RGFI –Revitalization of Fig Cultivation in Istria, 104 p(2008).
23. U,Aksoy, et al., Ege Bölgesi koşullarına uygun sofralık incir çeşit seleksiyonu. Türkiye I. Ulusal Bahçe Bitkileri Kongresi, 1: p. 545-548(1992).
24. I,Travers, Influence des conditions pédoclimatiques du terroir sur le comportement du pommier et la composition des pommes à cidre dans le Pays d'Auge, Caen(2004).

Effects of irrigation systems on yield and physical characteristics of date (*Phoenix dactylifera* .L) cv. 'Medjool'

Fouzia Alla[1*], *Kawtar* Jdaini[1], *Jamal* Mimouni[2], and *Mohammed Aziz* Elhoumaizi[1]

[1]Laboratory of Agricultural Production Improvement, Biotechnology and Environment (LAPABE), Faculty of Sciences, University Mohammed First, 717 Oujda 60000, Morocco.
[2]National Agency for the Development of Oasis and Argan Areas (ANDZOA), Bouarfa, Morocco.

Abstract. The shortage of irrigation water pooled with the effect of climate change and inadequate management in oases; affects the sustainability of the oasis ecosystem. Irrigation, as practiced in these oases, is not conducted according to the real water requirements of the crops, this management can induce water stress which affects the physical characteristics of the dates cultivated and consequently yield production, or on the contrary a significant water intake and a waste of the resource. This study investigated two irrigation systems; drip-irrigation, which is currently applied under date palms in the Moroccan oases, and the newly introduced underground irrigation (Micro-porous irrigation) .The aim of this work, was to assess the effects of different types of irrigation system on yield and physical quality parameters of 'Medjool' dates. Fruit quality assessment including fruit weight, length, breadth, pulp thickness, stone weight, length, breadth, number and average weight of bunches fruit per tree were evaluated, to identify the most effective irrigation system for date palm.

1 Introduction

Date palm (*Phoenix dactylifera* L.), one of the oldest known fruit crops, belonging to Arecaceae family. It is an important horticultural crop and it represents the heart of oasis cultivation systems [1]. Its cultivation around the world, mainly for its edible sweet fruits, has a considerable economic significance, which improves agricultural yields in arid environments [2]. The chemical composition of dates including the minerals, proteins, fats, carbohydrates, vitamins, and dietary fibers, are the main nutritional values that affect the quality of dates. Its richness plays a major role in humans health, showing a high antioxidant, anticancer, hepatoprotective, neuroprotective, nephroprotective, gastrointestinal protective, antidiabetic, antihyperlipidemic potentials [3].

*Corresponding author: fouzia.alla@gmail.com

Located in an arid and hostile environment, the oasis constitutes a unique ecosystem, based on the right balance of three elements: water, soil quality, and the presence of date palms whose parasol-shaped foliage creates a real microclimate. This tree is considered as one of the oldest fruit species, its cultivation was practiced by ancient civilizations for centuries, showing a great ecological and socio-economic importance due to its high plasticity, and adaptation to extreme pedo-climactic conditions [4] .The management of water resources is the major factor influencing the economic and social development of the Saharan oases. In Morocco, the date palm is the key species in oasis systems; it occupies a privileged place in oasis arboriculture and provides jobs for most of the local population. Moroccan date palm counts so far around 6.6 million palm trees spread on 51,000 hectares concentrated mainly in four regions, Tafilalet-Daraa, Souss-massa, Guelmil-Oued Noun and Figuig [5], composed of 453 cvs. including high proportion of Khalts (originating from natural palm seedlings) and other high value cultivars such as Boufeggous, Medjool, Bouskri, and Aziza bouzid [6] , with an significant average harvest of 129,562 tons in 2017[7]. However, these sub-desert areas are in a critical situation due to the scarcity of their water resources. The water shortage of irrigation water pooled with the effect of climate change and inadequate management in oases; pose a significant threat to the sustainability of the oasis ecosystem. The availability of water in quantity and quality associated with irrational use of irrigation water in the face of hostile agro-climatic conditions are the main factors causing negative repercussions in the oasis ecosystem. As a result of the poor management of water on the scale of farms and their environment, many consequences and negative effects on the economy of palm growers can be observed in these zones.

Therefore, finding ways to increase the quality of dates and production through irrigation systems is of great importance. In morocco, the most used method of irrigation is drip irrigation, this system has helped sustain the production of date palms and other crops, however, its use in the sub-desert areas has various problems, mainly for its inability to keep water safe from high evaporation rates, the spread of weeds around date palms, cracks in the channels leading to water leaks, and damage of canals by animals. As an alternative to this system, it would be favorable to use underground irrigation. This technique is defined as a slow release, sub-surface irrigation system. It is based on a unique micro-technology from 0-300μm, which significantly reduces water use with limited maintenance, at a reduced cost. This promising water saving technology enables increased irrigation efficiency by up to 70%, minimizing losses of evaporation, runoff and deep percolation.
Given the issue of shortage of water, it is important to examine the comparative performance of various irrigation systems predominantly in desert regions. This research will focus on analyzing the effects of underground irrigation systems on yield and physical characteristics of dates, as well as a comparative analysis of drip irrigation systems used in the Moroccan oasis.

2 Materials and methods

2.1 Experimental site location

This study was conducted out on a plantation of 50ha in Figuig (East of morocco), (N 32°09‘05"W 1°17'47") , for two successive years (2019-2020) .The climate of the study area is arid with an annual mean temperature of 32°C and annual mean precipitation of 92.4mm (2009-2019).

https://doi.org/10.1051/e3sconf/202233704009

2.2 Plant material

The variety used is'Medjool'. This variety of date palm is essentially characterized by its worldwide reputation expressed by the large size of its fruit and its excellent date palm quality generating better commercial added value. The date palms were planted in 2011 at a density of 6m x 6m, the area reserved for this test is 600 m^2.

The experiment was conducted in two blocks, comprising two different irrigation systems:

i) An underground irrigation system was installed with a circle of 6 m for each date tree, with a flow rate of 8 l/m*h.;
ii) A drip irrigation system was installed with two lateral lines for each row of date palm, with 6 drippers and a flow rate of 96 l/h. Each block consists of 208 trees and 48 trees per replicate was used (B1, B2, B3, and B4).

The annual irrigation water amount supplied to date palm (*Phoenix dactylifera* L.) following the farmer regime were 46,08 (m^3/tree) under drip irrigation system including 7,68 (m^3/tree) of water losses , and 20,736 (m^3/tree) for the underground irrigation system (micro-irrigation) including 1,536 (m^3/tree) of water losses. Fruits samples were harvested from each tree and fruit bunches at the 'Tamr' fruit maturity stage, employed simple random sampling technique. Dates were imported to the laboratory for analysis.

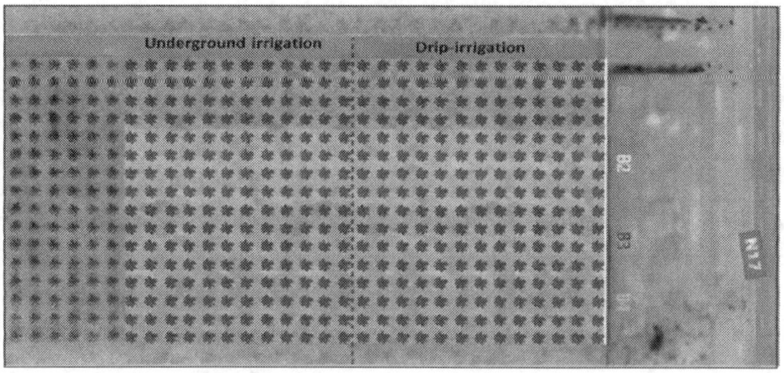

Fig. 1. Experimental plan

2.3 Physical characteristics

• Weight and stone weight were measured by using an electric balance with 0.01 gm sensitivity and average weight per fruit was calculated (RADWAG PS 360.R2).

• Pulp stone ratio was calculated by dividing the weight of pulp by stone weight.

• Fruit length, breadth, thickness was measured by using a digital Vernier caliper with 0.01mm accuracy.

2.4 Bunches fruit characterization

The evaluation of effect of two methods of irrigation, drip irrigation and underground irrigation, on bunches fruit characterization, is based in determination of the unit yield (in kg / tree): the total number and average weight of bunches fruit per tree.

2.5 Statistical analysis

The experimental data were analysed by using two-way analysis of variance (ANOVA) using SPSS statistics 20 edition. The effects of irrigation system and their interactions with different parameters were estimated and assessed using ANOVA, in particular least significant differences (LSD) along with using a significance level at ($P \leq 0.05$) using F-test.

3 Results

3.1 Physical characteristics

The analysis of the physical characteristics of date palms is presented below (Table 1).

Irrespective of the type of irrigation system, the fruit weight, breadth, length, and Pulp thickness did not significantly affected ($p \leq 0.05$). Underground irrigation resulted in the highest fruit weight, breadth, length, and pulp thickness (18.40±2.68g fruit^{-1},24.18±1.69 mm fruit^{-1}, 41.36±2.6 mm fruit^{-1}and 6.25±0.25 mm fruit^{-1}) compared to drip irrigation system (15.56 ±2.8g fruit^{-1} , 23±1.83mm fruit^{-1} , 38.25±2.43 mm fruit^{-1}, and 5,82±0,42 mm fruit-^{1}), respectively. The irrigation systems significantly affect only stone weight while stone breadth and length, were not significantly affected with irrigation systems. Underground irrigation resulted in highest stone weight, breadth and length (1.30±0.08 g fruit^{-1}, 9.09±0.48mm fruit^{-1}, and24.18±1.88mm fruit^{-1}) compared to drip irrigation system (1.14±0.14 g fruit^{-1}, 9.09±0.42mm fruit^{-1}, and 23.45±1.44mm fruit^{-1}) respectively (Table 2).However, there were great savings of irrigation water using underground irrigation systems compared to drip irrigation system. This is consistent with, [8-9]; and [10]; who reported that irrigation systems have no effect of on date fruit quality. Contrarily, research findings reported from the Middle East and North African countries claimed that irrigation systems influence date palm fruit quality [11, 12, 13, 14, 15, 16, 17]. Furthermore, several studies found the effect of different farming practices on physical characteristics of dates, such as fertilization treatments [18], pollination and pruning on fruit quality [19-20]. In our case, the two blocks of experiment were treated with the same fertilization treatments, the same farming practices and pollinated with the same mixture of pollen, however the mixture was different from 2019 to 2020.

Table 1.Effect of irrigation systems on physical characteristics of 'Medjool' dates per year

Irrigation system		Fruits			Pulps		Seeds			Ratio S/D	Ratio P/D
		Weight (g)	Breadth (mm)	Length (mm)	Thickness (mm)	Weight (g)	Weight (g)	Breadth (mm)	Length (mm)		
Drip-irrigation	2020	18,19±1,65	24,83±0,30	40,68±0,64	6,16±0,23	16,91±1,65	1,28±0,01	9,50±0,24	24,89±0,35	7,11±0,72	92,89±0,72
	2019	12,92±0,34	21,16±0,47	35,81±0,93	5,48±0,27	11,92±0,33	1,00±0,01	8,68±0,02	22,01±0,29	7,76±0,19	92,24±0,19
Underground irrigation	2020	19,98±1,28	25,87±0,50	43,78±1,37	6,36±0,22	18,61±1,22	1,38±0,10	9,57±0,09	26,06±0,72	6,90±0,47	93,10±0,47
	2019	16,83±2,87	22,48±0,38	38,94±1,46	6,13±0,22	15,61±2,86	1,22±0,03	8,61±0,13	22,29±0,57	7,47±1,02	92,53±1,02

Table 2.Effect of irrigation systems on physical characteristics of 'Medjool' dates

Irrigation system	Fruits			Pulps		Seeds			Ratio S/D	Ratio P/D
	Weight (g)	Breadth (mm)	Length (mm)	Thickness (mm)	Weight (g)	Weight (g)	Breadth (mm)	Length (mm)		
Drip irrigation	15,56±2,80	23,00±1,83	38,25±2,43	5,82±0,42	14,42±2,70	1,14±0,14	9,09±0,42	23,45±1,44	7,43±0,60	92,57±0,60
Underground irrigation	18,40±2,68	24,18±1,69	41,36±2,6	6,25±0,25	17,11±2,62	1,30±0,08	9,09±0,48	24,18±1,88	7,19±0,81	92,81±0,81

S: Seeds P: Pulps D: Dates

3.2 Bunches fruit characterization

The overall average production during two cycles of underground system differed significantly from the drip irrigation (5654.45 kg/year, 4739.4 kg /year) respectively. The average bunch mass and bunch fruit number were higher with a underground irrigation system compared to drip irrigation (8.65 kg/bunch fruit, 7.9 kg/bunch fruit, and, 652.5 bunch fruit , 595 bunch fruit)respectively (Table3). This indicates that the variability of these characters is not only genetic, but can be impacted by environmental parameters and management. The enhancement in both parameters was most likely due to the efficient use of water by the root system since it was straight provided to the absorbing functional zone. The yield parameter of date palm fruit of cv. 'Medjool' are improved with high reduction in irrigation water.

Table 3. Effect of irrigation systems on bunches fruit characteristics of 'Medjool' dates

Irrigation system	Bunch fruit number/ tree	Mean Production (kg/tree)	Mean overall production (kg)
Underground irrigation	5,68±1,13	49,22±10,09	4739,40±132,60
Drip irrigation	5,03±1,87	40,28±15,56	5654,45±181,85

4 Conclusion

Finding ways to increase date palm production through irrigation systems is of great importance. This study presents the effect of irrigation systems on yield and physical quality of 'Medjool' dates of Figuig oasis, Morocco. The information is most important in deciding the suitability of different irrigation systems for date palm especially in regions limited in water resources. The results showed that irrigation systems significantly affected only mean overall production of date and seed weight, whilst fruit weight ,breadth and length, pulp thickness, stone width and stone length, were not significantly affected by irrigation. However, there were great savings of irrigation water using underground irrigation systems compared to drip irrigation system. In conclusion, the underground irrigation system improved physical characteristics of 'Medjool' date, but the differences were not significant.

References

1. P. Munier. Le Palmier dattier. Techniques agricoles et productions tropicales, Paris, 221(1973)
2. B. Agoudjil, A. Benchabane, A. Boudenne, L. Ibos, M. Fois, ISO4 **43**, 491-497 (2011)
3. A.H. El-Far, H.M. Shaheen, M.M. Abdel-Daim, S.K. Al Jaouni, S.A. Mousa, Curr. Drug Discov. Technol **16**, 2-10 (2018)

4. My.H. Sedra. Le palmier dattier base de la mise en valeur des oasis du Maroc, techniques phoenicicoles et création d'oasis INRA, Editions 266 (2003)
5. FAO. The future of food and agriculture – Trends and challenges, Rome (2017)
6. My.H. Sedra. Date Palm Status and Perspective in Morocco, Date Palm Genetic Resources and Utilization, 257-323 (2015)
7. A. Hasnaoui, M.A. Elhoumaizi, M. Sindic, J. Sci. Res **3**, 139-149 (2010)
8. B. Ismail, I. Haffar, R. Baalbaki, Y. Mechref, J. Henry, Int. J. Food Sci **41**, 919-926 (2006)
9. A.D. Al-Qurashi, J. Food. Agric. Environ **8**, 404-408 (2010)
10. A. Aridah. *Effect of Type of Irrigation System on Productivity and Income of Date Palm Growers in Sebha, Libya*, thesis for the Degree of Doctor of Philosophy (2016)
11. I.A.R. Akasha. *Study of quality attributes of some libyan dates.* (Master thesis). Sebha University, Sebha, Libya (2005)
12. A. Al-Rawi, A.F. Al-Mohemdy. *Effect of water quality on the growth and yield of date palm,* in Proceedings of the Second International Conference on Date Palms, ICDP, 25-27 March 2001, Al-Ain, United Arab Emirates (2001)
13. F. Hussein, M.A. Hussein. *Effect of irrigation on growth, yield and quality of dry date palm fruits grown in Aswan area*, in Proceedings of the Second Symposium on Date Palm Trees, 27-30 May 2007, Alahsa, Saudi Arabia (1982)
14. A. Said, D. Kaoutrher, B. Ahmed, T. Mohammed, T. Brahim, Annu. Res. Rev. Biol **4**, 487-499 (2014)
15. M. Shaheen, Met., Env. & Arid Land Agric. Sci **18**, 23-39 (2007)
16. A.G. Tafti, M. Fooladi, Res. J. Biol. Sci **5**, 319-322 (2005)
17. I. Ullah, *Physical characteristics of fruit of different date palm cultivars during various maturity stages, in* Proceedings of the First International Scientific Conference for the Development of Date Palm and Dates Sector in the Arab World, Riyadh, Saudi Arabia (2011)
18. J.M. Basahi, Met., Env. & Arid Land Agric. Sci **17**, 26-42 (2006)
19. L. El-Juhany, Aust. J. Basic & Appl. Sci **4**, 3998-4010 (2010)
20. M. Shamsi, S. Mazloumzadeh, Agric. Sci. Technol **5**, 17-31 (2009)

Aqueous Enzymatic Extraction of safflower (*Carthamus tinctorius* L.) seed oil: A preliminary study

Chaymae Benkirane*[1], *Aymane* Allay[1], *Abdessamad* Ben Moumen[1], *Kamal* Belhaj[2], *Reda* Melhaoui[1], *Farid* Mansouri[1], *Malika* Abid[1], *Hana* Serghini-Caid[1], and *Ahmed* Elamrani[1]

[1] Laboratory of Agricultural Productions Improvement, Biotechnology and Environment (LAPABE), Faculty of Sciences, Mohammed I University, Oujda, Morocco.

[2] Laboratory of Sustainable Agriculture Management, Higher School of Technology, Sidi Bennour, University Chouaib Doukkali, El Jadida, Morocco.

Abstract. Reducing the pollution is a serious environmental challenge facing the oil producing industry worldwide. Conventionally, oil from oleaginous is extracted in large scale by mechanical expression or organic solvent extraction. In spite of its high yield, organic solvent extraction is time-consuming and very harmful to health and environment. Aqueous Enzymatic Extraction (AEE) is an eco-friendly extraction method using water as an extraction medium. Its application is a feasible alternative to traditional processing technologies without using organic solvents. However, its efficiency depends on many factors. This study aims to investigate the potential of AEE method to extract oil from safflower (*Carthamus tinctorius* L.) seeds using hemicellulase. Four extraction parameters with different levels that are incubation temperature, incubation time, initial pH, and enzyme concentration were evaluated. The optimum extraction conditions recorded were 65°C, pH 5, 4% w/w of enzyme and 3 hours of incubation. The obtained results showed that AEE reach a maximum of 70.45 % oil recovery. Therefore, Aqueous Enzymatic Extraction can be suggested as an environmentally cleaner alternative method to efficiently extract oil from safflower seeds.

1 Introduction

Safflower (*Carthamus tinctorius* L.), also called false saffron, is an herbaceous plant belonging to the Asteraceae family. It has been cultivated since antiquity for its ornamental and dyeing properties in several regions of the world notably in Indian and Egyptian civilizations [1]. Nowadays, it is mostly grown for its oilseeds containing 26 - 48% w/w of oil [2,3] depending on plant genotypes, environmental cultivation conditions and extraction methods.

Safflower seed oil has a yellowish color and a slightly nutty flavor [4]. It is considered as a versatile product since it is known for its culinary, medical, cosmetic, and technical

* Corresponding author: chaymae.1.benkirane@gmail.com

uses. Indeed, safflower oil is very rich in unsaturated fatty acids especially linoleic acid (approximately 80%) [5], which make it excellent for human consumption. It is also used in traditional medicine owing to its hypocholesterolemic, anti-diabetic, cardioprotective, antioxidant, and anti-inflammatory virtues. In addition, safflower oil fluidity and lightness properties make it ideal for cosmetic uses such as skin hydration and hair care. It also appears in the formulation of several dermatological and care body products. Moreover, there is other industrial and technical applications of safflower oil such as varnish and biodiesel production [1,4,6].

Several extraction methods are used conventionally for safflower oil extraction especially expeller pressing, solvent extraction, and prepress–solvent extraction [3]. Nowadays, the use of organic solvents (usually hexane) raises health, safety, and environmental concerns due to their toxicity and their flammability. Various researches showed that hexane is a dangerous air pollutant and have hazardous repercussions on the human neuronal and respiratory system when inhaled [7,8]. In this same context, studies have shown the existence of a high toxicity in animals fed with the defatted meal containing residual hexane [9,10]. These complications lead to seeking safe alternatives to preserve environment and health. Among the solutions found so far, the integration of "green" solvents and the use of "green" extraction techniques such as Aqueous Enzymatic Extraction (AEE).

Aqueous Enzymatic Extraction (AEE) is an environmentally friendly extraction technique wherein water is used as an extraction medium instead of organic solvents. This technique relies on the application of enzymes to break down cell walls and promote the release of targeted intracellular components. Various types of enzymes are used in AEE like cellulase, pectinase, hemicellulase, α-amylase, protease, and many others, depending on the structure of the seed [11,12]. Indeed, each enzyme is specific to a given substrate due to the complementarity between the shape of an enzyme's active site and the shape of its specific substrate [13]. AEE method has many advantages. It excludes or uses less organic solvents, provides oils that require less refining processes [10,11] and ensures encouraging yields without compromising the oil quality. Several studies reported that hexane-extracted and aqueous enzymatic-extracted oils were very similar in terms of physicochemical characteristics and fatty acid profile [14–16].

AEE has been successfully applied for oil extraction from several oilseed species, such as sunflower (Pakistan), almond (Iran), olive (Tunisia), and peanut (India), using either individual or combination of enzymes [14,17–19]. To the best of our knowledge, only one study has investigated the AEE of safflower oil using cellulase and protease [20].

The application of enzymatic extraction in different processes depends on a proper understanding of the factors influencing extraction. Several factors affect the performance of AEE such as pH, incubation time and temperature, agitation, type and concentration of enzyme, particle size of crushed seeds, and solid to liquid ratio [10]. Many authors have reviewed the main aspects relating to AEE [10–12].

The present study focused on evaluating the potential of AEE method, using hemicellulase enzyme, and determining extraction conditions that favor safflower oil recovery, which is a key step to optimize the overall process.

2 Material and methods

2.1 Plant material and reagents

The National Agricultural Research Institute (INRA, Regional Research Center, Meknes, Morocco) kindly provided safflower seeds of the *Cartamar* variety. The seeds were cleaned, finely ground in a laboratory grinder and stored at -20 °C in the dark until further use.

The Hemicellulase enzyme from *Aspergillus niger* (H2125) was purchased from Sigma Aldrich. Its standard activity is 1.5 units/mg solid using a β-galactose dehydrogenase system and locust bean gum as substrate. It is reported having also cellulase activity (0.03 units/mg solid) as per the supplier's specifications. Hexane was of analytical grade and obtained from Sigma Aldrich.

2.2 Solvent extraction of safflower seed oil

Oil was chemically extracted from 25 g of safflower seed grinding powder in a soxhlet apparatus using 300 ml of hexane at 60°C for 5 hours. Hexane was then eliminated under vacuum at 45 °C using a rotary evaporator. The oil yield was calculated.

2.3 Aqueous Enzymatic Extraction of safflower seed oil

The procedure for Aqueous Enzymatic Extraction (AEE) is summarized in Figure1. Finely comminuted seeds (3 g) were added to 15 ml of distilled water (ratio 1:5 w/v) in Erlenmeyer flasks. The pH was then adjusted with 0.1N HCl and 0.1N NaOH solutions. After adding the enzyme, the mixture was incubated in a water bath of adjustable temperature with constant shaking rate (120 rpm).

After incubation, the Erlenmeyer flasks were placed directly in ice to stop the enzymatic reaction. Centrifugation (at 3500 rpm for 20 min) was then carried out to separate the oily-aqueous phase from the solid phase containing the seeds and the insoluble proteins. Only the oily-aqueous phase was collected in another tube and 5 ml of hexane were added to facilitate the recovery of the oil contained in the emulsion. The hexane was then evaporated in a rotary evaporator at 45 °C and the oil recovery was calculated based on the yield obtained by the soxhlet extraction considered as control method.

$$\text{oil recovery } (\%) = \frac{\text{oil yield extracted by AEE}}{\text{oil yield extracted by soxhlet}} * 100 \quad (1)$$

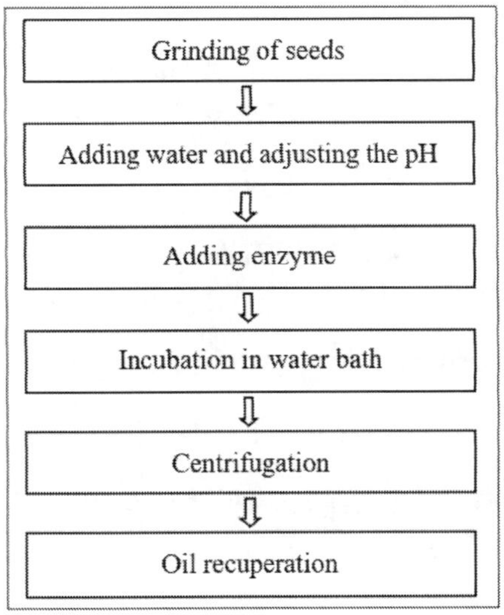

Fig.1. Schematic diagram of Aqueous Enzymatic Extraction of oil from safflower seeds

The extraction conditions were evaluated in several single factor experiments to know the operational range of the most important and effective variables. The factors studied are: Incubation temperature with five levels (40°C, 50°C, 60°C, 65°C and 70°C), initial pH with four levels (4.5, 5, 6 and 7), incubation time with five levels (1h, 2h, 3h, 6h and 8h) and enzyme concentration with eight levels (0 % w/w, 1% w/w, 2% w/w, 3% w/w, 4% w/w, 5% w/w, 7.5% w/w and 10% w/w).

3 Results and discussion

The conventional soxhlet method was used to determine the total amount of oil in safflower seeds. The oil extracted was 26.5±1.1 % which is in agreement with the oil content reported in the literature for the *Cartamar* variety from Morocco [5].

In Aqueous Enzymatic Extraction, the choice of the enzyme is crucial and must be depending on the structure of the seed and its predominant components [21]. In the current study, AEE was performed using hemicellulase which is a group of enzymes that catalyzes the breakdown of heteropolysaccharides. It includes glucanases, xylanases and mannanases which act on glucan, xylan, and mannan respectively.

After incubation, separating oil from other slurry components was the major problem encountered during the process. When centrifugation was carried out, four phases appear according to their density (Fig. 2). At the bottom, the centrifugation pellet consists of insoluble protein precipitate and heavy seed particles, above which there is an aqueous phase containing soluble compounds like soluble proteins and sugars. Then, we find the emulsion phase and a few droplets of free oil, sometimes forming a thin film on the surface. Indeed, in Aqueous Enzymatic Extraction, destabilizing the emulsion still remains a serious challenge because a considerable proportion of the total oil remains stuck in the emulsion [12,21]. There are some methods mentioned in the literature for destabilizing the emulsion like phase inversion, freezing and thawing, shearing, microfiltration, pH adjustment, and phospholipase/protease treatments [22]. But they not allow recovering all the amount of the emulsified oil. In our study, we chose to add 3 to5 ml of hexane to recover all the oil entrapped in the emulsion and determine the total amount of oil obtained by AEE method.

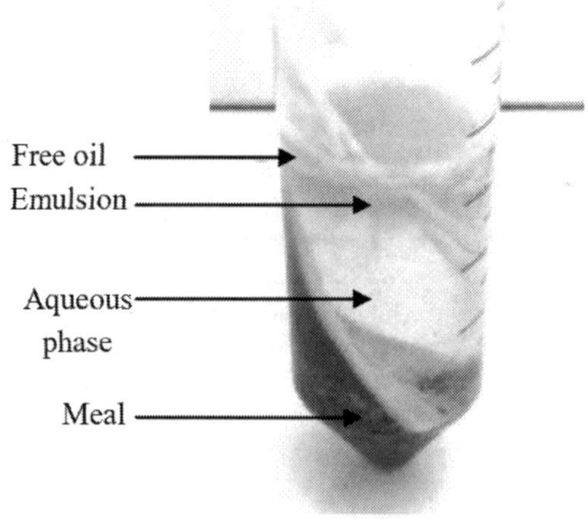

Fig.2. Different layers obtained following centrifugation after Aqueous Enzymatic Extraction of oil from safflower seeds

In the current study, the effect of different factors on oil recovery was evaluated. Each factor was independently varied, keeping other parameters fixed, to obtain the best conditions for extracting oil from safflower seeds.

3.1 Effect of temperature on safflower seed oil recovery

Temperature is a critical factor in the extraction process because it directly affects the enzyme activity. The figure 3(A) shows the variation of safflower oil recovery in response to different incubation temperatures. As the temperature increases from 40 to 65°C, so does the rate of oil recovery. The highest oil recovery is obtained at 65°C followed by a sharp decrease afterwards, due to the enzyme denaturation. Like all proteins, the thermal energy (high temperatures) disrupts the chemical bonds necessary to maintain the tertiary structure of the enzyme, which negatively affect the capacity of enzyme and substrate to fit together. On the other hand, at lower temperatures, enzyme and substrate have less kinetic energy and there are few interactions between them, which reduce or hinder the enzymatic activity. This characteristic pattern is reported in several works [19,20].

Optimum temperature differs from one study to other depending on the thermo-stability of the enzyme, the plant species, and the other extraction conditions. In this study, 65°C is the recorded optimum temperature value for the enzyme activity. At this temperature, the enzyme seems to be more stable and gives the highest oil recovery (70.45%). This result seems to be different from some studies which have shown that the optimum temperature for better enzyme activity is between 40 to 55 °C [19,20,23,24]. However, it is in agreement with several other authors who find that the stability of the enzyme is enhanced with respect to extreme temperatures [25,26], and this enhance oil extraction. Nevertheless, a high temperature might affect the oil quality, particularly, the oxidation state. Consequently, analyzes of the oxidation stability of these oils extracted at high temperature must imperatively be carried out.

3.2 Effect of pH on safflower seed oil recovery

Another important parameter that influences the extraction of seed oil by AEE is the pH of the incubation medium. Indeed, Rovaris et al. [27] during a comparable study on soybean, showed that the oil yield obtained by AEE was higher when controlling pH compared to tests where the pH was uncontrolled. Thus, the effect of four different pH values of the AEE incubation medium on the oil recovery was tested. The result presented in figure 3(B) shows that the optimum pH for hemicellulase is pH=5. Any incubation medium with a pH inferior or superior to this optimum value of 5, results in a remarkable reduction in oil recovery. In fact, each enzyme has a specific optimal value depending on its type (acidic, neutral or alkaline enzyme). Under more acidic or more alkaline conditions, the enzyme denatures and no longer allows the substrate to bind to its site of action. In general, hemicellulase is an acidic enzyme and almost all studies employing hemicellulase use a pH range from 4 to 5 [28].

Several authors emphasized the importance of extracting oil in a pH different from the isoelectric point of substrate's proteins. Otherwise, substrate-enzyme interaction could be compromised due to the insolubility of the proteins [11,21]. In this study, the highest oil recovery was obtained at pH 5 although it coincides with the isoelectric point of safflower proteins [29].

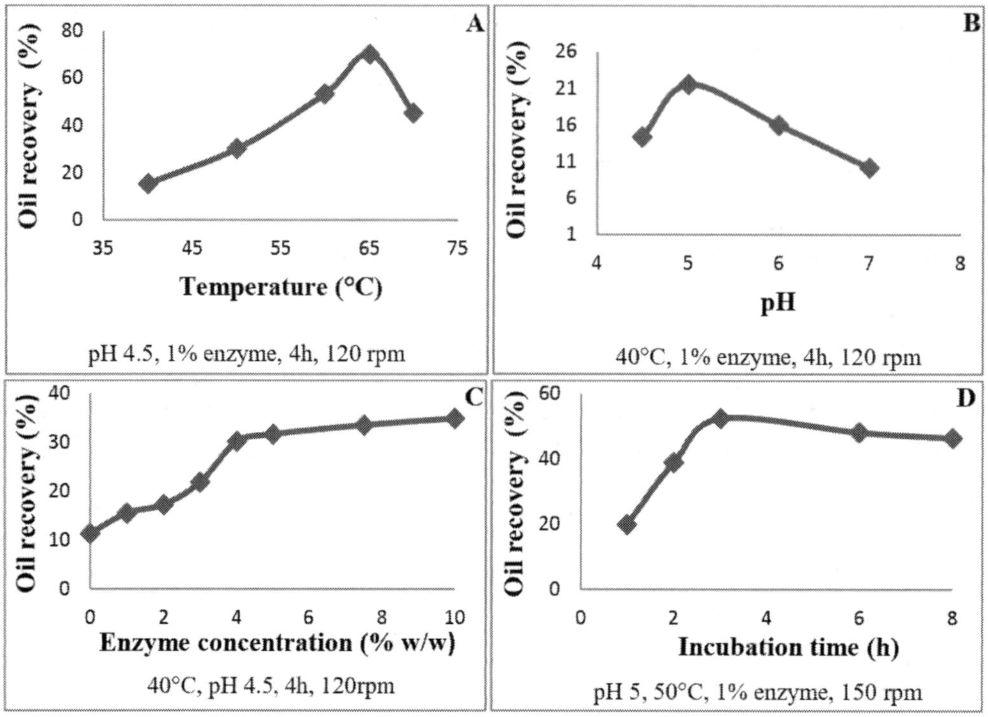

Fig.3. Effect of temperature (A), pH (B), enzyme concentration (C) and incubation time (D) on safflower seed oil recovery [3 g seeds, 1:5 solid: water ratio]

3.3 Effect of enzyme concentration on safflower seed oil recovery

The effect of enzyme concentration on oil recovery was investigated at eight different concentrations (Fig. 3(C)). The oil recovery initially increases with increasing enzyme concentrations, and then reaches a steady level with slight increase. An oil recovery of 11% is obtained for the control test (without addition of enzyme to the safflower seed oil extraction solution), while the addition of 10% enzyme (w / w) allows a clear improvement which corresponds to an oil recovery of 35%. This result is in agreement with several previously published results which concluded that the use of enzymes improves the oil extraction process [15,17,30]. According to Mansouri et al. [3], safflower seeds of *Cartamar* variety used in this study, have cells with pectocellulosic walls that contain 21.22% hemicellulose, 31.35% cellulose and 17.4 % lignin. Therefore, the use of hemicellulase enzyme (which has especially hemicellulase but also cellulase activity) allowed to destroy cell structural integrity and considerably facilitated the oil release comparably to the control.

The figure 3(C) shows that oil recovery hits a plateau at 4% w/w of hemicellulase, meaning that the gain in oil is no longer significant. That is, by exceeding 4% of enzyme concentration there is a very weak improvement, which does not economically justify the additional cost of the enzyme. Indeed, it has been reported that the high cost of enzymes is a serious challenge for the use of this extraction technique at the industrial level [10]. Hence, the interest in seeking a compromise between the enzyme quantity ensuring a high yield and its corresponding cost. Several studies have mentioned the importance of reaching the saturation phase of substrate active sites (by increasing enzyme concentration) in order to degrade the sample to the maximum and to promote oil recovery [20,31]. 4% is therefore suggested as the optimal hemicellulase concentration for safflower seed oil extraction.

Nevertheless, an abuse in increasing enzymatic concentrations, in some cases, may hinder the oil extraction process and trigger off flavors, bitterness and caramelization of sugars [10].

3.4 Effect of incubation time on safflower seed oil recovery

Over the tested range of incubation time (1-8h), the oil recovery increases significantly during the first three hours (from 19.85% to 39% respectively at 1 and 3 h), thereafter it decreases slightly (fig. 3(D)). In fact, the longer the incubation time increases, the greater the chances of promoting contact between the enzyme and the substrate. But, much longer extraction times do not necessarily result in higher yields, presumably due to the depletion of substrate. This pattern is supported by previously published studies on oil extraction from seeds of other safflower cultivars [20], from almond cake [32], *Pinus pumila* seeds [33] and rapeseeds [31].

The current study focused on the effect of four extraction factors mainly incubation temperature, pH, enzyme concentration, and incubation time. Nevertheless, other influential parameters relating to the process should not be neglected such as particle size of crushed seeds, agitation, and solid to liquid ratio. Generally, small particle size promote higher yield due to easier access of enzymes to their action sites [20]. The stirring rate helps mechanically disrupt the cell membranes and homogenize the components of the reaction mixture. Solid-liquid ratio is also a factor to consider. Low humidity can cause a thick suspension that hinders enzyme activity. However, high water amount leads to enzyme dilution and an oil yield decrease [10,11].

Overall, the optimum conditions for the aqueous enzymatic extraction of safflower oil using hemicellulase were 65°C, pH 5, 4% w/w of enzyme and 3 hours of incubation. It is worth mentioning that the combination of these optimal values gave a yield of 12% corresponding to 45% of oil recovery only. However, among the experimental tests carried out, a maximum oil recovery of 70.45% was obtained for a pH 4.5, 1% hemicellulase, 4h, and 65°C. This result implies that there is a strong interaction between the factors which is also in accordance with previous reports in the literature [14,30,34]. This finding clearly highlights the need for a statistical design of experiments to take into account the effect of the interaction.

In an earlier study, Gibbins [20] has also investigated the AEE of safflower oil from *Dincer* variety and found that the optimum conditions using the enzyme Celluclast 1.5L (a cellulase) were pH 4.84, 48.3°C, 0.25 mL enzyme/g substance, particles of <0.6 mm, giving a maximum oil recovery of 65% which is slightly lower than our result. The current study proves that AEE has a great potential in extracting oil from safflower seeds since high oil recovery has been achieved (70.45%) using a single enzyme. The combination of several enzymes like cellulase and protease could further significantly increase the oil yield [33]. Indeed, enzyme mixtures promote extraction because each one targets a well-determined site and cooperates together facilitating the oil release [13,14,33]. In general, combining at least one cellulase, hemicellulase, pectinase, and protease is more effective for maximum oil recovery. Cellulase, hemicellulase and pectinase act on cell walls by degrading them and facilitating the action of the protease which will degrade the proteins that bind to oil bodies.

According to the literature, the maximum oil recovery reached in this study was better than that obtained from sesame (57.4%) [16], soybean (29.48%) [27] and bayberry kernels (49.44%)[34]. Other studies have even been able to achieve an oil recovery greater than 90% in peanut [19], palm fruit [30], watermelon kernels [24] and corn germ [15] by combining several enzymes and optimizing the process.

4 Conclusion

Safflower oil was extracted by AEE method from *Cartamar* variety employing hemicellulase enzyme. The results obtained show that processing variables (duration and temperature of incubation, pH, enzyme concentration) have remarkable effects on the oil recovery. The applicability of the process was demonstrated since a maximum of 70.45% recovery was achieved. Aqueous Enzymatic Extraction (AEE) is therefore suggested as a promising green chemistry technique.

An optimization of safflower oil extraction by Response Surface Methodology (RSM) is envisaged. The present preliminary study, elucidating individual effects of extraction parameters, will therefore be of great interest in choosing the effective ranges for each independent variable in RSM.

References

1. S. Krist, in *Veg. Fats Oils,* 657–664 (2020)
2. B. Aydeniz, O. Güneşer, and E. Yılmaz, J. Am. Oil Chem. Soc. **91**, 99-101 (2014)
3. F. Mansouri, A. Ben Moumen, G. Richard, M. L. Fauconnier, M. Sindic, A. Elamrani, and H. Serghini Caid, OCL - Oilseeds Fats, Crop. Lipids **25**, 1-9 (2018)
4. J. Yeo, S. Fereidoon, and J. Smith, in *Bailey's Ind. Oil Fat Prod.*, edited by F. Shahidi, 7th ed. 533–576 (John Wiley & Sons, Ltd, 2020)
5. A. Ben Moumen, F. Mansouri, G. Richard, M. Abid, M. L. Fauconnier, M. Sindic, A. Elamrani, and H. Serghini Caid, Int. J. Food Sci. Technol. **50**, 804-810 (2015)
6. N. Khalid, R. S. Khan, M. I. Hussain, M. Farooq, A. Ahmad, and I. Ahmed, Trends Food Sci. Technol. **66**, 176-186 (2017)
7. A. Herskowitz, N. Ishii, and H. Schaumburg, N. Engl. J. Med. **285**, 82-85 (1971)
8. A. I. Rascu, Rom. J. Leg. Med. **23**, 279-284 (2015)
9. U.S. Environmental Protection Agency, *TOXICOLOGICAL REVIEW OF N-HEXANE: In Support of Summary Information on the Integrated Risk Information System (IRIS)* (Washington, DC, 2005)
10. S. P. J. Kumar, S. R. Prasad, R. Banerjee, D. K. Agarwal, K. S. Kulkarni, and K. V. Ramesh, Chem. Cent. J. **11**, 1-7 (2017)
11. P. W. Mwaurah, S. Kumar, N. Kumar, A. K. Attkan, A. Panghal, V. K. Singh, and M. K. Garg, Compr. Rev. Food Sci. Food Saf. **19**, 3-20 (2020)
12. J. J. Liu, M. A. A. Gasmalla, P. Li, and R. Yang, Innov. Food Sci. Emerg. Technol. **35**, 184-193 (2016)
13. G. Ricochon and L. Muniglia, OCL - Ol. Corps Gras Lipides **17**, 356-359 (2010)
14. M. Balvardi, K. Rezaei, J. A. Mendiola, and E. Ibáñez, JAOCS, J. Am. Oil Chem. Soc. **92**, 985-992 (2015)
15. R. A. Moreau, D. B. Johnston, M. J. Powell, and K. B. Hicks, J. Am. Oil Chem. Soc. **81**, 1071-1075 (2004)
16. S. Latif and F. Anwar, Food Chem. **125**, 679-684 (2011)
17. S. Latif and F. Anwar, JAOCS, J. Am. Oil Chem. Soc. **86**, 393-400 (2009)
18. N. Hadj-Taieb, N. Grati, M. Ayadi, I. Attia, H. Bensalem, and A. Gargouri, Biochem. Eng. J. **62**, 79-85 (2012)
19. A. Sharma, S. K. Khare, and M. N. Gupta, JAOCS, J. Am. Oil Chem. Soc. **79**, 215-218 (2002)
20. R. D. Gibbins, H. A. Aksoy, and G. Ustun, Int. J. Food Sci. Technol. **47**, 1055-1062 (2012)
21. M. Mat Yusoff, M. H. Gordon, O. Ezeh, and K. Niranjan, Food Chem. **211**, 400-408 (2016)

22. S. Jung, D. Maurer, and L. A. Johnson, Bioresour. Technol. **100**, 5340-5347 (2009)
23. J. Long, Y. Fu, Y. Zu, J. Li, W. Wang, C. Gu, and M. Luo, Bioresour. Technol. **102**, 9991-9996 (2011)
24. X. Sui, L. Jiang, Y. Li, and S. Liu, Procedia Eng. **15**, 4673-4680 (2011)
25. L. Wang, F. Liu, T. Li, D. Liu, Y. Xu, and Y. Yang, Appl. Sci. **8**, 1-14 (2018)
26. P. Maijala, N. Kango, N. Szijarto, and L. Viikari, Antonie Van Leeuwenhoek **101**, 905-917 (2012)
27. Â. A. Rovaris, C. O. Dias, I. P. da Cunha, R. M. C. Scaff, A. de Francisco, C. L. O. Petkowicz, and E. R. Amante, Ind. Crops Prod. **36**, 405-414 (2012)
28. K. Boutemak, N. Benali, and N. Moulai-Mostefa, Int. J. Chem. Mol. Eng. **9**, 1508-1511 (2015)
29. J. A. Ulloa, P. Rosas-Ulloa, and B. E. Ulloa-Rangel, J. Sci. Food Agric. **91**, 572-577 (2011)
30. C. B. Teixeira, G. A. Macedo, J. A. Macedo, L. Helena, M. Silva, A. Manoel, and C. Rodrigues, Bioresour. Technol. **129**, 575-581 (2013)
31. B. S. Zhang, Z. Wang, and S. Y. Xu, J. Am. Oil Chem. Soc. **84**, 97-105 (2007)
32. T. S. P. Souza, F. F. G. Dias, M. G. B. Koblitz, and J. M. L. N. Juliana, Processes **7**, 1-19 (2019)
33. F. Chen, Q. Zhang, H. Gu, and L. Yang, J. Chromatogr. A **1471**, 68-79 (2016)
34. Y. L. Zhang, S. Li, C. P. Yin, D. H. Jiang, F. F. Yan, and T. Xu, Food Chem. **135**, 304-308 (2012)

Removal of a non-steroidal anti-inflammatory drug (Piroxicam) in an aqueous medium by an agricultural by-product

Nora Samghouli[1,*], *Boutaina* Regraguy[1], *Fatima-Zahra* Abahdou[1], *Karima* Azoulay[1], *Imane* Bencheikh[1], *Jamal* Mabrouki[1] and *Souad* El Hajjaji[1,*]

[1]Laboratory of Spectroscopy, Molecular, Modeling, Materials, Nanomaterials, Water and Environment, (LS3MNWE), Department of Chemistry, Faculty of Sciences, Mohammed V University in Rabat, Av Ibn Battouta, B.P. 1014, Rabat 10000, Morocco

Abstract. The presence of pharmaceuticals in the environment has increased. These pollutants are toxic and non-biodegradable. Piroxicam (PRX) is a non-steroidal anti-inflammatory drug that ends up in wastewater via pharmaceutical industry activities and human being consumption. This work aimed to study the pharmaceutical pollutant removal from wastewater using agricultural by-products as low-cost adsorbent material. Different parameters were studied, such as time, initial adsorbate concentrations, and temperature. The study of the initial concentration-effect shows that the greatest amount of adsorbed is observed in low concentrations. The temperature has shown a negative effect in this study. The kinetics show that after 45min, the equilibrium is obtained; that means the exhaustion of all active sites. The representative model of this adsorption is the Langmuir isotherm according to the regression coefficient, which is equal to 0.99. The natural abundance of this material and the low cost of investment could offer a good alternative to other more expensive adsorbents such as activated carbon.

Keywords: Adsorption; Wastewater; agricultural by-product; Low-cost material; Piroxicam

1. Introduction

In recent years, Modernization and population growth have contributed to the contamination of water sources by new chemical products, which have led to a large number of environmental problems [1, 2]. Contamination of water by Pharmaceuticals is an increasing problem across the world [3–5]. The pollution of water by pharmaceuticals poses a serious threat to human health and the environment [6, 7], affecting a wide range of aquatic species in a variety of ways, including erratic behavior, reproduce and development abnormal, and a lack of metamorphosis [8]. The emergence of pharmaceutical substances and their metabolites in the environment occurs through the excretion of these pharmaceuticals through urine [9–11]. Therefore, the existence of a suitable technique for the removal of these pollutants from wastewater is a challenge. One of these pharmaceuticals is piroxicam, which is used as a non-steroidal anti-inflammatory drug to reduce symptoms of fever, rheumatic disease, and pain [12]. This medication is consumed in large quantities each year and is getting it in wastewater in the amounts of ng/L or μg/L, which indicate a high level of environmental toxicity according to the literature [13].

Several treatment techniques are used for to remove organic contaminants from wastewater, such as biological oxidation, coagulation, flocculation, and sedimentation [14], advanced oxidation processes (AOP), photo- Fenton [15], electrochemical oxidation / degradation [16, 17], membrane processes [18], and adsorption [19]. Despite the availability of all these techniques for the removal of organic pollutants, the adsorption process remains the best because of its high efficiency and effectiveness, low cost, easy to operate [20]. Moreover, adsorption can also remove pollutants without any generation of hazardous by-products [16, 21]. Adsorption is usually done with activated carbon, which has a big specific surface area but also comes at a high price. Other materials have been shown in several studies that are alternatives, more effective, and low-cost, such as orange peel [22], Rice Husk [23], Banana peel [24], and date palm leaves [25].

The present study aims to carry out an experimental investigation to remove piroxicam from an aqueous solution using treated artichoke waste as an adsorbent alternative. Kinetic and isotherm studies were performed to determine the equilibrium time and the adsorption type.

*Corresponding authors: nora_samghouli@um5.ac.ma or souad.elhajjaji@um5.ac.ma

2. Materiels and methods

2.1 Materiels

2.1.1 Adsorbent

In a flask containing NaOH (2 %) solution, the cleaned, artichoke waste was immersed and stirred for 2h at 50° C. The collected sample was filtered, rinsed with distilled water. The samples were then treated for 1 hour at 70°C with sodium hypochlorite (NaClO) (3.7 % w/v) in a ratio of (0.5: 3 v/v) NaClO / H_2O. The recovered sample was filtered and washed repetitively until it reached neutral pH, then it was dried in the open-air [26].

2.1.2 Adsorbate

Piroxicam is a non-steroidal anti-inflammatory, analgesic, and antipyretic drug having a long biological half-life [27]. The piroxicam was used as a model contaminant without any previous purification.
100 milligrams of piroxicam were dissolved in 100 mL of distilled water to make a stock solution. Then, four dilute solutions were made from the stock solution: 10 mg/L, 20 mg/L, 30 mg/L, and 40 mg/L.

2.2 Methods

2.2.1 Analysis method

The study of piroxicam adsorption onto treated artichoke waste (TAW) was carried out in a static system. A mass of TAW was introduced into a volume of 50 mL of piroxicam, the mixture was stirred (500 rpm/min) at room temperature. After liquid-solid (Piroxicam-TAW) separation by centrifugation (3500 rpm for 20 min). The residual concentration of the filtrates was analysed by UV-1600 spectrophotometer at 334nm [27].

The evolution in adsorption efficiency was carried out either by determining the adsorption yield of piroxicam (R(%)); or by calculating the adsorbed amount (Q (mg/g)). The following equations were used to calculate the adsorption yield and the amount of adsorption [28–30]:

$$R(\%) = \frac{Ci-Cf}{Ci} \times 100 \quad (1)$$

$$Qm = \frac{Ci-Cf}{m} \times V \quad (2)$$

Where: C_i is the initial and C_f is the final piroxicam concentrations (mg/L), m is the mass of treated artichoke waste (TAW) (g), and V is the volume (L).

2.2.2 Piroxicam adsorption onto TAW

2.2.2.1 Kinetic runs

The adsorption kinetics of piroxicam onto treated artichoke waste was studied by adding 1 g of TAW in 1 L of piroxicam solution. These suspensions were mixed continuously for 75 minutes at room temperature. At programmed time intervals, aliquots of the supernatant (3.6 ml) were taken, centrifuged, and analyzed by UV-VIS spectrophotometry.

2.2.2.2 Determination of adsorption isotherms

To study the piroxicam's concentration effect on the adsorption via treated artichoke waste, a piroxicam stock solution of 40 mg/L was prepared. To accomplish piroxicam adsorption isotherms, all the parameters were maintained constant except the concentration which was varied between 5–40 mg/L. The suspensions were shaken for 45 min (time sufficient to reach the equilibrium), then centrifuged (20 min at 3500 rpm) and analyzed with UV-Visible spectrophotometry.

3. Results and discussion

3.1 Adsorption of piroxicam onto TAW

3.1.1 Effect of contact time on piroxicam removal

Optimization of the adsorption time is one of the most important actions to get a high removal efficiency under specific operating conditions. This is also interesting from an economic standpoint because it suggests maximizing drug adsorption while minimizing time.
To study the adsorption kinetics of piroxicam onto TAW, the experiment was conducted under these operating conditions [31] (Adsorbent dose=2g/L, pH=3, T=20°C, and [Piroxicam]=10mg/L). Samples were taken at regular time intervals to determine the amount adsorbed quantity of piroxicam. The results obtained are shown in Figure 1.

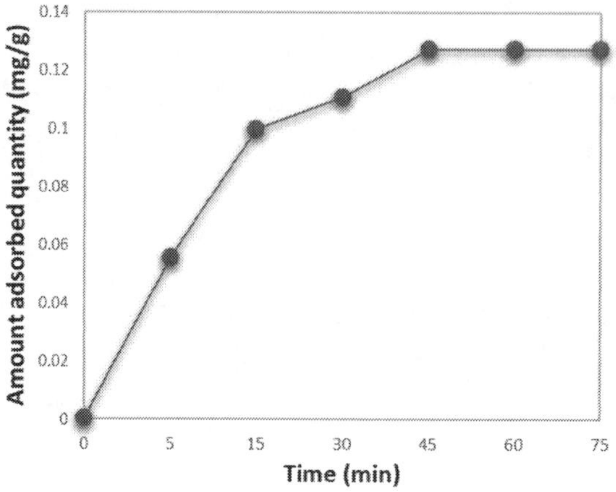

Fig. 1. Contact time effect on the piroxicam adsorption onto TAW.

The contact time effect on the piroxicam adsorption shows that during the first 15 minutes, the removal of piroxicam via adsorption onto TAW increases rapidly. It continues to rise at a slower rate than in the first minutes until it reaches equilibrium at 45 minutes. The high availability of active sites on the surface of TAW may explain the rapid increase in piroxicam removal in the first phase of treatment.
These phenomena can be explained by exhausting all active sites which exist on the surface of the studied material (TAW) [32, 33]. We may also deduce from the curve that the amount adsorbed quantity of TAW is 0.13 mg/g in 45 minutes. The TAW adsorbent is so effective to remove organic medicines from wastewater.

3.1.2 Adsorption kinetic models

The plots of the pseudo-first and pseudo-second-order kinetic models of piroxicam adsorption on TAW are shown in Figure 2.
The mathematical expressions of each model and the values of the regression coefficients, as well as the parameters of these kinetic models, are grouped in Table 1.

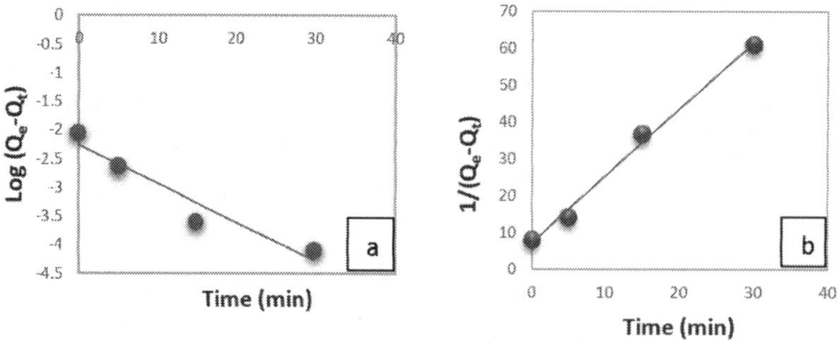

Fig. 2. Kinetic models of piroxicam adsorption onto TAW: pseudo-first-order (a); pseudo-second-order (b)

Table 1. Kinetic model formulas and parameters of the piroxicam adsorption onto TAW

Kinetic model formulas	**Parameters**	**Piroxicam/TAW**
Pseudo-first-order [34]	R^2	0.93
$\log(q_e - q_t) = \log(q_e) - K_1 \frac{K}{2.303} t$ (3)	k_1 (min^{-1})	0.16
	q_e (cal) (mg/g)	0.10

	q_e (exp) (mg/g)	0.13
Pseudo-second-order [35] $\frac{1}{Qe-Qt}=\frac{1}{Qe}+K_2t$ (4)	R^2	0.99
	k_1 (g/mg.min)	1.81
	$q_{e\ (cal)}$ (mg/g)	0.14
	$q_{e\ (exp)}$ (mg/g)	0.13

Where: q_e and q_t the adsorbed amount of piroxicam at a time (t) and at equilibrium time (mg/g) respectively; t: contact time (min); K_1, K_2: rate constants for pseudo 1st (1/min) and pseudo 2nd ($g.mg^{-1}.min^{-1}$) kinetic models respectively.

As shown in table 1, it can be seen that, in the case of first-order kinetics, the experimentally determined equilibrium adsorbed quantity is different from the calculated one. On the other hand, the experimentally determined equilibrium adsorbed quantity is adjacent to the one calculated using the second-order kinetics model. This model applies well in the case of the adsorbent/adsorbate systems studied, given the values obtained for the coefficients of determination R^2 which are very near to 1. This model implies chemisorption, which includes electron exchanges at the solid-liquid interface [36, 37].

3.1.3 Isotherm models

3.1.3.1 Initial concentration effect of piroxicam

The initial pollutant concentration can alter the efficiency of an adsorbent to remove pollutants, which is dependent on the combination of pollutant concentration and available binding sites on the adsorbent's surface [38, 39].

To study the effect of the initial concentration of piroxicam in an aqueous medium, different concentrations of piroxicam were changed with fixation of the remaining factors (adsorbent dose = 2 g/L, pH = 3, T = 20 °C), then, the solutions were mixed and stirred. These solutions were shaken until they approached equilibrium (time = 45 min). The effect of the piroxicam concentration onto TAW is shown in Fig. 3.

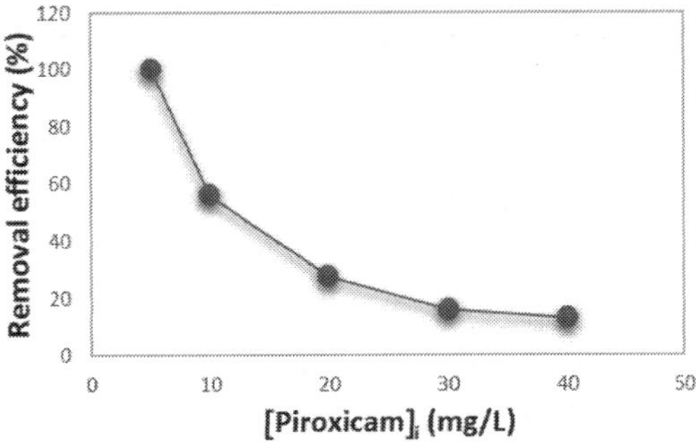

Fig. 3. Initial concentration effect on the adsorption of piroxicam onto TAW

As seen in Fig. 3, the initial concentration of piroxicam affects negatively the removal efficiency of TAW, so that the adsorption rate decreases with the increasing concentration of piroxicam. The adsorption rate of the TAW is decreased rapidly by 44.05% by increasing the concentration from 5 to 10 mg/L; meanwhile, the adsorption rate of the TAW adsorbent was decreased by 28.70% when the piroxicam concentration was increased from 10 to 20 mg/L, and then decreased slightly by 14.58% from 20 to 40 mg/L.

3.1.3.2 Adsorption isotherm models

Adsorption isotherms are often used to determine maximum pollutant adsorption capacities and to identify the type of adsorption. The results, processed according to the mathematical models of Langmuir and Freundlich, allowed us to determine the maximum adsorption capacity and the adsorption parameters [40].

The curves of the piroxicam adsorption isotherm models onto TAW are shown in Fig. 4.a and b respectively for the Langmuir and Freundlich models and the corresponding formulas and parameters are grouped in Table 2.

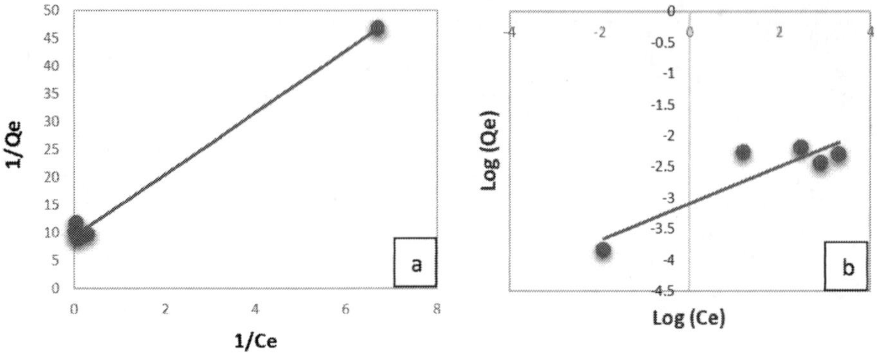

Fig. 4. (a) Langmuir and (b) Freundlich adsorption isotherms of piroxicam onto TAW

Table 2. isotherm model formulas and parameters of the piroxicam adsorption onto TAW

Adsorption isotherm model formulas	Parameters	Piroxicam/TAW
Langmuir [41] $\frac{1}{Qe} = \frac{1}{Qm} + \frac{1}{Qm.KL.Ce}$ (5)	R^2	0.99
	Q_m (mg/g)	20.33
	K_L	0.41
	R_L	0.11
Freundlich [42] $LogQe = LogKF + \frac{1}{nf} Log\ Ce$ (6)	R^2	0.98
	K_F	76.01
	1/n	0.41

Where: q_e and q_m are the adsorption quantity of piroxicam at equilibrium time and maximum adsorption quantity (mg/g) respectively; K_L and K_F are Langmuir constant ($L.mg^{-1}$) and Freundlich constant ($mg^{(1-n)}L^n.g^{-1}$) respectively; R_L: Separation parameter of equilibrium $RL = \frac{1}{1+b\times C0}$ [43]; $1/n_f$: Factor of heterogeneity;

The values of the regression coefficients indicate that the adsorption process of piroxicam by TAW is described favourably by the Langmuir isotherm (linear regression coefficient R^2 which is very close to unity). This finding suggests that adsorption occurs in a monolayer on homogeneous surfaces [44, 45].

3.1.4 Initial temperature effect

The temperature was altered between 20 and 60°C in this study to assess the temperature effect on the adsorption of piroxicam onto TAW.

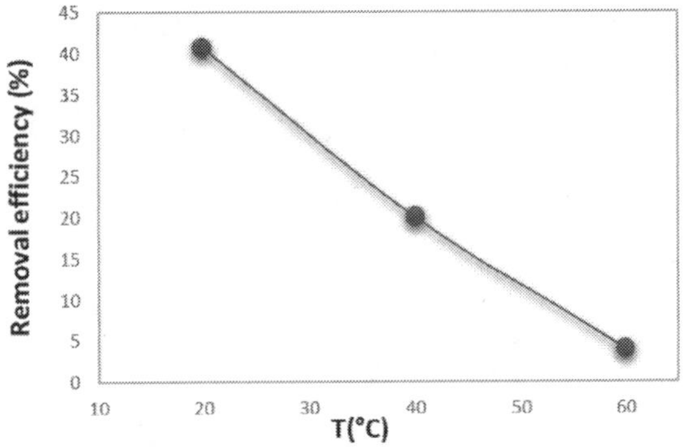

Fig. 5. Temperature effect on the adsorption of piroxicam onto TAW

The temperature effect on piroxicam adsorption onto the TAW (Fig. 5) reveals that the adsorption decreases rapidly (20.66%) between 20°C and 40°C and then decreases by 16.11% when the temperature rises to 60°C. Therefore, the temperature has a negative effect on the adsorption process in this study.

4. Conclusion

This work was intended as a contribution to the elimination of pharmaceutical products present in water. This study carried out on synthetic water gave results that prove the feasibility of adsorption of this residue (piroxicam) as well as the operating conditions of adsorption of this pharmaceutical product in water. It would be interesting to study the behavior of other pharmaceutical molecules and especially to apply it in real waters (waste from drug manufacturing plants, drinking water, etc.) to search for and eliminate residues of pharmaceutical products that can harm human beings.
In this study, different parameters were studied, the adsorption kinetics, the study of isotherms, the initial concentration effect of the adsorbent, and the effect of temperature. The results obtained allow the following conclusions:

- The study of the initial concentration-effect shows that the greatest quantity adsorbed is observed for low concentrations.
- The kinetics show that after 45min, equilibrium is reached; this means the exhaustion of all active sites that exist at the TAW interface.
- The representative model is the Langmuir isotherm according to the regression coefficient which is equal to 0.99.

The natural abundance of this material and the low capital cost could offer a good alternative to other more expensive adsorbents.

References

1. S. M. Siddeeg, M. A. Tahoon, and F. Ben Rebah, Mater. Res. Express **7**, 012001 (2019)
2. M. Kumar, B. P. Tripathi, and V. K. Shahi, J. Hazard. Mater. **172**, 1041 (2009)
3. E. Guinea, C. Arias, P. L. Cabot, J. A. Garrido, R. M. Rodríguez, F. Centellas, and E. Brillas, Water Res. **42**, 499 (2008)
4. E. Brillas, I. Sirés, and M. A. Oturan, Chem. Rev. **109**, 6570 (2009)
5. I. Sirés, E. Brillas, M. A. Oturan, M. A. Rodrigo, and M. Panizza, Environ. Sci. Pollut. Res. **21**, 8336 (2014)
6. B. Kasprzyk-Hordern, R. M. Dinsdale, and A. J. Guwy, Water Res. **43**, 363 (2009)
7. T. Mackuľak, L. Birošová, I. Bodík, R. Grabic, A. Takáčová, M. Smolinská, A. Hanusová, J. Híveš, and M. Gál, Sci. Total Environ. **539**, 420 (2016)
8. C. I. Kosma, D. A. Lambropoulou, and T. A. Albanis, Sci. Total Environ. **466–467**, 421 (2014)
9. A. Bavili Tabrizi and N. Seyyedeh Tutunchi, Adv. Pharm. Bull. **3**, 37 (2013)
10. S. Papoutsakis, Z. Afshari, S. Malato, and C. Pulgarin, J. Environ. Chem. Eng. **3**, 2002 (2015)
11. R. Zhang, P. Sun, T. H. Boyer, L. Zhao, and C.-H. Huang, Environ. Sci. Technol. **49**, 3056 (2015)
12. (n.d.)
13. S. M. Siddeeg, A. Amari, M. A. Tahoon, N. S. Alsaiari, and F. B. Rebah, Alex. Eng. J. **59**, 905 (2020)
14. C. Duran, D. Ozdes, A. Gundogdu, and H. B. Senturk, J. Chem. Eng. Data **56**, 2136 (2011)
15. J. M. Salman, V. O. Njoku, and B. H. Hameed, Chem. Eng. J. **174**, 41 (2011)
16. Md. J. K. Ahmed, M. Ahmaruzzaman, and R. A. Reza, J. Colloid Interface Sci. **428**, 222 (2014)
17. M. C. Somasekhara Reddy, L. Sivaramakrishna, and A. Varada Reddy, J. Hazard. Mater. **203–204**, 118 (2012)
18. A. Altınışık, E. Gür, and Y. Seki, J. Hazard. Mater. **179**, 658 (2010)

19. D. Smiljanić, B. de Gennaro, A. Daković, B. Galzerano, C. Germinario, F. Izzo, G. E. Rottinghaus, and A. Langella, J. Environ. Manage. **286**, 112168 (2021)

20. Y. Dai, Q. Sun, W. Wang, L. Lu, M. Liu, J. Li, S. Yang, Y. Sun, K. Zhang, J. Xu, W. Zheng, Z. Hu, Y. Yang, Y. Gao, Y. Chen, X. Zhang, F. Gao, and Y. Zhang, Chemosphere **211**, (2018)

21. I. Ali, Mohd. Asim, and T. A. Khan, J. Environ. Manage. **113**, 170 (2012)

22. I. C. Afolabi, S. I. Popoola, and O. S. Bello, Chemom. Intell. Lab. Syst. **203**, 104053 (2020)

23. T. D. Pham, T. T. Bui, T. T. Trang Truong, T. H. Hoang, T. S. Le, V. D. Duong, A. Yamaguchi, M. Kobayashi, and Y. Adachi, J. Mol. Liq. **298**, 111981 (2020)

24. M. Patel, R. Kumar, C. U. Pittman, and D. Mohan, Environ. Res. 111218 (2021)

25. F. Saremi, M. R. Miroliaei, M. Shahabi Nejad, and H. Sheibani, J. Mol. Liq. **318**, 114126 (2020)

26. I. Bencheikh, K. Azoulay, J. Mabrouki, S. El Hajjaji, A. Moufti, and N. Labjar, Chem. Data Collect. **31**, 100597 (2021)

27. A. Bachhav, S. Ahire, and A. Jadhav, Int. J. Pharm. Sci. Res. **10**, 811 (2019)

28. I. Bencheikh, K. Azoulay, J. Mabrouki, S. El Hajjaji, A. Moufti, and N. Labjar, Chem. Data Collect. **31**, 100597 (2021)

29. K. Azoulay, I. Bencheikh, J. Mabrouki, N. Samghouli, A. Moufti, A. Dahchour, and S. E. Hajjaji, Int. J. Environ. Anal. Chem. **0**, 1 (2021)

30. Y. Tan, Z. Cheng, Y. Liu, X. Gao, S. Liu, and Z. Shen, Chemosphere **280**, 130577 (2021)

31. N. Samghouli, I. Bencheikh, K. Azoulay, F-Z. Abahdou, J. Mabrouki, and S. El Hajjaji, EAI/Springer Innov. Commun. Comput., XIV,166 (2022)

32. S. Dawood, T. K. Sen, and C. Phan, Water. Air. Soil Pollut. **225**, 1818 (2013)

33. I. Bencheikh, K. Azoulay, J. Mabrouki, S. El Hajjaji, A. Dahchour, A. Moufti, and D. Dhiba, Sci. Afr. **9**, e00509 (2020)

34. S. Wang and H. Li, Dyes Pigments **72**, 308 (2007)

35. Y. S. Ho, D. a. J. Wase, and C. F. Forster, Environ. Technol. **17**, 71 (1996)

36. D. Mohan, K. P. Singh, and V. K. Singh, J. Hazard. Mater. **135**, 280 (2006)

37. Suhas, P. J. M. Carrott, and M. M. L. Ribeiro Carrott, Bioresour. Technol. **98**, 2301 (2007)

38. S. Sinha Ray, R. Gusain, and N. Kumar, in (2020), pp. 119–135

39. E. Malkoc and Y. Nuhoglu, J. Hazard. Mater. **127**, 120 (2005)

40. Z. Meçabih, S. Kacimi, and B. Bouchikhi, Rev. Sci. Eau J. Water Sci. **19**, 23 (2006)

41. J.-Z. Guo, B. Li, L. Liu, and K. Lv, Chemosphere **111**, 225 (2014)

42. A. Mittal, J. Mittal, A. Malviya, and V. K. Gupta, J. Colloid Interface Sci. **344**, 497 (2010)

43. N. Mohammadi, H. Khani, V. K. Gupta, E. Amereh, and S. Agarwal, J. Colloid Interface Sci. **362**, 457 (2011)

44. L. El Fakir, M. Flayou, A. Dahchour, S. Sebbahi, F. Kifani-Sahban, and S. El Hajjaji, Desalination Water Treat. 1 (2015)

45. V. K. Gupta and T. A. Saleh, Environ. Sci. Pollut. Res. Int. **20**, 2828 (2013)

E3S Web of Conferences **337**, 05002 (2022) https://doi.org/10.1051/e3sconf/202233705002
I2CNP 2021

Recycling of ashes from thermal stations for the production of an ecological mortar: New application in civil engineering

Redouane Mghaiouini [1,2]*, *Abdelhadi* Mortadi [2], *Dennoun* Saifaoui [3], *Mohammed* Salah [4], *Abderrazzak* Graich [2], *El ghaouti* Chahid [2], *Garmim* Taoufik [2], *Abderrahmane* Elmlouky [2], *Brahim* Mrani [5], *Hassan* Ghanim [5], *Mohamed* Monkade [2] *and Abdeslam* El Bouari[1]

[1] Department of Chemistry, Physical Chemistry Laboratory Applied Materials, Faculty of Sciences-Ben M'sik, Hassan II University, Casablanca Morocco.
[2] Department of Physics, Condensed Matter Laboratory, Faculty of Sciences, Chouaib Doukkali University, El Jadida Morocco.
[3] FSAC – UH2C, Laboratory for Renewable Energy and Dynamic Systems, Casablanca, Morocco
[4] Department of Chemistry, Molecular Modeling and Spectroscopy Research Team, Faculty of Sciences, Chouaïb Doukkali University, El Jadida, Morocco.
[5] Laboratory of Nuclear, Atomic, Molecular, Mechanical and Energetic, Faculty of Sciences, Chouaïb Doukkali University, El Jadida, Morocco.

Abstract. The main idea of this work is to investigate about the flexural and compressive forces of the mortar, which was developed from fly ash, hearth ash, and magnetized water with different times of magnetization. The results of this work reveal that the flexural and compressive force of mortars realized with magnetized water is higher than those prepared with non magnetized water. The best result of increasing the flexural and the compressive forces of the mortar are obtained when the water magnetization time is equal to 20 min. The growth in flexural and the compressive forces of the mortar elaborated with magnetized water is proportional to the magnetization time.

*Corresponding author: redouanemghaiouini@gmail.com

1 Introduction

The cement manufacturing industry is responsible for around 5% of global carbon dioxide (CO_2) emissions [1]. This alarming number has called for a deep reflexion to replace cement with other materials. Fly ash is a by-product made in thermal units. The country's dependence on coal for power generation is increasing and therefore the production of fly ash will also be greater. This widely available unused material can be used as a partial replacement for cement, as it has important cementations properties. According to the literature [2], the magnetic water ameliorate the strength of concrete by almost 50%.Thus, the use of magnetized water can reduce the use of cement in concrete. The change of characteristics of water by the action of magnetic source was cited in many papers within the literature [3-11]. In fact, water under a magnetic source undergoes many modifications in its physical properties (optics, electromagnetism, thermodynamics, and mechanics properties) compared with natural water. Xiao-Feng Pang et al. [12] have investigated the influences of the magnetic flux on physical properties of water (soaking effect, electric conductivity, thermal properties, hydrophobicity of water to materials and refraction index), as reported by them [12], these changes are caused by the modifications of micro-structures of water under the action of magnetic source. Nan Su et al. [13] have studied the effect of magnetized water on mortar and concrete containing fly ash, they found that magnetized water can improve the fluidity of fresh mortar and, at its solidification, and compressive forces. Huinan Wei et al. [14] have also worked on the influence of magnetized water in construction materials, they concluded that the utilization of magnetic water can improve the cracking resistance of materials. The amount of researches published on the influence of magnetized water on mortars remains insufficient, especially for the mortar laborite with the fly ash (pozzolanic material) or hearth ash. The recapture of those ashes from thermal installations will decrease the rejection of waste in nature and can improve the mechanical characteristics of the mortars [15-17]. Thus, the aim of this paper is to encourage the recuperation of the ashes and increase the standard of the mortar with a completely unique technology based on magnetized water. Thus, this work brings a more explanation of the mechanism of action of magnetized water on the properties of mortar. The results of this paper are going to be very helpful for engineers in building and construction to raise the use of magnetic water within the development of the new generation of mortars with good quality and cheap price.

2 Material and Preparation

2.1 Preparation of the mortar

Mortar is a blending of sand, hydraulic binder (cement or lime), and water. Herein, we focus in the valorization of the fly ash of thermal installations to formulate the mortar, which help to reduce the pollution fair. The fly ash (FA) is added as a replacement of the cement with different proportions of FA. The mass ratio, FA/cement adopted within the present work (Table 1) is adequate to 10% with reference (NFEN196-1). The mortars were arranged as a pieces of dimensions 4×4×16 cm (figure 1). The pieces were lifted in the open air, then they were tested after 28 days (mechanical resistance to bending and compression).

Fig. 1 Fly ash samples

Table 1 contains the different composition of the mortars. Mortar B0 is the reference. Mortar B10 was obtained by replacing part of the cement with fly ash (10% in masse).

Table 1 Mortar formulation B0 and B10.

Elements	Fly ash(g)	Cement(g)	Sand(g)	water(g)
B0	0	450	1350	225
B10	45	405	1350	225

2.2 Cement

The cement employed herein is CEM II (**CPJ45**: Moroccan standard NM 10.1.004). It's a hydraulic cement of 81.5% clinker, 12% limestone, the remainder is gypsum which regulates the setting. The chemical composition of CPJ 45 cement is presented in Table 2.

Table 2 Chemical composition of CPJ45 cement

Cement	$\%SiO_2$	$\%Al_2O_3$	$\%Fe_2O_3$	%CaO	%MgO	$\%SO_3$	$\%K_2O$
CPJ 45	17	5	3	63	2.3	3.33	1.2

2.3 Sand

The sand utilized comes from the coast (Oualidia Morocco). Its particles analysis was comply with standard NF P18-560 [16] with an electric sieve; the results obtained are summarized within the graph in Figure 2.

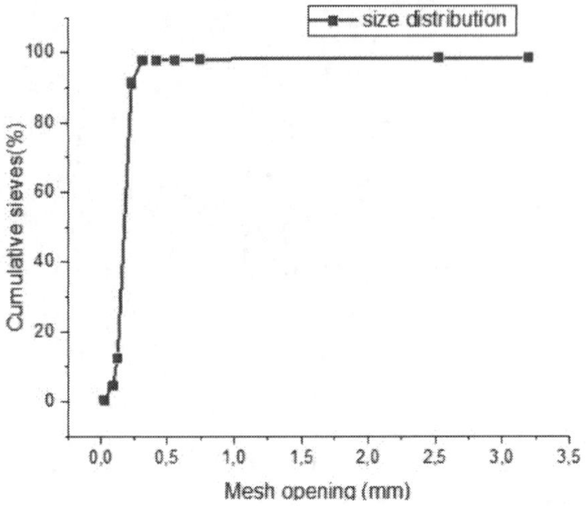

Fig. 2 Particles distribution of the sand

The grains curve indicates that the predominant grains is between 0.200 mm and 0.315 mm. Thus, the sand used is uniform and conforms to the fine grains. The apparent density of the sand used decided by considering a proportion of pores (41.5%) consistent with the reference [18]. The bulk density information's are presented in Table 3.

Table 3 Bulk density information's

Bulk density without compaction	Bulk density with compaction	Actual density
1,43g/cm^3	1,54g/cm^3	2,77g/cm^3

The actual density obtained, is equal to 2.77 g/cm^3, which comply with the specific weight of fine sand (2.65 g/cm^3).

2.4 Fly ash and hearth ash

2.4.1 Fly ash

Coal fly ash is a very fine powder obtained from the combustion of pulverized coal within the boilers of thermal power plants (Jorf Lasfar "JLEC"). The fly ash posses a gray color and have a smooth and glossy surface [19], like cement powder (Figure 3).

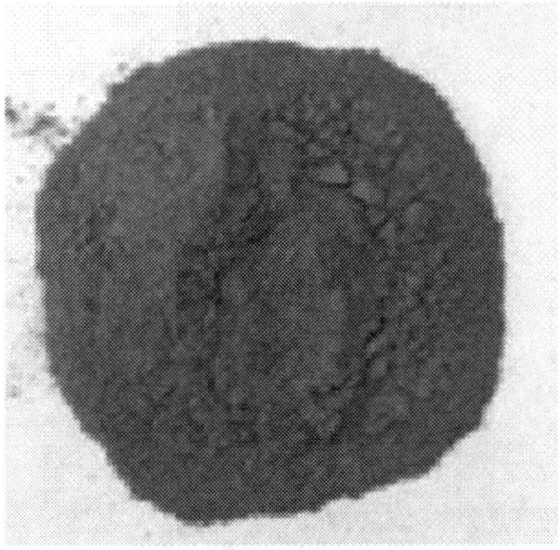

Fig. 3 The visual appearance of fly ash

Each sample was obtained by dividing several 5kg samples. Cutting consists of dividing the sample into four equal parts. Two opposite parts are recovered then homogenized. One among these last two parts is cropped then on. The operation is often repeated three or four fold to get a stratified sample [20]. Table 4 summarizes the composition of the fly ash.

Table 4 The chemical elements in the fly ash in percent %

SiO_2	Al_2O_3	Fe_2O_3	SiO_2+Al_2O_3+Fe_2O_3	CaO	MgO	SO_3	K_2O
57	34	3,4	94,4	5,05	0,02	0,5	0,03

The chemical constitution of the ash indicates that the sum of the percentages of the elements: SiO_2, Al_2O_3, and Fe_2O_3 is 94.4%. Thus, we classify them among the ashes of aluminous silica. The ash composition is presented in Figure 4. Two peaks of mullite and quartz were observed during this experiment. In combustion, these minerals change their structure and provides rise to a weak crystallized part within the sort of mullite and quartz and anamorphous part [16]. This result corresponds perfectly to the mineralogical composition of the bottom ash.

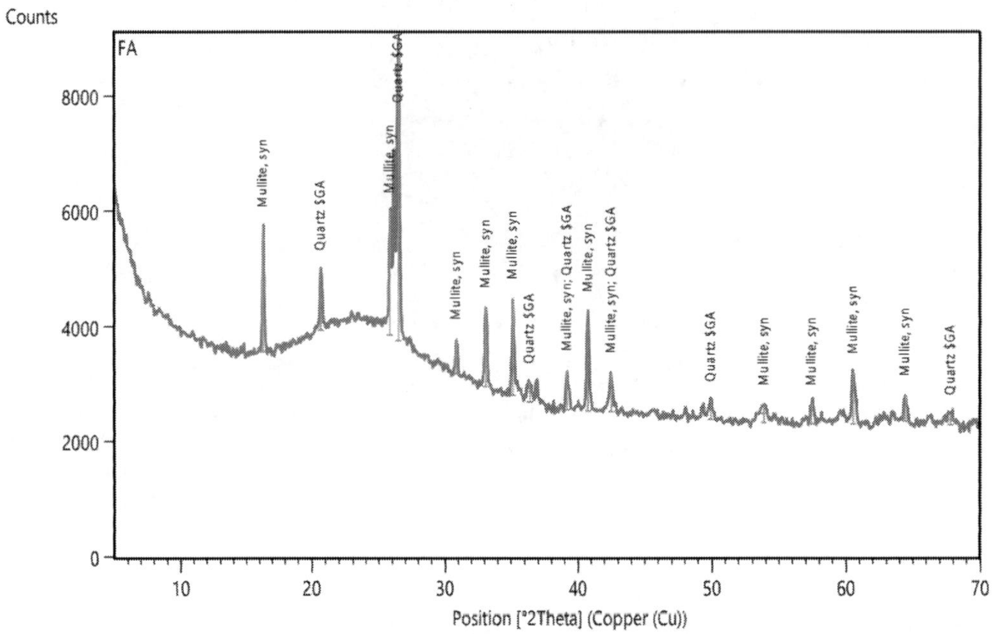

Fig.4 The mineralogical spectrum of the fly ash used

2.4.2 Bottom ash

The Bottom ash (BA)/cement mass ratios is equal to 10%. The process of preparation is the same as that of fly ash. The mortars prepared with Bottom ash are presented in Figure 5. **Table 5** present the different formulation of mortar elaborated from bottom ash.

Fig. 5 Samples of bottom ash mortar

Table 5 Mortar composition (gunity)

Elements	Bottom ash	Cement	Sand	Water
B0	0.00	457.07	1350	225
B10	52.07	405	1350	225

The outcomes of the analysis of the chemical composition of mortar produced from bottom ash (**table 6**), indicate that more than 80% of the chemical constitutions of the ashes were composed principally by SiO_2, Al_2O_3,and Fe_2O_3.

Table 6 Composition of the bottom ash used by %

CaO	SiO_2	Fe_2O_3	Al_2O_3	K_2O	Na_2O	ZnO	PbO	SO_3	MgO	CaO_{free}	LOI
1.92	52.07	8.86	23.34	1.9	0.4	0.01	0.01	1.87	1.09	0.29	8.24

The X-ray diffraction analysis indicates the presence of two peaks, the first one is for quartz (SiO_2) and the second one corresponds to mullite (**Figure 6**). This result can be explicated by the mineralogy of the coal used, which generally consists of crystalline silica in the form of quartz and phyllite minerals from the clay group (shales)[18].

https://doi.org/10.1051/e3sconf/202233705002

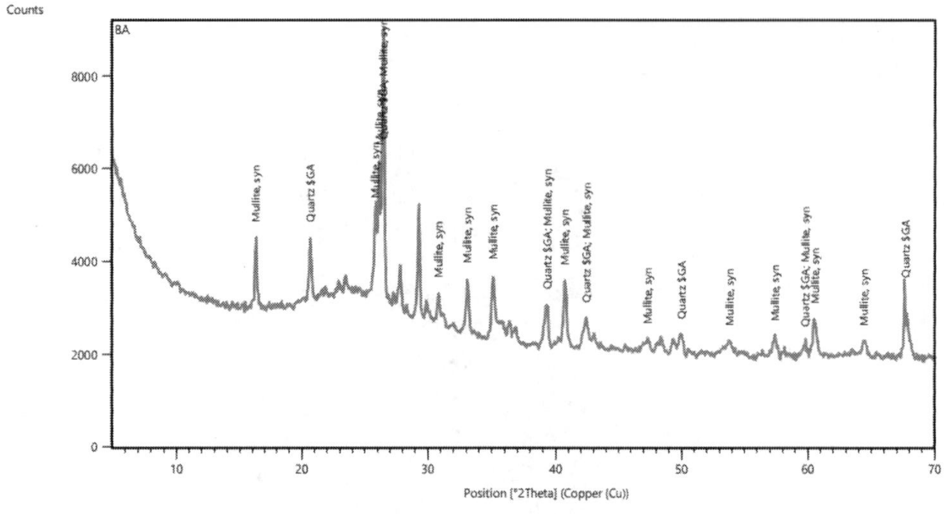

Fig.6 The mineralogical spectrum of bottom ash

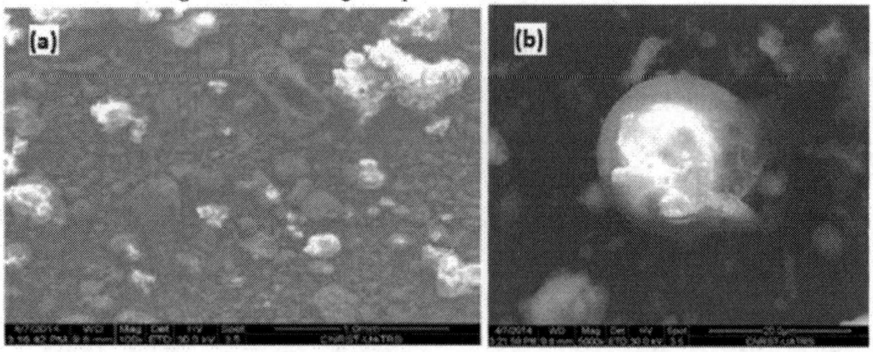

Fig.7 SEM images of Bottom ash morphology: (a) 1.0 mm level and (b) 20.0 μm.

Figure 7 shows the morphological structure of the crude ash obtained before filtration: they are in the form of spherical particles with irregular sizes that revealed the presence of porosity [21]. Most grains of cement are in the form of hollow spheres or spheres filled with smaller spheres. Micro crystals are also present on the surface of the particles, which reveal the presence of Mullite and Quartz. The elements Si, Al, Ca, Na and small amounts of Fe are contained in the plot spheres.

3 Experiment and materials description

3.1 Magnetization of water

The natural water was treated by an electromagnetic source of 40 nT, using a generator machine (**Figure 8**).

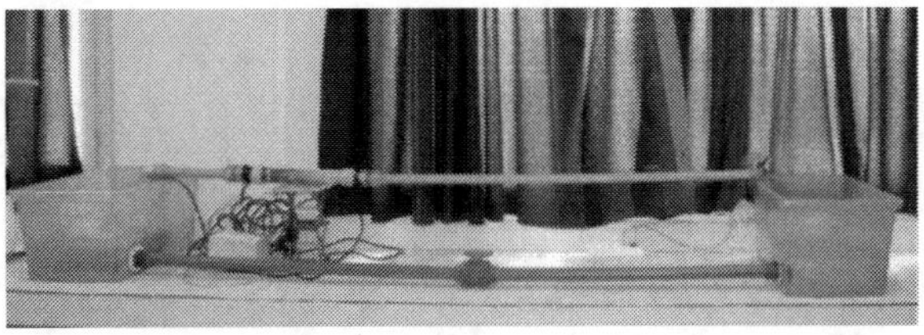

Fig. 8 Magnetic water production machine

The water flow can be adjusted by a differential regulator (PID) (**Figure 8**). When water crosses this electromagnetic field, it becomes magnetized.

3.2 Measurement of the mechanical properties

The test of the mechanical resistance to bending of mortars was realized on a material characterization machine (**figure 9**). The experimental protocol used is described by standard EN 196-1.

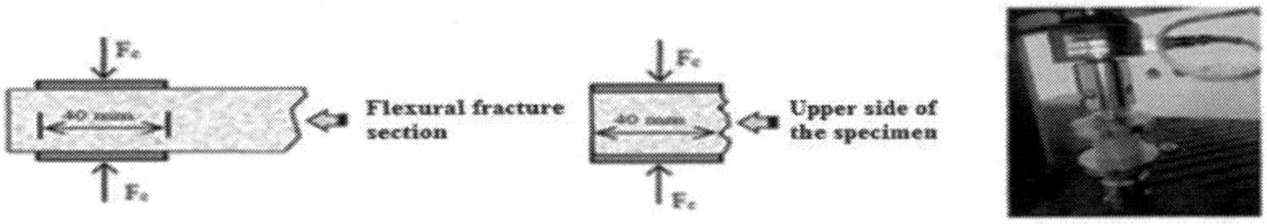

Fig. 9 Mechanical bending strength test instrument

The expression of resistance to the load applied to the center of the prism is expressed as

follows, $R_f = 3F_fL/2b^3$ (1)

F_f: load applied to the center of the prism at break (N)
L:distance between supports (mm)
b:side of the square section of the prism (mm)

After bending, each half-prism was reused to test the compressive strength (**Figure 6**)[8].

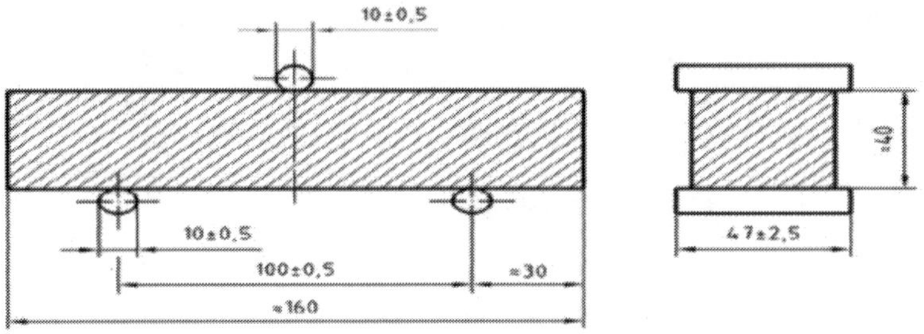

Fig. 10 Device for testing mechanical properties

A progressive force was on the cross-section (4×4cm) of the test piece with the same speed up to the breaking load from which the mechanical resistance to compression can be obtained in mega-pascals as follows.

$R_c = F_c/b^2$ (2)

Fc: Maximum load at break (N)
b: side of the square section of the prism (mm)

The authors produced a cement-based mortar with 10% fly ash and a water/cement ratio (W/CM) equal to 0.5.

4 Results and discussion

4.1 Physico-chemical parameters of tap and magnetized water

The different Physico-chemical parameters of natural water and magnetized water we represented in **table 7**, in order to observe the impact of the magnetic field on the water properties. Thus, the application of a magnetic field on water increases the value of pH from 7.69 to 8.25 [22] and decreases its conductivity from 1695 µS/Cm to 971 µS/Cm. The value of TDS is also decreasing dramatically from 848 to 485mg/l. For other parameters associated with the concentration of Ca, Cl and TH they stay the same.

Table7 Physico-chemical analysis of natural water and magnetized water

Parameters	Tap water	Magnetized water
pH	7.69	8.25
Conductivity(µS/Cm)	1695	971
TDS(mg/L)	848	485
Ca(mg/L)	78.5	79
Oxydability(mg/L)	0.41	0.49
Chloride(mg/L)	88.5	89
Temperature(°C)	17.8	17.8
Mg(mg/L)	23.83	23.9
Alkalinity(°F)	10.35	10.4
TH or title hydrometric(meq/l)	2.94	2.98
Ammonium(mg/L)	0.00	0.00
Fer(mg/L)	0.00	0.00

4.2 Results of mechanical tests on mortars

4.2.1 Flexural strength test (FS)

Figures 11 and 12, presents the variations in the average flexural strength (FS) mortars produced from fly ash and bottom ash, in place of cement, with a water/cement ratio equal to 0.5 at different times of water magnetization, 0 min, 10 min, 20 min, 30 min, and 40 min. We have observed that for all mortars containing magnetized water, a higher flexural resistance than those developed with natural water which doesn't exceed the value of 3.80 MPa for the FS of mortar based on 10% bottom ash and 3.44 MPa for the FS of mortar based on 10% fly ash. In summary, the extension of time of magnetization increases the FS of the mortar. The optimal time of magnetization of water is 30min which increases the value of FS to 4.5 MPa for a mortar based on 10% bottom ash or10 % of fly ash, after 30 min as a time of magnetization, the value of FS remains almost constant, which indicates the saturation phenomenon to magnetization.

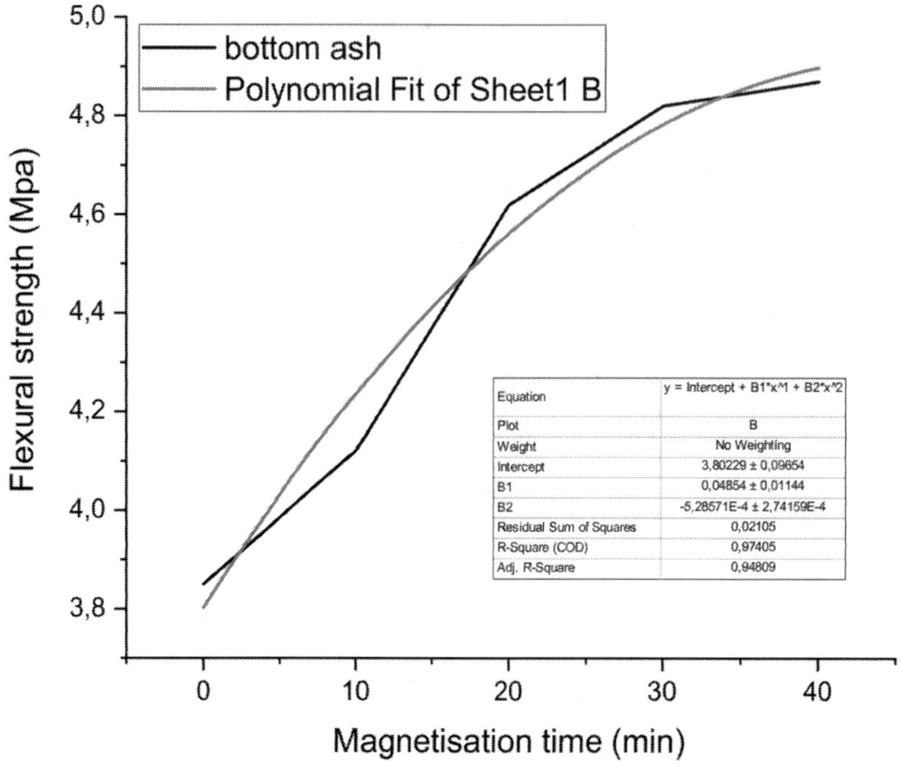

Fig. 11 Effect of EMF on the flexural strength of mortar based on10 % bottom ash

Table 8 ANOVA analysis of Flexural strength of bottom ash

		DF	Sum of Square	Mean Square	F Value	Prob>F
Flexural strength	Model	2	0,78987	0,39494	37,53136	0,02595
Flexural strength	Error	2	0,02105	0,01052		
Flexural strength	Total	4	0,81092			

From figure 11 we see that the most important result is to record with an R^2 (0.94). In this study, ANOVA analysis and F tests were performed to determine the statistically significant process parameters. If the critical F value is less than the calculated F value, there is a direct relationship between independent and dependent variables. The critical F-value is less than the calculated F-value (37,53136). As the P-value (0,02595) is much lower than 0.05, which proves the great significance of Flexural strength of bottom ash testing (table 8).
The figure 11 show the equation of the Flexural strength of bottom ash as a function of the magnetization time

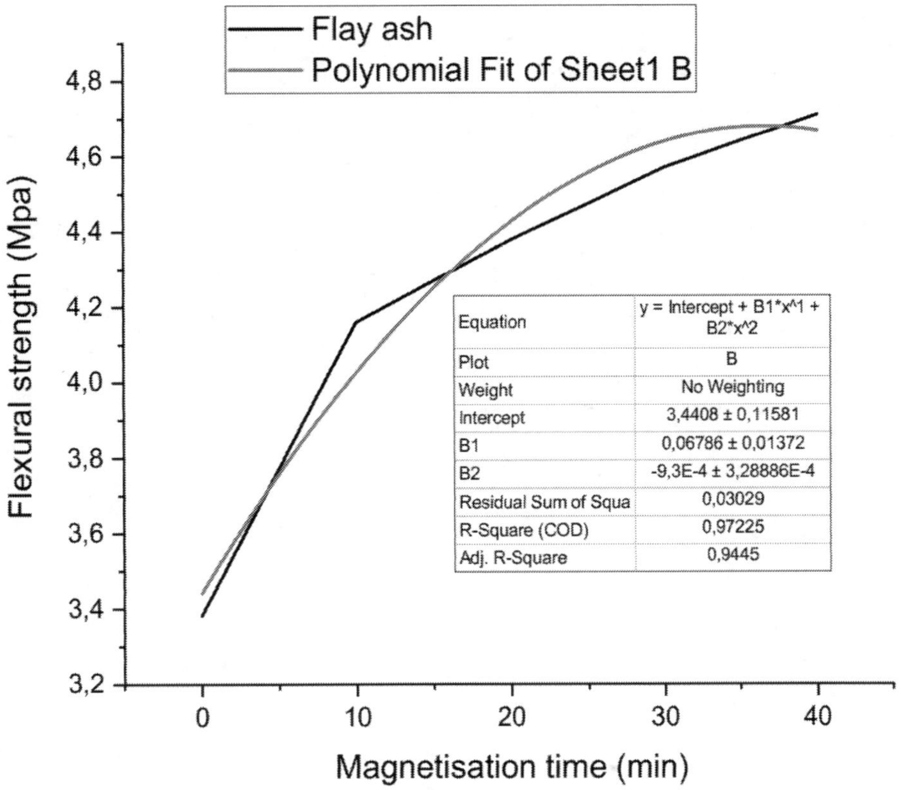

Fig. 12 Effect of EMF on the flexural strength of mortar based on 10% fly ash

Table 9. ANOVA analysis of Flexural strength of fly ash

	DF	Sum of Square	Mean Square	F Value	Prob>F
Flexural strength Model	2	1,06112	0,5305	65,03624	0,02775
Flexural strength Error	2	0,03029	0,01514		
Flexural strength Total		41,09141			

From figure 12 we see that the most important result is to record with an R^2 (0.94). In this study, ANOVA analysis and F tests were performed to determine the statistically significant process parameters. If the critical F value is less than the calculated F value, there is a direct relationship between independent and dependent variables. The critical F-value is less than the calculated F-value (65,03624). As the P-value (0,02775) is much lower than 0.05, which proves the great significance of Resistance to compression of bottom ash testing (table 9). The table in figure 12 show the equation of the Flexural strength of fly ash as a function of the magnetization time.

4.2.2 Compressive strength test

Figure 13 and **14** presents the variations in the average compressive force Mortars based on fly as hand bottom ash used in place of cement. With a water/cement ratio equal to 0.5 at different times of water magnetization, 0 min, 10 min, 20 min, 30 min, and 40 min, we note for all mortars containing magnetized water, higher compressive strength than those developed with tap water. For tap water, the value of compressive strength Mortar based on bottom ash is 7.5 MPa and 13.8 MPa for fly ash. In presence of magnetic water, we observe an increase of resistance to compressive strength. The optimum time of magnetization is equal to 30 min, after this value, the resistance to the compressive strenght remains constant (saturation to magnetization).

.

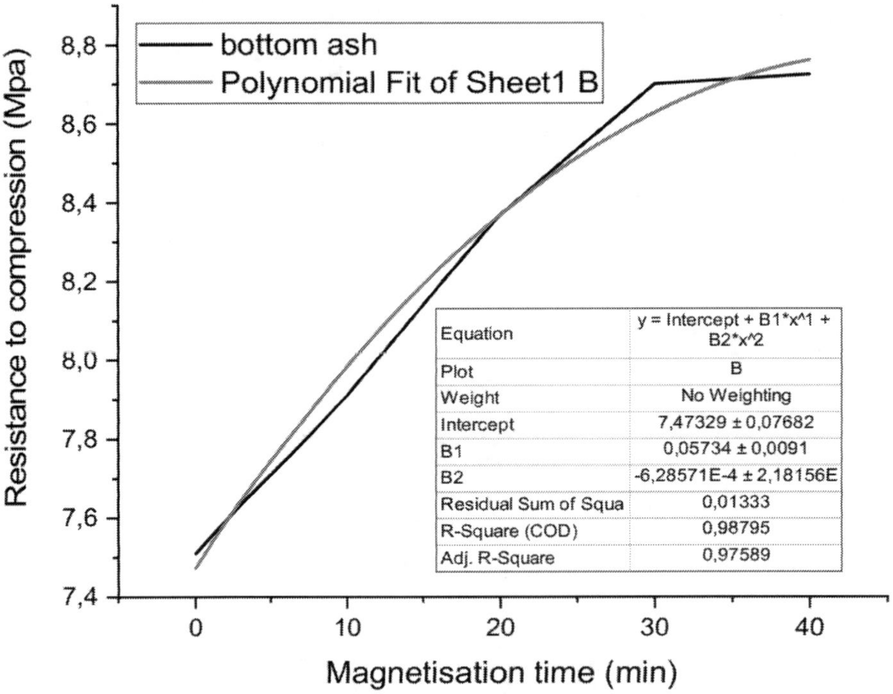

Fig. 13 Effect of EMF on the Resistance to compression of mortar based on 10% bottom ash

Table10. ANOVA analysis of Resistance to compression of bottom ash

		DF	Sum of Squares	MeanSquare	FValue	Prob>F
Resistance to compression	Model	2	1,09215	0,54608	81,9584	0,01205
Resistance to compression	Error	2	0,01333	0,00666		
Resistance to compression	Total	4	1,10548			

From figure 13 we see that the most important result is to record with an R^2 (0.98). In this study, ANOVA analysis and F tests were performed to determine the statistically significant process parameters. If the critical F value is less than the calculated F value, there is a direct relationship between independent and dependent variables. The critical F-value is less than the calculated F-value (81,9584). As the P-value (0,01205) is much lower than 0.05, which proves the great significance of Resistance to compression of bottom ash testing (table 10). The table in figure 13 show the equation of the Resistance to compression of bottom ash as a function of the magnetization time.

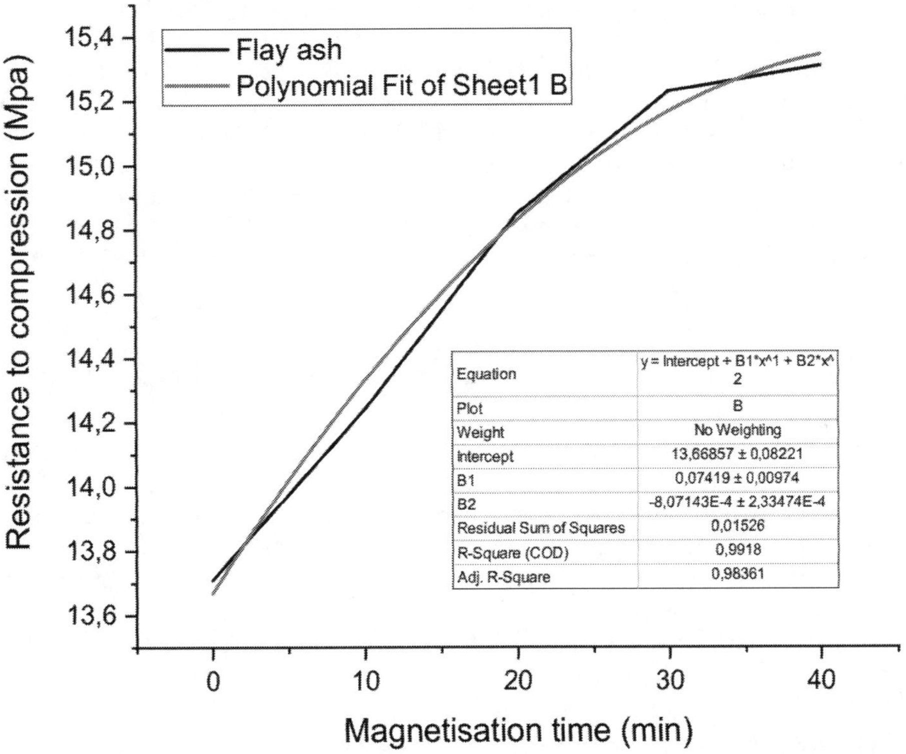

Fig. 14 Effect of EMF on the Resistance to compression of mortar based on10 % fly ash

Table 11 ANOVA analysis of Resistance to compression of fly ash

		DF	Sum of Square	Mean Squar	F Value	Prob>F
Resistance to compression	Model	2	1,84682	0,92341	121,00075	0,0082
Resistance to compression	Error	2	0,01526	0,00763		
Resistance to compression	Total	4	1,86208			

From figure 14 we see that the most important result is to record with an R^2 (0.98). In this study, ANOVA analysis and F tests were performed to determine the statistically significant process parameters. If the critical F value is less than the calculated F value, there is a direct relationship between independent and dependent variables. The critical F-value is less than the calculated F-value (121,00075). As the P-value (0,0082) is much lower than 0.05, which proves the great significance of Resistance to compression of fly ash testing (table 11). The table in figure 14 show the equation of the Resistance to compression of fly ash as a function of the magnetization time.

4.2.3 Interpretation of Flexural strength test and Compressive strength test

The interpretation of the previous results of flexural force and compressive force tests can be explained by the fact that the electromagnetic field change the structure of water molecules clusters (break of the macro clusters water molecules into small clusters of water, So they can easily penetrate the central part of the cement grains, increasing its resistance compared to mortar realized by natural water which forms larger clusters). Thus, with natural water we obtain incomplete hydration leading to bad mechanical properties [21,23]. Furthermore, the magnetized water can act as a dispersing agent, which improves the fluidity of the mortar increasing its quality in construction.

Conclusion

The magnetic field on water increases the value of pH and decreases its conductivity and its TDS. The concentration values of elements as Ca, Cl, and TH remain fixed after magnetization of water. For all mortars produced with magnetized water, we observed a higher flexural resistance than those with natural. The optimal time of magnetization of water is 30 min after this time of magnetization, the values of flexural and compression resistance remain constant, which indicates the phenomenon of saturation to magnetization. The electromagnetic field generated an electromagnetic force which can break the macro clusters water molecules into small clusters, thus they can easily penetrate the central part of the cement grains, increasing the resistance of mortar.

References

[1] N. Mahasenan, S. Smith, K. Humphreys, In Greenhouse Gas Control Technologies - 6th International Conference; Gale, J., Kaya, Y., Eds.; Pergamon: Oxford, 2,995(2003)
[2] P.Ubale, D. R. Pandit, P .A. Wadekar, Int. j. sci. technol. manag.,5,68(2016)
[3] M. C. Amiri, A. A. Dadkhah. Colloids Surf. A: Physicochem. Eng. Asp., 278 ,252 (2006)
[4] J. S. Baker, S. J. Judd. Water Res., 30, 247 (1996)
[5] A. Bubnov, V. Novotná, V. Hamplová, M.Kašpar, M. Glogarová, J. Mol. Structure, 892, 151 (2008)
[6] R. Cai, H. Yang, J. He, W. Zhu, J. Mol. Structure, 938, 15 (2009)
[7] A. C. Cefalas, E. Sarantopoulou, Z. Kollia, C. Riziotis, G. Dražic, S. Kobe, et al. J. Comput. Theor. Nanosci., 7, 1800 (2010)
[8] K. T. Chang, C. I. Weng, J. Appl. Phys., 100, 1 (2006)

[9] M. C. Chang, C. Y. Tai, Chem. Eng. J. 164, 1 (2010)
[10] E. Chibowski, L. Hołysz, A. Szcześ, Water Res., 37, 4685 (2003)
[11] E. Chibowski, A. Szcześ, Chemosphere, 203, 5 (2018)
[12] K. X. Zhou, G. W. Lu, Q. C. Zhou, J. H. Song, S.T. Jiang, H. R. Xia..J. Appl. Phys., 88, 1802 (2000)
[13] X. F. Pang, B. Deng, B. Tang, Mod. Phys. Lett. B, 26, 1 (2012)
[14] H. Wei, Y. Wang, J. Luo.Constr. Build. Mater,147,91(2017)
[15] R. Mghaiouini et al., Indian. J. Sci. Technol.,13, 1172, (2020)
[16] R. Mghaiouini et al., J. Ecol. Eng., 21, 245 (2020)
[17] R. Mghaiouini, M. Salah, M. Monkade, A. El Bouari, Arab. J. Sci. Eng. (2021)
[18] S. Ghorbani, M. Gholizadeh, J. D. B. Pavers, Mater., 11, 1 (2018)
[19] A. R. Esfahani, M. Reisi, B. Mohr, J. Mater. Civ. Eng.,30,1(2018)
[20] M. Gholhaki, A. kheyroddin, M. Hajforoush, M. Kazemi, Constr. Build. Mater.,158, 173 (2018)
[21]N. Su, C. F. Wu, Cem. Concr. Compos., 25, 681 (2003)
[22] R. Mghaiouini, A. Elmlouky, M. Salah, T. M. Al Antary, M. Monkade, A. El Bouari, A. Y. Ghidan Fresenius Environ. Bull. 29, 7922(2020)
[23] T. IM. Abdel-Magid, R. M. Hamdan, A. AB. Abdelgader, M. EA. Omer, N. M. R. Ahmed, Proc. Eng., 193, 494 (2017)

CNTs modified ZnO and TiO_2 thin films: The effect of loading rate on band offset at metal / semiconductor interfaces

Younes Ziat[1*], *Maryama* Hammi[1,2**], *Hamza* Belkhanchi[1***], *Ousama* Ifguis[3****], *Saloua* Rzaoudi[1], *Charaf* Laghlimi[4], *Abdelaziz* Moutcine[5] and, *Charaf* Lazrak[6]

[1]Engineering and Applied Physics Team (EAPT), Sultan Moulay Slimane University, Beni Mellal, Morocco
[2]University of Mohammed 5th — Rabat, Morocco
[3]Laboratory of engineering in chemistry and physics of material, Faculty of Sciences and Technology, Beni-Mellal, Morocco
[4]Applied Chemistry Team, Faculty of Sciences and Techniques of Al Hoceima, Abdelmalek Essaadi University, Al Hoceima, Morocco
[5]Molecular Electrochemistry and Inorganic Materials Team, Beni Mellal, Faculty of Science and Technology, Sultan Moulay Slimane University, Morocco
[6]Thermodynamics, Mechanics and Energetics: Experimentation, Modeling and Simulation: TME-EMS, Physical Department, Faculty of Sciences, Mohammed 5th University, Av Ibn Battouta, PO Box 1014, Rabat 10000, Morocco

Abstract. ZnO and TiO_2 are direct wide band gap semiconductors with intriguing properties. A wide range of applications makes it one of the most studied materials in the past decade, particularly when elaborated as nanostructures. In this work, we focus on synthesis of CNTs modified ZnO and TiO_2 thin films using sol-gel method. The morphological and optical characterizations of the based ZnO and TiO_2 films were carried out using scanning and transmission electron microscopy (SEM and TEM), XRD and UV spectroscopy. Electrical properties of the deposited ZnO/CNTs and CNTs /TiO_2 were studied using I-V measurements at room temperature in metal/semiconductor/metal configuration, by the use of an array of metallic micro-electrodes deposited on the surface of the elaborated thin films. This allows determining qualitatively the electrical conductivity of thin films and the different parameters of the Schottky junction between the composites nano-films and the substrate. This study is necessary for future applications in solar cell.

1 Introduction

Solar energy is the most inexhaustible and available renewable energy [1-5]. Therefore, the development of photovoltaic devices has attracted much interest in the scientific world.

Corresponding author: *ziat.estbm@gmail.com; **mar.hammi@yahoo.com; ***belkhanchihamza@gmail.com; ****ousama.if@gmail.com

Many materials are suitable for the design of photovoltaic devices, but many of them do not offer an optimal efficiency/cost ratio, thus limiting their expansion. Dye photovoltaic cells are commonly developed in the last decades as the 3rd generation of solar cells [6]. It is a simple and inexpensive technology according to the literature because it uses inexpensive and available semiconductor materials; for example: titanium dioxide (TiO_2) [7-11] or zinc oxide (ZnO) [12-15]. To make these cells more competitive it is important to study their characteristics and the possibility of adding (and/or replacing some components with) other materials that can improve photon absorption and charge transport in the active layer and limiting charge recombination in order to increase the conversion efficiency of the cell. Our attention was drawn to carbon nanotubes (CNTs) because of their extraordinary physical and mechanical properties and the fact that they are good candidates for improving the performance of solar cells. In addition, they have good electrical conductivity to nanocomposite metal oxides [16-18]. It is also possible to use CNTs in the active layer with titanium dioxide (TiO_2) or zinc oxide (ZnO) as in the present study.

Titanium dioxide (TiO_2) is a semiconductor with three polymorphs: anatase (quadratic), rutile (quadratic) and brookite (orthorhombic) [19]. According to the literature, anatase is recognized as the most photo-catalytically active phase due to the mobility of electron/hole pairs within the material and the ease of their transfer to the surface [12, 20].

The exceptional electronic properties of the anatase phase of TiO_2, especially the low rates of electron-hole recombination, make this polymorph a suitable candidate for application in solar cells. Dye Sensitized Solar Cells (DSSC) systems, incorporating pure or modified anatase TiO_2 and based on the photo-catalytic principle, have been developed [21]. Due to its remarkable functional properties, the range of applications for TiO_2 is very wide. For example, for the application of self-cleaning glasses [22], filtration of wastewater into drinkable water [23-24], Dye Sensitized Solar Cells (DSSC), integrating pure or modified anatase TiO_2 and based on the photo-catalytic principle, have been developed [25].

The ZnO is a wide band gap semiconductor widely used in short wavelength optoelectronic applications. It has also been used as a photoanode material in DSSCs due to its high charge carrier (electron) mobility, transparency to visible light, the possibility of making it highly conductive by doping and its greater synthetic flexibility, as well as its advantageous morphology [26].

In this study we present a simple and economic method to elaborate thin films and we make a comparative study between ZnO and TiO_2 thin films loaded with CNTs by different mass percentage and to see the effect of CNTs in the optical and structural properties and to see the most efficient metal oxide for photovoltaic use.

2 Synthesis

2.1 Synthesis of CNT

The carbon nanotubes are of considerable interest for both fundamental and applied research. CNTs are nano-structures of the fullerene family. Their typical diameter is in the nanometre range, while their length can reach a few millimetres, or even a few centimetres [27]. Some nanotubes are semiconductors, others are good electrical conductors. CNTs have remarkable physical, mechanical and electrical properties (excellent thermal and electrical conductivity and high mechanical strength: a CNT is 100 times stronger and 6 times lighter than steel) which have opened the field to numerous and promising applications. The applications of nanotubes are numerous: new, more resistant materials, new textiles (bullet-proof vests, etc.), nanoelectronics based on nanotubes, new generation flat screens, medical applications, etc. CNTs can thus be used to elaborate high-performance composite materials, conductive polymers or technical textiles. They are already used in the fields of sports equipment, aeronautics, automobiles, defence, medicine, etc. Nanotube powder also has multiple potential applications such as hydrogen storage, the manufacture of batteries for electric cars and photovoltaic [28]

The CNTs used in the experiments were prepared by chemical vapor deposition, and purified through the treatment of the concentrated mixed acid (nitric acid and hydrochloric acid), which was described in detail elsewhere [29-30].

2.2 Preparation of ZnO/ carbon nanotubes composite

ZnO/CNTs nanocomposites were synthesized using a sol-gel method based on Benyounes's work with slight modifications [12]. The solution-based strategies to synthesize ZnO nanoparticles typically require a precursor solution of zinc acetate-ethanol complex [31] or a long hydrolysis method [32]. In this work, we present a way to prepare small ZnO nanoparticles by directly mixing zinc acetate and CNTs. Dehydrated zinc acetate $[Zn(CH_3COO)_2\text{-}2H_2O]$ was chosen as the initial material to perform the experiment. Two-methoxy ethanol was used as solvent and monoethanol amide (MEA) as stabilizer. The zinc acetate dihydrate and CNTs were initially dissolved in a mixture of solvent (MEA) and stabilizer (two-methoxyethanol) at room temperature. The zinc acetate concentration was approximately 0.35 M while the molar ratio of MEA to zinc acetate $Zn(CH_3COO)_2$ was 1.0. To get a clear and homogeneous solution, this mixture was stirred at 60 °C for 2 h. After the mixture was cooled at room temperature, it was used as a coating solution. After the mixture was prepared for 24 h, the coating was performed. The solution was poured onto glass substrates, with spin cycles of 1000 rpm for only 30 s. When the spin coating was done, the films obtained were dried at a temperature of 400 °C for 20 min on a hot plate. The objective of this process was to evaporate the solvent and to remove the organic residues. This whole procedure was repeated 12 times to get a thickness of the sintered films close to 800 nm.

2.3 Preparation of carbon nanotubes/TiO_2 composite

The coating solution for titanium isopropoxide: $C_{12}H_{28}O_4Ti$ is got by dissolving 2 ml of Aldrich 98% $C_{12}H_{28}O_4Ti$ and 5 ml of isopropanol solution: C_3H_8O, and then different weight percentages (%) of CNTs powders ranging from 0.05 wt% to 0.9 wt% CNTs had been dispersed in the resulting mixture. This mixture was agitated at 63 °C during 11 minutes. Next, it was added 5.5 ml of acetic acid: CH_3COOH. After that, the new mixture was stirred for 18 minutes. Thereafter, after two hours, the final mixture was agitated after adding 12ml of methanol CH_3OH to make the solution of sol-gel. The latter solution was spin-coated onto plain ITO glass substrates which had been previously soaked in methanol and acetone at different spinning speeds (2800, 3000 and 5000 rpm) for 30s, followed by a drying process at 100-110°C for 10 min to evaporate the solvent. At last, the annealing temperature was subsequently raised from (100 to 500°C).

3 Characterization

After the TiO_2/ carbon nanotubes and ZnO/ carbon nanotubes thin films were prepared, the structural and morphological characterizations were characterized using AFM (atomic force microscopy) and SEM (scanning electron microscopy).UV-Visible transmission spectroscopy was used to study the optical properties. The optical transmission spectra were measured in the visible and near UV range. As a result, this method permits to determine the transmission T (%) of a material from a given wavelength λ that has been precisely selected.

The optical transmission spectra of the thin films developed were recorded using a Spectrophotometer UV-Visible: (Jasco V-530) on the range of 300-800 nm wavelength.

4 Results and discussion

4.1 Optical Properties

In this work, we focus on synthesis of CNTs modified ZnO and TiO_2 thin films using sol-gel method, and make a comparative study of the effect of CNTs on ZnO and TiO_2 thin films.

The transmittance spectrum (T) of thin films versus wavelength at room temperature in the spectral range of 300-800 nm is presented in Fig. 1. The spectrums show a highly transmittance up to 89% in the visible domain for the uncharged thin film of TiO_2 (Fig. 1(a)) and 76% for the uncharged ZnO thin film (Fig. 1(b)) , which indicates that TiO_2 and ZnO are highly transparent materials, it is emphasized that the transmittance in the overall wavelength range is significantly reduced with increasing CNT filler content ranging from 0.05 to 0.9 wt.% CNT up to 64% for CNT/TiO_2 and 30% for CNT/ZnO. Decreased optical transmittance is due to the fact that CNTs are not transparent and absorb incoming light and increase light scattering.

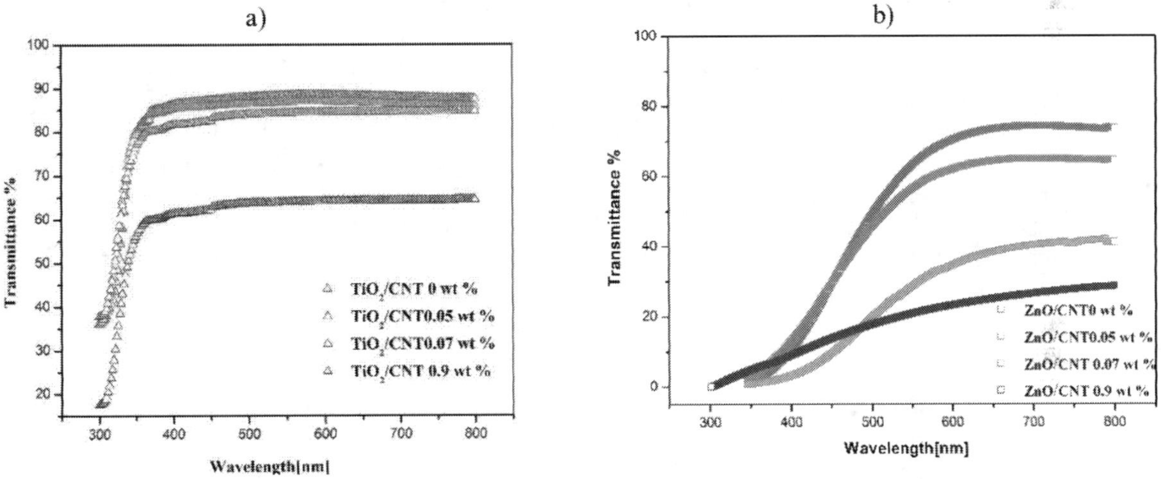

Fig. 1. Plot of transmittance (T) versus wavelength of doped thin films, a) TiO_2/CNT, b) ZnO/CNT

The value of the optical gap energy (Eg) of the film, the refractive index of the films, the optical constants and their thickness is determined from the spectrophotometer, which can record the optical transmission of the films according to the wavelength. This technique is based on the double relation light-matter. When the films under study absorb ultraviolet or visible light, the absorbed energy leads to the excitation of electrons that move to a higher energy level.For this purpose, we will use the following formulas given by the method of Manifacier and Swanepoel [33-34]. The thickness of the nanostructures is determined from the following equation:

$$d = \left(\frac{\lambda_1\lambda_2}{2(\lambda_1 n_2 - \lambda_2 n_1)}\right) \quad (1)$$

With n_1 and n_2 present, respectively, the refractive indices of the films for the wavelengths λ_1 and λ_2 adjacent. The refractive indices in the spectral region of the transparent, low and medium absorption zones can be calculated as follows:

$$n_1 = \left[N_1 + (N_1^2 - S^2)^{1/2}\right]^{1/2} \quad \text{and} \quad n_2 = \left[N_2 + (N_2^2 - S^2)^{1/2}\right]^{1/2} \quad (2)$$

In addition, the Swanepoel coefficient (N) in the transparent spectral region can be calculated by the following expression:

$$N_1 = 2S\left(\frac{T_{max1} - T_{min}}{T_{max1} \times T_{min}}\right) + \left(\frac{S^2+1}{2}\right) \quad \text{and} \quad N_2 = 2S\left(\frac{T_{max2} - T_{min}}{T_{max1} \times T_{min}}\right) + \left(\frac{S^2+1}{2}\right) \quad (3)$$

Where S is the refractive index of the glass, $T_{max}(\lambda)$ and $T_{min}(\lambda)$ represents the maximum and minimum values for the transmission curve.

Using the equation 2 and 3 to determine equation 1 and determine the thickness d to determine the absorption α of the of the Epoxy/Silicone N-CNT nanocomposite films of which it is made using the Bouguer-Lambert-Beer relation [34]:

$$T = \exp(-\alpha d) \quad (4)$$

If transmittance T is expressed in (%), the absorption coefficient is shown by:

$$\alpha = \frac{1}{d} \ln\left(\frac{100}{T}\right) \quad (5)$$

We can from the transmittance spectra (T) calculate the optical Gap value of semiconductors from the Tauc formula (Eg(eV)) defined by the following equation using [35-36]:

$$(\alpha h\nu) = B\,(h\nu - Eg)^n \quad (6)$$

The relation can be rewritten in a logarithmic form such as:

$$\ln(\alpha h\nu) = \ln B + n \ln(h\nu - Eg) \quad (7)$$

Where α is the absorption coefficient, ν is the absorption frequency, B is constant, h is Planck's constant and n is dependent on the type of optical transition, The constant n depends on the nature of the optical Gap, it is 1/2 for a direct optical Gap and 2 for an indirect optical Gap. To determine whether the electronic transition that occurred in the studied samples is direct or indirect, "ln(αhν-Eg)" is plotted as a function of "ln(hν)" that has been fitted by equation 7 in the linear region in the plot, employing the means squares method, where B, n, and Eg are the fitting parameters, with this method being able to indicate the type of the optical transition of the TiO_2/CNT and ZnO/CNT thin films are direct.

The band gap energy has been obtained using the extrapolation of the linear part of the absorption edge $(\alpha h\nu)^2$

Table 1. Eg values versus loading rates of TiO_2/CNT @ wt % and ZnO/CNT @ wt %

TiO_2/CNT wt %.	0	0.05	0.07	0.9
Eg(eV)	3.98	3.48	3.46	3.2
ZnO/CNT wt %.	0	0.05	0.07	0.9
Eg(eV)	3.45	2.76	2.71	2.31

From the transmission spectra, we deduced the optical gap of ZnO/CNT and TiO_2/CNT films using the relation (6). based on the plot of $(\alpha h\nu)^2$ versus incident photon energy (hν) shown in Table 1, we can estimate the optical band gap by extrapolating the linear region in the x-axis (αhν)2. The obtained optical gap is high for the uncharged thin films of TiO_2 and ZnO (3.98 eV for TiO_2 and 3.45 for ZnO), because it is optically transparent and it means that no absorption is possible in the visible, they cannot be excited without being charged with another element to decrease the Eg, so we charge with CNT and we see in Table 1, that there is a decrease of Eg from 3.98 to 3.2 eV for CNT/TiO_2 and from 3.45 to 2. 31eV for ZnO/CNT, this decrease in optical gap with the increase in charge rate is mainly caused by the addition of CNTs which influences the overall carrier concentration and the increase in free electron concentration which decreases the band gap (approximation between the valence band and the conduction band). The band gap reduction effect has been clearly observed in ZnO films doped with CNTs than TiO_2. Therefore, the system becomes progressively more conductive with the addition of CNTs, which seems to slightly change the electronic band structure of the ZnO and TiO_2 film. Furthermore, the optical gap shift with increasing concentration of charged CNTs can also be correlated with the surface roughness and density of the film which explains the small Eg decrease of the CNT/TiO_2 thin film from 3.98 to 3.2 eV due to the film roughness. Absorption shift of ZnO/CNT and TiO_2/CNT nanocrystal line films can be explained by the Burstein-Moss effect [37-38]. In fact, the reduction of the band gap is beneficial for the photovoltaic efficiency of broadband semiconductors such as ZnO, as it is more efficient to use visible light.

4.2 Microstructural insight

4.2.1 Transmission electron microscopy (TEM) of CNT

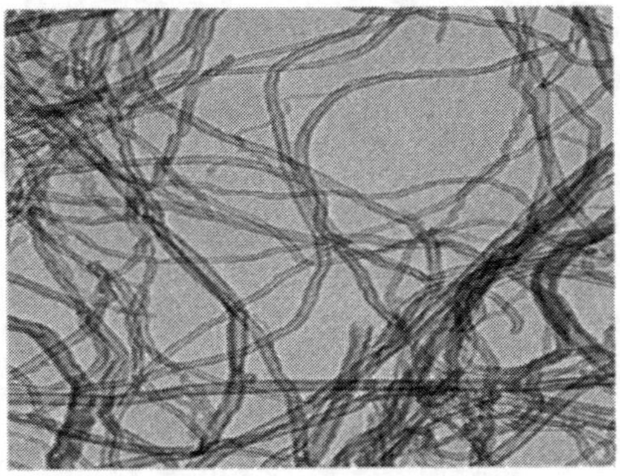

Fig. 2. TEM images of CNTs

The CNTs were synthesized by CVD by using ethylene as the carbon source (CNT). Transmission electron microscopy (TEM) images of the CNT sample indicate the existence of uniform multi-walled CNTs, see Fig. 2.

4.2.2 Atomic force microscopy (AFM)

The surface condition of the films was observed by AFM microscopy. Fig. 3 shows the morphology, in three dimensions, associated with the TiO_2 and ZnO films.

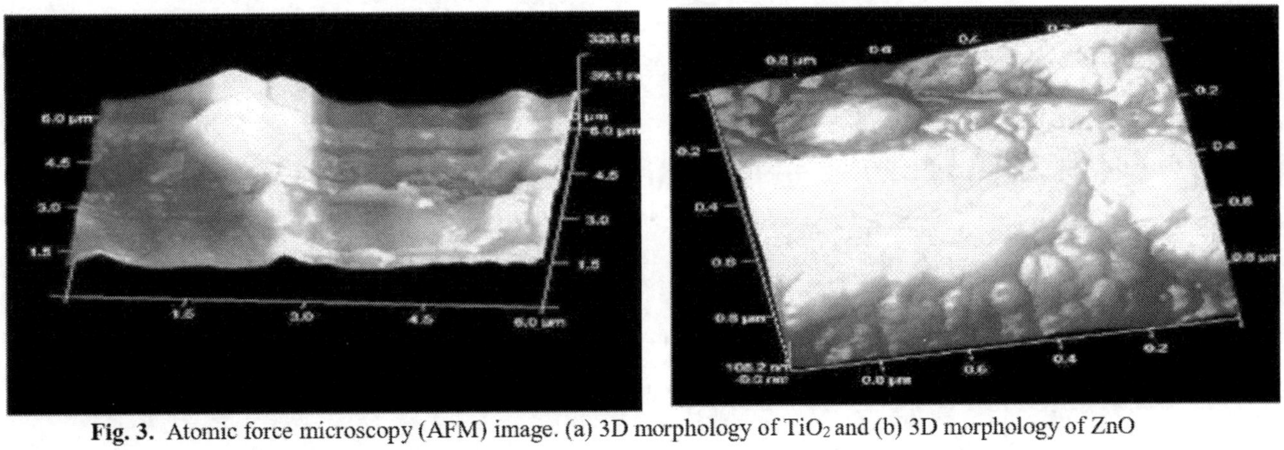

Fig. 3. Atomic force microscopy (AFM) image. (a) 3D morphology of TiO_2 and (b) 3D morphology of ZnO

https://doi.org/10.1051/e3sconf/202233705003

Indeed, the roughness depends on the chosen method of fabrication. This leads to different thicknesses of the films, which suggests different grain sizes of the ZnO and TiO_2 layer. Similarly, the observation of AFM images clearly shows a layer thickness of 650 nm for the TiO_2 sample and a value of 500 nm for ZnO

4.2.3 SEM micrographs of TiO_2/CNT and ZnO/CNT

Morphological study was performed on the surface of TiO_2/CNT and ZnO/CNT thins to understand the surface quality and size distribution of representative TiO_2 and ZnO thins films on top of ITO and show the homogeneous surface of the substrates, the characterization of each of the TiO_2 and ZnO nanoparticles samples was carried out with SEM (Scanning Electron Microscopy) microscope respectively.

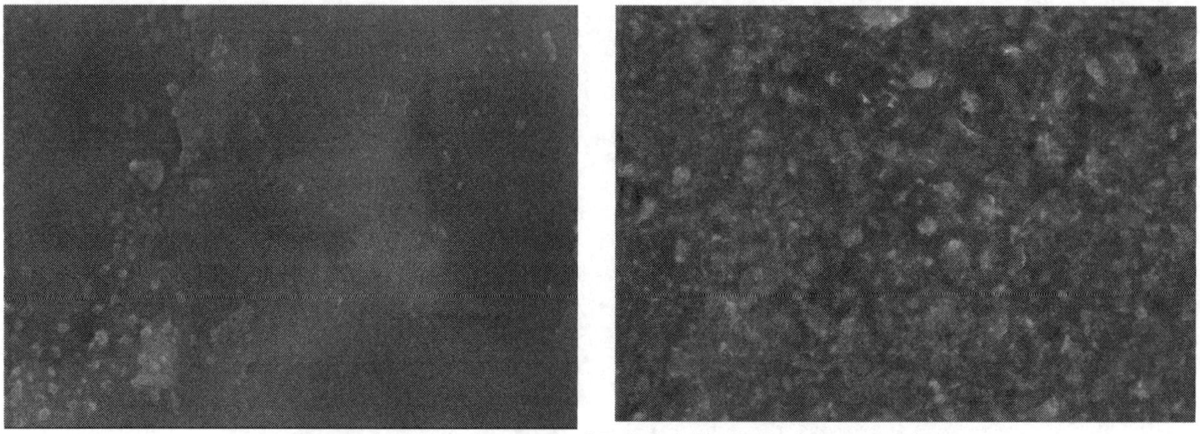

Fig. 4. (a) SEM micrograph of ZnO and (b) SEM micrograph of TiO_2

Fig 4 (a) and (b) show SEM images of the fabricated ZnO/CNT and TiO_2/CNT thin films respectively. The images show a clear granular surface morphology that reveals the formation of crystallized TiO_2 and ZnO.

The hardened composites showed a very dense and with good relative smoothness on the surface. The SEM images showed that the CNTs particles were uniformly dispersed in the ZnO and TiO_2 films. As a result, it is clear that there is a good miscibility of the particles between the phases, and that CNTs are well inserted in the porous layer of TiO_2 and ZnO, that CNT surfaces are covered by the TiO_2 and ZnO particles, which mean that TiO_2 and ZnO is well grafted on CNTs. The SEM allows us to clearly see the connection between the CNT particles and the TiO_2 and ZnO films will increase the efficiency of the charge transport in the device since they have very good electrical conductivity.

5 Charge carrier effect on Schottky contact barrier

The Schottky junction is the simplest way to construct the photovoltaic cell. It is a single layer of semiconductor deposited between two electrodes. One electrode provides ohmic contact with the thin film, while the second should provide current rectifying properties. It is the electric field generated at the blocking surface which allows the dissociation of the excitons by forming a potential barrier. The mobilities of the carriers are generally low in such materials, which tends to increase the value of the series resistance.

The both studied oxides (ZnO and TiO_2) based nanocomposites thin films were investigated on optical point of view. Thin films showed high transmission property (up to 80%) in the visible wavelength region. The optical band gap narrowing is more likely caused by light absorption.

The obtained nanocomposites thin films are likely to be investigated for the feasibility of Schottky junction. To the best of our knowledge, this is the first time that such kind of composites is proposed for a photovoltaic application. In order to verify the origin of the photovoltaic effect, The energy band diagram metal-semiconductor is drawn. The depicted junction (Fig.5) is of a typical metal- ZnO/CNTs and metal-TiO_2/CNTs.

Metal is characterized by fermi level below which there is a pool of electrons. It could be due to two overlapping bands generally. The work function of metal is greater than the work function of the semiconductor. Within this study we made the ohmic contact using gold (Au).

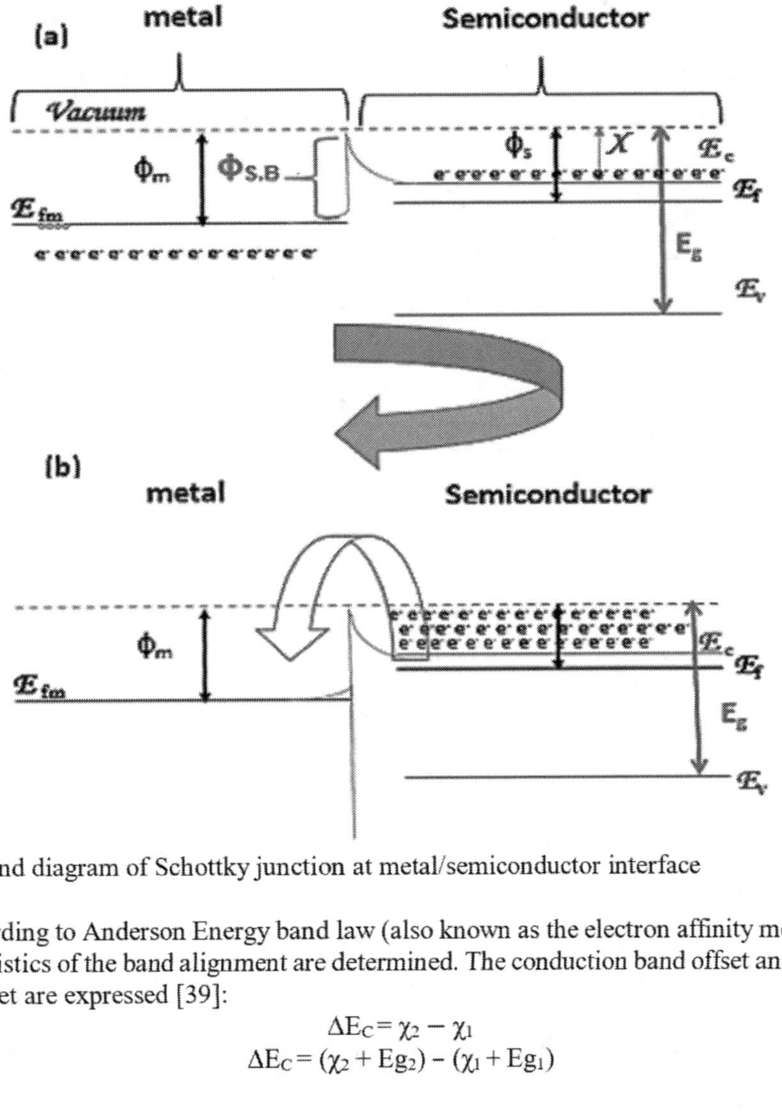

Fig.5. Band diagram of Schottky junction at metal/semiconductor interface

According to Anderson Energy band law (also known as the electron affinity model), the characteristics of the band alignment are determined. The conduction band offset and valence band offset are expressed [39]:

$$\Delta E_C = \chi_2 - \chi_1 \quad (8)$$

$$\Delta E_C = (\chi_2 + Eg_2) - (\chi_1 + Eg_1) \quad (9)$$

Where χ is the electron affinity, Eg is bandgap, and subscripts 1 and 2 correspond to metal and ZnO or TiO_2, respectively. In our case, the work function is about 4.5 eV for ZnO [40], 1.39 eV for anatase TiO_2 [41] and 5 eV for Au. The studied interfaces produce Schottky barrier, this barrier is not allowing the charge carrier transfer from semiconductor to the metal as indicated in Fig. 5 (a).The solution to this problem is the incorporation of more charge careers and forward bias application (Fig.5 (b)) so as to make higher level E_C offset leading to the occurring of fluent charge transfer, this is in a good consistence with the band gap narrowing, which is considered as an important key factor of the good conversion yield in solar cells. This effect was highlighted by some researchers who have used n-ZnO thin films to make potentially high-efficiency and low-cost solar cells [42-44]. Also in some other work, a nickel (Ni)-doped ZnO film was prepared by spray pyrolysis, and the optical band gap was reduced from 3.47 eV for un-doped ZnO film to 2.87 eV for 15% Ni-doped film [45].

Sundaram et al. evidenced that when ZnO is highly doped, the fermi level almost overlaps with the conduction band edge giving a rise to band alignment. Hence, the work function of ZnO is considered the same as electron affinity. Thus, in our case, these values result in ΔE_C and ΔE_V of ~0.5 eV and 1.81eV, respectively.

6 Current-voltage characteristics

Two distinct junctions, Au-TiO_2/CNTs and Au-TiO_2/CNTs, were made to verify the cause of the photovoltaic effect.

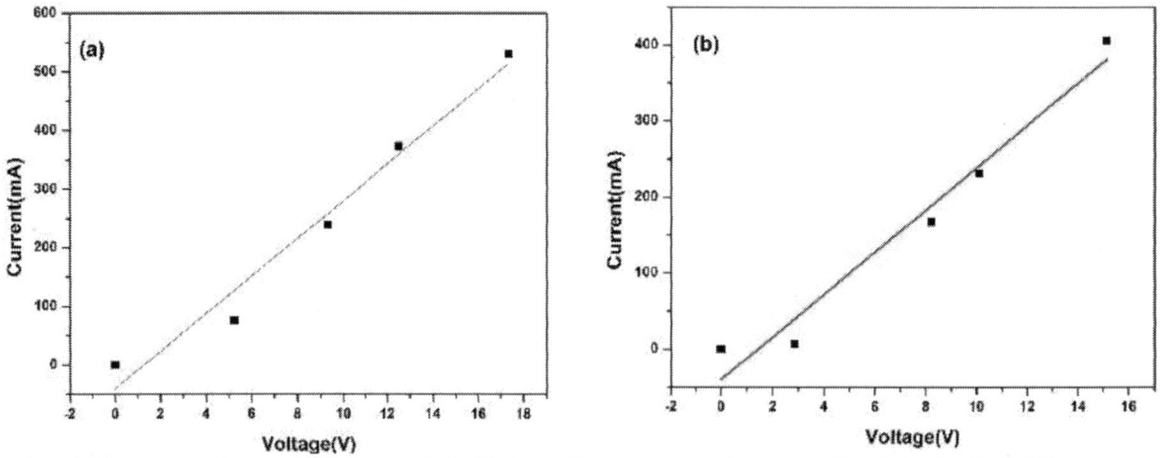

Fig.6. (a) Current–voltage profile of Au-TiO_2/CNTs, (b) Current—voltage profile of Au-ZnO/CNTs

After obtaining the highly filled thin films, we evaluated the current–voltage characteristics of Au-ZnO/CNTs and Au-TiO_2/CNTs two junctions, Fig. 6; this clearly indicates two ohmic connections.

7 Conclusion

In this study, we succeeded to elaborate TiO_2/CNT and ZnO/CNT thin films by sol-gel method and deposited on ITO glass substrates. TiO_2 and ZnO are known for their applications in the fields of photovoltaic and photoactivity, the choice of the substrates is based on these applications. Indeed, the indium oxide coated glass substrate (ITO) is used in solar cells. Structural and morphological characterizations were determined by radiation diffraction, scanning electron microscopy (SEM) and atomic force microscopy (AFM), UV-vis transmission spectroscopy was used to study the optical properties, the results obtained suggest that the films are of fairly high optical quality.

We were able to see the effect of increasing the dopant concentration of our CNT material as absorption elements, the average transmittances in the visible region are decreased due to the induction of CNT which went from 89% in the visible region for the uncharged TiO_2 thin film to 64% for CNT/TiO_2 0.9 wt.% CNT and 76% in the visible region for the uncharged ZnO thin film to 30% for ZnO/CNT 0.9 wt.% CNT, the increase in the amount of carriers leads to an increase in the visible absorption.

The results indicate that the film exhibits a direct optical transition with an Eg of 3.98 (uncharged TiO_2) to 3.2 eV for CNT/TiO_2 and 3.45 (uncharged ZnO) to 2.31 eV for ZnO/CNT as a function of loading rate which is in good agreement with the similar gap of the semiconductors. These results are very encouraging, they show an improved opto-electronic property compared to undoped materials and deserve to be further explored for the development of advanced devices especially for loaded ZnO that has very favorable optical character and their Eg decreases more than TiO_2 when loaded with CNT. These results also show that it was possible to modify the charged ZnO and TiO_2 films by inserting a charge and suggest that these thin films are good candidates for optoelectronics and solar cells applications especially for ZnO based thin films since the highly loaded samples exhibited band alignment at Au/ZnO-CNTs Schottky interface.

Acknowledgements: The authors would like to thank the support of Ms. ***Khadija Ziat***. Our thanks are extended to Mr. ***Mohammed Ali Errachid*** from "Economy Management, Mohammed 5th University of Rabat, Faculty of Law, Economics and Social Sciences-Souissi, Rabat, Morocco" and Ms. ***Maryem Errachid*** from "Universität Paderborn, Warburger Straße 100, 33098 Paderborn, Germany" to accept the language revision. We are warmly grateful the professor ***Mohamed Ghaleb*** from "*Lycée Oulad Youssef, Délégation Béni Mellal, Région Béni Mellal—Khénifra, Maroc*" to discuss the DATA of this work. Also, we are grateful to professors ***Y. Lakhal, F-Z. Baghli, M. Benchagra, A. Balouki, M. Aqil*** and ***A. Boudaoud*** who are the founders of "Engineering and Applied Physics Team (EAPT) of Sultan Moulay Slimane University, Beni Mellal, Morocco" for the fruitful discussion and the provided expertise. A special thanks is addresses to professors ***Mohamed ES-SEQALLY*** from "*Faculté d'Economie et de Gestion, Université Sultan Moulay Slimane, Béni Mellal, Maroc*" and ***Abderrahmane Amhirik*** from "Faculty of Arts and Humanities, Beni Mellal, Morocco" for their valuable proposals.

References

1. H. Kim, E. Park, S. J. Kwon, J. Y. Ohm, H. J. Chang, Renew. Energ **66**, 523 (2014).
2. M. Hosenuzzaman, N.A. Rahim, J. Selvaraj, M. Hasanuzzaman, A. B. M. A. Malek, A. Nahar, Renew. Sust. Energ. Rev **41**, 284 (2015).
3. Z. Lan, J. Li, TELKOMNIKA Indones J. Electr. Eng **12**, 4419 (2014).
4. G. Liu, Inter. J. Elect. Power Energ. Syst **56**, 168 (2014).
5. N.S. Lewis, Chem. Rev **115**, 12631 (2015).
6. N. T. R. N.Kumara, A. Lim, C. M. Lim, M. I. Petra, P. Ekanayake, Renew. Sust. Energ. Rev **78**, 301 (2017).
7. H. Belkhanchi, Y. Ziat, M. Hammi, C. Laghlimi, A. Moutcine, A. Benyounes, Optik **229**, 166234 (2021).
8. Z. Zarhri, Y. Ziat, O. El Rhazouani, A. Benyoussef, A. Elkenz, A, J. Phys. Chem. Solids **94**, 12 (2016).
9. Z. Zarhri, Y. Ziat, O. El Rhazouani, A. Slassi, A. Benyoussef, A. El Kenz, J. Magn. Magn. Mater **406**, 212 (2016).
10. H. Belkhanchi, Y. Ziat, M. Hammi, C. Laghlimi, A. Moutcine, A. Benyounes, In E3S Web. Confer **183**, 05002 (2020).

11. Z. Zarhri, M. Ángel Avilés Cardos, Y. Ziat, M. Hammi, O. El Rhazouani, D. Avellaneda Avellaneda, J. Alloys Compd **819**, 153010 (2019).
12. A. Benyounes, M. Hammi, Y. Ziat, A. Slassi, N. Zahra, Appl. Phys. A **124**, 1 (2018).
13. A. Slassi, Y. Ziat, Z. Zarhri, M. Abdellaoui, and A. Fakhim Lamrani, Phys. Scr **90**, 085801 (2015).
14. A. Slassi, N. lakouari, Y. Ziat, Z. Zarhri, A. F. Lamrani, E. K. Hlil, A. Benyoussef, Solid State Commun **218**, 45 (2015).
15. M. Hammi, Y. Ziat, O. El Rhazouani, A. Slassi, Optik **131**, 399 (2017).
16. J. Arranz-André, W. J. Blau. Carbon **46**, 2067 (2008).
17. T. Sawatsuk, A. Chindaduang, C. Sae-kung, S. Pratontep, G. Tumcharern, Diam. Relat. Mater **18**, 524 (2009).
18. K. Lee, C. Hu, H. Chen, K. Ho, Sol. Energ0 Mater. Sol. Cells **92**,1628 (2008).
19. T. Esch, I. Gadaczek, and T. Bredow, Appl. Surf. Sci **288**, 275 (2014).
20. N. Sakai, A. Fujishima, T. Watanabe, K. Hashimoto, J. Phys. Chem. B **105**, 3023 (2001).
21. M. Nazeeruddin, E. Baranoff, M. Grätzel, Sol. Energy **85**, 1172 (2011).
22. N. Mufti, I. K. R. Laila, H. Fuad, A. Fuad, Int. Conf. Phys. Instrum. Adv. Mater **853**, 1 (2017).
23. R. Bergamasco, F. Da Silva, F. Arakawa, N. Yamaguchi, M. Miranda Reis, C. Tavares, M. T. De Amorim, C. R. Tavares, Chem. Eng. J **174**, 102 (2011).
24. S. Lee and S. Park, J. Ind. Eng. Chem **19**, 1761 (2013).
25. M. Nazeeruddin, E. Baranoff, M. Grätzel, Sol. Energ **85**, 1172 (2011).
26. K. Keis, L. Vayssieres, H. Rensmo, S. E. Lindquist, A. Hagfeldt, J. Electrochem. Soc **148**, A149 (2001).
27. F. A. Carey, R. J. Sundberg, Ad. Org. Chem: Part A: Struct. Mech, **187** (2000).
28. M. Kar, S. Saha, R. Sarkar, S. Pal, P. Sarkar, P, J. Phys. Chem. C **124**, 7652 (2020).
29. Y. Yu, L. Ma, W. Huang, F. Du, J.C. Yu, J. Yu, J. Wang, P.K. Wong, Carbon **43**, 670 (2005).
30. L.P. Zhu, W.Y. Huang, L.L. Ma, S.Y. Fu, Y. Yu, Z.J. Jia, Acta Phys. Chim. Sin **22**, 1175 (2006).
31. L. Spanhel, M. A. Anderson, J. Am. Chem. Soc **113**, 2826 (1991).
32. D. W. Bahneman, C. Kormann, M. R. Hoffmann, J. Phys. Chem **91**, 3789 (1987).
33. Y. Ziat, H. Belkhanchi, M. Hammi, O. Ifguis, Int. J. Photoenergy **2021** (2021).
34. R. Swanepoel, J. Phys. E Sc. Instru **16**, 1214 (1983).
35. Younes Ziat, A. Abbassi, M. Hammi, O. El Rhazouani, Opt. Quantum Electron **48**, 511 (2016).
36. J. Tauc, A. Menth, Non-Crystalline 8, 569 (1972).
37. E. Burstein, Amerc. Phys. Society 93, 632 (1954).
38. T. S. Moss, Proc. Phys. Soc 67, 775 (1954).
39. H. Solar et al., "Electron Affinity and Bandgap Optimization of Zinc Oxide for Improved Performance of ZnO / Si," pp. 1–8, (2019).
40. Sundaram, K.B. Work function determination of zinc oxide films. J. Vac. Sci. Technol. A: Vac. Surf. Films,**15**, 428–430 (1997).
41. S. Kashiwaya, J. Morasch, V. Streibel, T. Toupance, W. Jaegermann, and A. Klein, The Work Function of TiO_2, Surfaces **1**, 73–89, (2018).
42. Baturay, S.; Ocak, Y.S.; Kaya, D. The effect of Gd doping on the electrical and photoelectrical properties of Gd:ZnO/p-Si heterojunctions. J. Alloys Compd, **645**, 29–33 (2015).
43. Ren, X.; Zi, W.; Ma, Q.; Xiao, F.; Gao, F.; Hu, S.; Zhou, Y.; Liu, S.F. Topology and texture controlled ZnO thin film electrodeposition for superior solar cell efficiency. Sol. Energy Mater. Sol. Cells, **134**, 54–59 (2015).

44. Zeng, X.; Wen, X.; Sun, X.; Liao, W.; Wen, Y. Boron-doped zinc oxide thin films grown by metal organic chemical vapor deposition for bifacial a-Si:H/c-Si heterojunction solar cells. Thin Solid Films, **605**,257–262 (2016).
45. Das, S.C.; Green, R.J.; Podder, J.; Regier, T.Z.; Chang, G.S.; Moewes, A. Band Gap Tuning in ZnO Through Ni Doping via Spray Pyrolysis. J. Phys. Chem. C, **2013**, 12745–12753 (2013).

Isolation and characterization of bacteria producing hydrolase enzymes from two moroccan hypersaline environments

*Abdessamad*El berkaoui[1, 2*],*Aicha*Ait alla[2], *Abdellatif*Moukrim[2], *Faissal*El filali[1], *Said*Hanoune[1], *and**Naima* Boum'handi[1].

[1]Laboratory of Microbiology, Seafood Processing and Technology Centre (CSVTPM), National Research Institute of Marine Fisheries (INRH), BP: 1050, Agadir- Morocco.
[2]Laboratory " Aquatic Systems: Marine and Continental Environment, University Ibn Zohr, Faculty of Sciences, BP: 8106, Agadir, Morocco.

Abstract. Halophilic bacteria are a group of microorganisms living in saline environments and in some cases need the salinity to survive. Furthermore,these bacteria species have the potential for interesting and promising applications. In fact, this is due not only to their ability to adapt to extreme physico-chemical conditions but also because many of them they produce interesting substances for the industry such as extracellular hydrolytic enzymes.The present study aims to isolate and characterize halophilic bacteria producing hydrolase enzymes from twosalinesin Sidi Moussa-Oualidia complex (Morocco) and to screen their potential to produce hydrolase enzymes. In this study, 15 halophilic bacteria were identified and analyzed for their ability to produce extracellular enzymes. The phylogenetic analysis based on 16S rDNA sequencing indicated that the 15 isolated strains belong to the genera *Bacillus*, *Oceanobacillus* and *Virgibacillus*. The study of enzymatic activity revealed that all isolates studied were capable of producing at least one extracellular hydrolytic enzyme of which 9 produced amylase, 6 cellulase, 13 DNase, 11 lipase, and 8 protease. In addition to their ability to produce extracellular hydrolytic enzymes, our isolates have demonstrated their potential adaption to extreme Physico-chemical conditions. These properties could allow them to be used for many industrial applications.

1 Introduction

Halophilic and halotolerant bacteria are a group of microorganisms living in saline environments and in some cases need the salinity to survive. To maintain cell structure and function in high salt concentration, these bacteria have developed various biochemical strategies [1].Indeedmost halophilic and halotolerant bacteria studied to date are attracting interest for their capacity to produce compounds with high potential in industrial processes,

*Corresponding author:a.elberkaoui@yahoo.fr

such as extracellular hydrolytic enzymes[2, 1], including protease, amylase, lipase, cellulase, and DNase. These enzymes have very diverse potential uses in different biotechnological and industrial applications such as food, biomedical, pharmaceutical, detergents, cosmetic and textile industries[3-5]. Moreover, research on the biotechnological potential of halophilic bacterial enzymes is increasing because of their stable activity at a high salt concentration as well as at high temperature and alkaline pH.These properties would allow halophilic enzymes to be used in industrial processes requiring harsh conditions[6-8]. However, to date,very few halophilic enzymes have been used in biotechnological applications or industrial processes compared to enzymes from alkalophilic and thermophilic bacteria[9]. The Sidi Moussa-Oualidia complex is an important Moroccan wetland including several hypersaline environments: Sidi El Abed salines,Sidi M'bareksalines, Sidi Moussa lagoon and salines, andOualidia lagoon and salines. It is located on the Atlantic coast of Morocco, in the province of El Jadida. Accordingly,it was designated as a wetland of international importance under the Ramsar Convention in 2005. However,until now, no environmental study has focused on the phylogenetic diversity of halophilic bacteria in the Sidi Moussa-Oualidia complex. Therefore, the present study aimed to isolate and characterize the halophilic bacteria producing hydrolase enzymes from the Sidi El Abed and Sidi Moussa salines.

2 Materials and methods

2.1 Study area

Sediment samples were collected from two sites in the Sidi El Abed (33°02'N, 08°42'W) and Sidi Moussa (33°01'N, 08°43'W) salines located on the Atlantic coast of Morocco(Figure 1).

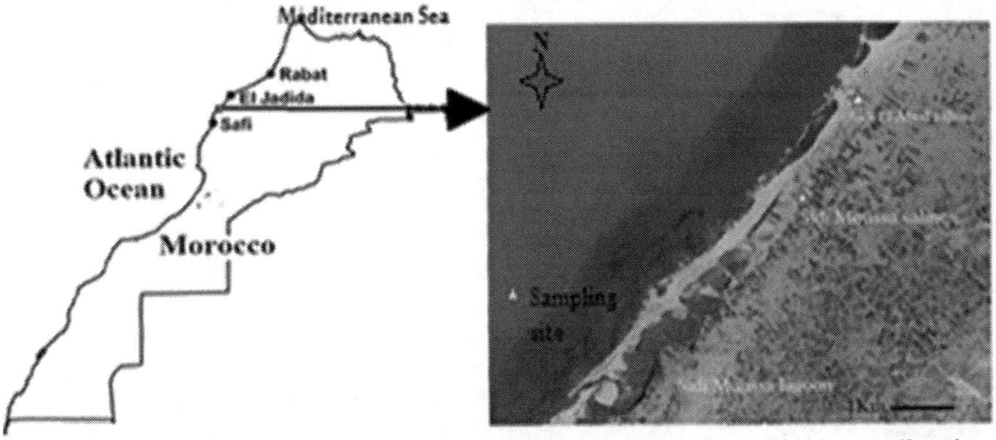

Figure 1: Geographical location of the Sidi El Abed and Sidi Moussa salines with the sampling sites (Google Earth image).

2.2 Isolation of bacterial strains

15 g of each sediment sample was collected and homogenized in 15 mL of sterilized saltwater with 5% NaCl and serially diluted up to10^{-4}dilution. For the isolation of the halophilic bacteria, 0.1 ml of each serially diluted sample was plated on one of five different salt concentrations (2%, 5%, 10%, 15% and 20%) in the Columbia medium containing the following composition (g per L): polypeptones (17); pancreatic heart peptone (3); corn starch (1); sodium chloride (5); yeast extract (3); agar (13.5); pH 7.3 +/-

0.2 (Biokar Diagnostics, Beauvais, France) and incubated at 30°C for 48 h. 192 Distinct colonies were selected and subcultured several times on Columbia agar to obtain high purity. The isolates were stored in 20% of glycerol at -80°C for further analysis.

2.3 16S rDNA sequencing and phylogenetic analysis

Fifteen strains were randomly selected according to their phenotypic, biochemical and physicochemical characteristics and to their capacity to synthesize extracellular hydrolytic enzymes. The genomic DNA of these strains was extracted and purification by Genomic DNA Mini Kit (Invitrogen, USA) following the procedure suggested by the manufacturer.The 16S rRNA was amplified using PCR Supermix Kit (Invitrogen, USA) and the universal primers FD1 (AGAGTTTGATCCTGGCTCAG) and rP2(ACGGCTACCTTGTTACGACTT)synthesized by Bio Basic Inc[10].The amplification procedure was performed according to the manufacturer's recommendations on the MiniAmp thermal cycler (Applied Biosystems, USA). Sanger sequencing was performed at National Center for Scientific and Technical Research (CNRST, Morocco). Preliminary identifications were realized based on sequence assembly and by query in the NCBI database. Phylogenetic analysis was performed using the software MEGA 7 [11]. The partial sequences of the 16S rDNA gene have been deposited in the NCBI database under the following accession numbers MZ317300-MZ317314.

2.4 Morphological and physiological characterization

The morphological characteristics of the isolates were determined by standard methods [12]. The salt tolerance of the bacterial strains was tested on Columbia agar with 5%, 10%, 15% and 20% NaCl and incubated at 30°C. The pH effect on isolates was examined by growing them on Columbia agar with a three pH value (4.5, 7.5 and 9.2) and incubating them at 30°C. Growth at different temperatures was evaluated on Columbia agar after incubation at 10, 30 and 50°C. the growth was measured by visual observation after incubation for 24 to 72 hours.

2.5 Determination of extracellular hydrolytic activities

Screening of strains for extracellular hydrolytic activities was performed as follows: Extracellular cellulase activity was revealed as described by Sadfi-Zouaoui et al [13]. DNase activity was tested according to Jeffries et al [14]. Amylolytic, lipolytic and proteolytic activities were assayed as described by Cowan [15], Sierra [16] and Sánchez-Porro et al [8] respectively. The halo diameter of hydrolytic activity (Hd) and colony diameter (Cd) of each strain were measured and the difference between Hd and Cd was calculated and considered as an estimate of hydrolytic activity (Figure 2).

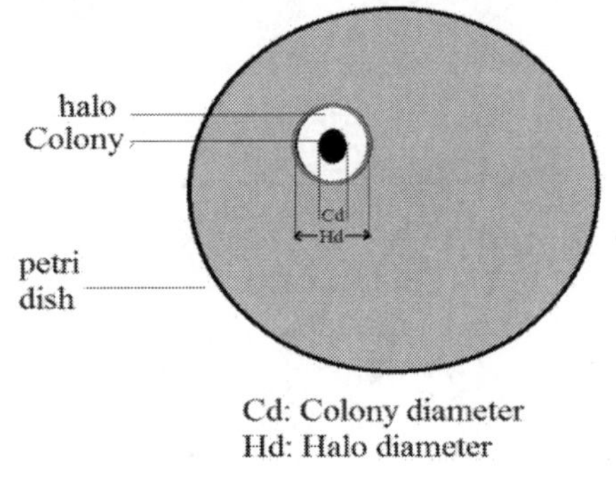

Cd: Colony diameter
Hd: Halo diameter

Figure 2:Method used to measure hydrolytic activity.

3 Results and discussion

3.1 Identification of bacterial strains.

Comparative analysis of the 16S rRNA partial sequences of the strains isolated from the sediments of Sidi El Abed and Sidi Moussa salterns revealed the following distribution: *Bacillus subtilis* has four strains, *Bacillus licheniformis* has three strains, *Cytobacillusoceanisediminis*(The new name for *Bacillus oceanisediminis*[17])and *Bacillus halotolerans* each have two strains, and *Bacillus mojavensis*, *Bacillus aquimaris, Virginibacillussalarius* and *Bacillus spizenii*(Previously considered a subspecies of *Bacillus subtilis* but was promoted to species status byDunlap et al. [18]) each have one strain (Table 1).The phylogenetic tree constructed byUPGMAmethod, presented in Figure 3, showed similar results.

This result reveals that *Bacillus* is the most dominant genus, with 12 strains representing 6 species; the other two genera, *Cytobacillus* and *Virgibacillus*, were represented by two and one species, respectively. Therefore,*Bacillus* is very adapted to saline sediments, as proven by this study.

Table 1: Affiliation of the 16S rRNA gene sequences and accession numbers of isolated halophilic bacteria.

Representative isolate	Tentative identification based on nearest neighbor	16S similarity (%)	GenBank accession number
HBFO11	*Virgibacillussalarius*	99.88%	MZ317305
HBSW27	*Cytobacillusoceanisediminis*	100%	MZ317311
HBFW7	*Cytobacillusoceanisediminis*	100%	MZ317310
HBFW2	*Bacillus subtilis*	100%	MZ317309
HBSO12	*Bacillus subtilis*	100%	MZ317302
HBFW4	*Bacillus subtilis*	100%	MZ317313
HBFW17	*Bacillus subtilis*	100%	MZ317308
HBSO7	*Bacillus spizizenii*	100%	MZ317301
HBSW4	*Bacillus mojavensis*	100%	MZ317307
HBSO2	*Bacillus licheniformis*	100%	MZ317300
HBSO20	*Bacillus licheniformis*	99.69%	MZ317304
HBFW20	*Bacillus licheniformis*	100%	MZ317314
HBSO19	*Bacillus halotolerans*	100%	MZ317303
HBSW29	*Bacillus halotolerans*	100%	MZ317312
HBFO12	*Bacillus aquimaris*	100%	MZ317306

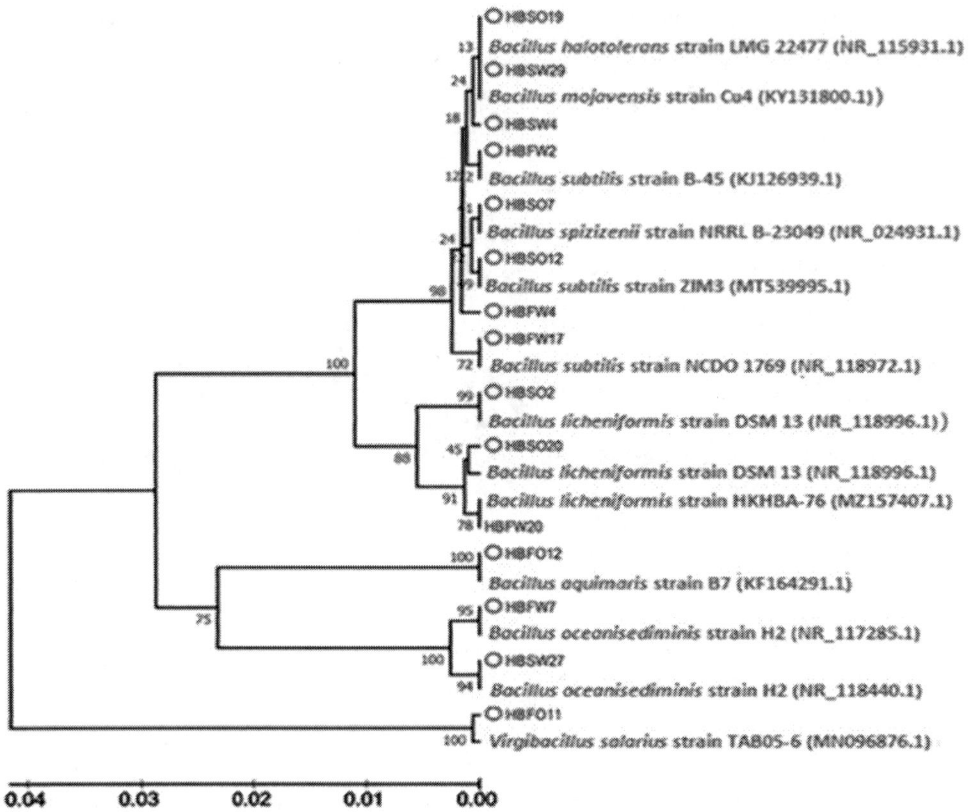

Figure 3: Tree of phylogenetic based on16S rRNA gene sequences analysis from the halophilic bacteria isolated from Sidi El Abed and Sidi Moussa salterns. The red circle indicates the 15 strains analyzed. Bar, 0.01 substitutions per nucleotide position.

3.2 Effect of physico-chemical conditions on bacterial strain growth

The growth of bacterial strains under different physico-chemical conditions is reported in Table 2. All strains were able to grow optimally on medium with 5% NaCl,thirteen of them were able to grow on medium with 10% NaCl,only two strains belonging to *B. subtilis* were able to tolerate a concentration of 15% and no strain tolerate 20% NaCl. These results indicated that these strains are moderately halophilic as classified byKushner and Kamekura[19].

The results of the Influence of temperature of incubation showed that all strains were able to grow at 30°C and 50°C except three strains, that could not tolerate 50°C, and only two strains: *C.oceanisediminis* HBSW27 and *B.aquimaris* HBFO12 were able to grow at 10°C. Hence, the strains identified in this study are designated as mesophilic and, in some cases, thermotolerant. Moreover, all isolates studied were tolerant of neutral pH, with 8 strains designated as acid-tolerant and 8 strains also considered alkaline-tolerant [20]. The pH tolerance of *V.salarius* HBFO11 and *B.subtilis* HBSO7 was quite wide (4.5 to 9.4).

The physico-chemical characterization of the isolated bacteria proved the great biotechnological potential of our strains, due to their ability to tolerate other extreme conditions than salt.

Table 2: Growth of halophilic bacteria identified at different physico-chemical conditions.

Species	Strains	[NaCl]				pH			Temperature		
		5%	10%	15%	20%	4.5	7.2	9.4	10 °C	30 °C	50 °C
V.salarius	HBFO11	++	+	-	-	+++	++	++	-	+++	++
C. oceanisediminis	HBSW27	++	+	-	-	-	++	-	-	++	+
	HBFW7	+++	-	-	-	-	+++	+	+	+++	-
B.subtilis	HBFW4	+++	++	-	-	++	++	-	-	++	++
	HBFW17	+++	++	-	-	+++	+++	-	-	+++	+++
	HBSO12	+++	++	-	-	++	+++	+	-	+++	-
	HBFW2	+++	++	+	-	+++	+++	-	-	+++	+++
B.spizizeni	HBSO7	++	++	+	-	+++	++	++	-	++	++
B.mojavensis	HBSW4	+++	+	-	-	-	+++	++	-	+++	+
B.licheniformis	HBSO20	++	++	-	-	-	++	++	-	++	+++
	HBSO2	+++	++	-	-	-	+++	++	-	+++	+++
	HBFW20	++	++	-	-	+	++	-	-	++	+++
B.halotolerans	HBSO19	+++	++	-	-	+	++	++	-	++	++
	HBSW29	++	-	-	-	-	++	-	-	++	+
B.aquimaris	HBFO12	++	++	-	-	-	++	-	+	++	-

Symbols: + growth; - no growth.

3.3 Hydrolytic activities of halophilic bacteria strains

The enzymatic activity results revealed that all identified strains were capable of producing at least one extracellular hydrolytic enzyme. Indeed, nine strains produce amylase, six strains produce cellulase, thirteen strains produce DNase, eleven strains produce lipase, and eight strains produce protease. The results of the combined hydrolytic enzyme activities show that seven strains (HBFO12, HBSO20, HBSW4, HBSO12, HBFW4, HBFW17, HBFO11) were able to produce four hydrolytic activities, four strains (HBSO2, HBFW2, HBFW7, HBSW27) showed three hydrolytic activities, three strains (HBSO19, HBFW20, HBSO7) displayed two hydrolytic activities and one strain (HBSW29) had one hydrolytic activity (Table 3).

The current study indicated that DNase was the most frequent activity, followed by lipase, amylase, protease and cellulase activities respectively. In a similar study, Moreno et al. [21]investigated the ability of halophilic bacteria to produce various extracellular hydrolytic enzymes (lipase, protease, DNase, amylase, pullulanase and xylanase) in the Atacama Desert, and also found that DNase producers were the most abundant isolates.In contrast, Kaitouni et al.[22] found that cellulase was the most represented enzyme activity among halophilic bacteria isolated from hypersaline environments of the pre-Rif region (Morocco). Indeed, the difference in bacterial diversity between the hypersaline environments of northern [22]and northwestern Morocco (current study) could be attributed to the fact that the two studies have only one species in common(*B. subtilis*).

Table 3: Hydrolytic activity of halophilic bacteria strains.

Species	Strain Number	Hydrolytic activity (mm)				
		Lipase	Cellulase	Protease	Amylase	Dnase
V.salarius	HBFO11	1	8	0	4	2
C.oceanisediminis	HBSW27	0	4	0	1	14
	HBFW7	0	6	0	4	10
B. subtilis	HBFW4	3	0	5	4	3
	HBFW17	3	0	4	4	1
	HBSO12	0	8	8	8	4
	HBFW2	12	0	1	0	8
B.spizizenii	HBSO7	4	0	0	0	4
B.mojavensis	HBSW4	10	4	0	1	8
B.licheniformis	HBSO20	6	12	0	4	2
	HBSO2	3	0	1	0	4
	HBFW20	0	0	2	0	1
B.halotolerans	HBSO19	4	0	1	0	0
	HBSW29	2	0	0	0	0
B.aquimaris	HBFO12	1	0	6	2	12

Note: After incubation, the hydrolytic activity of each strain was measured as follows: Colony diameter - Halo diameter (mm).

The enzyme profiles for each species are presented in Table 3. All four strains belonging to *B.subtilis*were capable of producing protease and DNase, three were lipase and amylase producers, and only one showed cellulase activity. Furthermore, *B. subtilis* HBSO12 showed great biotechnological potential, due to its high capacity to produce all the enzymes studied except lipase, and it is also the most protease-producing strain. Moreover, several studies have already proven that *B. subtilis* is a source of extracellular protease [23-26]. In addition, *B. subtilis*has been previously reported as an excellent producer of different extracellular hydrolytic enzymes, including cellulase, amylase, lipase and protease, with satisfactory characteristics for many industrial applications such as industrial fermentation of sugars, soybeans and tobacco[27-30]. Indeed, the *B. subtilis* strains isolated in this study were able to produce all the extracellular enzymes already mentioned.

B. licheniformis, the second most represented species in this study, displayed low enzymatic activity, except for strain HBSO20 which showed the highest productivity of cellulase. This strain is a promising producer of other enzymes besides cellulase compared to previous studies [31, 32]. It also showed the ability to grow significantly at temperatures of 50°C and tolerance to alkaline pH.Accordingly, these characteristics give the strain HBSO20 the potential to produce thermostable and halo-alkaline enzymes for use in industries that require extreme conditions.

C.oceanisediminis,B.aquimaris, B.halotolerans, B.mojavensis, and *B.spizizenii* are among the species identified in this study and have already demonstrated technological interest in previous studies. Firstly,*C.oceanisediminis* the species that produced the most Dnase in this study has been previously identified as a source of these extracellular enzymes.[33, 21].secondly,the *B.aquimaris* species, which was shown to be capable of producing amylase, protease, Dnase, and lipase in this study, has previously demonstrated its biotechnological importance in several studies[34, 35]. Thirdly, both strains identified as *B.halotolerans*were capable of producing a lipase. Moreover, in a recent study, a lipase from this species was characterized and showed properties suitable for various industrial applications [36]. Fourthly, *B.mojavensis* HBSW4, the second most lipase-producing strain in this study, has also shown the ability to secrete Dnase, cellulase and amylase.Similarly,Moreno et al [21] had previously reported the potential of this species to produce lipase and Dnase. Fifthly, *B.spizizenii* has shown the ability to produce only lipase and Dnase.On the other hand,a recent study reported the ability of this species to produce cellulase and protease, which has not been demonstrated by our strain [37].Lastly, HBFO11 identified as a member of *V.salarius* showed the ability to produce all hydrolases studied, except protease. As well, according to a similar study, this species was reported as a producer of cellulase and nuclease[38].

The species isolated in this study were previously isolated from several environmental extremes and showed high enzymatic potential as listed in table 4.

Table 4: Previous studies that isolated the same species as the present study with hydrolase activities

Isolation Site	Number of identified strains	Species similar to the species isolated in this study. (Number of strains)	Hydrolytic Activity	References
Debagh Hot Spring (Algeria)	41	*B.mojavensis (16)* *B.subtilis (3)* *B.licheniformis (11)*	Not determined	[39]
Mahala salt lake in Fars province (southern Iran)	13	*B.aquimaris(1)* *B.subtilis (2)*	Lipase	[40]
Maharloo hypersaline lake located in the south of Shiraz, Iran.	16	*B. subtilis (3)*	Protease	[23]
saline lakes ecosystems (Algeria)	74	*C.oceanisediminis (1)* *B.subtilis (1)* *B.spizizenii (1)* *B.halotolerans, (3)* *V.salarius (4)*	Amylase Cellulase Esterase Gelatinase Inulinase , Nuclease Pectinase Protease Xylanase .	[41]
Natural and artificial hypersaline environments in the pre-Rif region (northern Morocco)	56	*B. subtilis (6)*	Cellulase Amylase Protease Pectinase Inulinase	[22]
Marsh and two salterns from Lower Loukkos(west of Morocco)	122	*C.oceanisediminis (8)* *B.licheniformis(1)* *B.aquimaris (32)*	Protease Cellulase Amylase DNase	. [33]
Atacama Desert in northern Chile	25	*C.oceanisediminis (1)* *B.mojavensis (1)* *B.licheniformis (1)* *B.aquimaris (1)*	Protease Lipase Amylase DNase Pullulanase Xylanase	[21]
This study: The Sidi Moussa-Oualidia complex, Morocco	15	*V.salarius (1)* *C.oceanisediminis(2)* *B.subtilis (4)* *B.spizizenii(1)* *B.mojavensis(1)* *B.licheniformis (3)* *B.halotolerans(2)* *B. aquimaris(1)*	Protease Lipase Amylase DNase Cellulase	

4 Conclusion

The study of the biodiversity of halophilic bacteria producing extracellular enzymes from the salines of Sidi Moussa and Sidi El Abed allowed the identification of 15 strains belonging to eight species (*V.salarius, C. oceanisediminis B. subtilis, B. spizizenii, B. mojavensis ,B. licheniformis B. halotolerans and B. aquimaris*) and showed the presence of higher hydrolase activity in all identified halophilic species. Furthermore, the growth of bacterial strains under different physicochemical conditions showed a great diversity within the same species. Therefore, these properties make our isolated strains potentially useful in industrial processes where extreme conditions would inhibit ordinary enzymes. Indeed, to explore the diversity of halophilic bacteria in Morocco, other research is currently underway to screen halophilic bacteria producing extracellular hydrolytic enzymes in other Moroccan hypersaline environments.

References

1. A. Ventosa, J.J. Nieto, and A. Oren. Microbiol. Mol. Biol. Rev. **62**(2): p. 504-544 (1998).
2. R. Margesin and F. Schinner. Extremophiles. **5**(2): p. 73-83 (2001).
3. N. Kulkarni, A. Shendye, and M. Rao. FEMS microbiology reviews. **23**(4): p. 411-456 (1999).
4. F. Niehaus, C. Bertoldo, M. Kähler, and G. Antranikian. Applied microbiology and biotechnology. **51**(6): p. 711-729 (1999).
5. A. Pandey, P. Selvakumar, C.R. Soccol, and P. Nigams. Current science: p. 149-162 (1999).
6. J. Gomes and W. Steiner. Food technology and Biotechnology. **42**(4): p. 223-225 (2004).
7. S. Mechri, K. Bouacem, M. Amziane, A. Dab, F. Nateche, and B. Jaouadi. BioMed Research International. **2019** (2019).
8. C. Sánchez-Porro, S. Martin, E. Mellado, and A. Ventosa. Journal of applied microbiology. **94**(2): p. 295-300 (2003).
9. A. Oren. Environmental technology. **31**(8-9): p. 825-834 (2010).
10. W.G. Weisburg, S.M. Barns, D.A. Pelletier, and D.J. Lane, *16S ribosomal DNA amplification for phylogenetic study*. Journal of bacteriology. **173**(2): p. 697 (1991).
11. S. Kumar, G. Stecher, M. Li, C. Knyaz, and K. Tamura,. Molecular biology and evolution. **35**(6): p. 1547 (2018).
12. L. Prescott, J. Harley, and D. Klein. *McGraw-Hill Higher education Companies.* Inc., New York (2003).
13. N. Sadfi-Zouaoui, B. Essghaier, M. Hajlaoui, M.-L. Fardeau, J. Cayaol, B. Ollivier, and A. Boudabous. Journal of Phytopathology. **156**(1): p. 42-52 (2008).
14. C.D. Jeffries, D.F. Holtman, and D.G. Guse. Journal of Bacteriology. **73**(4): p. 590 (1957).
15. D.A. Cowan. Biotechnology, the Science and the Business: p. 311-340 (1991).
16. G. Sierra. Antonie van Leeuwenhoek. **23**(1): p. 15-22 (1957).
17. S. Patel and R.S. Gupta. International journal of systematic and evolutionary microbiology. **70**(1): p. 406-438 (2020).
18. C.A. Dunlap, M.J. Bowman, and D.R. Zeigler. Antonie van Leeuwenhoek. **113**(1): p. 1-12 (2020).
19. D.J. Kushner and M. Kamekura. Halophilic Bact. **1**: p. 109-140 (1988).
20. H. Saito and H. Kobayashi. Science progress. **86**(4): p. 271-282 (2003).
21. M. Moreno, F. Piubeli, M. Bonfa, M. García, L. Durrant, and E. Mellado. Journal of applied microbiology. **113**(3): p. 550-559 (2012).

22. L.B.D. Kaitouni, J. Anissi, K. Sendide, and M. El Hassouni. Annals of Microbiology. **70**(1): p. 1-15 (2020).
23. Y. Ghasemi, A.S. RSOUL, A. Ebrahiminezhad, A. Kazemi, M. Shahbazi, and N. Talebnia, *Screening and isolation of extracellular protease producing bacteria from the Maharloo Salt Lake.* (2011).
24. W. Kim and S. Kim. Journal of food biochemistry. **29**(5): p. 591-610 (2005).
25. T. Maruthiah, P. Esakkiraj, G. Prabakaran, A. Palavesam, and G. Immanuel,. Biocatalysis and Agricultural Biotechnology. **2**(2): p. 116-119 (2013).
26. N. Nisha and J. Divakaran. African Journal of Biotechnology. **13**(16) (2014).
27. Y. Bakri, S. Elkhouri, M. Harba, and Y. Akeed. Biological Sciences-PJSIR. **63**(2): p. 113-118 (2020).
28. J. Dai, A. Dong, G. Xiong, Y. Liu, M.S. Hossain, S. Liu, N. Gao, S. Li, J. Wang, and D. Qiu. Frontiers in Microbiology. **11**(1539) (2020).
29. T. Eggert, U. Brockmeier, M.J. Dröge, W.J. Quax, and K.-E. Jaeger. FEMS Microbiology Letters. **225**(2): p. 319-324 (2003).
30. T. Peláez, R. Guillermo, and C. Yuliet. Acta Microscopica. **29**(5) (2020).
31. F. Azadian, A. Badoei-Dalfard, A. Namaki-Shoushtari, and M. Hassanshahian,. Molecular Biology Research Communications. **5**(3): p. 143 (2016).
32. S. Shajahan, I.G. Moorthy, N. Sivakumar, and G. Selvakumar. Journal of King Saud University-Science. **29**(3): p. 302-310 (2017).
33. I. Berrada, A. Willems, P. De Vos, E. El Fahime, J. Swings, N. Bendaou, M. Melloul, and M. Amar. African Journal of Microbiology Research. **6**(10): p. 2419-2434 (2012).
34. P. Shivanand and G. Jayaraman. Process Biochemistry. **44**(10): p. 1088-1094 (2009).
35. N. Trivedi, V. Gupta, M. Kumar, P. Kumari, C. Reddy, and B. Jha. Chemosphere. **83**(5): p. 706-712 (2011).
36. S.C. Podeti, Ramya & Alpula, Nagaraju & Lapaka, Suresh. International Journal of Pharmacy and Biological Sciences.: p. 1-16 (2020).
37. F. Masmoudi, S. Tounsi, C.A. Dunlap, and M. Trigui. Plant Cell Reports: p. 1-15 (2021).
38. T. Menasria, M. Monteoliva-Sánchez, L. Benammar, M. Benhadj, A. Ayachi, H. Hacène, A. Gonzalez-Paredes, and M. Aguilera. World Journal of Microbiology and Biotechnology. **35**(9): p. 132 (2019).
39. M. Arab, S. Bakour, R. Lalaoui, N. Aissaoui, F. Nas, A. Hoceini, P.-E. Fournier, and N. Klouche-Khelil. Geomicrobiology Journal. **36**(2): p. 137-147 (2019).
40. Y. Ghasemi, S. Rasoul-Amini, A. Kazemi, G. Zarrini, M. Morowvat, and M. Kargar, *Isolation and characterization of some moderately halophilic bacteria with lipase activity.* Microbiology. **80**(4): p. 483 (2011).
41. T. Menasria, M. Monteoliva-Sánchez, L. Benammar, M. Benhadj, A. Ayachi, H. Hacène, A. Gonzalez-Paredes, and M. Aguilera. World Journal of Microbiology and Biotechnology. **35**(9): p. 1-16 (2019).

Effect of quince (*Cydonia oblonga Miller*) fruit extract against oxalocalcic crystallization

Ibtissam Elhadri[1*], *Latifa* Baddade[1,2], *Moulouda* El mouftari[1], and *Mohamed* Berkani[1]

[1] Laboratory of Applied Spectro-Chemometry and Environment, University Sultan Moulay Slimane, Faculty of Science and Technology, Beni Mellal, Morocco.
[2]Transdisciplinary Team of Analytical Sciences for Sustainable Development, University Sultan Moulay Slimane, Faculty of Science and Technology, Beni Mellal, Morocco.

Abstract. Kidney stone or Urolithiasis is a solid particle that forms in the urinary tract. In Morocco, as in many countries, most patients use medicinal plants as an alternative therapy for many diseases, including lithiasis. The fruit of *Cydonia oblonga Miller.* is traditionally used for the prevention and treatment of several diseases. The present study aims to explore the effects of aqueous extracts of pulp and peel quince on oxalocalcic crystallization. These results show that the aqueous extracts of *C. oblonga M* can inhibit calcium oxalate crystallization.

1 Introduction

Urinary calculi affect a large part of the population, it is evident that urine supersaturation is an important factor in the formation of these urinary concretions. Indeed, when the over-saturation is sufficient, it leads to the nucleation of crystals and their growth, then followed by aggregation, which lead to the formation of stones [1]. In general, kidney stones are composed of calcium salts, uric acid, cystine and struvite, depending on the main etiological entity. Each type of stone has its own group of causes, so the management of each entity is specific [2].

According to the literature, approximately 70% of human urinary stones contain calcium phosphate and calcium oxalate. Calcium oxalate monohydrate (COM) or Whewellite ($CaC_2O_4.H_2O$), calcium oxalate dihydrate (COD) or Weddellite ($CaC_2O_4.2H_2O$), and calcium oxalate trihydrate (COT) or CaOxite ($CaC_2O_4.2H_2O$) are three forms of hydrated calcium oxalates, which can present in urinary stones [3].

Urine contains many molecules able to inhibit crystallization. Dietary factors appear to have an effect on the ability of urine to inhibit calcium oxalate crystallization[1]. The potential use of natural substances to inhibit crystallization is more widespread, we therefore tested the inhibitory effects of *Cydonia oblonga Miller.*, the common Quince, belonging to the Rosaceae family, genus *Cydonia*[4].

In this paper, we report the effects of aqueous extracts of quince fruit on experimental calcium oxalate crystals.

* Corresponding author: ibtissam.elhadri00@gmail.com

2 Material and Methods

2.1 Fruit collection and extract preparation

Healthy quince (*C. oblonga Miller*) fruit samples were collected in October 2019 from the region of Beni Mellal - khenifra, fruits were separated into peel, pulp and seeds, packed into plastic bags and frozen at -20 °C until used.
Aqueous extract of *C. oblonga Miller* was obtained by maceration of homogenized fruit material in water for three days in a refrigerated chamber at 4°C in the dark, then the mixture was centrifuged at 10000g for 15 min at room temperature, the supernatants were collected by filtration using Whatman Filter Paper (pore size 6 mm). Extracts were kept at 4°C in the dark, the extract's concentration was between 1 and 10 mg/ml.

2.2 Crystallization of CaOx

The inhibition of the crystallization was studied according to the method described by Beddade et al [5], In this respect, stock solutions of calcium chloride ($CaCl_2$) and ammonium oxalate $(NH_4)_2C_2O_4$ were prepared at a final concentration of 4 mmol/L, in a buffer containing sodium acetate (0.2 mol/L) and NaCl (0.15 mol/L) at pH 6. Before being used in crystallization experiments, both solutions were filtered through an 0.22 mm filter and warmed to 37 ◦C in order to approximate urinary physiological conditions.

2.3 Crystallization without inhibitor

$CaCl_2$ solution of 4 mM was transferred into in a beaker, to which we added an equal volume (10 ml) of equimolar solution of $(NH_4)_2C_2O_4$, to trigger the formation of the crystals. After incubating the mixture for 2 h under stirring in the water bath at 37 °C, the crystals were analysed morphologically under light microscope.

2.4 Crystallization with inhibitor

The inhibitor effect on crystallization was studied with several concentrations of the extract ranging from 1 to 10 mg/ml. Nevertheless, it should be mentioned that the inhibiting substance was added to the sodium acetate solution to adjust the pH value to 6. The experiments were carried out with 10 mL of $CaCl_2$ and 10 mL of the extract to which 10 mL of $(NH_4)_2C_2O_4$ was added, which promotes both the formation of the crystals.

2.5 Microscopic observation

The solutions were maintained at 37 °C in a water bath under constant stirring for 2 hours, then the number and the morphology of crystals in each concentration of extract was carried out using a polarized optical microscope. Each test was performed in triplicate.

2.6 Crystal characterization

The crystals were recovered through a 0.22 μm microfiltration membrane and dried under vacuum at 37 °C for 24 hours. The obtained solid samples were characterized by X-ray diffraction (XRD), also the chemical composition of the crystals was determined by Fourier transform infrared spectra (FTIR) between 4000 and 450 cm^{-1} with a resolution of 4 cm^{-1}.

3 Results

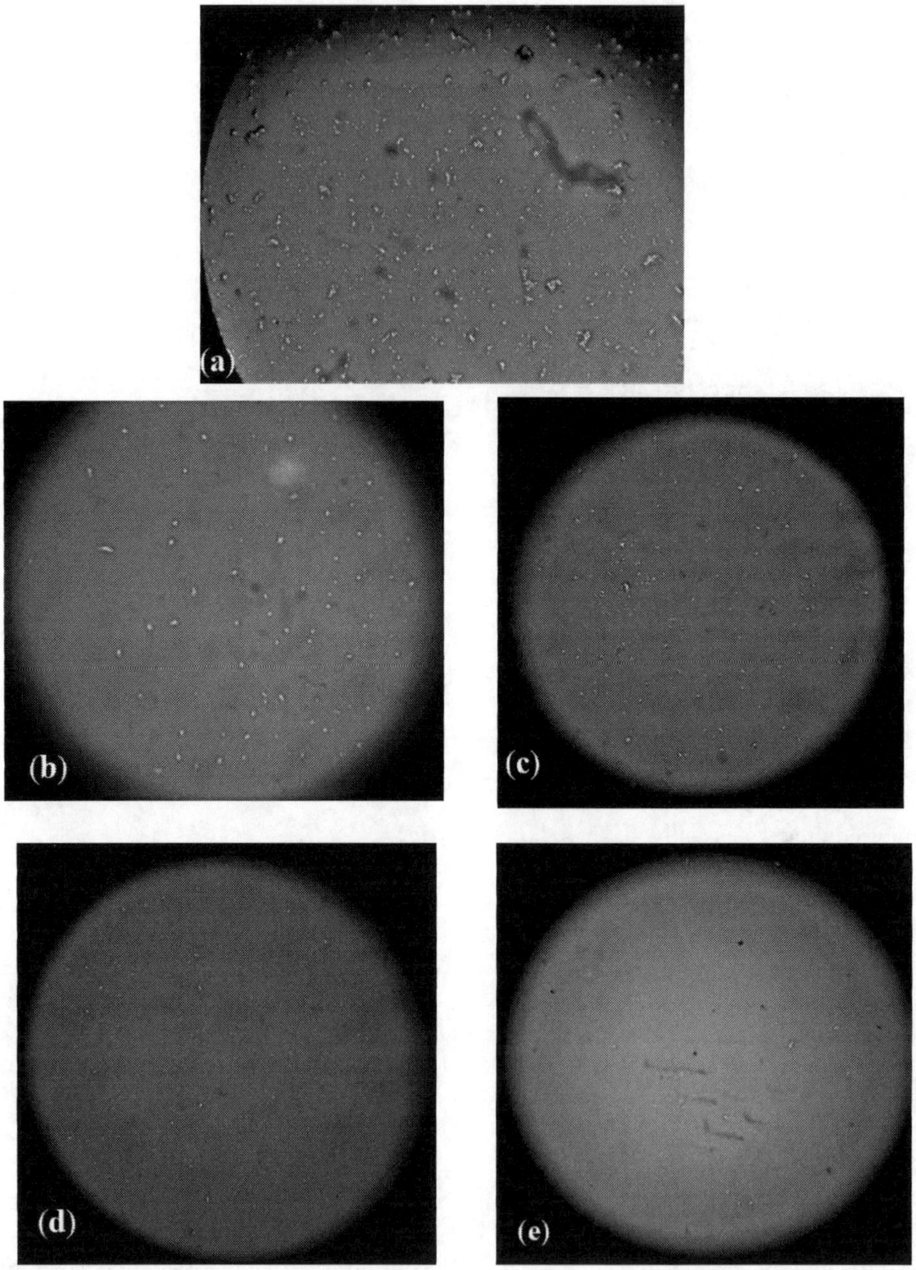

Fig. 1 Micrographs of calcium oxalate crystals in the presence of aqueous extract of *C. oblonga Miller.* pulp: (a)= 0 mg/ml, (b)= 1 mg/ml, (c)= 5 mg/ml, (d)= 8 mg/ml and (e)= 10 mg/ml (Gross×200)

3.1 Microscopic observation

The preventive effect of the aqueous extract of quince pulp and peel against the calcium oxalate's crystallization was investigated in vitro at a constant temperature of 37 °C using synthetic super saturated urine. The latter was prepared with calcium dichloride $CaCl_2$ and ammonium oxalate $(NH_4)_2C_2O_4$. However, the development of calcium oxalate crystals was observed under a microscope after magnetic stirring of a synthetic urine according to the content of the aqueous extract of *C. oblonga Miller* pulp and peel in order to detect the minimum concentration of the aqueous extract which can inhibit the crystallization of calcium oxalate.

The light microscopic photographs taken of the control group reveals the presence of aggregates in addition to a predominance of the crystals as shown in figure 1(a). Furthermore, the incubation of the calcium oxalate crystals in the presence of the different concentrations fruits extracts caused significant decreases in the number and size of the latter compared to control group in a dose-dependent manner. Also, the analysis of the pictures taken for different samples in figure 1 and figure 2 shows that treatment with the *C. oblonga Miller* peel extract had a more notable effect than *C. oblonga Miller* pulp extract on CaOx crystals inhibition, significantly.

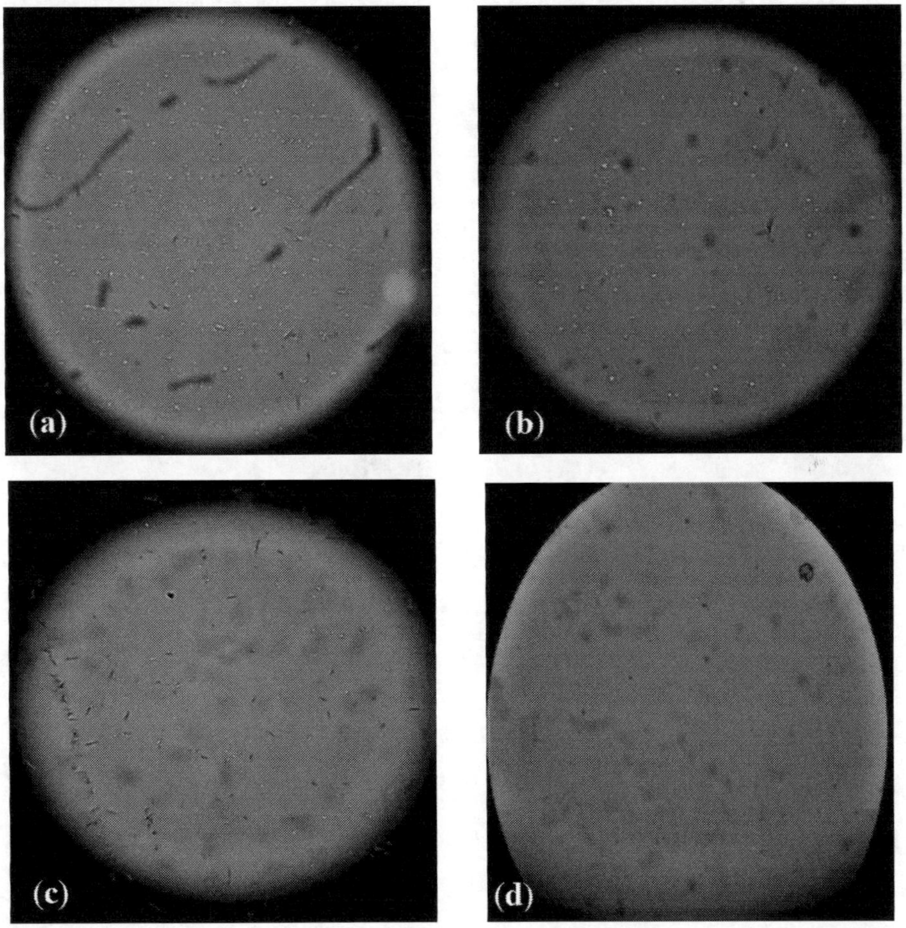

Fig. 2 Micrographs of calcium oxalate crystals in the presence of aqueous extract of *C. oblonga Miller*. peel: (a)= 1 mg/ml, (b)= 5 mg/ml, (c)= 8 mg/ml and (d)= 10 mg/ml (Gross×200)

Comparison of the images in both figure 1 and figure 2 shows that *C. oblonga Miller* peel and pulp extracts at a dose of 1 mg/mL decreases the number and size of crystals by reducing supersaturation. Increasing the extracts concentration shows a significant effect on crystal size and number. This result can be demonstrated by the inhibition of growth and crystal

 https://doi.org/10.1051/e3sconf/202233705005

aggregation by the polyphenols contained in the plant or by the dissociation of the aggregates by the bioactive molecules of the extracts.

3.2 XRD analysis of CaOx crystal's

The XRD pattern of the synthesized crystals was fitted by Rietveld refinement using Full prof software. Figure 3 shows the XRD experimental pattern presented by the black line, while the red line shows the calculated data and blue line provides the differences between both. It can be observed that the crystal phase composition of the solid shows diffraction peaks at 14.919°, 24.381°, 30.072°, 35.978°, 38.142°, 40.588°, and 43.478° which correspond to the (-101), (020), (-202), (112), (130), (202), and (321) [6], [7] (PDF #21-0838) respectively, revealing the presence of COM. In addition, we detected the peaks at 11.019° and 16.113° representing the index of reflective planes for (011) and (110) [8], suggesting the presence of DOC and TOC respectively. These data indicate clearly the monoclinic structure of the synthesized crystals.

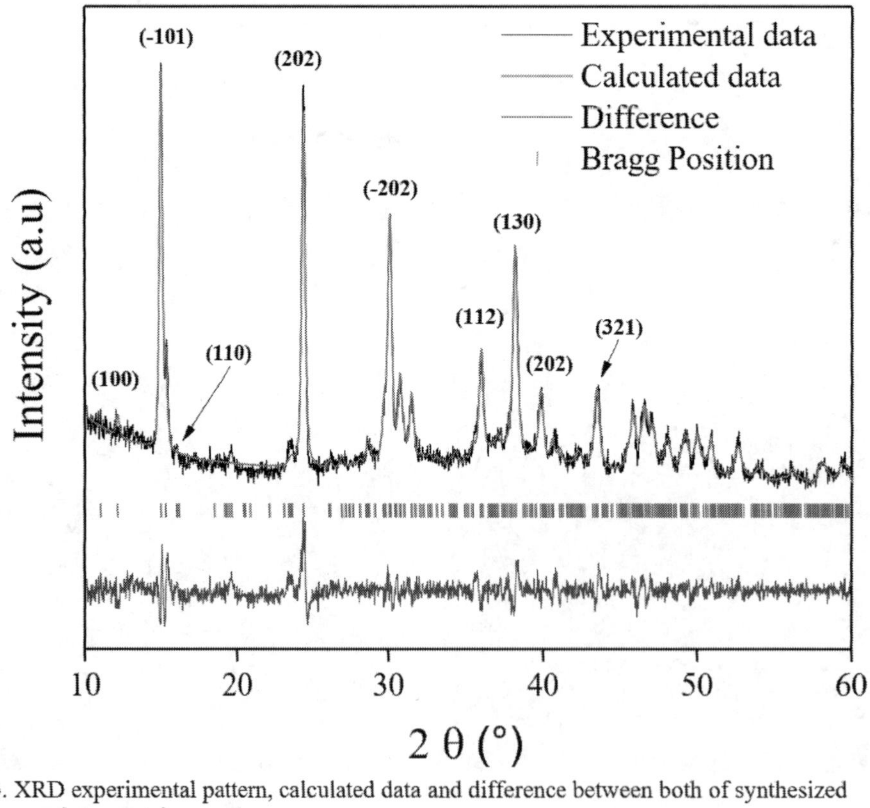

Fig .3. XRD experimental pattern, calculated data and difference between both of synthesized calcium oxalate crystals.

3.3 Characterization by Fourier Transform Infrared Spectroscopy (FTIR).

The synthesized crystals were also examined by FTIR measurements in order to verify their chemical constitution. The obtained FIIR spectra of sample is illustrated in figure 4.

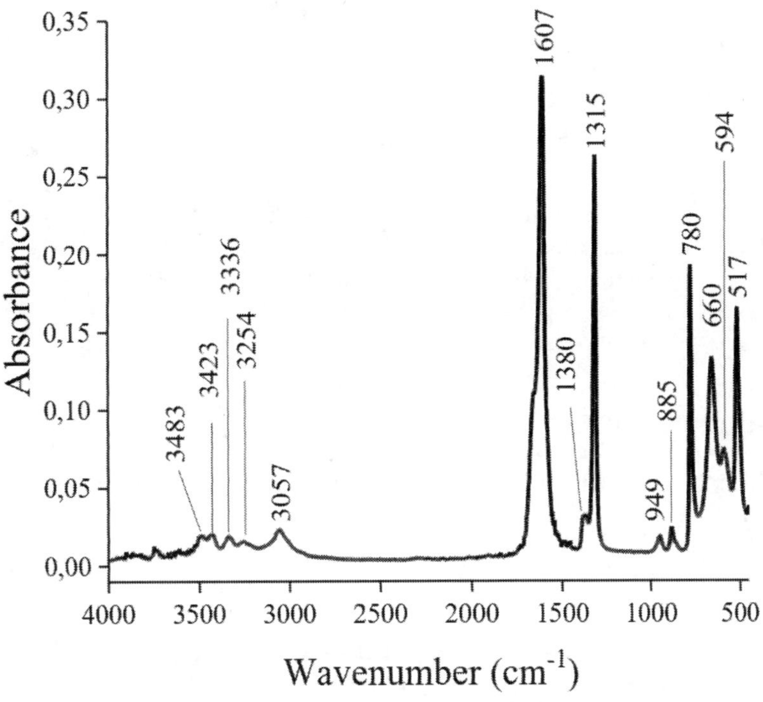

Fig. 4. FTIR spectra of the synthesized calcium oxalate crystals

According to Figure 4, multiple staircase bands at 3483 to 3057 cm^{-1} critical to a valence vibration of the OH stretching of water and corresponding to COM crystals. While the COD crystals show only one absorption peak at 3423 cm^{-1} [9], [10]. In addition, the absorption band due to the balancing vibration of the coordinated water plane appears at 780 cm^{-1}, also two bands around 949 cm^{-1} and 660 cm^{-1} appear, characterizing the formation of COD [7]. The absorption band at 885 cm^{-1} results from the C=C stretching mode and the band at 517 cm^{-1} is caused by the in-plane bending of O=C=O [10]. While absorption bands were observed at 1607 cm^{-1} and 1380/1315 cm^{-1}, the former are attributed to the antisymmetric carbonyl stretching band (vas (COO-)) and the latter are consistent with the symmetric metal carboxylate stretching band (vs (COO-)); these bands correspond to COT crystals [11]. XRD and FTIR measurements confirmed the presence of mixed crystals between COM + COD + COT.

4 Discussion

Numerous studies have shown that the species *Cydonia oblonga Miller* is an excellent natural source of phenolic acids and flavonoids. In this study, the in vitro inhibitory effect of aqueous extract of C. oblonga Miller peel and pulp on crystallization was evaluated at different concentrations of the extract. In order to valorise and verify the effectiveness of antilithiatic activity of *C. oblonga Miller*. The results of this investigation provide evidence that *C. oblonga*. fruit can have a litholytic effect. In addition, our results are similar to those of Beddade *et al* [5], [12] have studied the effect of aqueous extract of *Z. Lotus* and *A. Unedo* on the crystallization of calcium oxalate in aqueous solution and in the Human in vitro, and in a recent study conducted by Kachkoul *et al* [13] two extracts of *Punica granatum L.* fruit peel, it was found that they had an anticrystallization capacity. Another study suggested that the methanol soluble fraction isolated from *Hamulus lupulus L.* was the most effective on inhibition and dissolution of calcium oxalate crystals model [2]. The study carried out by Zaki *et al* [14] showed that the hydroalcoholic extract of *C. zeylanicum* has a significant inhibitory effect on the crystallization of CaOx, but not in a concentration-dependent manner.

5 Conclusion

In conclusion, we found that quince fruit extracts, particularly the pulp and peel, had significant effects. To our knowledge, this is the first report concerning the anticrystallization activity of this fruit. Nevertheless, such in vitro results need to be confirmed in vivo in order to develop a potent antilithic drug. In addition, the mechanism by which the drug exerts its effect in the animal model of lithiasis needs to be further studied.

Acknowledgements

The authors thank the National Scientific and Technological Research Center (CNRST) for financial support.

References

1. S. Kulaksizoğlu, M. Sofikerim, and C. Çevik **40** 589–594 (2008)
2. A. Frckowiak, T. Kozlecki, P. Skibinski, W. Gawe, E. Zaczynska, A. Czarny, K. Piekarska, R. Gancarz. J. Cryst. Growth **312,** 3525–3532 (2010)
3. V. Tszzorl and C. DOTVTBNECGBRRT. Am. Mineral. **65,** 327–334 (1980)
4. W. Zhou, A. Abdurahman, A. Umar, G. Iskander, E. Abdusalam, B. Berké, B. Bégaud, N. Moore. J. Ethnopharmacol. **154,** 163–169 (2014)
5. M. El Bir, M. Oubenali, M. Echajia, S. Rabi, and M. Berkani. **23,** 583–589 (2018)
6. Y. Zhang, Y. Tang, J. Xu, D. Zhang, G. Lu, and W. Jing. J. Solid State Chem., 231, 7–12 (2015)
7. J. Gao, J. F. Xue, M. Xu, B. S. Gui, F. X. Wang, and J. M. Ouyang. Int. J. Nanomedicine **9,** 4399–4409 (2014)
8. Y. Shen, W. Yue, A. Xie, S. Li, and Z. Qian. Colloids Surfaces B Biointerfaces. **45,** 120–124 (2005)
9. R. Kachkoul, T. Sqalli Houssaini, Y. Miyah, M. Mohim, R. El Habbani, and A. Lahrichi. Prog. en Urol. 28, 156–165 (2018)
10. S. Bhattacharyya, A. K. Mandal, and S. K. Singh. J. Postgrad. Med. Educ. Res. **48,** 128–131 (2014)
11. R. Kachkoul, T. Squalli Housseini, M. Mohim, R. El Habbani, Y. Miyah, and A. Lahrichi. J. Integr. Med. **17,** 430–437 (2019)
12. L. Baddade, I. Elhadri, and M. Berkani. Journal of Biomedical Research and Health Economics. **(2),** 1–5 (2020)
13. R. Kachkoul, T. S. Houssaini, M. Mohim, R. El Habbani, and A. Lahrichi. Evidence-based Complement. Altern. Med., **2020**, 1-14 (2020)
14. S. Zaki, N. Jahan, M. Kalim, and G. Islam. J. Integr. Med., **17,** 273–281 (2019)

E3S Web of Conferences **337**, 05006 (2022) https://doi.org/10.1051/e3sconf/202233705006
I2CNP 2021

Green synthesized nanomaterials as an alternative to restricted chemicals in postharvest preservation of horticultural crops: a review

Kwanele A. Nxumalo[1] *and Olaniyi A.* Fawole[1]*

[1]Postharvest Research Laboratory, Department of Botany and Plant Biotechnology, University of Johannesburg, P.O. Box 524, Auckland Park, Johannesburg 2006, South Africa.

Abstract. Synthetic chemicals have been successfully used to control economic postharvest pathogens such as Botrytis cinerea, Penicillium spp., Rhizopus spp., etc., thus reducing postharvest losses. However, their active ingredient and co-formulants have resulted in their use being restricted internationally due to health concerns and consumers' requests for safe and natural alternatives. As an alternative to chemicals, green synthesized nanomaterials incorporated with edible coatings in a polymer matrix have been successfully applied on horticultural crops to improve their shelf-life Due to their known antimicrobial activities, plant extracts and nanoparticles such as silver and zinc oxide have been successfully used in food processing applications. Green sources are used as stabilizer and reducing agents to synthesize shape and size-controlled metal and metal oxide nanoparticles. Thus, eco-friendly nanoparticles are produced without using toxic materials and are cost-effective. Edible coatings form a thin layer that prevents oxidation, maintains the product quality, can be safely eaten as part of the product, and does not add unfavourable properties. Studies showed that edible coatings with or without green nanomaterials have been successfully used to control postharvest losses and improve the shelf-life of fruit crops such as plums, raspberries, etc.

*Corresponding author: nxumalokwaneleandy@gmail.com

1 Introduction

Postharvest losses of fruits and vegetables are mainly caused by decay brought about by fungal plant pathogens [1]. According to the Food and Agricultural Organisation [2], one of the most feasible options for meeting future food needs is reducing postharvest losses. Postharvest losses pose a serious threat to the storage and marketability of fruits and vegetables, leading to nutrient loss and the spread of microorganisms responsible for degradation, leading to huge fresh produce losses[3-4]. Therefore, innovative technologies such as edible coatings (EC)[5] and controlled atmosphere (CA) are suggested as possible solutions for maintaining the quality of agricultural products during storage [3]. The spread of harmful microorganisms and the use of chemicals in postharvest preservation of horticultural crops poses a severe threat to the environment, crop quality and human health [6]. The application of green synthesized nanomaterial to control postharvest losses of horticultural crops has been generally regarded as safe [GRAS] to both the environment and human health [7-8]. The exceptional properties of green nanomaterials make them a practical choice for sustainable horticulture in general andin particular postharvest diseases of fruit and vegetable at the scientific and industry level[9]. The objective of this review was to document theapplication of green synthesized nanomaterials as an alternative to restricted chemicals in postharvest preservation of horticultural crops.

2 Economic importance of postharvest losses of horticultural crops

In the developing world, fruit and vegetable losses significantly impact poor and vulnerable people [2]. This comes in the form of social (health and livelihoods), environmental and economic costs[10]. According to FAO [11], about one-third of the edible parts of food produced for human consumption gets lost or wasted globally, which is about 1.3 billion tons per year. Food waste at the consumer level in industrialized countries (222 million tons) is almost as high as the total net food production in sub-Saharan Africa (230 million tons). Losses and waste in fruits and vegetables are the highest categories of losses among all types of foods, with up to 60 percent of all fruits and vegetables making their way to the landfill [2]. Horticultural losses impact food security for poor people, food quality and safety, economic development, and the environment [11]. Economically, postharvest losses of fruits and vegetables have a direct and negative impact on the income of both farmers and consumers [2], thus making food insecurity a question of access (purchasing power and prices of food) than a supply problem[10]. Improving the efficiency of the horticultural supply chain and reducing postharvest losses could reduce the high costs of fruits and vegetables and help bring down their cost and increase access for the consumers [10-11]. Given the high perishability of horticultural crops, the use of green nanomaterials, as an alternative to control postharvest losses, could be one way of reducing the economic cost of horticultural crops, improving their availabilityand increasing farmincome [9].

 https://doi.org/10.1051/e3sconf/202233705006

3 Green nanomaterials as a viable alternative to control postharvest losses of horticultural crops

The application of nanotechnology is largely promoted to develop active food packaging materials with excellent performance to protect horticultural crops and extend their shelf-life [12]. Promising performances of antimicrobial nano-coatings have been reported when they were applied on fresh fruits and vegetables [13-14]. Traditionally, nanoparticles (NPs) have been synthesized using physical and chemical methods and have high toxic residues, are expensive to produce, and are environmentally unfriendly [15]. Recently, green methods to synthesize metal-oxide nanoparticles have attracted much attention because they are simple, low cost and environment friendly [16-17]. These methods involve the use of plant, fruits, and vegetable extracts, which are used as stabilizers and reductant agents to control the crystal growth, and for scale-up and industrial production of NPs [18]. These plant extracts have organiccompounds, such as polyphenols that act as reductants and amino and carboxylic groups that may help as stabilizers, favouring the conversion of metal compounds into specific NPs [19]. The application of green nanomaterial to control postharvest losses of horticultural crops is summarised in Table 1.

Table 1. Application of green nanomaterial to control postharvest losses of horticultural crops

Green material	Synthesized nanoparticles	Fruit/vegetable	Focus of the study	Key findings	Reference
Citrus sinensis peel extract	Zinc-oxide NPs	Strawberries	Application of Citrus sinensis nano-coatings on fresh strawberries to evaluatethe preservation effect	Strawberries coated with citrus peel nanoparticles incorporated into carboxymethylcellulosesodium (CMC) had decreased weight loss, rate of rotting, and the storage quality was improved	[20]
α-pinene	Chitosan	Bell pepper	α-pinene chitosan nanoparticles in controlling to Alternaria alternate on bell pepper	The chitosan nanoparticles and edible coating inhibitedA. alternata during the cold storage period of bell pepper and preserved the physicochemical quality.	[21]
chitosan/gum arabic (CH/GA)	ZnONPs	Banana	ZnO NPs incorporated into chitosan/gum Arabic (CH/GA) for the preservation of bananas	Banana coated with the CH/GA/ZnO maintained freshness for more than 17 d in comparison with the less than 13 d for the control banana at 35 °C and 54 % RH.	[22]
Olive and marjoram leaf extract	ZnO-NPs	Bell pepper	Evaluating disease management of gray and black mold of pepper fruits (artificially inoculated with the tested two fungi)	Results showed significant reduction to the appearance of both diseases of pepper fruits treated with any of the two green synthesized ZnO nanoparticles compared with chemically synthesized ZnO-NPs and control.	[23]
Bergamot essential oil (BEO)	Titanium-NPS	Mango	Effects of polylactic acid (PLA)/BEO/nano-TiO2 film and PLA/BEO/nano-TiO2 + nano-Ag in maintainingthe quality of mangoes.	Results with PLA and PLA/BEO films indicated that PLA NC films successfully restricted adverse changes in total acidity, vitamin C content, and color. At the same time, firmness retention in the mangoes was enhanced. Results showed that the shelf life of these fruits was extended up to 15 d.	[24]
Carrageenan	ZnO-NPS	Mango	The effect of carrageenan/ZnO-NPs (CZ) coating in maintaining the storage quality of mango	On d 13 of storage, the decline in the total acidity of mangoes exposed to CZ 0.5 and CZ 1 NP coating treatments was significantly lower than that in the control samples. The CO2 production peak occurred after 7 d of storage, where the mangoes in the control sample produced the highest level of CO2. Until day 33, minimal degradation was evident in the	[25]

				mangoes with CZ 0.5 coatings compared to samples without coatings.	
Beeswax	beeswax solid lipid nanoparticle s (BSLN)	strawberry	Nano-coatings with BSLN-with xanthan gum (XG) as an alternative in the shelf life of strawberries stored in refrigeration.	The coating with 10 g/L of BSLN, 4 g/L of xanthan gum, and 5 g/L of propyleneglycol contributed positively to preservation while also decreasing decay rates; reflected in less fungal growth, weight loss, and physiological damage to the end of storage.	[26]

4 Antimicrobial and mechanism of action

The efficacy of NPs in reducing microbial growth on fruits/vegetable surface is mainly dependent on the type, concentrationand properties of coating films, as well as the type, size, shape, concentration, chelation, release and photocatalytic properties of the applied NPs [27-28].The synergistic effect of coating materials and NPs in the barrier, the types and structure of tested microorganisms and the tested conditions could also influence antimicrobial activity of the applied green nanomaterial edible coatings [27, 29]. A study by Cano *et al.* [30] revealed that Ag-loaded films exhibited antimicrobial activity against postharvest pathogens such as Aspergillus niger and Penicillium expansum. However, the result appeared to be concentration dependent. Jia *et al.* [31] reported that silver/CS NPs exhibited a high effect on Botrytis cinerea fungi by suppressing the growth and germination at a concentration of 0.02 mg/mL in vitro and in vivo. Similarly, El-Sherbiny *et al.* [32] indicated that 100 μl/mlAg-NPs and the in-situ development of Ag-NPs on the surface of poly(ε-caprolactone)/curcumin/grape leaf extract-Ag hybrid NPs controlled Aspergillusflavus. These antimicrobial activities could significantly contribute to controlling the decay and quality loss of horticultural crops during storage or when exported [3,14]. Otoni *et al.* [33] and Goudarzi *et al.* [34]reported that nanoparticles and edible coatings (EC) act directly on the pathogen by inhibiting spore germination and mycelium growth and altering morphological structures and reactive oxygen species (ROS) generation, spore mitochondrial membrane potential (MMP) and adenosine triphosphate (ATP) content.The preservation mechanism includes gas modification and induction defence properties of EC with NPs by forming a thin coating film on the surfaces of fresh products after being treated and dried, thus, modifying the concentrations of O2, CO2, and ethylene produced by horticultural crops [3, 14, 35].

Gao *et al.* [36] reported that strawberries coated with carboxymethylcellulose (CMC) enriched with ZnO nanoparticles synthesized from Citrus sinensis peel extract (CMC/CSP) had higher efficacy in controlling Botrytis cinerea compared to CMC enriched with commercial ZnO nanoparticles (CMC/CZN). Strawberries coated with CMC/CSP showed a lower rate of rotting and increased the storage period by 2 d compared to the other treatments. The higher antifungal activity of the CMC/CSP damaged the DNA and protein of the postharvest pathogen resulting in their cells death, thus reducing its severity and preserving the quality of the fruit for a longer time. An in vitro study by Saekow *et al.* [37] found that ZnO-NPs synthesized from pineapple peel extracts in CMC coating delayed the severity of Alternaria alternata persimmon and tomato fruit.After 3d, results showed that control fruit had a progressive higher disease development in the early stage of the storage period. Contrary to the control, CMC+ZnONPs coated persimmon and tomato hada lower level of disease development started showing signs of diseases development at 6 d of storage. Sing [38] reported that zinc oxide nanoparticles have a strong good antifungal activity against selected strains of bacteria and fungus compared to conventional zinc oxide particles.

5 Conclusion and future prospects

Green nanomaterials have emerged as an alternative to fungicides and have proven cost-effective and have high efficiency in maintaining fruit/vegetable quality. The low cost, environmentally benign, low-toxicity, and good performance green nanomaterials show a great potential to be applied in postharvest preservation of horticultural crops. For this reason, research efforts are needed to enhance our knowledge in the application of nanotechnology in postharvest technology. Currently, there is a dart of knowledge on how the synthesis of green nanomaterials could be optimized for high yield and efficacy in the protection of horticultural crops. In addition, long termrisks and hazards of the use of nanomaterial remain unknown. There is a need to investigate this area and be cautious with the use of certain metaloxides such as TiO2and X, which have been studied to cause detrimental adverse factors such as colon cancer in humans. Finally, factors, including dose, kind of nanomaterials, application timing, and fruit/vegetable physiology, can affect the efficacy of nanomaterials; therefore, this area also requires further investigation.

References

1. P. Tripathi, N. Dubey. Postharvest Biol. Technol. 32:235-245 (2004)
2. Food and Agriculture Organization of the United Nations (FAO). Agricultural markets and sustainable development: Global value chains, smallholder farmers and digital innovations. Rome, FAO. https://doi.org/10.4060/cb0665en (2020)
3. Y. Xing, W. Li, Q. Wang, X. Li, Q. Xu, X. Guo, X. Bi, X. Liu, Y. Shui, H. Lin, H. Yang. Molecules. 24: 1695 (2019)
4. O.A. Fawole, S.C. Riva, U.L. Opara. Agronomy. 10:1-21 (2020)
5. S.C. Riva U.O. Opara, O.A. Fawole. Scientia Horticulturae. 262: 109074 (2020)
6. Food and Agriculture Organization (FAO) and the World Health organisation (WHO) of the United Nations. WHO, Geneva, Switzerland (2013)
7. L.A. Castillo, O.V. López, J. Ghilardi, M.A. Villar, S.E. Barbosa, M.A. García. Food Hydrocoll. 51: 432-440. (2015)
8. I. Shahabi-Ghahfarrokhi, F. Khodaiyan, M. Mousavi, H. Yousefi. Int. J. Biol. Macromol. 77: 85-91 (2015)
9. S. Ruffo Roberto, K. Youssef, A.F. Hashim, A. Ippolito. Nanomaterials. 9:1752(2019)
10. J. Gastavsson, C. Cederberg, U. Sonesson. Food and Agriculture Organization (FAO) of the United Nations: Rome, Italy (2011)
11. Food and Agriculture Organization of the United Nations (FAO). Accessed 17 November 2020 (2017).
12. D. Bhuyan, G.W. Greene, R.K. Das. Critic. Rev. Biotech.39(6):759-778 (2019)
13. B.J. Arroyo, A.C. Bezerra, L.L. Oliveira, S. J. Arroyo, E.A.D. Melo, A.M.P. Santos. Food Chem.125566(2019).
14. K.A. Nxumalo, A.O. Aremu, O.A. Fawole. Sustainability 13:5897 (2021)
15. O.J. Nava, C.A. Soto-Robles, C.M. Gómez-Gutiérrez, A.R. Vilchis-Nestor, A. Castro-Beltrán, A. Olivas, P.A. Luque. J. Molec. Struc. 1147:1-6 (2017).
16. A.K. Mittal, Y. Chisti, U.C. Banerjee. Biotech. Adv.31(2):346-356 (2013)
17. K. Elumalai, S. Velmurugan. Appl. Surf. Sci. 345: 329-336 (2015)
18. F. Davar, A. Majedi, A. Mirzaei, J. Am. Ceram. Soc. 98(6):1739-1746 (2015)
19. M. Stan, A. Popa, D. Toloman, A. Dehelean, I. Lung, G. Katona. Mat.Scie.Semicond. Proc. 39: 23-29 (2015)
20. Y. Gao, D. Xu, D. Ren, K. Zeng, X. Wu. Lwt.126:109297 (2020)
21. G. Hernández-López, R.I. Ventura-Aguilar, Z.N. Correa-Pacheco, S. Bautista-Baños, L.L. Barrera-Necha. Int. J.Bio.Macrom.165:1881-1888 (2020)

22. D.D. La, P. Nguyen-Tri, K.H. Le, P.T.M. Nguyen, M.D. Nguyen, A.T.K. Vo, M.T.H. Nguyen, S.W. Chang, L.D.Tran, W.J. Chung, D.D. Nguyen. Prog. Org. Coat.151:106057 (2021).
23. M. Hassan, M.A. Zayton, S.A. El-Feky. J. Biotech.Bioengin.3(4):8-15 (2019)
24. H. Chi, S. Song, M. Luo, C. Zhang, W. Li, L. Li, Y. Qin. Sci. Hortic. 249: 192–198 (2019)
25. B. Meindrawan, N.E. Suyatma, A.A. Wardana, V.Y. Pamela.Food Packag.Shelf Life.18:140-146 (2018).
26. M.L. Zambrano-Zaragoza, D. Quintanar-Guerrero, A. Del Real, R.M. González-Reza, M.A. Cornejo-Villegas, E. Gutiérrez-Cortez. Coatings. 10:253 (2020)
27. S.S. Soni, G.S. Dave, M.J. Henderson, A. Gibaud. Thin Solid Films. 531: 559-565 (2013)
28. Y. Xing, Q. Xu, L. Jiang, D. Cao, H. Lin, Z. Che, Y. Ma, X. Li, Y. Cai. RSC Adv. 5: 22825-22837 (2015).
29. Q. Xu, Y. Xing, Z. Che, T. Guan, L. Zhang, Y. Bai, L. Gong. J. Food Saf. 33: 179-189 (2013)
30. A. Cano, M. Cháfer, A. Chiralt, C. González-Martínez. Shelf Life. 10:16–24 (2016).
31. R. Jia, H. Jiang, M. Jin, X. Wang, J. Huang. Food Res. Int. 78: 433–441 (2015)
32. I.M. El-Sherbiny, A. El-Shibiny, E. Salih. J. Photochem. Photobiol. B.160:355-363 (2016)
33. C.G. Otoni, R.J. Avena-Bustillos, H.M.C. De Azeredo, M.V. Lorevice, M.R. de Moura, L.H.C. Mattoso, T. McHugh. Compr. Rev. Food Sci. Food Saf. 16: 1151–1169 (2017)
34. V. Goudarzi, I. Shahabi-Ghahfarrokhi. Int. J. Biol. Macromol. 106: 661-669 (2018)
35. H. Song, W. Yuan, J. Peng, W. Wei, X. Wang, L. Yang, Y. Zhang. Postharvest Biol. Technol. 119: 41-48 (2016)
36. Y. Gao, D. Xu, D. Ren, K. Zeng, X. Wu. Lwt, 126:10929 (2020)
37. M. Saekow, M. Naradisorn, W. Tongdeesoontorn, Y. Hamauzu. J. Food Scie. Agric. Technol. 5 (Spcl. Iss.): 41-4 (2019).
38. S. Sing. Toxicol. Mechan. Met. 29(1):37 (2018)

AUTHOR INDEX

AUTHOR INDEX